Hydrogen Production from Nuclear Energy

Greg F. Naterer • Ibrahim Dincer •
Calin Zamfirescu

Hydrogen Production from Nuclear Energy

Springer

Greg F. Naterer
Faculty of Engineering and
Applied Science
Memorial University of Newfoundland
St. John's, Newfoundland
Canada

Ibrahim Dincer
Calin Zamfirescu
Faculty of Engineering and Applied Science
University of Ontario Institute of Technology
Oshawa, Ontario
Canada

ISBN 978-1-4471-4937-8 ISBN 978-1-4471-4938-5 (eBook)
DOI 10.1007/978-1-4471-4938-5
Springer London Heidelberg New York Dordrecht

Library of Congress Control Number: 2013934979

Printed on acid-free paper

Springer is part of Springer Science+Business Media (www.springer.com)

To Veronica, Julia, Jordan and Josie Naterer
To Ibrahim Emir, Zeynep, Ibrahim Eren, Miray, Meliha and Gulsen Dincer
To Ioana and Iuliana-Maria Zamfirescu

Preface

Hydrogen is the most abundant element in the universe. It was first identified in the 1850s, during the industrial revolution, when a new type of society had emerged with the invention of the steam engine and its rapid transformation of the world's standard of living. In this mechanization era, the energy from coal was converted to mechanical power by steam engines, while the world shifted from a feudal system to widespread industrialization and capitalism. A second industrial revolution followed in the twentieth century—called the electrification era—when mechanical power from steam produced by large-scale power plants was used to supply national and regional grids with electricity. Industrial water electrolysis and fossil fuel reforming technologies to generate hydrogen for many types of industrial processes became well established since the early 1900s. A new self-conscience of society developed towards a concern for sustainable development whereby anthropogenic activities would neither harm the environment nor jeopardize the access of future generations to valuable materials and energy resources for their needs. In this context, the development of a hydrogenation era emerged in this current century, with respect to sustainable hydrogen generation at a large scale and its use as a clean energy carrier or chemical feedstock for industrial processes.

Hydrogen is the basic constituent of many chemical compounds with high industrial and societal value. Combined with nitrogen from air, hydrogen can be converted to ammonia—a fertilizer, refrigerant, and key compound for a large variety of pharmaceuticals and chemicals. Combined with carbon dioxide, ammonia gives urea, another valuable fertilizer which is generally produced today in a polluting and non-sustainable manner. Hydrogen combined with carbon dioxide can yield methanol. Also, other synthetic fuels such as diesel or gasoline can be produced by hydrogenation of carbon dioxide, making it possible to develop a cleaner transportation sector for road vehicles. It is envisaged that fuel cell vehicles will be become widely available in the coming decades. This will require a major increase in hydrogen demand, for example, by hundreds of millions of tonnes annually. Upgrading of heavy oil and oil sands in Alberta, Canada, is another need for hydrogen in vast quantities. Sustainable hydrogen production is also needed for cleaner steel production, plastics, construction materials, etc.

The key challenge of the emerging hydrogen era is to generate hydrogen at a large scale in a sustainable manner. This is a key question which has preoccupied scientists and engineers for decades. One needs a feedstock resource containing hydrogen such as water, hydrogen sulfide, or fossil hydrocarbons. Furthermore, one needs a source of energy which is "clean(er)" to extract hydrogen from the selected resource. In addition, the method should generate hydrogen as efficiently as possible. Nuclear-based hydrogen production is an important pathway to sustainable hydrogen production at a large scale. In principle, nuclear energy to generate hydrogen can lead to: (1) better efficiency, (2) better cost-effectiveness, (3) better resource use, (4) better knowledge through design and analysis, (5) better energy security, and (6) better environment. These are six main pillars for sustainable energy development of humankind.

The use of nuclear energy to generate hydrogen as an energy carrier was first proposed for military applications in the 1960s when a small, mobile nuclear reactor was proposed to generate hydrogen from water by electrolysis which could then supply fuel for heat engines. The main benefit of nuclear energy is the compactness of the energy source. A nuclear plant can generate 1 MW per acre of footprint at a high capacity factor, whereas other sustainable sources are much less dense (for example, concentrated solar power generates 25 kW per acre with a capacity factor smaller than 0.3). Nuclear fuels are extensively available worldwide, especially with the new breeding reactor technology, which makes use of thorium.

The first chapter of this book presents a general perspective on hydrogen as an energy carrier including its historical evolution and related concepts and technologies, as well as hydrogen's applications for sustainable energy in two major sectors: residential and transportation.

The second chapter discusses the role of nuclear energy for sustainable hydrogen generation at a large scale. The capacity factor of a nuclear power plant can be increased to high values (over 95 %) if hydrogen is produced locally, either by electrolysis or other processes linked to the nuclear reactor. The use of hydrogen for heavy oil and bitumen upgrading, as well as its positive environmental impact, is also discussed. In addition, hydrogen use in road vehicles, rail, and air transport is analyzed for various case studies.

Chapter 3 summarizes the main nuclear hydrogen production programs worldwide over the last ~40 years. The Generation IV International Forum (GIF) is as an international organization that coordinates the development of the next generation of nuclear reactors for electricity, hydrogen, and process heat production. There are six main reactor concepts in GIF and a roadmap established for a synergistic and cooperative development among the international partners. These partners include the U.S. Department of Energy, Euratom (European Union), Japan Atomic Energy Agency, International Nuclear Energy Association, Commissariat of Atomic Energy (France), Korean Atomic Energy Research Institute, Natural Resources Canada, China Atomic Energy Authority, and Paul Scherrer Institute (Switzerland), among others. Other non-GIF concepts that have been developed in countries such as Russia, India, South Africa, and Brazil are also presented. A case study of a recent international consortium for copper–chlorine cycle development is highlighted.

Chapter 4 presents water electrolysis technology including the fundamental science and technology. Alkaline, proton exchange membrane and solid oxide electrolyzers are discussed in detail. In addition, other emerging hybrid electrolysis systems are discussed. The main systems of nuclear-electrolytic hydrogen production from water are outlined.

Thermochemical water splitting cycles are comprehensively examined in Chap. 5. Fundamentals of thermochemical and electrochemical reactions are introduced. The general methodology of thermochemical and hybrid cycle construction and assessment is described. The selected chemical reactions must be thermodynamically feasible, but also having satisfactory equilibrium yield. A down-selection procedure of thermochemical and hybrid cycles is presented. The main development aspects of three of the most promising cycles are discussed in detail. These include the sulfur–iodine cycle, the hybrid sulfur cycle, and the copper–chlorine cycle.

Chapter 6 focuses on the promising copper–chlorine cycle, which has experienced rapid development and progress in recent years.

Chapter 7 involves integrated systems for nuclear hydrogen production. This chapter explains all six nuclear reactor concepts selected in the GIF roadmap. Further sections are focused on a linkage of the very high temperature reactor (VHTR) with steam electrolyzers, integration of VHTR with the sulfur–iodine cycle, integration of a nuclear plant with the hybrid sulfur cycle, and integration of a copper–chlorine thermochemical cycle with a supercritical water cooled reactor (SCWR). A case study is presented for integrated nuclear, desalination, and hydrogen production technologies.

The concluding Chap. 8 presents future trends and emerging opportunities with nuclear hydrogen production. A series of newly developed concepts of systems and high-temperature components and materials is presented. Also, various advanced power cycles coupled to fission reactors and thermochemical water splitting plants are discussed. In addition, nuclear fusion technology for hydrogen generation is briefly examined. This is a long-term prospective technology, which is expected to be developed mostly beyond the 2050s.

This book provides extensive analyses and compilations, as well as the fundamentals of the main processes, technological problems, and design challenges. Nuclear hydrogen generation is an emerging technology which links the physics of nuclear reactors with the chemistry of hydrogen production processes, as well as the engineering of integrated systems that perform many simultaneous functions which are complex because they generate multiple outputs (e.g., electricity and hydrogen).

The authors gratefully acknowledge the support of a large number of individuals who made this work possible through their efforts of advancing the knowledge of this subject. In particular, we acknowledge the researchers and staff of the Clean Energy Research Laboratory (CERL) at the University of Ontario Institute of

Technology (UOIT) for all of their contributions and assistance, especially Dr. Zhaolin Wang, Dr. Kamiel Gabriel, Dr. Marc Rosen, Edward Secnik and Dr. Venkata Daggupati, as well as all ORF (Ontario Research Fund; Cu–Cl cycle) team members. Special thanks also to colleagues and coworkers who have provided a wealth of information, materials and support for this book, particularly at Atomic Energy of Canada Limited (AECL) - Dr. Sam Suppiah, Dr. Lorne Stolberg, Dr. Alistair Miller, Dr. Romney Duffey, and the Argonne National Laboratory (ANL) - Dr. Michele Lewis, Dr. Magali Ferrandon.

Greg F. Naterer, Ibrahim Dincer and Calin Zamfirescu
Oshawa, Ontario
March, 2013

Contents

Chapter 1
Hydrogen as a Clean Energy Carrier

Abstract In this chapter, the role of hydrogen as a clean energy carrier is discussed. In the first part of the chapter, a historical perspective on hydrogen exploration, production, and use is presented. It appears that the modern world evolved from a mechanization period (when heat engines were developed to generate motive force from fossil fuels) to an electrification era (when the motive power developed by power plants is converted to electricity to supply large national or regional grids) and to a hydrogenation era (when the energy from primary sources is converted by clean processes to hydrogen as an energy carrier). It is argued that the hydrogenation era just started and will be implemented during the first century of the current millennium. Furthermore, this section reviews the main national/regional research programs deployed in the last 20 years on hydrogen research. Various uses of hydrogen including as energy carrier in transportation sector, energy storage medium, and greenhouse gas reduction agent are discussed in detail. It is shown for a case study of comprehensive life cycle analysis that among six vehicle types hydrogen and fuel cell vehicles show the greatest potential. Also, the use of ammonia as synthetic fuel produced from clean hydrogen and nitrogen represents an attractive alternative to conventional fuels for vehicles.

1.1 Historical Perspective

Modern society requires many types of services to maintain a good standard of living: electricity, hot water, space heating, air-conditioning, fuels, various chemicals and materials, etc. Traditional methods of producing these commodities are primarily driven by combustion of fossil fuels which is a major contributor to air pollution. The global population and its demand for these services are rapidly increasing, so the rising use of fossil fuels has a major impact on climate change.

G.F. Naterer et al., *Hydrogen Production from Nuclear Energy*,
DOI 10.1007/978-1-4471-4938-5_1,

This scenario explains the worldwide motivation to develop cleaner, more efficient production methods, which will be eventually able to replace or supplement traditional processes by incorporating sustainable energy sources and hydrogen technologies. Hydrogen is a clean energy carrier because it can be generated by low- or zero-carbon sources (e.g., nuclear, water, biomass, solar). Subsequently hydrogen can be converted to electricity, synthetic fuels, or heat with little or no environmental pollution.

Additionally, hydrogen is an important chemical that enters the constituency of many materials used in industry and society: plastics, foods, pharmaceuticals, fertilizers, metallic materials, construction materials, etc. For cleaner generation of hydrogen, there is less pollution from the production of such hydrogen-containing materials. Therefore, hydrogen itself can be viewed as a critical commodity for future society, which can replace or at least complement fossil fuels (oil, coal, natural gas) that are extensively used at present. In this section, we follow the historical evolution of the concept of "hydrogen as clean energy carrier," emphasizing the main issues and achievements to date.

Hydrogen was discovered in the Middle Age by Paracelsus (sixteenth century), who reacted iron with sulfuric acid, a reaction that releases gaseous hydrogen, at that time called "inflammable air." Cavendish observed in 1766 that hydrogen combustion generates only water. In 1783, Lavoisier, Laplace, and Meunier demonstrated that water is composed from two parts of hydrogen and one part of oxygen. The name of hydrogen was proposed by Lavoisier to indicate that it is a constituent of water ("hydro"). In 1784, the steam-iron process has been patented by Lavoisier and Meunier as the first technology to generate hydrogen in industrial quantities. Many other researchers—e.g., Gay Lussac, von Humboldt, and Charles—contributed by the end of the eighteenth century to advances in knowledge, use, and production of hydrogen. Charles demonstrated the first hydrogen flying balloon in 1783. In Paris, a system for night lighting was installed in 1803, which uses a syngas comprising 40–60 % hydrogen. The use of hydrogen as a town gas operated in many towns since the nineteenth century until the middle of the twentieth century. This is one of the first uses of hydrogen as an energy carrier.

The technological development of society followed together with the evolution of hydrogen as an energy vector. Consider the history of primary energy that society uses for supplying its needs. There are three types of energies to be considered, namely, the energy of coal, fossil hydrocarbons (petroleum, oil shale, oil sands, etc. and natural gas), and renewable/sustainable sources which are "zero carbon." The renewable/sustainable energies are those that do not release greenhouse gas (GHG) emissions upon their use (although some GHGs are emitted indirectly—for example due to the construction of an energy distribution infrastructure). Typical examples of "zero carbon" energies are biomass, hydroelectric, wind energy, solar, nuclear, and others. The authors compiled data from past literature (Bareto et al. 2009) to illustrate the chart in Fig. 1.1 representing the share of "zero-carbon," coal, and fossil fuel energy use worldwide since 1800 until the present. The chart is extrapolated until 2100 based on a prediction model by Bareto et al. (2009).

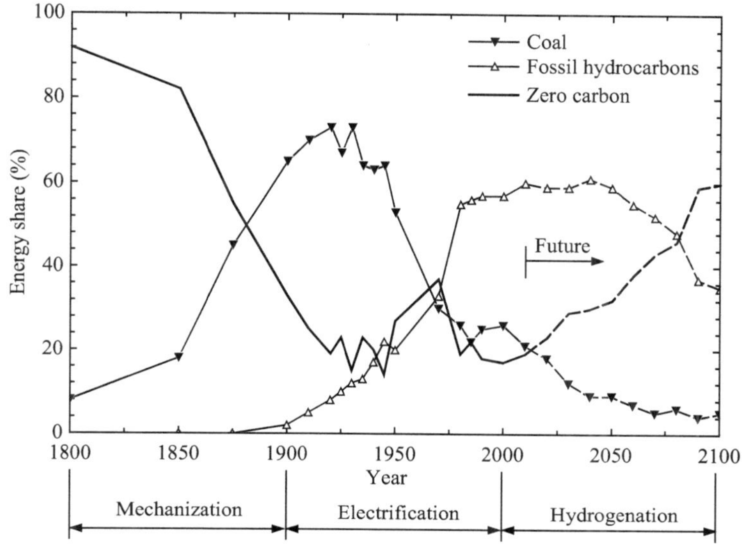

Fig. 1.1 Shares of world energy supply since the start of the industrial revolution

It is interesting that before the industrial revolution, the world energy supply comprises only "zero carbon" sources: wood, wind, hydro, and animal power. Starting with the industrial revolution and the introduction of the steam engine, the consumption of coal increased sharply. One can speak about a "mechanization" period of extended about 100 years during which Western civilizations implemented an industrial society. There was remarkable technical and scientific progress in this period, especially in England and France. Regarding hydrogen, the discovery of water electrolysis was made by Nicholson in 1800 and the invention of the fuel cell by Grove in 1839. By the end of the nineteenth century, Wroblewski realized the liquefaction of hydrogen, a process that was further perfected by Dewar and made commercially by Claude.

By the time of 1900, another source of energy started proliferating in the society, namely, petroleum. In the same time, advances in electric power generation and the electrical grid propelled a second industrial revolution of around 100 years, which is called the "electrification" period (see Winter 2009). With the help of electricity, devices like pumps, compressors, conveyers, mixers, etc. became more effective, which encouraged the sharp development of the process industry. Germany was noted in the beginning of the twentieth century by many and important inventions in the chemical and process industry. One of the most important processes ever invented is the Haber–Bosch ammonia synthesis process (1913). Manufacturing of ammonia by this process became a major consumer of hydrogen in the world. Further evolution of the process led to synthesis of urea at an industrial scale, which is a major fertilizer produced from ammonia and carbon dioxide. In the Ruhr region of Germany, a high development of coal mining and metallurgy occurred at the beginning of the twentieth century. Hydrogen was highly needed by the metallurgical sector. Consequently, by 1938, a hydrogen pipeline of 240 km was installed to deliver 250 × 10^6 Nm^3/year of hydrogen obtained from coal. Presently in Europe, there are about 1,500 km of hydrogen pipelines, while in the USA, there are about 900 km of hydrogen pipelines.

The geopolitical context in the first half of the twentieth century encouraged the development of engines and fuels for transportation. Gasoline gained in proportion due to its high octane number, its high energy density, and cheap process of extraction from crude petroleum. Currently, the road transportation sector uses mostly gasoline and diesel and the marine transport industry uses heavy petroleum or diesel. Hydrogen has found uses in space transport as liquefied rocket fuel.

It is observed that by the end of the twentieth century, science became more and more the dominant driving force influencing the development of society. A self-conscience was developed (the so-called Noosphere, see Goltsov et al. 2006) that made humans aware of possible catastrophic scenarios related to the unsustainable use of fossil energy resources. The intellectual activity and the acquired technological and scientific knowledge created the start of a new 100-year era, which will follow the "mechanization" and "electrification" eras. This has been often cited as the "hydrogenation" of society (see Winter 2009). According to predictions suggested in Fig. 1.1, during the next 100 years, the "zero-carbon" energy sources will increase in share and a "hydrogen civilization" will emerge (see Goltsov et al. 2006). The rise of a hydrogen civilization is induced by preemptive actions founded on scientific analyses regarding the negative impact of GHG emissions on climate and the depletion of fossil fuel resources.

The chart in Fig. 1.2 shows how a basic scientific analysis predicts the decarbonization of the world energy supply, from one based mostly on carbon (molecular mass 12) to one that is based on hydrogen (molecular mass 2). This reflects a trend in the direction of "hydrogenation" of the world energy supply, which is an evolution toward worldwide use of hydrogen as an energy carrier. The plot is made under the assumption that the average carbon content of fossil hydrocarbons (mainly natural gas and petroleum) is around 1:3, while 2:3 is the hydrogen content (the carbon-to-hydrogen ratio for methane is 1:4 while for gasoline it is ~1:2). Therefore, the carbon content of the global energy supply is obtained from Fig. 1.1 by addition of the coal curve to one-third of the fossil hydrocarbon curve. The non-carbon share of the energy supply is obtained by adding 2/3 of the fossil hydrocarbon curve with the zero carbon curve. Figure 1.2 shows a historical representation of the carbon share and non-carbon share in the energy mix of the world supply. Note that the non-carbon share is represented by hydrogen as part of fossil hydrocarbons plus hydrogen from any other sustainable source, plus electricity. Since hydrogen and electricity are corresponding energy vectors—which can be converted from one to the other—it is useful to associate the non-carbon energy and molecular mass of 2 (hydrogen), while carbon energy has a molecular mass of 12. With these facts, one can plot, as indicated in Fig. 1.2—the right y-axis—the historical trend of the equivalent molecular mass of the worldwide energy supply (calculated as a weighted sum of the corresponding atomic masses of

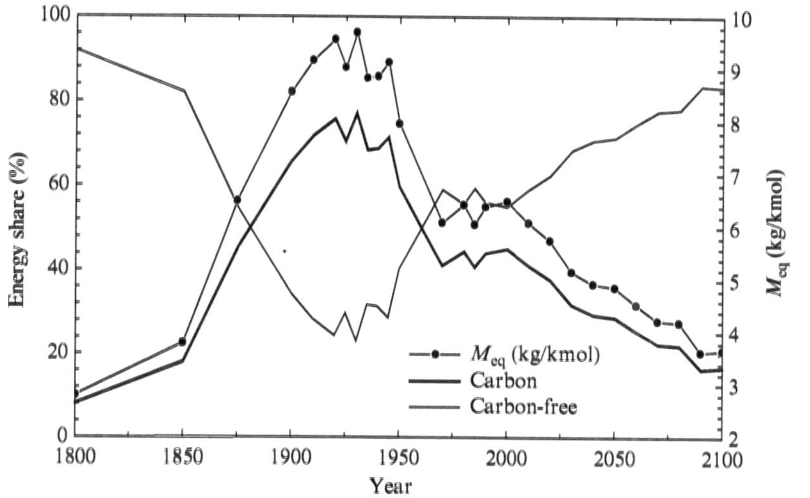

Fig. 1.2 Illustration of the historical trend of energy decarbonization

C and H). The curve (M_{eq}) shows an interesting feature: it sharply rises during the coal-dependent "mechanization era," decreases with fluctuations during the "electrification" era, and steadily decreases during the future "hydrogenation" era. This downward character toward energy decarbonization is a result of societal-will demonstrated during the last 40 years.

The concept of switching to an economy based on hydrogen as an energy carrier was shaped by the end of the 1960s and beginning of the 1970s. According to Bocris (2002), it started with the quest of finding a clean energy carrier that can replace or complement electricity. It was argued that hydrogen can be stored in large scales in depleted oil fields or in balloons under the sea at pressures of 20–50 bars. In the initial views—characterizing the years of the 1970s—hydrogen stored in large scales would be transported at long distances (~1,000 km) for which the economics shows that its transportation is better than electricity. In more recent years, another concept appeared, namely, hydrogen storage at small scales to satisfy power generation needs in vehicles or small power generators.

The energy crisis from the beginning of the 1970s influenced positively the consideration by the scientific community of a hydrogen economy. A key event impacting the progress of the hydrogen economy concept was the THEME conference held in Miami (FL) in 1974. The International Association of Hydrogen Energy (IAHE) was established at that time, which became a world leader in promoting the hydrogen economy concept. Several other associations were formed since that time; Table 1.1 contains a sample list of hydrogen associations worldwide. One of the most known and prolific conferences—initiated and maintained by IAHE—is the World Hydrogen Energy Conference (WHEC), the first being held in Miami, 1976. By 2005, the Kyoto Protocol was signed, which represents an

Table 1.1 Worldwide associations which are active in hydrogen energy

Name	Year[a]	Description
CHFCA	2009	Canadian Hydrogen and Fuel Cell Association collaborates with IAHE to promote hydrogen adoption in Canada. It was formed in 2009 as a result of a merger of the Canadian Hydrogen Association (CHA) and Hydrogen and Fuel Cells, Canada. Through the former CHA, the CHFCA has been active in the field of hydrogen energy for over 25 years.
NHD	2009	Nuclear Hydrogen Division of IAHE organizes and promotes activities that foster development of nuclear systems for hydrogen production, including thermochemical cycles, electrolysis, and related elements such as safety and materials science.
NH_3FA	2003	Ammonia Fuel Association promotes ammonia as a source of hydrogen for better and more compact storage. It indirectly sustains a hydrogen economy. It started as a series of annual conferences since 2003, named the Annual Ammonia Alternative Fuel Conferences.
ISO	1990	International Organization for Standardization formed ten subcommittees to implement international standards for hydrogen energy technologies.
IAHE	1974	International Association of Hydrogen Energy is the main pillar in the hydrogen movement and its initiator. It governs the World Hydrogen Energy Conference and it is responsible for publishing the International Journal of Hydrogen Energy.

[a]Year since the organization became involved in hydrogen energy activities

engagement of many countries to develop a cleaner energy system; hydrogen is an important part of this goal.

In Table 1.1, the Nuclear Hydrogen Division (NHD) of IAHE is an organization that promotes and fosters collaborations amongst countries pursuing nuclear production of hydrogen. In the Generation IV International Forum (GIF), 12 countries around the world are laying the groundwork for the next generation of nuclear reactors, which will produce hydrogen as well as electricity. In addition to organizing and promoting nuclear hydrogen meetings and activities, NHD aims to serve as public outreach, and advocacy to policy makers on the nuclear expansion is under way at this time around the world. Technology roadmaps are being developed that will foster the emergence of hydrogen as a key commercial product of nuclear plants in the future.

By the end of the 1980s, the hydrogen economy concept became well known. Russia started publication of Nuclear-Hydrogen Energy Technology periodicals showing a systematic scientific effort for developing a hydrogen economy from nuclear energy. By the end of the 1990s, many of the fundamental problems of hydrogen production, transportation, distribution, storage, utilization, materials, and safety were solved. As a result of the sustained effort in promoting hydrogen—the future energy carrier—many governments initiated research programs on the topic; some relevant of these programs are listed in Table 1.2. These led to many accomplishments, some of them very spectacular like the first flight demonstration of a large-size hydrogen passenger plane—Tupolev 155—in Moscow in 1988, marketing of the first 1 MW SOFC with 70 % efficiency by Westinghouse in

Table 1.2 Sample national or international hydrogen programs

Region	Program	Year	Description
Japan	WE-NET	1993	"Clean Energy Systems Utilizing Hydrogen" is a program aiming to realize the "World Energy NETwork." The project term is 1993–2020.
Texas, USA	ARISE	1990	NSF Center for H_2 research. Electrolysis and photo-electrolysis of H_2S to generate hydrogen. Photo-catalysis of water.
USA	Freedom Car and Hydrogen	2005	DOE program to promote hydrogen economy in the USA.
Europe	Shell Hydrogen	2000	$500M for hydrogen programs.
	CUTE		Clean Urban Transport for Europe.
	ECTOS	2000	Ecological City Transport System.
	Quickstart	2003	Hydrogen production from renewables to use and propel vehicles.
	HyCHAIN	2006	Over 160 FC vehicles in Germany, Spain, and Italy.
Canada–Europe	Euro-Quebec	1992	European-Canadian project to study the transport of hydropower from Quebec to Hamburg in the form of cryogenic hydrogen moved on barges.
Canada	NRC—Vancouver	2010	Around ten hydrogen-fuelling stations were installed in Vancouver during the Olympic games.
Ontario, Canada	Cu-Cl cycle	2011	International consortium of five countries developing an integrated plant of the copper-cycle of thermochemical water splitting for nuclear hydrogen production.
Germany–Saudi Arabia	HySolar	1986	Promoting hydrogen as an energy carrier for long distances. Development of hydrogen production methods.
France	PAN'H	2005	National Programme for Hydrogen Action.
Iceland	Iceland	1998	Aim to convert Iceland to a complete hydrogen economy by 2030.
Turkey	UNIDO-ICHET	2004	International Centre for Hydrogen Energy Technologies formed as a UNIDO project funded by the Ministry of Energy and Natural Resources of Turkey with the role to promote the hydrogen economy.

2001, and the orbiting aerospace plane "Venture Star" by Boeing and Lockheed-Martin that uses a mixture of cryogenic liquid–solid H_2 for propulsion (Veziroglu, 2000). The Japanese hydrogen submarine Urashima 600 was equipped with fuel cells and was launched in 2000. A French submarine with hydrogen and fuel cells with installed power of 300 kW each was built in recent years; it stores the hydrogen in FeTi metal hydrides that contain 1.8 t of fuel onboard. The weight of the storage tank is about 90 times heavier (Bose and Malbrunot, 2007).

One promising option is nuclear hydrogen production. Although nuclear energy is discouraged in some countries, there are fundamental, technical, economical, and sociopolitical advantages in favor of nuclear energy. In Europe, Germany is

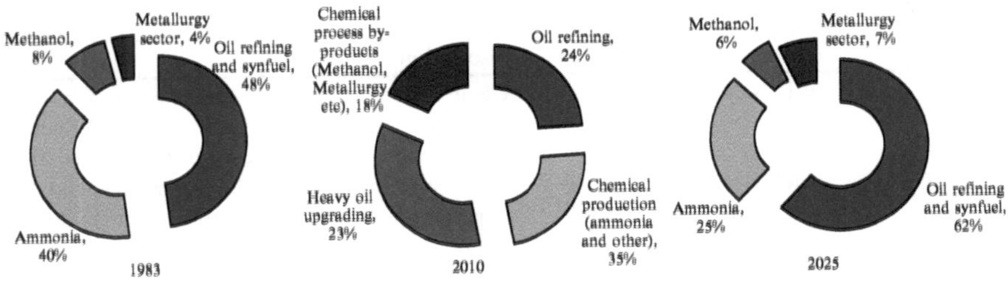

Fig. 1.3 Past and forecasted hydrogen utilization in Canada adapted from Dincer and Zamfirescu (2011)

focusing its efforts on the development of a nuclear-free energy infrastructure involving production of hydrogen from renewable sources such as wind, solar, and biomass. It is interesting to note that a wind power plant must occupy about 100 times more land area than a nuclear power plant site. On the contrary, France bases its energy supply on nuclear power plants. France exploits the only commercial breeder reactor.

Potential pathways to implementing nuclear hydrogen production, where hydrogen has the role of an energy carrier and storage medium, were developed initially in the 1970s. These include the Gas Research Institute and General Atomics in the USA, which identified a number of thermochemical cycles for water splitting, which can be coupled to nuclear reactors to generate hydrogen. More recently, the interest in thermochemical cycles was renewed in 2003 at the Argonne National Laboratory (ANL; see Lewis 2003). A large-scale international program to develop the copper–chlorine cycle in conjunction with generation IV nuclear reactors is currently in progress. It involves a collaboration between Canada, USA, Argentina, Czech Republic, Romania, and Slovenia, led by the University of Ontario Institute of Technology (UOIT), together with Atomic Energy of Canada Limited (AECL) and ANL, and collaborating with the Generation IV International Forum. This program and its accomplishments will be described in detail in this book.

The present worldwide production of hydrogen is around 50 Mt per year. In Canada, which is very active in hydrogen development, hydrogen is utilized as a feedstock for various chemical processes in industries and refineries. Canada produces more than 3 Mt of hydrogen per year (see Dincer and Zamfirescu 2011). Presently there is a significant use of hydrogen in Canada for upgrading heavy oil, particularly for the oil sand sites in Alberta. In the past, hydrogen has been used in Canada for oil refining, ammonia production, methanol production, and process gas in metallurgical sectors. The growth of hydrogen utilization trends in Canada is reported in Fig. 1.3. The oil refining sector and ammonia production accounted in 1983 for 48 % and 40 %, while at present it is observed that the oil refining and heavy oil upgrading has a total share of hydrogen utilization of 47 %. Based on forecasts, by 2025, the oil refining share for hydrogen demand will reach 62 %.

Table 1.3 Thermophysical properties of hydrogen

Property	Description	Property	Description
Melting point (MP)	Transformation solid–liquid at 1 atm, MP = 14.01 K	Boiling point (BP)	Transformation liquid-vapor at 1 atm, BP = 20.3 K, 99.8 % *para*-hydrogen
Critical point	32. 97 K, 12.9 bar	Liquefaction	14.1 MJ/kg theoretical energy
Flame	Flame speed[a] 2.75 m/s, adiabatic flame temperature 2,400 K	Ortho-para percent	At standard temperature, 75 % *ortho* and 25 % *para* hydrogen
Vapor density	0.1 kg/m^3 at 1 atm and 298 K 1.34 kg/m^3 at 1 atm and 20.3 K	Liquid density	70.79 kg/m^3 at 1 atm and 20.3 K
LHV	Lower heating value, 119.9 MJ/kg	HHV	Higher heating value, 141.8 kJ/kg
Flammability	4–18 % and 59–75 % mix with air, ignition temperature 858 K	Detonation	13–65 % mix with air with 2 km/s, suppression of shock wave at 1.47 MPa
Diffusivity	In air, 0.61 cm/s	Conductivity	0.177 W/m K

[a]Six times faster than that of gasoline

1.2 Physical Properties

Hydrogen is the most abundant chemical element in the universe (atomic number $Z = 1$). Having the simplest molecular structure, hydrogen can easily lose valence electrons and therefore it is very reactive. Therefore, hydrogen generally cannot be found as an individual element on earth, but rather in chemical compounds. Water is the most abundant source of hydrogen on earth; also, hydrogen is part of most fossil fuels and biomass. On earth, hydrogen can also be found in the form of hydrogen sulfide (H_2S). Hydrogen has a high calorific value, the lowest molecular weight, the highest thermal conductivity among all gases, and the lowest viscosity. Table 1.3 summarizes the main thermophysical properties of hydrogen, while Fig. 1.4 shows its T–s diagram.

The heating value per unit of weight of hydrogen is much higher than that of conventional fuels. However, hydrogen cannot be kept easily in a condensed phase because its density is too low. Therefore, regardless of its storage method, the heating value of hydrogen per unit of volume is the lowest as compared to conventional fuels. Nevertheless, the efficiency of power generation systems fuelled with hydrogen is much higher than that obtained with conventional fuel. This compensates for the low storage density problem of hydrogen. An important advantage is that hydrogen combustion is completely clean, producing only water vapor in the exhaust gas. Heat recovery can be applied to hydrogen combustion to obtain liquid water which may be recycled back to produce hydrogen.

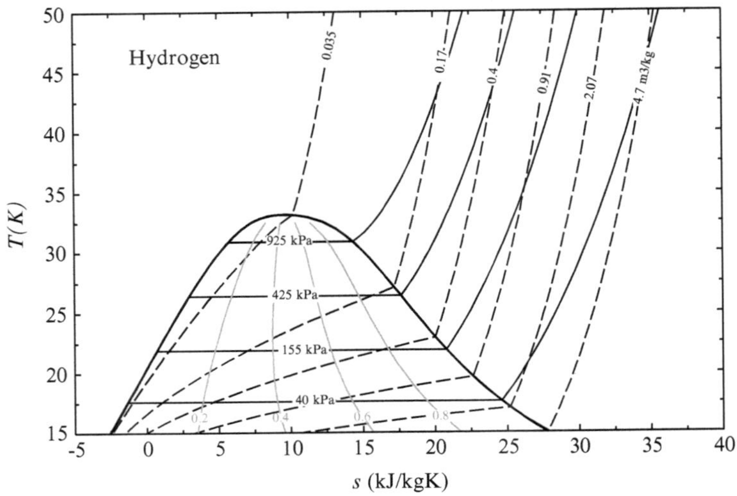

Fig. 1.4 *T–s* diagram of hydrogen

1.3 Transportation

There are relatively few difficulties related to hydrogen transport in pipelines. As mentioned previously, hydrogen in the form of town gas was used for at least 150 years to supply networks of many metropolitan cities. There are however energy costs due to hydrogen transportation and distribution. The approximate orders of magnitude of energy losses from hydrogen production to hydrogen delivery are listed as follows (see Dincer and Zamfirescu 2011).

- Production: 10 % if produced via fossil fuel reforming and 25 % if electrolysis is used.
- Conditioning: purification and compression or liquefaction. Purification is done to 4 ppm for compressed hydrogen or 1 ppm for liquefaction. Conditioning totals 10–15 % losses at compression (to 20–80 MPa, respectively), and 30–150 % losses at liquefaction (for 1 t/h and 0.01 t/h, respectively).
- Transport: Via pipeline 1.4 % per 150 km.
- Distribution at gas station: 7.5–40 % losses due to electrical energy to pump H_2 (for 100 and 2,000 cars per day, respectively).

Apart from pipelines, there are many other alternatives for hydrogen transportation. One possibility is to transport hydrogen via the existing infrastructure of natural gas pipelines. In order to pursue this without modifications, hydrogen is mixed with natural gas in a proportion of 1:9. Hydrogenation of natural gas pipelines is an excellent way of reducing GHG emissions.

Pipelines dedicated to hydrogen transport exist in Europe and USA (mentioned previously). It is found that for practical reasons, the diameter of the pipeline is

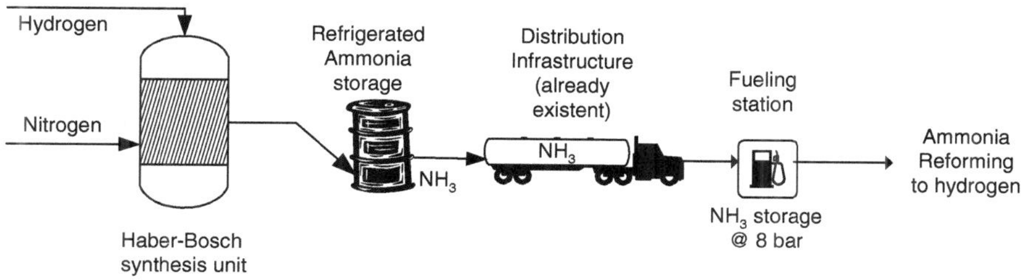

Fig. 1.5 Hydrogen through an ammonia transportation scheme

ideally in the range of 100–200 mm and the operating pressure varies between 0.34 and 10 MPa. Material embrittlement poses some significant technical problems; however there are various technical solutions available (see Bocris 2002).

Road transport is typically done in canisters of compressed hydrogen carried by trucks. Due to the large weight of the canisters, a typical truck of 40 t can transport only ~700 kg hydrogen. The typical pressure of compressed hydrogen is 20 MPa with a storage density of 14 kg/m^3 (gas). The mass of gas with respect to the storage tank is only 6 %. There is technology of hydrogen compression and storage up to 70 MPa which improves the hydrogen packing up to 8 % from the storage tank weight. The longest distance for hydrogen transport in a compressed gas phase is around 500 km, when the cost of transport balances the cost of carried hydrogen.

A more compact transportation method of hydrogen is in its cryogenic liquid state. There are cryogenic cisterns for road transport which can carry 3.5 t of hydrogen in a 40 t heavy truck. The cryogenic mode of transportation is applicable to rail and road transport too.

Other means of hydrogen transport involve chemical transformations. Two main possibilities exist: (1) hydrogen is converted into methanol based on an established technology which uses carbon dioxide recovered from any type of process, and (2) conversion of hydrogen in ammonia via the mature Haber–Bosch process. There is an energy cost in both cases with the intermediate carrier synthesis, but this is compensated by cheaper transportation and distribution. Retrieving hydrogen from ammonia or methanol at a distribution point is straightforward. It requires relatively little energy, which can be derived onsite from renewables.

Figure 1.5 illustrates the hydrogen through ammonia transportation scheme. A simplified cost analysis of this scheme in comparison to road transport in cryogenic and compressed gas states was performed by Dincer and Zamfirescu (2011). It shows that although converting hydrogen from ammonia introduces a 25 % cost penalty, due to its cheaper distribution and reformation, transportation of hydrogen via ammonia competes well with road transport of compressed and liquefied hydrogen.

The last segment of hydrogen transport is its distribution (or delivery) to the users. Two types of fuelling stations exist up to now for hydrogen applications (vehicles): compressed gas or cryogenic liquid stations. Two versions exist for vehicles fuelling with compressed hydrogen: transfer via a special hose from the high-pressure reservoir of the fuelling station to the fuel tank of the vehicle, or using interchangeable fuel tanks. The latter solution can be done with a certain degree of automation; it can be fast and adapted to light vehicles. For transfer of liquid hydrogen, the only practiced solution is to convey the fuel via a liquid pump.

1.4 Storage

The issue of hydrogen storage can be discussed only in conjunction with the storage scale. One can have very large storage capacities of order of hundreds of thousands of cubic meters or more, or one can have intermediate storage in large-size reservoirs, small-scale storage in road transport, or even microscale storage in portable power applications (e.g., powering laptops). For each case, it corresponds a specific set of solutions.

Very-large-scale storage can be applied in the vicinity of industrial producers of hydrogen, including future nuclear hydrogen facilities. In this case, hydrogen can be injected into exhausted oil wells, underground caverns, or salt domes. There is good experience of hydrogen storage in salt caverns. Praxair has a hydrogen pipeline in the Gulf of Mexico that is integrated with a storage facility in caverns. Linde stores hydrogen underground in porous rocks at pressures up to 40 atm.

For medium-size storage, an interesting possibility is hydrogen storage in the form of ammonia or urea. Ammonia can be stored seasonally as refrigerated liquid at atmospheric pressure in tankers of 60 kt. This strategy is widely applied because ammonia is a product produced throughout the year and consumed mostly in summers (as fertilizer). A typical sealed reservoir is a building size of 50 m diameter and 30 m height.

Figure 1.6 illustrates a storage facility of ammonia. Since a molecule of ammonia embeds three atoms of hydrogen, it is relatively straightforward to retrieve hydrogen from ammonia under an energy cost of 15 % from the lower heating value of the extracted hydrogen. The amount of hydrogen generated from 60 kt of ammonia is about 10 kt.

For a smaller scale, hydrogen is difficult to store. It is very difficult to store hydrogen in a satisfactory dense phase so that the packed energy density is similar to that of common fuels used today for vehicles. On a higher heating value base, 1 kg of gasoline is equivalent to 0.3 kg of hydrogen. Gasoline has 1.3 L/kg, while the equivalent liquefied hydrogen contains 3.5 times more volume. Compressed hydrogen at 400 bar and 298 K occupies 9.6 times more volume than gasoline. The following storage methods are considered for hydrogen for small-size storage:

- Cryogenic liquid at standard pressure and 20 K
- Compressed gas at standard temperature and very high pressure
- Physical binding of hydrogen molecule in a solid material matrix
- Chemical binding to synthesize a denser chemical that can later release hydrogen

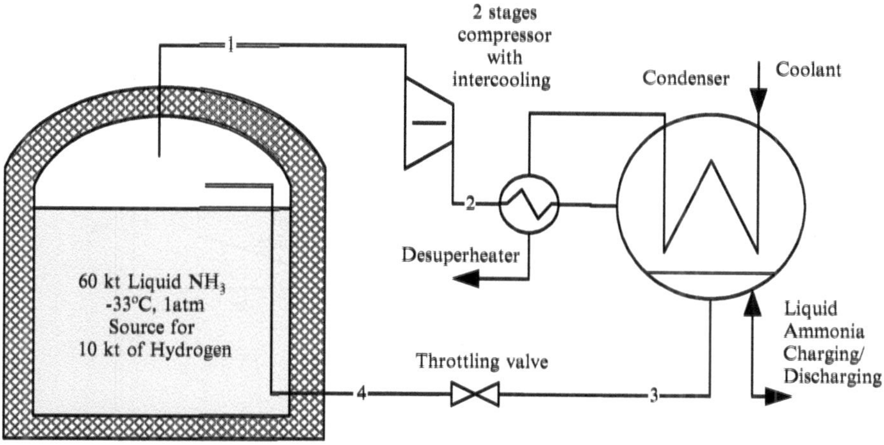

Fig. 1.6 Seasonal storage of hydrogen, chemically embedded in ammonia

Two processes can be used for storing hydrogen in a solid matrix: physisorption and chemisorption. In general, it requires some thermal energy at the discharging phase and some cooling is applied at charging. Sodium, lithium, magnesium, and boron can create chemical hydrides that store hydrogen more densely than possible in metal hydrides via physisorption and chemisorption. Decomposition of chemical hydrides is irreversible. Therefore, sophisticated processes must be put in place to recycle the chemical compounds after hydrogen release. Consequently the devices that generate hydrogen out of chemical hydrides are voluminous.

If the volume of the hydrogen storage and that of the system that extracts hydrogen from it are included, many chemical hydrides appear non-satisfactory for vehicular applications. For example, based on a recent assessment, the US Department of Energy (DOE) assessed boron-based hydrides as not applicable for road vehicles. However, ammonia borane is a good choice for portable applications. It is one of the most promising hydrogen storage systems with 120 g/L in a weight concentration of 19 % (see Dincer and Zamfirescu 2011). Ammonia borane releases hydrogen by thermal cracking according to the following reaction occurring at successively higher temperatures:

$$\left.\begin{array}{r}2NH_3BH_3 \xrightarrow{<120\,^\circ C} (NH_2BH_2)_2 + 2H_2 \\ (NH_2BH_2)_2 \xrightarrow{\sim 150\,^\circ C} (NHBH)_2 + 2H_2 \\ (NHBH)_2 \xrightarrow{>500\,^\circ C} BN + 2H_2\end{array}\right\}.$$

It is interesting to note from the above reactions that ammonia borane can release hydrogen at temperatures lower than 120 °C; if properly conducted, the decomposition reaction can release hydrogen even at room temperature. Releasing hydrogen

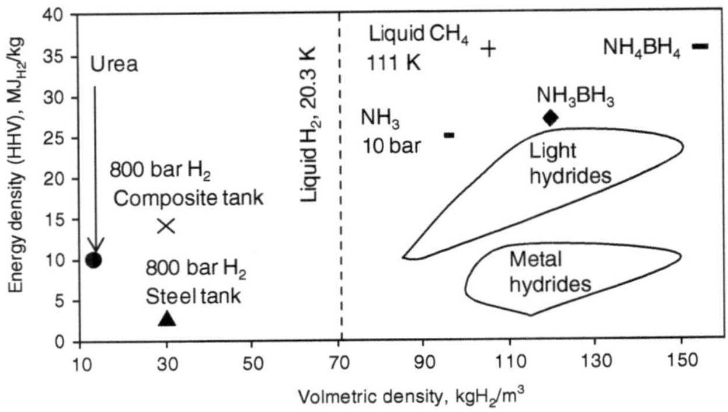

Fig. 1.7 Density of hydrogen storage with various technologies

at low temperature is a crucial characteristic for storage systems with regard to cold start-up of small engines.

Metal amines were proposed as an alternative solution for hydrogen storage via ammonia intermediates, e.g., magnesium chloride which forms $Mg(NH_3)_6Cl_2$. This molecule embeds hydrogen at 109 g/L and 92 g/kg and releases ammonia by thermal desorption.

Ammonia can also be used to store hydrogen "chemically" in small-scale applications. An alternative to ammonia is urea, which is a source of ammonia that can be obtained by hydrolysis. Urea is very stable, can be stored for very long time and it is allowed on passenger cars. When synthesized from biomass, urea is a "zero carbon" fuel and hydrogen source. For extraction of hydrogen from urea, a sustainable thermal energy source can be used, for example heat recovery from exhaust gases. Ammonia-based compounds—as shown in the above example urea—have good capability of hydrogen storage and release.

Figure 1.7 presents the hydrogen storage density with various technologies in terms of volumetric storage (*x*-axis) and gravimetric storage (*y*-axis). The ammonia–boron compound with a chemical formula NH_4BH_4 appears to be the most "dense" hydrogen storage medium.

1.5 Applications

1.5.1 Reduction of GHG Emissions in the Residential Sector

Consider a case study of the analysis of the potential to reduce GHG emissions in the residential sector of Canada by using hydrogen as an energy carrier. This example is based on the work of Lubis et al. (2009). The residential sector in Canada consumes mainly fossil fuels and requires energy according to the graph of

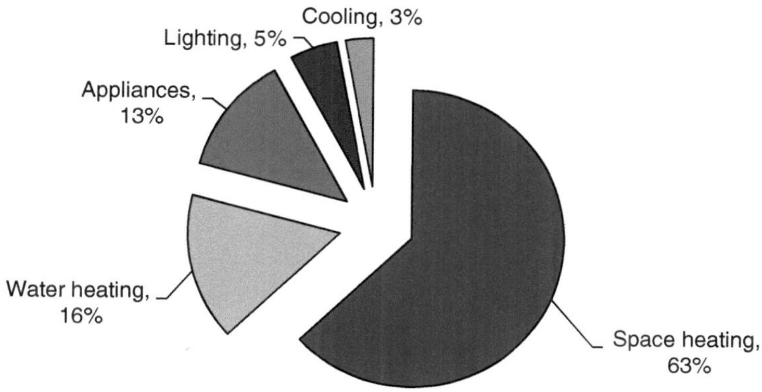

Fig. 1.8 Shares of energy usage in residential sector of Canada [data from Lubis et al. (2009)]

Table 1.4 Breakdown by task and province of energy consumption (in TJ) in Canada in 2005

Service	Energy carrier	Province									
		NFL	PEI	NS	NB	QUE	ON	MAN	SAS	AB	BC
Space heating	Fuel	68.4	60.8	58.7	87.2	95.5	90.5	68.4	76.0	116.0	63.0
Water heating	Fuel	11.6	19.3	15.7	12.2	13.5	25.6	19.0	22.1	29.6	23.6
Operating appliances	Electricity	17.7	7362	18.4	18.1	18.4	17.1	20.6	16.5	15.9	19.3
Lighting	Electricity	6.0	2.5	6.9	6.0	7.2	7.3	7.5	6.6	6.7	8.8
Space cooling	Electricity	0.02	0.05	0.22	0.29	2.0	9.2	4.2	0.84	0.08	0.50

NFL Newfoundland, *PEI* Prince Edward Island, *NS* Nova Scotia, *NB* New Brunswick, *QUE* Quebec, *ON* Ontario, *MAN* Manitoba, *SAS* Saskatchewan, *AB* Alberta, *BC* British Columbia

Fig. 1.8. The objective is to analyze the reductions in GHGs achievable through the use of hydrogen technologies, as a substitute for electricity provided by the grid and space heating from natural gas furnaces for single-family detached houses.

It is assumed that electricity is required for lighting, space cooling, and appliances, while heating and water heating requirements are provided by a furnace (see Table 1.4). The alternative system delivers the required electrical and thermal energy using hydrogen in fuel cells and furnaces. The calculation of GHGs for each case considers the emissions at the consumption of energy as well as production of fuels. For the alternative energy system, hydrogen is combusted in the furnace to generate heat and supplied to the fuel cell to produce electricity. Therefore, at the end use, there are no direct emissions of GHG. For the reference system, the furnace is supplied with natural gas, which is modeled as methane.

The furnace efficiency is assumed to be 60 %, and then the rate of fuel consumption is determined according to

$$\eta = \frac{\dot{Q}_{\text{useful}}}{\dot{m}_{\text{fuel}}\text{LHV}_{\text{fuel}}}. \tag{1.1}$$

Table 1.5 Greenhouse gas emission factors for the electricity of each Canadian province

Province	NFL	PEI	NS	NB	QUE	ON	MAN	SAS	AB	BC
GHG (kg/GJ)	234	235	243	221	1	251	3	334	143	104

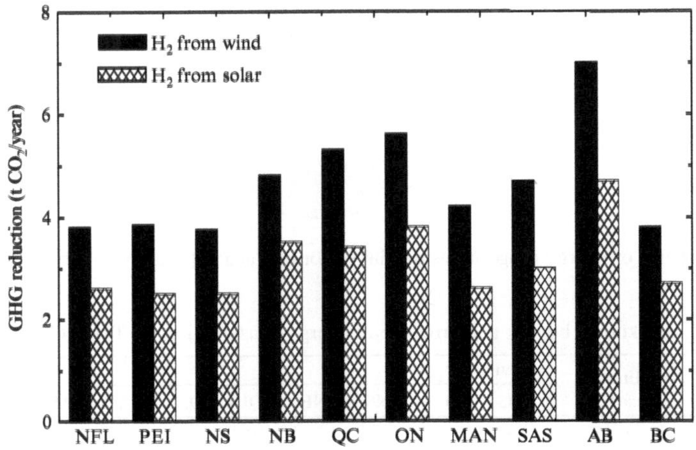

Fig. 1.9 Reductions in greenhouse gas emissions associated with heating systems in the residential sector in Canada, for the alternative case considered

For the hydrogen furnace system, the efficiency is assumed within the practical range of 40–50 %. From (1.1), the hydrogen fuel consumption is determined for any given heat demand. For the fuel cell system, it is assumed that the availability factor is 91.3 %. The evaluation of GHG emissions of hydrogen production considers some indicators as 20.5 kg/GJ when hydrogen is generated from wind energy, and 30.6 kg/GJ if photovoltaic arrays are used. It results in the specific grid emission in each province according to Table 1.5.

The GHG emission reductions for heating systems when hydrogen technologies are used in place of natural gas are shown in Fig. 1.9. The GHG emission reductions for electricity consumption systems when hydrogen technologies are used in place of grid-based electricity are shown in Fig. 1.10. It shows that the greatest annual reduction, 7.2 t CO_2, occurs in Alberta when hydrogen from wind energy is used, mainly because it is the province with the largest thermal energy requirements for heating as well as a large population. It can be concluded that the heating system in Canada must be improved to reduce air pollution.

Regarding the electrical energy supply, the results shown in Fig. 1.10 demonstrate that GHG emissions are reduced in most Canadian provinces when solar- or wind-based hydrogen technologies are used in place of grid-based electricity. In two provinces, Quebec and Manitoba, the hydrogen technologies are not attractive for electrical power generation because they increase the GHG emissions with respect to the reference case. The greatest annual reduction in GHG emissions occurs in Saskatchewan mainly because the GHG emission factors for the

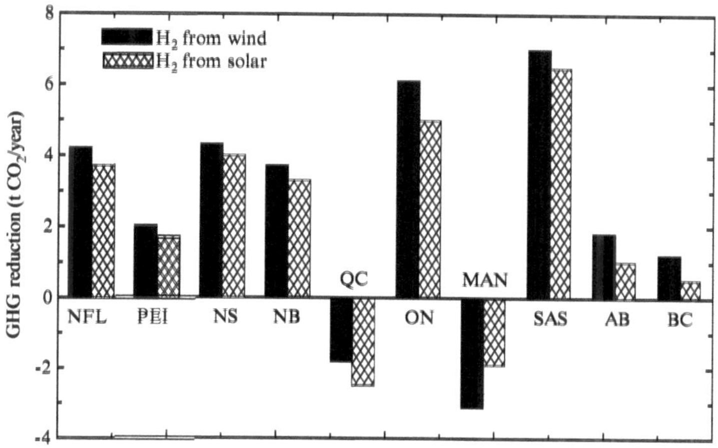

Fig. 1.10 Reductions in greenhouse gas emissions associated with electricity use in the residential sector in Canada, for the alternative case considered

electricity grid in Saskatchewan are higher than other provinces. Although the differences in the reductions or increases in GHG emissions are dependent on whether the hydrogen is derived from wind or solar energy, this dependence is not noticeable when hydrogen supplies heat.

The analysis suggests that the hydrogen technologies considered in the alternative cases have significant potential for reducing emissions of GHGs in Canada's residential sector. Noting that the residential sector consumes a 20 % share of Canada's overall energy supply, it indicates that the potential of hydrogen energy is significant for reducing GHG emissions nationally.

1.5.2 Comparative Life-Cycle Assessment of Hydrogen and Conventional Vehicles

This case study demonstrates the advantages of using hydrogen as energy carrier in passenger vehicles. This example is detailed elsewhere (Dincer and Zamfirescu 2011); this section provides an overview of main results and analysis. Sustainable development requires environmentally benign vehicles characterized by little or no atmospheric pollution during their life cycle. In Canada, 70 million tons of GHGs are emitted per year from vehicle driving.

Six types of vehicles are considered in Table 1.6. The material inventory for batteries and fuel cell stacks is indicated in Table 1.7. The inventory includes analyses regarding the fuel processing stages (both gasoline and hydrogen cases). The hydrogen production efficiency from natural gas is 78 % while that of gasoline from crude oil is 85 %. The analysis also considered the case of hydrogen

Table 1.6 Main characteristics of vehicles

Vehicle	Fuel type	Engine type	Fuel consumption (MJ/100 km)	Driving range (km)
Conventional	Gasoline	ICE	237	540
Hybrid	Gasoline	ICE + electric	138	930
Electric	Electricity	Electric	67	164
Fuel cell	Hydrogen	Electric	129	355
H_2-ICE	Hydrogen	ICE	200	300
NH_3-H_2-ICE	Ammonia	ICE	175	430

Source: Dincer and Zamfirescu (2011)

Table 1.7 Inventory of main construction materials

Component	Mass	Remarks	
Battery hybrid vehicle	53 kg	1.8 kWh capacity; 1.96 MJ electricity and 8.35 MJ petroleum per kg	
Battery electric vehicle	430 kg	27 kWh capacity; 1.96 MJ electricity and 8.35 MJ petroleum per kg	
PEMFC stack	Electrode	4.44 kg	Contains platinum, ruthenium, carbon paper
	Membrane	5.64 kg	Nafion
	Bipolar plate	53 kg	Contains polypropylene, carbon fibers, carbon powder
	End plate	2.8 kg	Aluminum alloy
	Current collectors	1.14 kg	Aluminum alloy
	Tie rod	2.05 kg	Steel
Vehicle	Breakdown of life-cycle energy consumption (GJ/unit)		
	Vehicle type	ICE	PEMFC
	Material production	50	55
	Vehicle assembly	25	24
	Vehicle distribution	2.1	2.1
	Vehicle use	819	195
	Disposal	0.30	0.3

Source: Dincer and Zamfirescu (2011)

production from wind energy and from photovoltaics. Figure 1.10 presents the life-cycle GHG emissions for 1 MJ of mechanical work generated at the vehicle shaft.

The emissions are correlated with the vehicle efficiency. Taking into account that the efficiency of the hydrogen vehicle is high (>40 %) and that of the gasoline vehicle is low (~20 %), these results show that the gasoline vehicle emits the most during its lifetime. The result from Fig. 1.11 is further used to determine the GHG emissions for the vehicles. For an assessment of environmental impact, the air pollution indicator (AP) is calculated and both the AP and GW (global warming parameter in kg CO_2 equivalent per 100 km) are correlated with the curb mass of car. The formulae used for calculations and the amounts of GHGs and AP are presented in Table 1.8.

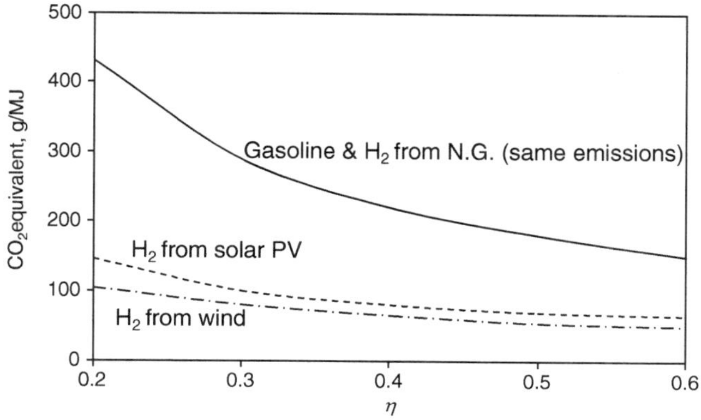

Fig. 1.11 GHG emissions for the entire life cycle, during fuel processing and utilization

Table 1.8 Environmental impact for life cycle of case-study vehicles

Vehicle	Air pollution (AP) and global warming (GW) equations	AP (g/100 km)	GW (kg CO_2 equiv/100 km)
Conventional	$AP = m_{curb}AP_m$; $GW = m_{curb}GWP_m$	3.6	1.5
Hybrid	$AP = (m_{curb} - m_{bat})AP_m + m_{bat}AP_{bat}$; $GW = (m_{curb} - m_{bat})AP_m + m_{bat}GWP_{bat}$	4.2	1.7
Electric	$AP = (m_{curb} - m_{bat})AP_m + m_{bat}AP_{bat}$; $GW = (m_{curb} - m_{bat})AP_m + m_{bat}GWP_{bat}$	6.2	2.0
Fuel cell	$AP = (m_{curb} - m_{fc})AP_m + m_{fc}AP_{fc}$; $GW = (m_{curb} - m_{fc})AP_m + m_{fc}GWP_{fc}$	17.8	4.1
H_2-ICE	$AP = m_{curb}AP_m$; $GW = m_{curb}GWP_m$	4.0	1.5
NH_3-H_2-ICE	$AP = m_{curb}AP_m$; $GW = m_{curb}GWP_m$	3.0	1.4

Source: Dincer and Zamfirescu (2011)

This analysis considers all phases of a product (system, technology) life cycle, starting with the extraction of resources, manufacturing of materials and parts, assembly, use, maintenance, and disposal. It shows that the lowest global warming impact is that of hydrogen—internal combustion engine (either directly supplied by hydrogen or indirectly, via ammonia—the NH_3-H_2-ICE case). Also H_2-ICE vehicles show lower air pollution indicators (see Table 1.8). In Table 1.8, the fuel cell vehicle has high air pollution due to the manufacturing phase of the vehicle which involves many air polluting processes. In conclusion, hydrogen-fuelled vehicles are a good choice for the future hydrogen economy, although nowadays, due to high costs and a lack of infrastructure, they do not yet compete with the existing gasoline technology.

1.6 Conclusions

This chapter discussed the role of hydrogen as a potential energy carrier for better environmental and sustainable development. A focus on hydrogen is given with respect to historical perspectives, physical properties, and comparisons with other fuels and options, as well as its use in transportation, and options with respect to storage and other applications.

Nomenclature

AP	Air pollution indicator, g/100 km
GW	Global warming indicator, kg CO_2/100 km
GWP	Global warming potential, kg CO_2/100 km
LHV	Lower heating value (MJ/kg)
$\dot{m}$	Mass flow rate (kg/s)
$\dot{Q}$	Heat flux (kW)

Greek Letters

η Efficiency

Subscripts

Fuel Natural gas or hydrogen

References

Bareto L, Makihira A, Riahi K (2009) The hydrogen economy in the 21st century: a sustainable development scenario. Int J Hydrogen Energy 28:267–284

Bocris JO'M (2002) The origin of idea of a hydrogen economy and its solution to the decay of the environment. Int J Hydrogen Energy 27:731–740

Bose T, Malbrunot R (2007) Hydrogen. Facing the energy challenges of the 21st century. John Libbey Eurotext, Esher, UK

Dincer I, Zamfirescu C (2011) Sustainable energy systems and applications. Springer, New York

Goltsov VA, Veziroglu TN, Goltsova LF (2006) Hydrogen civilization of the future—a new conception of the IAHE. Int J Hydrogen Energy 31:153–159

Lewis D (2003) Hydrogen and its relation with nuclear energy. Prog Nucl Energy 50:394–401

Lubis L, Dincer I, Naterer GF, Rosen MA (2009) Utilizing hydrogen energy to reduce greenhouse gas emissions in Canada's residential sector. Int J Hydrogen Energy 34:1631–1637

Veziroglu TN (2000) Quarter century of hydrogen movement 1974–2000. Int J Hydrogen Energy 25:1143–1150

Winter C-J (2009) Hydrogen energy—abundant, efficient, clean: a debate over the energy-system-of-change. Int J Hydrogen Energy 34:S1–S52

Chapter 2
Nuclear Energy and Its Role in Hydrogen Production

Abstract In this chapter the role of nuclear energy in hydrogen production at large scale is discussed. In the first part of the chapter various routes of hydrogen generation using nuclear energy are described. Five routes are identified for hydrogen generation by water splitting, among which four are based on thermal energy derived from nuclear reactor, while the fifth is based on the radiolytic effect (that is, disintegration of water molecule under the impact of nuclear radiation). The role of hydrogen as energy storage medium for load levelling of the regional electrical grid is extensively discussed. It is shown that hydrogen production when electricity demand is low, storage and its use in fuel cell for power generation when electricity demand is high, represents a very attractive method for effective generation of electricity in regional grids, which reduces the costs and decreases the environmental production when nuclear energy is the primary source. Large-scale hydrogen production is also essential for petrochemical operations and heavy (nonconventional) oil upgrading, or oil-sand extraction/processing procedures. Hydrogen option represents a potential solution for transportation sector where it can be used either directly (hydrogen is stored onboard of vehicles) or indirectly (hydrogen is converted in a synthetic fuel such as gasoline, diesel, methanol, or ammonia). All means of transportation can benefit from hydrogen as energy carrier; in this chapter the road, rail, and air transport are analyzed in detail.

2.1 Introduction

The constant rise of energy demand in the world imposes an increasing supply of power generation capacity, while efforts must be made to reduce the greenhouse gas (GHG) emissions associated with higher capacity. In addition to clean electricity generation, there is a need for high-temperature thermal energy supply by methods with reduced pollution. As a result, worldwide research efforts are focusing on the development of renewable energy technologies, such as wind, concentrated solar thermal, ocean, biomass, geothermal, among others.

G.F. Naterer et al., *Hydrogen Production from Nuclear Energy*,
DOI 10.1007/978-1-4471-4938-5_2,

Although the generation potential from renewable sources is sufficient to satisfy the energy needs of humankind, their relatively high costs have limited their large-scale adoption. A major challenge with solar energy is its day–night intermittency and intensity fluctuation. Solar-to-hydrogen technology can reach an efficiency of a few percent at an elevated cost. Wind power fluctuates and it is characterized by a small capacity factor. Massive use of biomass energy has ethical issues related to the dilemma of "crops for energy or crops for food." Exploitable geothermal sources are limited.

In the energy mix of the future, there is a need for a stable source of energy that can assure a base load without GHG emissions. For this goal, nuclear energy must be considered. All nuclear power stations presently in operation are designed to steadily generate electricity to satisfy the base load of an electrical grid. There are an increasing number of countries that are adopting nuclear energy programs, showing a growing worldwide effort to develop a fourth generation of nuclear reactors for combined power, high-temperature heat, and hydrogen. In the future, all energy systems are expected to be hybrid systems, which are a combination of various energy resources and energy conversion methods to operate as one system, to maximize efficiency and to reduce environmental impact and waste heat. Hydrogen is a promising energy carrier to link between renewable and nuclear energy sources in those hybrid systems which are sustainable and environmentally benign. A non-exhaustive list of advantages of nuclear hydrogen production is given as follows.

- It generates a new market—hydrogen—which can be sold either as electricity (via fuel cells) during peak periods when the price is higher, or as fuel for transportation, or as a chemical feedstock to industry, whichever is more advantageous.
- It levels the electricity demand profile of the grid.
- It allows for more efficient power production as the generators (other than nuclear) can operate mostly near to their nominal load.
- It reduces pollution associated with power production because the generators operate with higher efficiency.
- It aids development of renewable energy sources which need temporary energy storage (such as hydrogen) to smooth the fluctuating availability profile.
- It makes nuclear power plants more cost competitive and secure by integration with high-temperature electrolysis and thermochemical cycles.

In this chapter, the role of nuclear energy for hydrogen production is discussed and related issues are analyzed. The objective is to introduce the main methods of hydrogen production from nuclear energy and to study three major roles of nuclear hydrogen, namely: (1) energy storage medium for load levelling of an electrical grid, (2) promoting cleaner petrochemical operations, and (3) providing an environmentally benign fuel for the transportation sector.

2.2 Nuclear Hydrogen Production

Nuclear energy is generated in fission reactors using uranium-based fuel (or thorium in the near future). The generated nuclear radiation is converted to high-temperature heat which is transported by a heat transfer fluid for further use. The regular use of nuclear heat is power generation in large-scale power plants. It is possible to divert some of the high-temperature heat from nuclear reactors to supply chemical processes that eventually generate hydrogen from water splitting. Moreover, nuclear radiation present in nuclear reactors and during all phases of nuclear fuel processing can be used directly to generate hydrogen from water.

Five general methods can be envisaged to generate hydrogen from nuclear energy through water decomposition: (1) radiolysis, (2) electrolysis, (3) high-temperature steam electrolysis, (4) hybrid thermochemical water splitting, and (5) thermochemical water splitting. Method (1) uses nuclear radiation to directly split the molecule of water into hydrogen and oxygen; method (2) uses electricity derived from nuclear energy to electrolyze water; methods (3) and (4) are called hybrid because they use both electricity and high-temperature heat to split water; and method (5) directly uses high-temperature heat. All possible paths of water splitting using nuclear energy are represented in Fig. 2.1. As shown in the figure, nuclear energy is converted into nuclear radiation (either in the reactor or during the fuel processing cycle). Nuclear radiation is converted further to high-temperature heat. Both the radiation and heat are used to generate hydrogen, as indicated.

The radiolytic (also known as chemo-nuclear) water splitting method uses high-energy radiation or kinetic energy of fission products to excite water molecules and generate hydrogen and oxygen. Experiments of Carty et al. (1981) show that water can be split with fission fragments and steam can be split by alpha particle irradiation. An experiment of alpha irradiation of steam splits six water molecules with 100 eV radiation, which corresponds to 15 % efficiency (Carty et al. 1981). However, with current commercial reactors, a configuration for radiolytic water splitting is not practical because of requirements of containment of the radioactive material, although radiolysis can, in principle, be implemented at spent fuel pools of existing nuclear power plants.

The alternative option is to use high-temperature heat in four ways mentioned above (2–5). In current nuclear reactors, nuclear particles are thermalized (slowed by collisions) and nuclear radiation is transferred with the purpose to convert nuclear energy into heat. A thermal fluid is used to carry the heat from the confined nuclear reactor. Using this heat to split water implies a requirement to supply the required Gibbs energy and enthalpy to each molecule.

Figure 2.2 shows the variation of Gibbs free energy and enthalpy of water dissociation with temperature. The Gibbs energy for water dissociation can be calculated from the reaction enthalpy and entropy of $H_2O(l) \rightarrow H_2(g) + 0.5O_2(g)$ with the equation $\Delta G = \Delta H - T \times \Delta S$, where T is the temperature at which the reaction occurs.

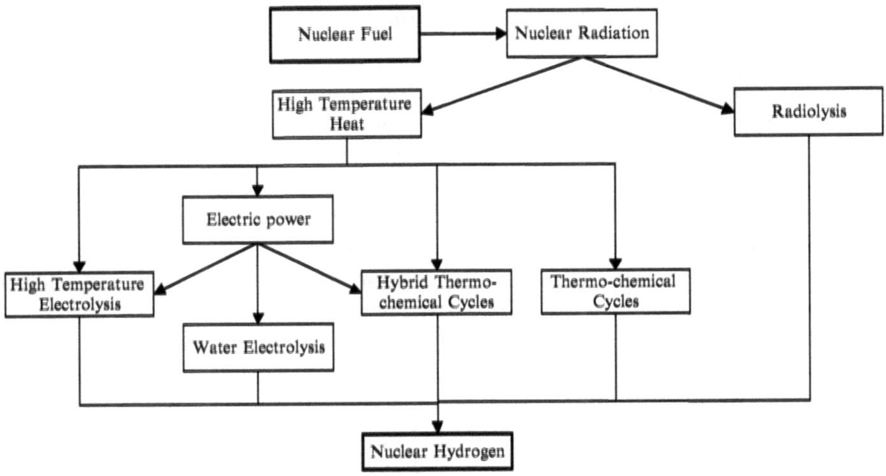

Fig. 2.1 Nuclear water splitting pathways for hydrogen production

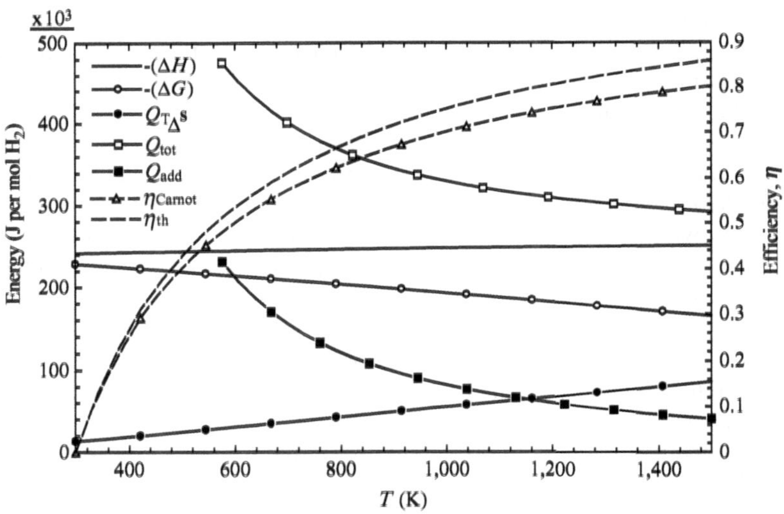

Fig. 2.2 Thermal energy to split 1 mol of water molecule and its required conversions

Although the reaction enthalpy increases with temperature, the Gibbs energy decreases. The Gibbs energy cannot be transferred as heat; it requires an organized form of energy such as work, electricity, high-energy radiation, etc. Method (2) mentioned above uses electricity only to split water molecules by electrolysis. A part of the electrical energy supplied at the electrodes is received by the molecules of water as the Gibbs free energy. The other part, $Q_{T\Delta S} = -T \times \Delta S$, is transmitted as heat, although it is provided in electrical form at the electrodes of the electrolyzer. The $Q_{T\Delta S}$ term originates from internal conversion of electricity into heat, by Joule effects within the electrolyte. The total energy supplied fully electrically to the electrolyzer is therefore $\Delta H = \Delta G + Q_{T\Delta S}$. It can be observed from Fig. 2.2 that the ratio between ΔG and ΔH decreases with temperature. At 298 K, ΔG is 98 % of ΔH; at 1,500 K, this ratio reaches 66 %. This suggests that water electrolysis (method 2) is applicable at temperatures close to the standard value. However, at higher temperatures, other hybrid electrothermal methods (3, 4) or thermal-only methods (5) are applicable.

The energy delivered by the future generation of nuclear reactors will be in the form of high-temperature heat. The level of temperature will be increased with respect to reactors today. For example, the planned CANDU-SWCR reactor generates heat at 650 °C, as compared with the present CANDU-6 which delivers steam at 330 °C. For a water splitting reaction that is conducted at a given temperature T, the heat needed to supply the required Gibbs energy to the reaction can be calculated from the following equation:

$$\Delta G = Q_{\Delta G} \times (1 - T_0/T), \tag{2.1}$$

where T_0 is the reference temperature of the environment. The amount of heat required to drive the water splitting reaction becomes $Q_{\text{tot}} = Q_{\Delta G} + Q_{T\Delta S}$. The heat flow $Q_{\text{add}} = Q_{\text{tot}} - |\Delta H|$ is the minimum additional heat that must be available over the heat corresponding to water dissociation in order to be able to deliver to the molecules the required Gibbs energy and the energy necessary to compensate for the irreversibilities $(-T\Delta S)$. Figure 2.2 indicates the variation of Q_{tot} and Q_{add} with reaction temperature. The maximum thermal efficiency of the water splitting process can be determined by

$$\eta_{\text{th}} = \frac{\Delta H}{Q_{\text{tot}}}. \tag{2.2}$$

According to Fig. 2.2, the maximum thermal efficiency at 825 K (maximum temperature within the Cu-Cl water splitting cycle) is 0.68, and at 1,500 K (viz. high-temperature steam electrolysis) is 0.86. For these temperatures, the Carnot efficiency is 0.64 and 0.80, respectively. The Carnot efficiency curve is also indicated in Fig. 2.2. From this analysis, it is thermodynamically advantageous to generate hydrogen from high-temperature sources because a certain amount of transferred heat (namely, $Q_{T\Delta S}$) can be used directly as thermal energy for the process. This shows the advantage of nuclear hydrogen production by hybrid or thermal-only methods.

The production of hydrogen by water electrolysis does not require the electrolyzer to be placed at the nuclear plant site. It is possible to use distributed electrolyzers connected to the electrical grid. This method is beneficial for the entire grid, including the nuclear power generator, because it allows for load levelling. Section 2.3 of this chapter discussed the importance of delocalized

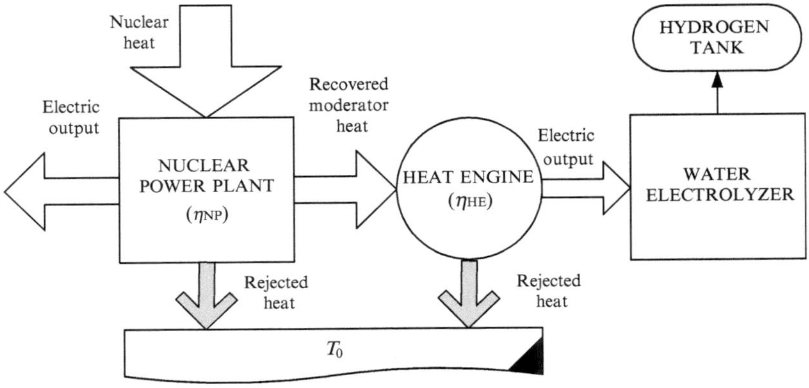

Fig. 2.3 Cogeneration of hydrogen at nuclear power plants from recovered moderator heat

water electrolysis in synergy with thermochemical production at a nuclear power plant site. Another method for nuclear-based water electrolysis is indicated in Fig. 2.3; it couples the moderator to a heat engine which produces more power to operate an electrolyzer. Water electrolyzers, discussed in Chap. 4, have a practical electricity-to-hydrogen efficiency of about $\eta_E = 60\,\%$. Nuclear reactors of the present generation have an efficiency of typically $\eta_{NP} \cong 30\,\%$. The efficiency is over 45 % with a fourth-generation modular helium reactor (MHR) and advanced gas reactor (AGR). If heat rejected by moderator of nuclear power plant is recovered and used to drive a heat engine, then power generation efficiency limited by a Carnot factor of the order of 12 %, thus an achievable $\eta_{HE} = 6\,\%$. It results in a practical recovered heat-to-hydrogen efficiency of $6\,\% \times 60\,\% = 3.6\,\%$. With the assumption that moderator heat represents 5 % of the heat rejected by a power plant, it results in generated hydrogen that represents an equivalent energy of maximum 1 % from the generated power. At a scale of production of 700 MW, the hydrogen production would be significant, approximately 600 kg/day.

The thermal-to-hydrogen efficiency of high-temperature electrolyzers, hybrid and direct thermochemical water splitting cycles is limited as indicated by the η_{th} curve in Fig. 2.2. High-temperature steam electrolyzers operate at ~1,000 °C. The electricity-to-hydrogen efficiency can be over 90 %, as discussed in Chap. 5. With a 30 % efficiency of power generation, it results in 27 % nuclear heat-to-hydrogen efficiency. In high-temperature electrolysis, at least a portion of energy equal to the Gibbs energy of the reaction must be given as electricity. If a hybrid thermochemical cycle is used instead, the required electricity is typically much smaller than the Gibbs energy for water splitting at that temperature. On the other hand, the heat requirement is higher than $Q_{T\Delta S}$. Direct thermochemical cycles do not require electrical energy input. In Fig. 2.4 the layout of a system for coupling nuclear reactor with thermal of hybrid water splitting systems is suggested. A part of the heat from nuclear reactor and a part of the power generated by the nuclear power plant are supplied to water splitting system.

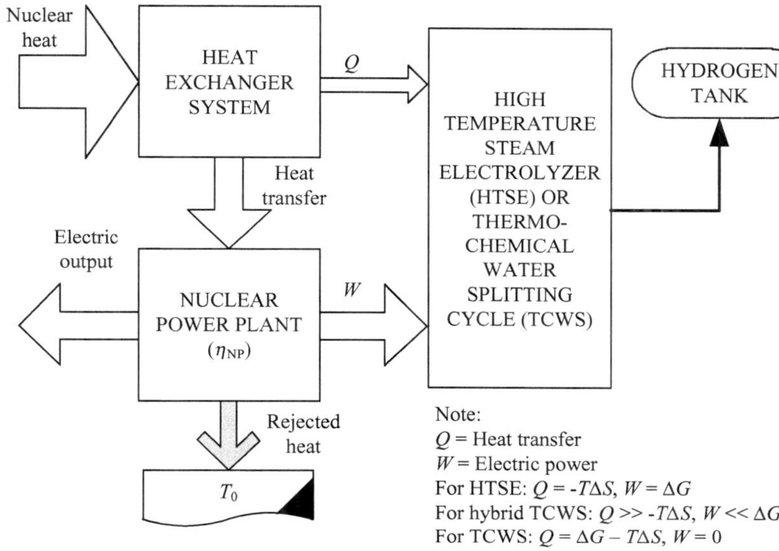

Fig. 2.4 Configuration for coupling high-temperature steam electrolysis or thermochemical water splitting cycle with a nuclear reactor

In the following sections, thermochemical cycles that can be adapted to nuclear reactors of the fourth generation are described. Thermochemical cycles aim to reduce the amount of electricity and the temperature for water splitting reactions via a number of intermediate steps in which additional chemical reactions occur. The overall reaction is $H_2O \rightarrow H_2 + 0.5O_2$. In a multistep process it is, in theory, possible to decrease the work requirement to as low as zero by performing reactions with positive entropy changes at high temperatures and reactions with negative entropy changes at low temperatures.

The sulfur–iodine cycle (S–I) comprises three steps and requires heat input at 800–1,000 °C. The first step is the Bunsen reaction according to $I_2 + SO_2 + 2H_2O \rightarrow 2HI + H_2SO_4$ in which exothermic sulfur dioxide (SO_2) gas absorption takes place in the liquid phase at 20–100 °C. Gaseous SO_2 reacts with iodine (I_2) and water (H_2O) to generate an aqueous solution of hydriodic acid (HI) and sulfuric acid (H_2SO_4). Then the two types of acids produced at the end of the reaction are separated by liquid–liquid-phase separation in the presence of excess iodine. The HI decomposition reaction $2HI \rightarrow I_2 + H_2$ generates hydrogen with a low endothermic heat of reaction at 300–500 °C in the gas phase. The H_2SO_4 decomposition reaction occurs according to $H_2SO_4 \rightarrow H_2O + SO_2 + 0.5O_2$ in two stages; at first, gaseous H_2SO_4 decomposes into H_2O and SO_3 at 400–500 °C; secondly, SO_3 decomposes into SO_2 and O_2 at about 800 °C with the help of a solid catalyst. The sulfur–iodine cycle is discussed in detail in Chap. 6.

A variation of the S–I cycle is the sulfuric acid cycle developed initially by the Japan Atomic Energy Agency (JAEA). It includes a low-temperature electrolytic process at 450 °C of sulfuric acid decomposition, $H_2SO_4 \rightarrow H_2O + SO_3$, using a solid electrolyte that conducts oxygen ions. The sulfur trioxide is decomposed thermally, $SO_3 \rightarrow SO_2 + 0.5O_2$ at 550 °C, and the sulfur dioxide is combined electrochemically with water at 80 °C to release hydrogen according to $2H_2O + SO_2 \rightarrow H_2SO_4 + H_2$.

The ISPRA Mark 9 thermochemical cycle requires 650 °C as a maximum temperature to drive a hydrolysis reaction of iron dichloride according to $6FeCl_2 + 8H_2O \rightarrow 2Fe_3O_4 + 12HCl + 2H_2$. It follows a chlorination step of magnetite occurring at 150 °C, and a decomposition step of iron tri-chloride $6FeCl_3 \rightarrow 3Cl_2 + 6FeCl_2$ conducted at 430 °C.

The lithium-nitrite cycle proposed by the Argonne National Laboratory (ANL) comprises three steps, namely, (1) hydrogen iodide production, $LiNO_2 + I_2 + H_2O \rightarrow LiNO_3 + 2HI$ by oxidation of lithium nitrite at about 25 °C; (2) hydrogen production through thermal decomposition of hydrogen iodide, $2HI \rightarrow H_2 + I_2$ at 425 °C; and (3) lithium nitrate decomposition, $LiNO_3 \rightarrow LiNO_2 + 0.5O_2$ at 475 °C.

The magnesium-chloride cycle was studied at ANL and the Idaho National Laboratory. It comprises three steps, one of which is the electrolysis of anhydrous hydrochloric acid, $2HCl \rightarrow H_2 + Cl_2$. The second step occurs as chlorination of magnesium oxide, $MgO + Cl_2 \rightarrow MgCl_2 + 0.5O_2$ at 500 °C, while the third step is the hydrolysis of magnesium chloride, $MgCl_2 + H_2O \rightarrow 2HCl + MgO$ at 450 °C.

Atomic Energy of Canada Limited (AECL) and ANL have been developing a low-temperature thermochemical cycle called the copper–chlorine (Cu–Cl) cycle, which is currently under development and scale-up at the University of Ontario Institute of Technology (UOIT). It has a number of versions depending on the number of steps: three, four, and five steps. The maximum temperature required with the five-step cycle is 530 °C, which occurs as follows:

- Steam at 400 °C and solid copper chloride ($CuCl_2$) at 400 °C enter a fluidized bed reactor, where an endothermic chemical reaction occurs, $2CuCl_2(s) + H_2O(g) \rightarrow CuO{\cdot}CuCl_2(s) + 2HCl(g)$.
- Copper oxychloride decomposes thermally at 500 °C and releases oxygen according to the reaction $CuO{\cdot}CuCl_2(s) \rightarrow 2CuCl(l) + 0.5O_2(g)$.
- Electrolytic copper is produced, $4CuCl(s) \rightarrow 2CuCl_2(aq) + 2Cu(s)$ at 80 °C in an aqueous solution.
- The aqueous $CuCl_2$ is dried, $CuCl_2(aq) \rightarrow CuCl_2(s)$.
- Solid copper is combined with hydrochloric acid at 400 °C to produce hydrogen according to $2Cu(s) + 2HCl(g) \rightarrow 2CuCl(l) + H_2(g)$.

Orhan (2011) found through ASPEN simulations that the Cu–Cl cycle reaches a heat-to-hydrogen efficiency of over 40 %. This cycle is analyzed in detail in Chap. 7, including its variations with regard to the number of steps. Because of its required temperature level of 530 °C, the Cu–Cl cycle is the most promising cycle to be linked with the Generation IV CANDU-SWCR reactor.

2.3 Energy Storage for Load Levelling

In any market, the price is determined based on the economics of supply and demand. In the free electrical energy market, the price is determined by supply, demand, and regulation. The market comprises at least four entities: the electricity generators, the grid, the consumers, and the regulatory body. The electricity generators represent a multitude of power plants (nuclear, hydro, coal, natural gas, etc.), which supply electricity to the grid; this is called the "supply side" of the energy system. The consumers of electricity are those who determine the electricity demand (or the grid load) at any particular moment; they are denoted generically as the "demand side." The grid has a role to transport the electricity from producers to consumers. Additionally, the grid is connected to electricity storage systems able to smooth short-term fluctuations in the demand. The grid is also connected to grids of neighboring jurisdictions to allow for electricity trade by exports and imports. The regulatory body establishes the rules of economic transactions for electricity; thus it influences the market price.

There is a potential to reduce the price of electricity generated by the grid system if nuclear power plants are coupled with hydrogen production facilities. In this section, the importance of energy storage in the form of hydrogen for levelling the grid load is discussed. Moreover, the synergy between centralized base load hydrogen generation at nuclear stations and off-peak hydrogen generation by electrolysis at decentralized locations is discussed with respect to grid economics and ecological factors.

Figure 2.5 illustrates—as a typical case—the supply and demand profile for mid-July 2011 in Ontario. From these curves, the supply meets the demand only incidentally. The supply is generally higher than the demand, except for short periods when there is more demand than supply, indicating that electricity imports from neighboring provinces may be required. The normalization in Fig. 2.5 is made with respect to peak generation for the day. The grid is supplied by a mix of electrical energy generators: nuclear power stations; coal and natural power plants; and hydro, wind, solar, and other generators. Nuclear power plants operate steadily at their optimum to generate the grid base load, while other generators operate at part load.

The load factor, defined as the ratio of generated power vs. the production capability of the generator, quantifies the degree of part load. Figure 2.6 illustrates the load factor of Ontario's grid generators for mid-July 2011. Similar profiles exist for other national/regional grid systems. The nuclear power load factor is quasi-constant and around 0.98. For the coal power plants, it varies from close to 0 to 0.75; for natural gas it varies from about 0.20 to 0.60; for hydropower it starts from 0.40 in the mornings, and reaches 0.60 in the evenings and 0.05 in the night; solar power is only available during the day and it reaches its maximum load factor at about noon. Operation at part load of any power plant cannot be optimal. Therefore, any load variation can lead to efficiency losses; more fuel is needed to generate a unit of electrical energy. This is reflected in the electricity price fluctuations.

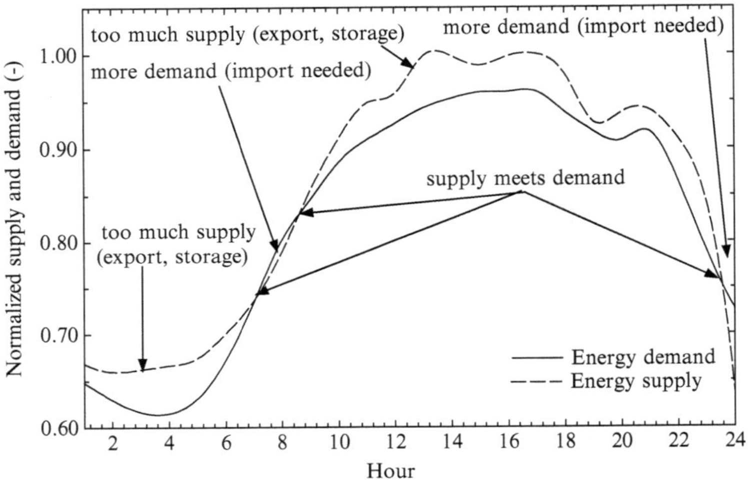

Fig. 2.5 Daily supply/demand profiles in mid-July 2011 in Ontario, Canada [data from IESO (2011)]

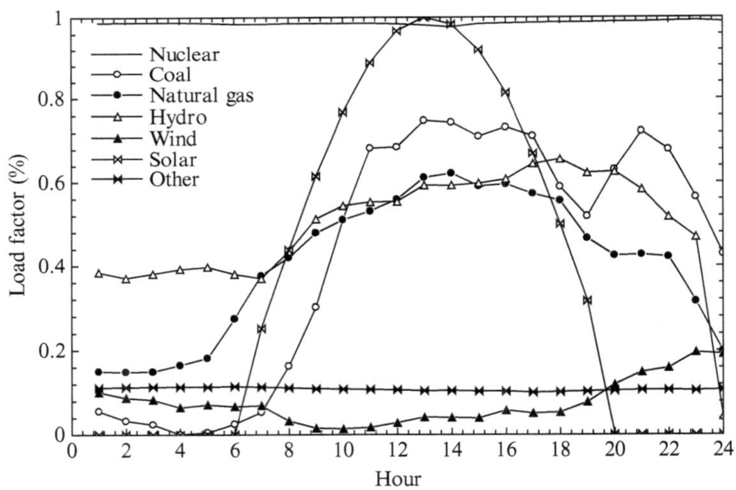

Fig. 2.6 Load factors of grid power generators by type for mid-July in Ontario [data from IESO (2011)]

In the electricity market, there is always a difference between the generated power and the generation capacity of the suppliers. The capacity is larger than the generation. The electricity price is directly influenced by the difference between the generation capability of the electrical energy system and the demand.

The pricing mechanism of electricity is further explained using actual data from an electrical grid. Figure 2.7 shows the demand and generation capability profiles of

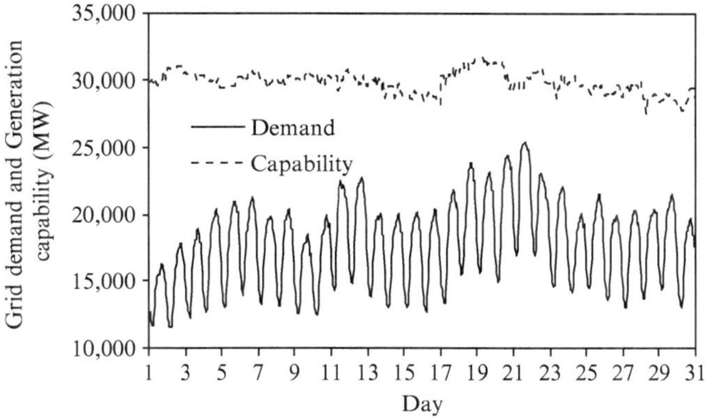

Fig. 2.7 Ontario's grid demand and generation capability in July 2011 [data from IESO (2011)]

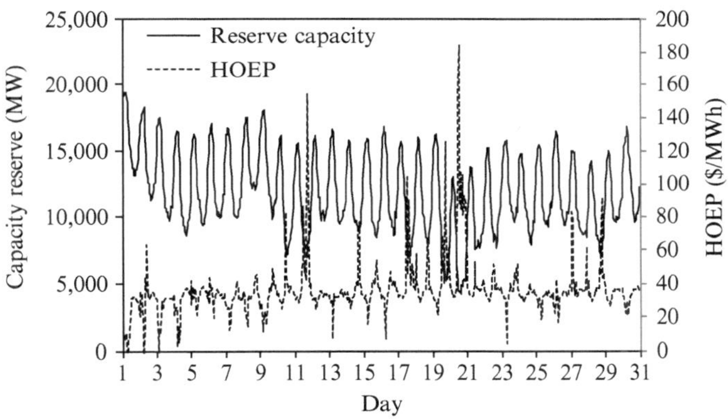

Fig. 2.8 Electricity price correlated with the reserve capacity of Ontario grid in July 2011 [data from IESO (2011); *HOEP* hourly ontario electricity price)]

Ontario's grid in July 2011. The difference between generation capability and the demand is called the "reserve capacity" of the generators. The reserve capacity of the day can be determined by subtracting the "demand" profile from the "capacity" profile. Figure 2.8 correlates the reserve capacity of generators with the hourly electricity cost. Prices are generally higher when the reserve capacity is lower. In addition, the electricity price tends to be higher at peak demand, which typically falls in early evenings when people arrive from work. In off-peak (low demand) hours (e.g., nights) and days (e.g., weekends), the electricity price is low. The low price of electricity in off-peak periods creates the opportunity to generate relatively cheap hydrogen.

The cost of grid-produced hydrogen correlates with three factors: (1) the price of electricity; (2) additional electrolysis capacity, above a continuous level that does not vary with electricity prices throughout the day; and (3) the required storage of

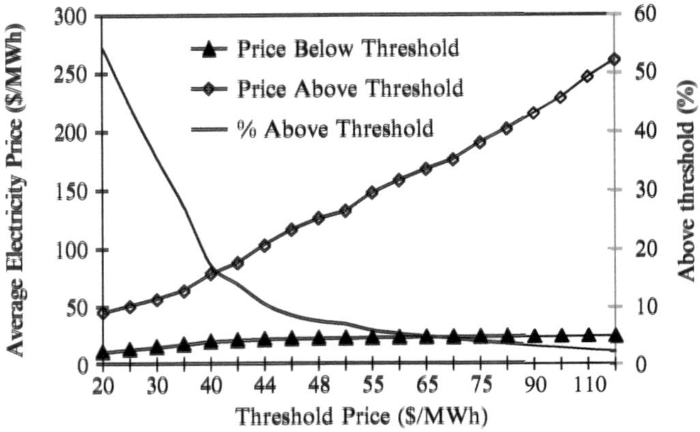

Fig. 2.9 Electricity prices above and below a threshold in Alberta, 2002

hydrogen. Note that the percentage of electricity supply above the price threshold of electricity decreases when the electricity price rises (see Fig. 2.9). If off-peak electricity is used to generate hydrogen by electrolysis, then a threshold price is established, wherein hydrogen is only produced below the threshold. When this threshold price is lowered, more capacity of electrolysis is needed to produce a given quantity of hydrogen over a shorter duration. Also, more storage capacity is needed due to longer periods without generation of hydrogen. Therefore centralized hydrogen generation should be applied.

Generation of hydrogen during the off-peak hours has many advantages. Ideally, for the best economics, the entire difference between the generation capacity and grid demand can be converted to hydrogen, as an energy carrier and storage medium. The grid can generate hydrogen in a delocalized or localized manner. Grid-connected water electrolyzers of capacity up to hundreds of kW or 1 MW can be distributed over a territory—especially in urban regions—to generate hydrogen in a delocalized manner. High-temperature electrolysis systems and thermochemical water splitting cycles can be coupled to heat-rejecting power plants to generate large quantities of hydrogen at their locations. The plants can be equipped with fuel cells to generate back electricity when the grid needs it, or the hydrogen can be sold to local transportation or industrial sectors.

In the subsequent part of this section, the additional benefits of integrating electrolysis and thermochemical hydrogen production are examined. Their unique synergy of coupling decentralized off-peak electrolysis with centralized base load thermochemical production is a significant advantage. The results are derived from work of Naterer et al. (2008). Four cases are analyzed here for hydrogen production.

- System I (benchmark): Decentralized production of hydrogen by water electrolysis.
- System II: Hydrogen production by steam-methane reforming (SMR).

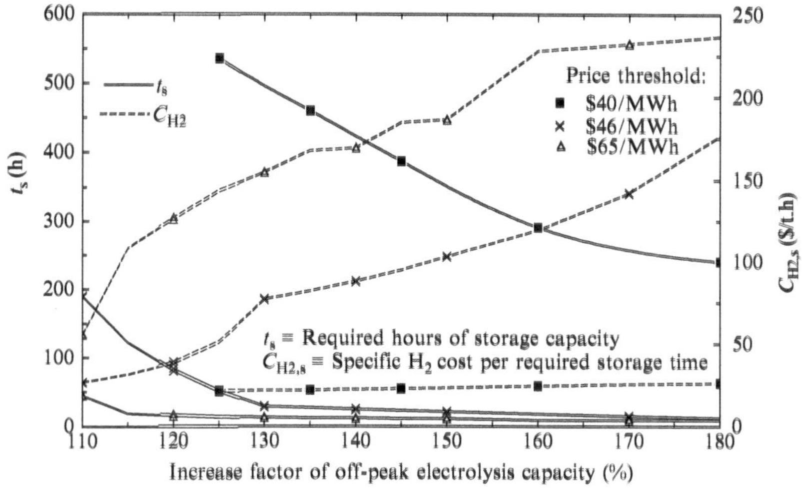

Fig. 2.10 Required hours of hydrogen storage capacity and the specific cost of hydrogen correlated to the increase factor of off-peak electrolysis capacity

- System III: Thermochemical Cu–Cl cycle (TCCC) coupled to an SWCR nuclear reactor.
- System IV: TCCC coupled to a CANDU-III reactor assisted by natural gas combustion.

For System I (the benchmark case), a capital cost of the electrolyzer of \$300/kW is assumed with 15 %/year return on investment and a 10-year amortization period. The electrolysis hydrogen is stored in tubes with an assumed storage cost of \$0.8M/t.

The capacity and storage of hydrogen rise when the price threshold is reduced. For a given electrolysis capacity, if the units operate only when electricity prices are the lowest, then the electrolysis units remain idle for a longer period of time, thereby increasing the required hours of storage. More storage is needed to meet demand, which increases the overall cost.

The electrolyzers are never shut down: during the off-peak hours, the electrolyzers run to a capacity much higher than that corresponding to peak hours of operation. Figure 2.10 shows the influence of the increase factor of off-peak electrolysis capacity with respect to the peak period on the storage time and specific hydrogen cost. If the off-peak capacity factor is higher, then the required hours of hydrogen storage time are reduced. The cost of hydrogen tonnage follows the same trend; however, the cost per unit mass of hydrogen per hour of storage time increases with the capacity factor of the electrolyzer for off-peak periods. Consequently, this factor must be established based on a trade-off analysis that considers the specific storage time and the costs. Moreover, one notes from Fig. 2.10 that the chosen price threshold that defines on- and off-peak periods is a parameter that influences the analysis to a large extent. An optimum exists at a particular electrolysis installation capacity, above which the price per kg of hydrogen produced becomes excessive.

System II comprises a number of centralized plants of large capacity that generate hydrogen by SMR. Consider a capacity of SMR plant of 10 t hydrogen/day (1,420 GJ/day) which was found a suitable choice for integration with Ontario's grid. The capital cost of the SMR plant is approximated by scaling laws according to

$$C_{SMR}/(\dot{m}_{H_2})^{0.66} = C_{ref}/(\dot{m}_{ref})^{0.66}, \tag{2.3}$$

where $\dot{m}$ is the rate of hydrogen production per day and it represents the plant scale; C is the capital cost; the subscript "ref" indicates a reference case; and the exponent 0.66 is a typical scale factor for chemical plants.

For reference values in (2.3), one notes that an SMR plant of 230 t/day has a capital cost of \$65M. Thus for 10 t/day, it results in \$8M investment. The storage facility for hydrogen at a centralized location comprises large-size pressurized reservoirs. The associated investment cost for those is taken to be a typical value of \$0.2M/t (which is lower than tube storage used for System I). The resulting SMR capital cost is \$3.1/GJ, assuming a 20 % annual capital charge. If one assumes that hydrogen is transported from the centralized production facility to distributed users by trucks, there is about \$0.8M additional capital reflected in truck operations; consequently the total capital in the System II capital cost is \$3.9/GJ.

The operational cost of System II depends on the fluctuating price of natural gas. In addition, one has to account for natural gas distribution charges and the SMR conversion efficiency (about 79 %). With such assumptions, the estimated hydrogen price—as indicated in Naterer et al. (2008)—becomes \$8.4/GJ. The cost associated to carbon dioxide sequestration is around \$1.6/GJ.

System III uses the Cu–Cl cycle to split water into hydrogen and oxygen through intermediate copper and chlorine compounds. This cycle comprises a set of chemical reactions that form a closed internal loop that recycles all chemicals on a continuous basis, without emitting any GHGs externally to the atmosphere. The same (2.3) is used to determine the capital cost of a 10 t/day Cu–Cl plant. In this case, the reference values taken from a study by the Argonne National Laboratory—as cited in Naterer et al. (2008)—are \$124 M for a Cu–Cl plant that produces 125 t/day of hydrogen. Accounting to capital discharge and a 3 % inflation rate up to 2015 when the cycle is assumed to become a commercial technology, it yields a contribution of \$7.7/GJ for the capital portion of the hydrogen cost.

A thorough analysis with Aspen Plus software shows that a heat-to-hydrogen efficiency of the Cu–Cl cycle of 54 % is theoretically possible, as compared to 30 % which is the nuclear-to-hydrogen efficiency obtained with Canada's next generation of reactors coupled to electrolysis. Thus there is a significant margin of superior overall conversion efficiency in favor of the Cu–Cl technology. To be conservative and more realistic in the estimates, it is assumed that the heat-to-hydrogen efficiency of the Cu–Cl cycle is 43 %. This cycle allows for a high degree of heat recovery and internal heat recycling between hotter and colder streams, thereby leaving a relatively smaller portion of between 20 and 30 % for the remaining high-temperature heat source at about 530 °C; this heat amounts to about 66 GJ/t of hydrogen. Canada's supercritical water reactor of Generation IV is designed to supply steam at 25 MPa and 625 °C, which is sufficient to supply the required heat to a Cu–Cl plant. It is assumed that a Cu–Cl plant is coupled with a medium-size nuclear reactor of 700 MW. Sample calculations show that for 10 t of hydrogen per day produced with 43 % efficiency, the amount of thermal power needed is ~40 MW, which represents 6 % of the nuclear plant capacity. There are required inputs of about 4.2 kg/s of water and 7.6 MW of heat assuming that the plant operates steadily. It results in 1 % of steam that must be extracted from the power plant and that the high-pressure working fluid (steam) can be cooled from 625 to 618 °C to deliver heat through a heat exchanger to the Cu–Cl cycle. The margin of efficiency improvement is more than one-third higher than the overall heat-to-hydrogen efficiency of electrolysis.

The Cu–Cl cycle includes an energy-intensive spray drying process that can be driven by utilizing waste heat from within or external to the Cu–Cl cycle. Factoring in the additional efficiency gains realized by this waste heat recovery, it yields for System III a similar range of operational cost as the previous natural gas estimate of between $6/GJ and $7/GJ.

Note also that this approach can be adapted with the existing CANDU reactors with a top-up heat source like natural gas. This is the situation considered in System IV, introduced above. The operational cost estimate of this version of System IV is in line with the previous cost estimate, namely, at about $6.3/GJ.

The transportation costs of hydrogen from the centralized Cu–Cl plant consider truck transport at $4.6/GJ for an average distance of up to 16 km. In hydrogen transport, there lies a key synergy between off-peak electrolysis and thermochemical production of hydrogen because the trucking costs rise with distance from the Cu–Cl plant. However, decentralized electrolysis (which does not require hydrogen transportation) has lower costs further away in surrounding towns. Thus, similarly to centralized base load electricity with nuclear power, together with decentralized supply from wind or solar power, the electrolysis and thermochemical production methods have synergistic roles with each other. Similar economic trade-offs exist between thermochemical production and electrolysis with pipeline hydrogen transport as an option. Decentralized electrolysis becomes more economically attractive in towns that surround a nuclear power plant with hydrogen generation capability, because shorter pipelines are needed between fueling stations in cities.

The above systems have a unique potential to serve both decentralized needs in off-peak hours and centralized base load production from a nuclear station, respectively. This potential indicates that water-based production of hydrogen can become more competitive against the predominant existing technology provided that there is a method to effectively link these systems. A first result from the study by Naterer et al. (2008) is presented in Fig. 2.11 where the cost per GJ of hydrogen (delivered to users) for three of the systems considered in the analysis is shown: I,

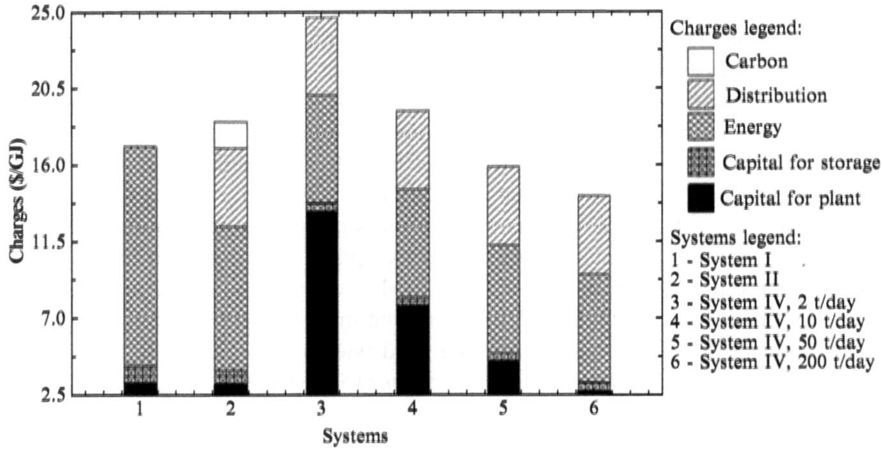

Fig. 2.11 Cost comparison of various production systems for hydrogen [data from Naterer et al. (2008)]

Table 2.1 Cost details for compared Systems I, II, and IV

			IV			
System	I (10 U × 1 t/d)	II (10 t/d)	2 t/d	10 t/d	50 t/d	200 t/d
Capital for plant ($/GJ)	3.2	3.1	13.2	7.7	4.4	2.7
Capital for storage ($/GJ)	1.0	0.8	0.5	0.5	0.5	0.5
Energy ($/GJ)	12.8	8.4	6.3	6.3	6.3	6.3
Distribution ($/GJ)	0	4.6	4.6	4.6	4.6	4.6
Carbon ($/GJ)	0	1.6	0	0	0	0
Total ($/GJ)	17.0	18.5	24.6	19.1	15.8	14.1
Total ($/kg)	2.41	2.67	3.49	2.71	2.24	2.00

II, and IV. System III is excluded in this comparative chart because the specific aim is to compare only the technology of today.

The Generation IV SCWR is still under development; therefore System III will be analyzed later with the purpose to study the potential of improvement of hydrogen production technology. The detailed structure of the cost components is indicated in Table 2.1. For capacities below ~10–20 t/day, electrolysis from off-peak electricity has a lower unit cost of hydrogen production, although the advantage reverses at higher capacities.

The next part of the study considers various scenarios of technology improvement for the Cu–Cl plant and increased benefits of hydrogen production with thermochemical cycles during off-peak periods and applications of the SCWR. Regarding technology improvement, it is assumed that in a relatively short time horizon, the capital cost of a Cu–Cl plant can be reduced by 25 %. System IV with a 25 % capital charge reduction is denoted in the next comparative graph, illustrated as Fig. 2.12, with IV-25 %. Regarding System III (SCWR), which is included in the

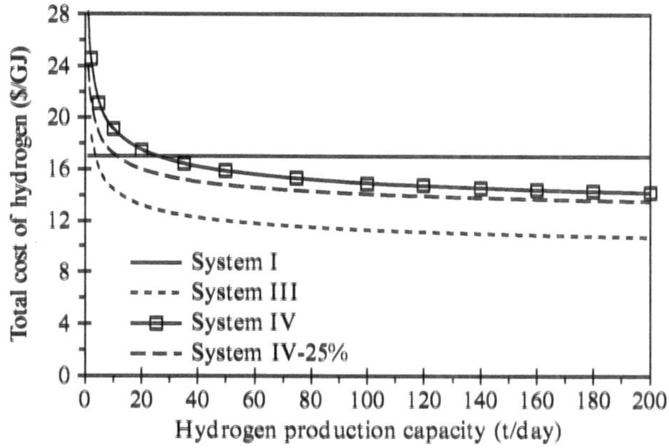

Fig. 2.12 Cost comparison of Systems I, III, III-25 %, and IV with respect to hydrogen production capacity [data from Naterer et al. (2008)]

analysis reported in Fig. 2.12, this uses the same 25 % reduced capital cost of the Cu–Cl plant.

Furthermore, System III uses off-peak times to generate more hydrogen by diverting the high-pressure/high-temperature steam from the turbine to the Cu–Cl plant using a bypass heat exchanger. During the on-peak period, the thermochemical cycle maintains a base load production of hydrogen. Discounting the energy cost by an equivalent amount as the off-peak electrolysis, a further potential cost reduction of $2.8/GJ is illustrated in Fig. 2.12 for System III. A key result is the predicted crossover point between electrolysis and thermochemical production in Fig. 2.12, at about 10–20 t/day. It should be kept in perspective that various idealizations, simplifications, and uncertainties are used in this cost analysis, so the results are intended more to provide qualitative trends, rather than precise costs.

System III appears to be the most effective with respect to Systems I and IV and IV-25 % for large-scale applications. A further discount of the hydrogen cost by System III is obtained by selling the oxygen produced from water splitting. At a market price of about $0.21/kg of oxygen, this effectively represents a discount on the hydrogen cost. In Fig. 2.13, the cost of hydrogen produced by System III if the oxygen is sold as a by-product is indicated; the system is identified on the graph with the legend "System III + oxy." This system is compared to SMR, with System II on the same chart (no carbon capture is assumed for System II). For more expanded comparisons, two systems II with CO_2 capture are considered (see Naterer et al. 2008):

- "System II + $30-carbon" using a deep geological aquifer located 300 km from the SMR plant with associated capture and sequestration costs of $30/t of CO_2.
- "System II + $100-carbon" which assumes the application of a carbon tax of $100/t of emitted CO_2.

In Fig. 2.13, it is predicted that the SMR/Cu–Cl plant crossover occurs at 70 t/day if no carbon sequestration is applied, whereas if it does apply the crossover, it

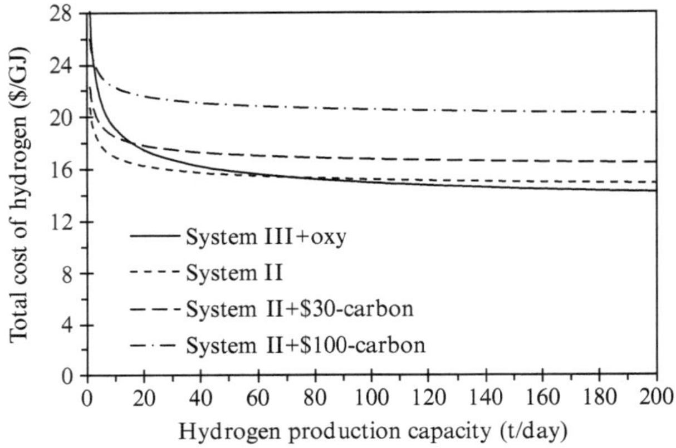

Fig. 2.13 Cost comparison between hydrogen production Systems II (SMR) and III (Cu–Cl + SCWR) for a range of production capacities [data from Naterer et al. (2008)]

occurs at about 20 t/day. Moreover, if carbon is taxed, the thermochemical production becomes competitive vs. SMR at around 5 t of hydrogen per day.

The effect of distributed nodes that leverage the synergy between decentralized hydrogen production and local community infrastructure needs can extend the production capacity of electrolyzers beyond 10 t/day of hydrogen production. Thermochemical methods have significantly higher thermal efficiency, but electrolysis can take advantage of low electricity prices during off-peak hours, local use of by-products of heat and oxygen, as well as intermittent and decentralized supplies of electricity like wind, solar, or tidal power. By effectively linking these systems, production of hydrogen from nuclear and/or renewable energy can become more competitive against SMR.

2.4 Heavy Oil Upgrading and Petrochemical Operations

The world dependence on crude oil and petroleum products will not be eliminated in the current century. The crude oil consumption in recent years increased from 11 millions of barrels per day (mbd) in 1950, to 57 mbd in 1970, to around 85 mbd at present. Crude oil (or petroleum) is a naturally occurring hydrocarbon-based liquid or solid (e.g., bitumen in oil shale, oil sands), found in underground rock formations. The main hydrocarbons included in petroleum are alkanes, cycloalkanes, aromatics, asphaltics, naphthalenes, and paraffins. From the total world crude oil reserves, 30 % is conventional oil, 15 % heavy oil, 25 % extra heavy oil and 30 % is found in forms of bitumen in oil shale and oil sands.

Historical trends of crude oil production and the forecast of future demand are indicated in Fig. 2.14. This figure has been compiled using data derived from previous studies by Dincer and Zamfirescu (2011) and Wang and Naterer (2010).

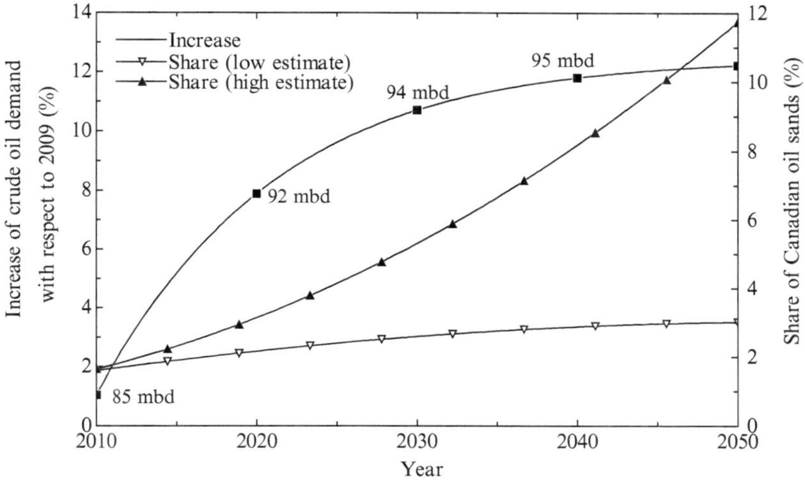

Fig. 2.14 Forecast of crude oil demand increases until 2050 and Canadian oil sands share

It is estimated that until 2040, the world crude oil demand will increase by10 % with respect to the present and afterwards will likely remain quasi-stable at around 95 mbd. Other oil reserves except conventional oil will start to be exploited in a larger degree than today. Figure 2.14 indicates the low and the high estimates of crude oil demand from Canadian oil sands. These estimates are expressed in Fig. 2.14 in percentages from the total crude oil demand of the world. By 2050, the share of Canadian oil sands—which is essentially very heavy oil—will represent between 3 and 12 % of the total crude oil demand. From this figure, one can derive that the technologies of crude oil upgrading and other petrochemical operations (distillation, extraction, fractionation, purification, cracking, etc.) must be advanced to reach higher efficiency, more cost-effectiveness, and lower pollution.

Three main commodities are mainly consumed by petrochemical refineries in order to extract and process crude oil, namely, mechanical power, heat, and hydrogen. In addition, for some processes (oil sand extraction), large quantities of water are consumed, especially in the form of steam. Mechanical power is needed to operate machines and equipment for crude oil extraction and processing. It can be found in the form of direct shaft work, electricity to operate electric motors, or a fuel (diesel, gasoline, or hydrogen in the future) to run machines. The heat is needed by many chemical processes like extraction of hydrocarbon fractions, distillation of crude oil, thermo-catalytic cracking, etc. Hydrogen is needed to upgrade crude oil to lighter fractions to fabricate diesel, gasoline, methanol, etc. With today's technology, the power, heat, and hydrogen needed in refineries and crude oil extractions from the oilfield are highly polluting with large quantities of GHGs released. Two examples of inefficient resource utilizations with today's technology include:

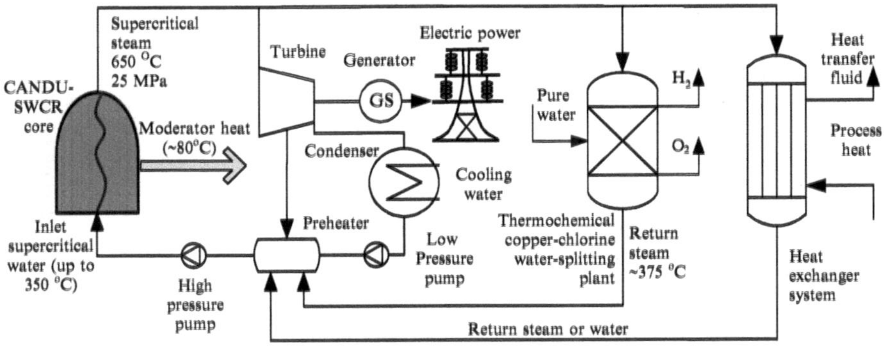

Fig. 2.15 General layout of a coupled CANDU-SWCR + Cu–Cl plant for power, heat, and hydrogen generation

- Using high-quality fuels to heat crude oil in refineries for distillation
- Using natural gas to generate hydrogen for upgrading bitumen from oil sands

This section examines the role of nuclear hydrogen generation for improving the environmental performance of heavy oil upgrading and petrochemical operations. It focuses on three major operations: extraction of bitumen from oil sands, upgrading bitumen to crude oil, and crude oil distillation to produce a range of refinery fuels. The results can be extended for oil shale extraction and upgrading without major modifications. Also, the results can be applied to refineries of traditional crude oil supplied with power, heat, and hydrogen derived from nuclear heat and water, for better cost-effectiveness and reduced pollution.

Many of the petrochemical operations require high temperatures (more than 300 °C). Therefore, the envisaged nuclear reactor is a Canadian CANDU-SCWR (Supercritical Water Reactor) of Generation IV which produces high-pressure steam at around 650 °C and 25 MPa. This steam can be used as a valuable source of power, heat, or hydrogen. Figure 2.15 presents the conceptual coupling of a CANDU-SWCR with a copper–chlorine thermochemical water splitting plant and a heat exchanger system for general-purpose petrochemical operations (including traditional oil or oil sand exploitation and bitumen upgrading). The aim of this system is to provide the required energy and material needs to petrochemical processes such that transportation fuels (gasoline, diesel, jet fuel, kerosene) and refinery by-products (fuel oil, coke, asphalt, methanol, sulfur) can be obtained without burning additional high-quality fuels, reducing thus the pollution and increasing the cost competitiveness.

In order to establish an analysis basis and compare various technologies, consider first the traditional petrochemical operations including those from oil sand extraction and upgrading—which are relatively novel developments in the oil industry. Crude oil must first be extracted from the oilfield (oil sand, oil shale field) and then it is further processed. With the current technology, there are specific materials and energy requirements for each of these two phases. The extraction work is laborious and requires costs (and energy) mainly for locating the reservoir, drilling, oil extraction and recovery, pipelines, and truck transportation to a refinery. For the oil sands case, high energy is spent by surface mining (which is applied up to 75 m deep deposits). The deeper deposits are more economically exploited via in situ methods. Very large quantities of steam are needed to extract the bitumen from oil sands. It requires soaking, softening, bitumen slurry pumping, and separation from sand. This needs roughly 1 barrel of clean water per barrel of crude oil; for heating purposes, on average 17 cubic meters of natural gas are combusted to produce one barrel of crude oil.

Typically more than 90 % of bitumen in the oil sands is recovered while the rest of the mined material is deposited in tailing ponds which measure more than 2 m^3 for each barrel of crude oil. Fresh water must be used with the in situ extraction method (steam-assisted gravity drainage (SAGD)). A major problem is that water becomes contaminated with many pollutants including mercury, heavy metals, and naphthenic acids. In Alberta, the area covered by contaminated water in tailing ponds increased about three times in the last five years (as of 2011, it extended to around 150 km^2). The estimated GHG emission associated with oil sand extraction is 130 kg CO_2 per barrel of synthetic crude oil. The energy and GHG emissions and environmental pollution due to tailing ponds associated with oil sand exploitation are among the most significant of all oil extraction technologies.

There are some major technologies currently applied to extract bitumen from oil sands in Alberta. The most widely applied is "SAGD" which involves drilling to horizontal wells, one above and one below. Steam is injected in the above well, which soaks and softens the bitumen that flows in the bottom well. A bitumen slurry is formed, which is pumped at the surface where bitumen is separated mechanically. The process requires steam at 100–300 °C and around 17 m^3 of natural gas to be combusted. "Cyclical steam stimulation" is another method that uses only one well of steam injection and bitumen slurry extraction after soaking and softening. The "toe-to-heel air injection" method produces a combustion process in the bitumen well by injecting air; the generated heat soaks and softens the bitumen which becomes fluid and then can be extracted. "Cold heavy oil production with sand" is a method extensively applied in Alberta to unconsolidated sand stones; the method uses sand to enhance the bitumen extraction. The SAGD method is also applied with other working fluids than steam. One version is to use hydrocarbon solvents to dilute and slightly upgrade the bitumen; this method is called "vapor extraction."

Another promising version is to use supercritical carbon dioxide ($ScCO_2$). It is shown by Fang (2010) that the optimum pressure and temperature of supercritical carbon dioxide extraction for oil sands are ~25 MPa and 60 °C. This creates a large potential to use low-grade heat recovery. Moreover, the bitumen separation process is very much facilitated by an application of pressure swing principles and recycling of carbon dioxide. After the mechanical separation of bitumen from sand and other impurities at optimal pressure and temperature, a further separation is required: separation of bitumen from carbon dioxide. Pressure swing separation implies a sharp reduction of pressure to 7 MPa (slightly below the critical pressure). Thus, carbon dioxide transforms in the gas phase and what remains is a hot liquid bitumen

Table 2.2 Required inputs for selected petrochemical processes for 1 barrel of crude oil

Process		Inputs
Bitumen extraction from oil sands[a]	SAGD	Steam at 100–300 °C; 27 m^3 natural gas as heat for 1 barrel of crude
	$ScCO_2$	CO_2 at 25 MPa, 60 °C
Bitumen upgrading to crude oil		Hydrogen
Crude oil distillation		2 kg steam at 2 bar 400 °C for 1 barrel of crude oil 102 MJ of heat at 400 °C for 1 barrel of crude oil

[a]Includes oil sand extraction from the well

Table 2.3 Bitumen upgrading temperatures and process description

Upgraded oil type	*T* (°C)	Brief process description
Athabasca bitumen	350–530	Thermal cracking, fluidized bed
Canada bitumen	380–460	Hydrothermal visbreaking, in supercritical water
Fort McMurray (Alberta) oil sands	400	Natural zeolite cracking
Heavy oil	450–550	Treated zeolite catalytic cracking
Alberta bitumen	500	Delayed coking
Alberta bitumen	300–400	Hydro-treating
Heavy oil	450	API 618 standards upgrading
Alberta bitumen	525	Hydrocracking
Canada bitumen	300–400	Fixed bed, removal of S, N, and O
Canada bitumen	410–420	Single cracker, increase H/C ratio
Canada bitumen	470–510	Fluid catalytic cracker

at low pressure. The captured carbon dioxide is recompressed to around 8 MPa, condensed at ~30 °C, repressurized to 25 MPa, heated, and reinjected in the well. In contrast, with $ScCO_2$ extraction in SAGD, water cannot be reinjected due to contaminants. Moreover, the contaminated steam cannot be released into the atmosphere. Therefore, with SAGD, steam is condensed and the resulted contaminated water is accumulated in tailing ponds, as also mentioned above.

Table 2.2 lists the input requirements for a bitumen extraction process, and two other petrochemical processes: bitumen upgrading to crude oil and crude oil distillation. The data from Table 2.2 is derived from the studies of Wang et al. (2010), Dincer and Rosen (2007), and Fang (2010). Regarding bitumen upgrading to crude oil, the required temperatures and process parameters are listed in Table 2.3. In Alberta, most of the hydrogen is produced in SMR plants along Edmonton's "Upgrader Alley." The temperature required by the upgrading depends strongly on the bitumen properties, cracker type, processing method, and upgrading degree, which can be synthetic crude oil or lighter oil, depending on the amount of hydrogen addition.

The crude oil distillation requires steam and heat in the approximate amounts listed in Table 2.2 for 1 barrel of crude oil. By distillation, a large pallet of refinery fuels is obtained: wash oil, heavy gas oil, light gas oil, heavy-to-light diesel, kerosene, naphtha, and liquefied petroleum gas (LPG). Steam is needed in the

Table 2.4 GHG emissions of bitumen extraction from Canadian oil sands using gas for heating

Year	2015	2020	2025	2030
Bitumen production scale (10^6 barrels/day)	2.2	3.4	4.1	4.3
Natural gas volume for heating (10^6 m^3/day)	59.5	90.6	110.4	116.1
Heat supplied by natural gas (MW_{th})	25,603	39,379	47,487	49,803
GHG emissions from combustion for heat (10^6 t/year)	42.6	65.6	79.1	83.0
Compared with 2006 CO_2 emissions (% of 2006)	7.8 %	12.0 %	14.5 %	15.2 %

Note: Canada's total CO_2 emissions in 2006 were 544.68×10^6 t

process of distillation; some 5 % of the supplied steam is recovered as water. As indicated in Dincer and Rosen (2007), the overall efficiency of double-stage distillation plants is around 55 %. A part of the fuel products are combusted to generate flue gas at 1,100 °C for heating.

In the situation, when a system sketched in Fig. 2.15 is used to provide power, heat, and hydrogen to the operations (extraction, upgrading, distillation, etc.) there are no direct GHG emissions. For a comparative analysis purpose, various common methods can be considered to supply these petrochemical processes with heat, power, and hydrogen. For instance, heat can be supplied by combustion of natural gas or by electric heating. In both cases, there are associated GHG emissions if the electricity is derived from the local grid. In Alberta and Saskatchewan—where oil sands are exploited—84 and 60 % of the power generated to the grid, respectively, are due to fossil fuel combustion.

Assume an evolution of Canada's production of oil sand bitumen according to Table 2.4, which falls in the range predicted in Fig. 2.10 (discussed above). It is possible to calculate the GHG emissions due to natural gas combustion for bitumen extraction. Natural gas is used for methane, in which one generates 1 mol of CO_2 per combusted mole of fuel. If bitumen extraction continues to be driven by natural gas combustion, the associated GHG emissions by 2030 would represent a large 15.2 % portion of total CO_2 emissions of Canada in 2006.

Regarding hydrogen production, the traditional methods that compete with the solution proposed in Fig. 2.15 (CANDU SWCR + Cu–Cl cycle) are SMR, coal gasification, and water electrolysis using electricity from the grid. Other alternative methods include the use of the copper–chlorine cycle to split water. The energy to drive the Cu–Cl process can be derived from natural gas combustion or nuclear heat. Each case will be considered.

SMR proceeds with syngas production followed by a water gas shift reaction. It consumes more methane than the stoichiometric balance to compensate for the need of reaction heat; this heat is obtained by combustion of additional methane. For full conversion, the overall reaction is $1.315CH_4 + 1.37H_2O + 0.63O_2 = 1.315CO_2 + 4H_2$. In practice, the yield of hydrogen is 65 % on average; therefore one obtains

$$\text{Emissions (SMR)} = 0.51CO_2/H_2(\text{combustion of methane as the heat source}). \qquad (2.4)$$

However, if the heat needed to supply SMR is derived from a clean source such as solar energy (required 800 °C), then no direct GHG emissions are associated with it. Therefore, the net GHG emission of SMR becomes

$$\text{Emissions (SMR)} = 0.39\text{CO}_2/\text{H}_2(\text{clean heat source, } T > 800\,^\circ\text{C}). \tag{2.5}$$

Coal gasification occurs at 800–1,800 °C according to the endothermic reaction $2H_2O + C = CO_2 + 2H_2$. If heat is supplied to the reaction by combustion of additional coal, the stoichiometric equation becomes $2H_2O + 1.453C + 0.453O_2 = 2H_2 + 1.453CO_2$. The practical conversion efficiency of coal to hydrogen is 60 % and the thermal efficiency of the process is assumed to be around 70 %. Therefore,

$$\begin{aligned}\text{Emissions (coal gasification)} = 1.21\text{CO}_2/\text{H}_2(&\text{combustion of coal as the}\\ &\text{heat source)}\end{aligned} \tag{2.6}$$

and

$$\begin{aligned}\text{Emissions (coal gasification)} = 0.84\text{CO}_2/\text{H}_2(&\text{clean heat source,}\\ &T > 800 - 1{,}800\,^\circ\text{C}).\end{aligned} \tag{2.7}$$

The GHG emissions of water electrolysis are correlated with the emissions of electricity generation. In order to make comparisons with SMR and coal gasification on a common basis, assume, conservatively, that electricity is generated from natural gas with 30 % efficiency, and water is dissociated through electrolysis with 100 % efficiency. This process occurs according to the following overall reaction: $1.187(CH_4 + 2O_2) + H_2O = 1.187(CO_2 + 2H_2O) + H_2 + 0.5O_2$. Therefore,

$$\text{Emissions (water electrolysis, } 100\,\% \text{ fossil fuel power)} = 1.19\text{CO}_2/\text{H}_2. \tag{2.8}$$

For Alberta and Saskatchewan, the emission equations—considering 84 and 60 %, respectively—for GHG emissions due to combustion of fossil fuels in power generation (see above) and assuming that natural gas is mainly used in these provinces for power generation yield

$$\left.\begin{aligned}\text{Emissions (electrolysis, } 84\,\% \text{ fossil fuel power, Alberta)} = 0.99\text{CO}_2/\text{H}_2\\ \text{Emissions (electrolysis, } 60\,\% \text{ fossil fuel power)} = 0.72\text{CO}_2/\text{H}_2\end{aligned}\right\}. \tag{2.9}$$

Note that the emissions of water electrolysis appear higher than SMR. To lower them to the SMR level, the clean power fraction of the grid must be in excess of 70 %. Next, consider hydrogen production with the Cu–Cl cycle when the heat needed to sustain its reaction is derived from natural gas combustion.

To produce one mole of hydrogen, the total required heat by all endothermic reactions is 555 kJ and the total released heat of all exothermic processes is 232 kJ within the Cu–Cl cycle. Assume that 50, 70, 80, and 90 % of the heat released by exothermic processes in the Cu–Cl cycle is recycled internally within the cycle. Then the required high-temperature heat input is 220 MJ/kg, 197 MJ/kg, 185 MJ/kg, and 173 MJ/kg H_2, respectively. Assume that there are no heat losses from the combustion chamber and that natural gas can be modeled as methane. All of the combustion heat of natural gas is transferred to the cycle. Therefore, the overall chemical equation to split water using the Cu–Cl cycle with heat from the combustion of methane is

$$x \times (CH_4 + 2O_2) + H_2O = x \times (CO_2 + 2H_2O) + H_2 + 0.5O_2, \qquad (2.10)$$

where x is the molar fraction of methane that must be added to the cycle to make itself sustainable.

The values for x are 0.548, 0.489, 0.460, and 0.432 when the released heat is 50 %, 70 %, 80 %, and 90 % reused internally, respectively. Because methane burning generates steam, this can be used in the Cu–Cl cycle as a portion of feed H_2O; the other portion will be liquid water that must be boiled. Together with steam, CO_2 is introduced into a hydrolysis reaction of the Cu–Cl process, hence a need for separating HCl from CO_2 in the downstream units. However, the introduction of CO_2 also has significant advantages: (a) heat carried in high-temperature CO_2 of the methane combustion exhaust gases can be recovered back to the hydrolysis reaction; and (b) CO_2 is inert to $CuCl_2$ and can reduce the decomposition of copper oxychloride. Therefore, if steam and carbon dioxide are recycled to provide part of the heat (and feed) to the Cu–Cl plant, then the molar fraction x from (2.10) becomes 0.47, 0.42, 0.40, and 0.37, when the released heat by exothermic processes of the Cu–Cl cycle is 50, 70, 80, and 90 % reused internally.

The Cu–Cl cycle requires electricity to drive the electrolytic process of CuCl/HCl aqueous system. Assume that this electricity is derived from the local grid (for Alberta and Saskatchewan cases). Therefore, the corresponding GHG emissions are

$$\left.\begin{array}{r} \text{Emissions (electrolysis Cu–Cl, 84 \% fossil fuel power, Alberta)} = 0.07CO_2/H_2 \\ \text{Emissions (electrolysis Cu–Cl, 60 \% fossil fuel power)} = 0.05CO_2/H_2 \end{array}\right\}. \qquad (2.11)$$

The results regarding GHG emissions per mole of hydrogen produced with all considered methods are summarized in Tables 2.5 and 2.6. Table 2.5 shows those methods that use a fossil fuel heat source. Table 2.6 indicates the hydrogen production methods with anon-carbon heat supply. In both cases, the required electricity is supplied from the local grid. It is found that the Cu–Cl cycle has the lowest CO_2 emissions.

When fossil fuels are used as an electricity and heating supply to the Cu–Cl cycle, then the cycle has equivalent or slightly lower emissions than SMR,

Table 2.5 GHG emissions of H_2 production using fossil fuel heat source with various methods

Method	GHG[a]	Description of the system
SMR	0.51	The burning of methane supplies all of the required enthalpy of endothermic processes.
Coal gasification	1.21	The burning of coal supplies all of the required enthalpy of endothermic processes.
Water electrolysis	1.19	100 % of the electricity is supplied from the combustion of fossil fuels.
	0.99	84 % of the electricity is supplied from the combustion of fossil fuels (AB).
	0.72	60 % of the electricity is supplied from the combustion of fossil fuels (SK).
Cu-Cl cycle	0.44 (AB) 0.42 (SSK)	90 % The burning of methane supplies all of the net heat input to the Cu–Cl cycle. The electricity for the electrolytic step of Cu–Cl cycle is supplied from the power grid: Alberta, 84 % fossil fuel generated; Saskatchewan, 60 % fossil fuel generated. The percentages are the internally reused portions of the heat released by exothermic processes of the Cu–Cl cycle after the steam in the exhaust gases produced from the methane combustion is recovered.
	0.46 (AB) 0.44 (SK)	80 %
	0.48 (AB) 0.46 (SK)	70 %

[a]Given in mol CO_2 emitted per mole H_2 produced

Table 2.6 GHG emissions of H_2 production using non-carbon heat supply with various methods

Method	GHG[a]	Description of the system
SMR	0.39	The enthalpy of endothermic processes is completely supplied from clean heat sources.
Coal gasification	0.84	The enthalpy of endothermic processes is completely supplied from clean heat sources.
Cu–Cl cycle	0.07 (AB) 0.05 (SK)	The net heat input to the Cu–Cl cycle is completely supplied from non-carbon heat sources, e.g., solar or nuclear, but the electricity for the electrolytic step of the Cu–Cl cycle is supplied from the power grid: Alberta, 84 % fossil fuel generated; Saskatchewan, 60 % fossil fuel generated.
	0	If the electricity for the electrolytic step of the Cu–Cl cycle is completely clean from a non-carbon source (SWCR + Cu–Cl system).

[a]Given in mol CO_2 emitted per mole H_2 produced

depending on the steam and heat recovery percentages. When non-carbon heat sources are used, the level of CO_2 emissions of SMR is about eight times higher than the Cu–Cl cycle. If both heat and electricity supplies are clean, then SMR still has significant emissions while the Cu–Cl cycle produces none.

With the current SMR technology used to generate hydrogen for bitumen upgrading of Canadian oil sands, the GHG emissions increase in the near future according to the forecast shown in Table 2.7. It will reach around 10 % of Canadian CO_2 emissions in 2006. These emissions come on top of emissions due to bitumen

Table 2.7 GHG emissions of hydrogen production for bitumen upgrading with SMR technology

Year	2015	2020	2025	2030
Bitumen production scale (10^6 barrels/day)	2.2	3.4	4.1	4.3
Hydrogen for upgrading (t/day)	7,480	11,560	13,940	14,620
GHG emissions from SMR process (10^6 t/year)	30.0	46.4	56.0	58.7
Compared with 2006 CO_2 emissions (% of 2006)	5.5 %	8.5 %	10.3 %	10.8 %
Heat supplied by SWCR to produce H_2 ($MW_{thermal}$)[a]	19,046	29,435	35,495	37,227
Number of SCWR + Cu–Cl plants required[b]	1.4	2.2	2.7	2.8

Note: Canada's total CO_2 emissions in 2006 were 544.68×10^6 t
[a]Hydrogen is produced by CANDU-SWCR and Cu–Cl plant linkage according to system from Fig. 2.15
[b]Assumed six nuclear reactors per plant, for equivalent 1,000 $MW_{electric}$ each unit
Data from Wang and Naterer (2010)

Table 2.8 GHG emissions of oil sand extraction, upgrading, and distillation with current technology (SAGR, SMR, and natural gas combustion at refinery)

Year	2015	2020	2025	2030
GHG emissions for bitumen extraction (10^6 t/year)	42.6	65.6	79.1	83.0
GHG emissions from SMR process (10^6 t/year)	30.0	46.4	56.0	58.7
GHG for crude oil distillation (10^6 t/year)	10.0	15.2	18.2	19.2
Total GHG emissions (10^6 t/year)	82.6	127.2	153.3	160.9
Compared with 2006 CO_2 emissions (% of 2006)	15 %	23 %	28 %	30 %

Note: Canada's total CO_2 emissions in 2006 were 544.68×10^6 t

extraction via SAGD (see Table 2.4). By 2030, the total emissions for bitumen extraction and upgrading to crude oil are predicted to represent 25 % of total Canadian CO_2 emissions in 2006. Herein lies a benefit of using nuclear hydrogen production with CANDU-SWCR and the Cu–Cl cycle (Fig. 2.15). Based on the amount of required hydrogen, the number of nuclear plants to be installed in the oil sand locations is indicated in Table 2.7. If such an investment is pursued, the GHG emissions remain at the level of 15 % with respect to the year 2006, corresponding to the bitumen extraction process, only.

For distillation of one barrel of crude oil, one needs 2 kg of steam at 2 bar and 400 °C, and 102 MJ of heat (see Table 2.2) with 55 % efficiency. Assume that the heat to generate steam is derived from natural gas combustion; moreover, the heat needed to drive the distillation process is derived also from natural gas combustion (this scenario models today's refinery technology). It results in roughly 200 MJ of natural gas that must be combusted to satisfy these heating needs. Therefore, 4.2 kg of methane must be combusted to distillate 1 barrel of crude oil. The corresponding GHG emissions are12 kg CO_2 per distilled barrel of crude oil. The total carbon emissions with current technology are shown in Table 2.8. From Table 2.8, by applying SWCR and Cu–Cl technology, at least 30 % of GHGs are mitigated for oil sand extraction, upgrading, and distillation. This can be achieved by installing three nuclear plants.

For the year 2030, the GHG mitigation in terms of tons of CO_2 equivalent of Canadian oil sand processing is presented in Fig. 2.16. It can be observed that the

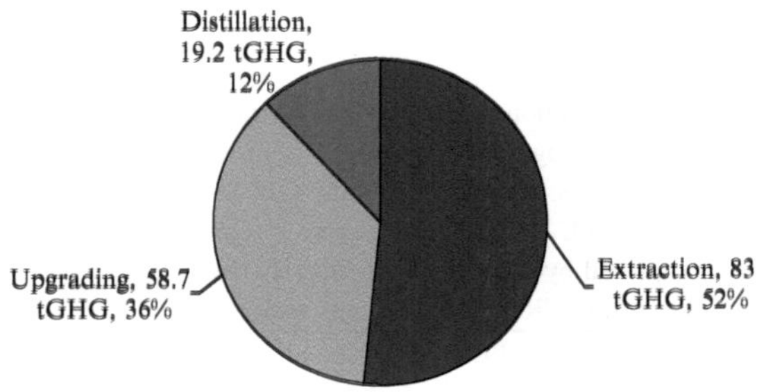

Fig. 2.16 GHG mitigation by 2030 if nuclear hydrogen with CANDU SCWR and Cu–Cl water splitting cycle is applied to Canadian oil sand processing

majority of GHG mitigation is due to improving the extraction process, whereas the heat needed to drive the process is replaced with carbon-free nuclear heat. Upgrading of bitumen is the second major segment of environmental improvement, achievable by providing heat and hydrogen from nuclear reactor energy. The distillation of crude oil in a refinery accounts for 12 % mitigation of GHG emissions.

Thus, by using nuclear hydrogen, in conjunction with nuclear heat and power in the petrochemical sector—more specifically oil sand processing—major GHG pollution reductions can be achieved. Additional gains of environmental performance are achieved by improving the extraction technology for bitumen. Application of $ScCO_2$ extraction—currently in development—reduces the water and energy requirements, as well as creates the opportunity to sequester CO_2 in situ, at oil sand locations.

2.5 Nuclear-Produced Hydrogen for the Transportation Sector

2.5.1 Transportation Sector Outlook

The transportation sector is characterized by intense energy demand in any country. Taking Canada as a sample case, Fig. 2.17 indicates the distribution of energy demand and GHG emissions for the transportation sector of this country. In Canada, the transportation sector accounts for about 32 % energy demand and 35 % GHG emissions. It can be observed that road transport is responsible for 78 % of GHG emissions in this sector.

Figure 2.18 shows the fuel consumption in the Canadian transportation sector, as per fuel type. Motor gasoline is the major fuel in Canada; its consumption represents 54 % in energy terms from all fuels. Diesel fuel follows gasoline with 33 %, and next is aviation fuel with 10 %. Figure 2.19 details the energy and GHG emissions for all kinds of transport vehicles in Canada. According to this data, the most pollution is due to passenger cars (25.3 %) and heavy trucks (23.3 %). It is expected that in the near future the global demand for transportation of goods and people will significantly increase. If the transportation demand increases at the same pace with global population, the level of pollution rises significantly. In this section, the role of nuclear hydrogen for a cleaner and better transportation sector is studied.

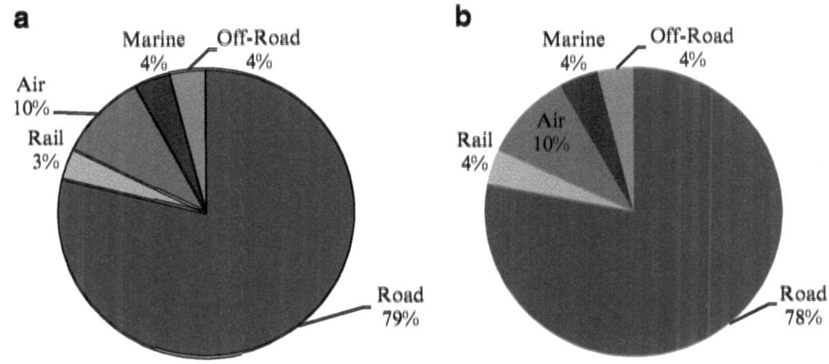

Fig. 2.17 Energy (**a**) and GHG emissions (**b**) by the transportation sector of Canada for each major transportation mode in 2008 [data from NRCAN (2010)]

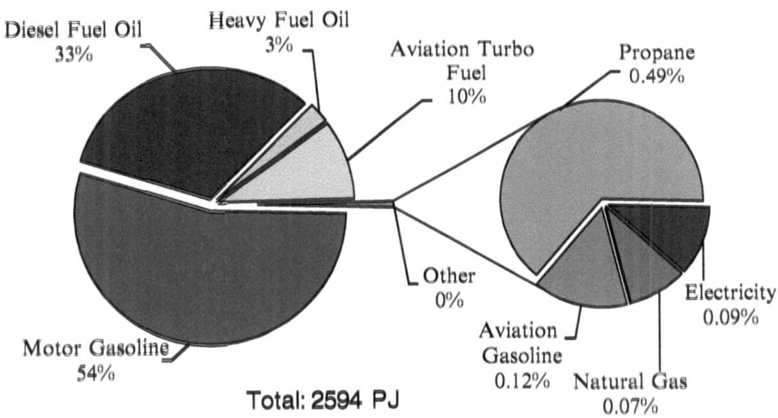

Fig. 2.18 Fuel consumption in Canadian transportation sectors, per fuel type [data from NRCAN (2010)]

Hydrogen is a potential energy carrier for powering all kinds of vehicles for road, rail, marine, air, and space transport. In Sect. 2.3 it is shown how hydrogen—as an energy carrier—can be generated using the power grid and base load nuclear energy with positive consequences on the electricity price, pollution reduction due to better operation of the grid, and additional pollution reduction due to promotion of hydrogen as an energy carrier for fuel cell vehicles.

2.5.2 *Hydrogen-Fuelled Road Transport*

The general layout of a fuel cell vehicle and hydrogen supply side is outlined in Fig. 2.20. Hydrogen is generated at partly centralized locations at the site of nuclear

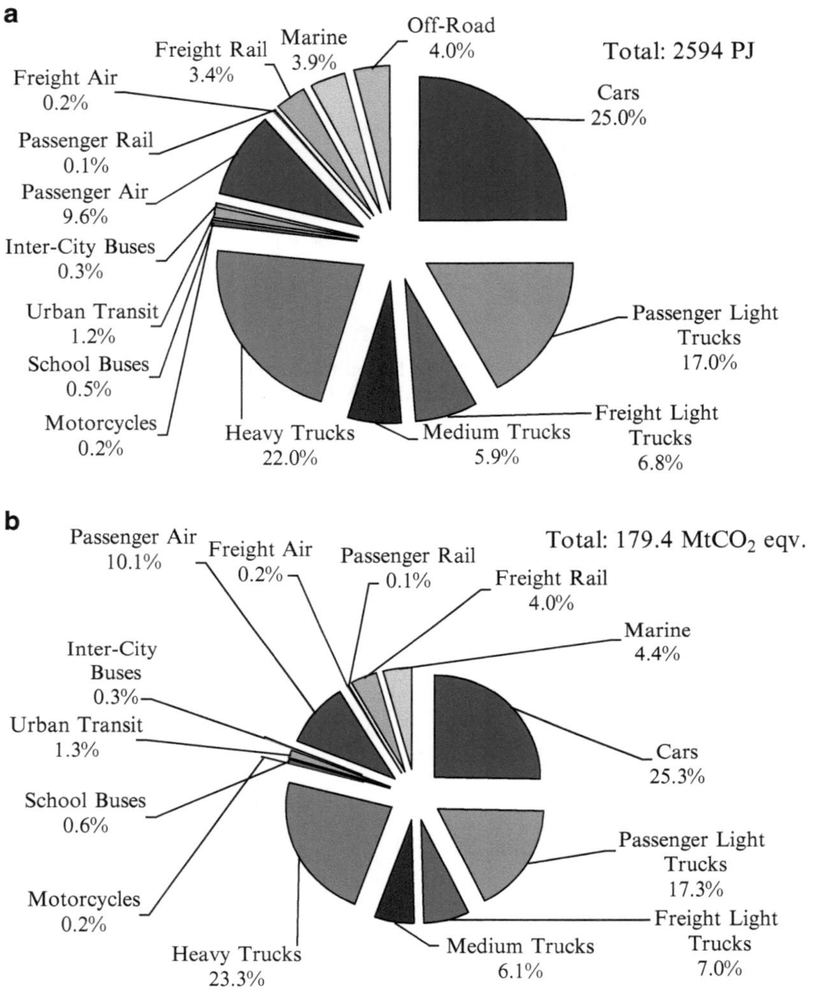

Fig. 2.19 Energy consumption (**a**) and GHG emissions (**b**) of transportation vehicles in Canada as per 2008 (*source*: NRCAN 2010)

power plants, and partly generated at decentralized locations using grid electricity and electrolysis. The energy utilization in the overall system from Fig. 2.20 (including fuel cell vehicles and the hydrogen supply system) can be estimated based on past studies; Dincer et al. (2010) and Dincer and Zamfirescu (2011). Figure 2.21 indicates the energy utilization in fuel cell cars.

Note that the results reported in Fig. 2.21 are estimates and refer only to typical fuel cell cars. Most of the losses are from hydrogen production and supply and due to hydrogen conversion back to electricity (overall totalling 65 %). In fuel cell cars, the shaft work for propulsion represents around 24 % of the primary energy. Using

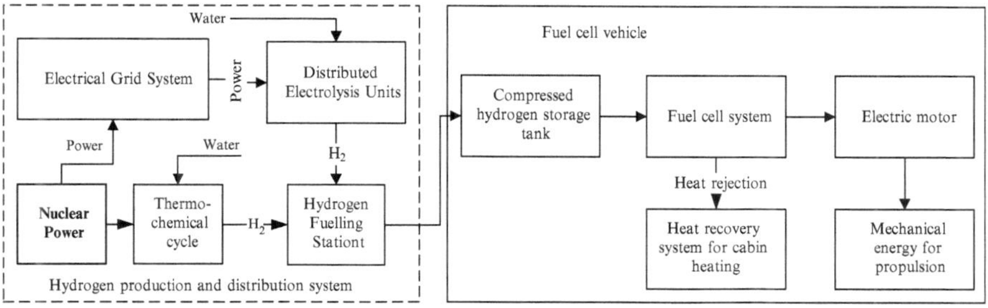

Fig. 2.20 Main subsystems of a fuel cell vehicle and hydrogen supply

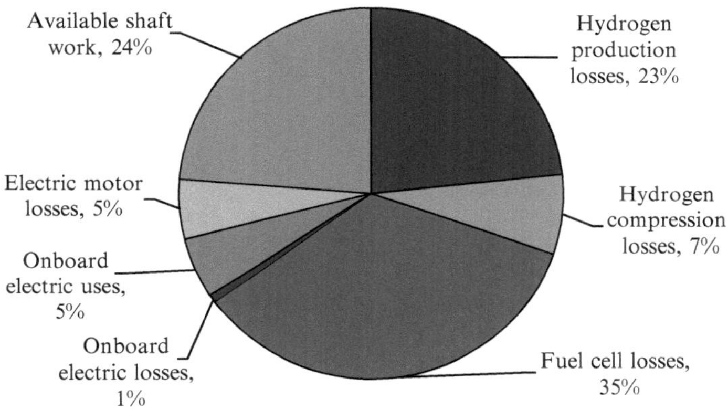

Fig. 2.21 Energy utilization in a typical fuel cell vehicle with respect to primary energy

actual electrical grid data from Environment Canada (2010), estimates are made based on energy utilization of fuel cell vehicles (Fig. 2.21) for the GHG emissions of electricity-driven passenger cars for each province. For comparison purposes, two types of electricity-driven cars are considered, namely, the fuel cell vehicle and electric vehicle. It is assumed that the vehicles are supplied by the electricity grid.

For the electric vehicle, it is assumed, based on data from Dincer et al. (2010), that there is a 54 % electricity to shaft power efficiency. The results presented in Fig. 2.22 show that the application of electric vehicles in Canada is beneficial except in three provinces where electrical grids are characterized by high GHG pollution due to their fossil fuel-based power sources. Future grid improvements by (potentially) adding nuclear-hydrogen capacity in Alberta, Saskatchewan, New Brunswick, and Nova Scotia can foster growth of fuel cell vehicles together with electric vehicles in all Canadian provinces. Fuel cell vehicles have a longer driving range than electric vehicles, which also cannot be used reliably yet in severe climatic conditions when cabin heating or cooling is needed, because in such cases, the driving range of electric vehicles may become significantly reduced in many situations.

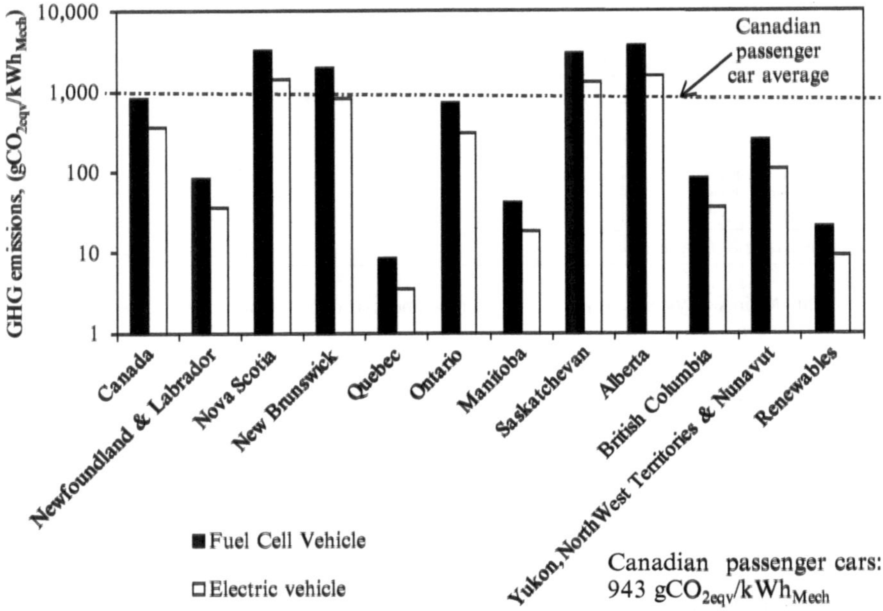

Fig. 2.22 Estimates of GHG emissions of electricity-driven passenger cars in Canada

Apart from electric and fuel cell vehicles, the hybrid powertrains appear as a promising solution for sustainable transportation. Figure 2.23 compares the GHG emissions and air pollution indicators for hybrid, electric, fuel cell, and conventional passenger cars. The air pollution indicator quantifies the environmental impact of exhaust gases of automobiles other than GHGs, which include the following airborne pollutants: CO, NO_x, SO_x, and volatile organic compounds (VOCs). According to Dincer et al. (2010), the indicator uses the following weighting coefficients: 0.017, 1, 1.3, and 0.64, for each airborne pollutant, respectively.

In the study reported in Fig. 2.23, four real vehicles are considered: Toyota Corolla (conventional), Toyota Prius (hybrid), Toyota RAV4EV (electric), and Honda FCX (fuel cell). It is assumed that hydrogen is generated by grid electricity assuming average worldwide pollution indicators associated with electricity generation; these indicators are taken from the study of Dincer et al. (2010), namely, 77.5 g GHGs per MJ of electricity and 0.296 g/MJ as the air pollution indicator. All three alternative vehicle options reduce the life cycle pollution to about half, with respect to a conventional vehicle. However, fuel cell vehicles show the smallest air pollution among all, during the vehicle utilization stage. The electric vehicle is characterized by air pollution during the utilization stage due to its battery recycling, mainly. If the GHGs of hydrogen production decrease, the fuel cell vehicle has the best likelihood to have the lowest carbon footprint.

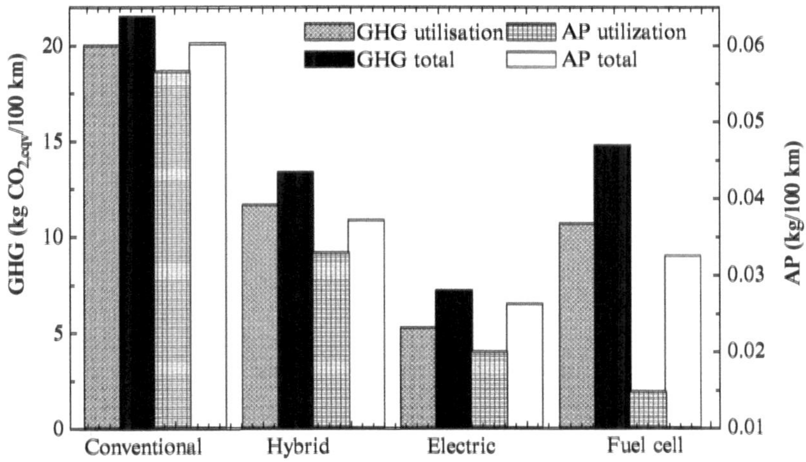

Fig. 2.23 GHG and air pollution (AP) related to the fuel utilization stages and total environmental impact for different types of vehicles [data from Dincer et al. (2010)]

2.5.3 *Hydrogen-Fuelled Railway Transport*

For large-scale transportation (trains, ships, airplanes), it appears feasible to store hydrogen onboard in its pure form. One of the first applications of hydrogen for widespread transportation vehicles may be implemented at a large scale in the near future for train locomotives, which offer an attractive mode of transportation in terms of energy consumption per passenger kilometer of travel. An option for cleaner rail transport is electrification with an electricity supply from clean generators (such as hydro, nuclear, wind, etc.). Electrified trains transfer more than 85 % of the electricity input to the wheels, making it one of the most efficient transportation systems, when compared with private automobiles, buses, or airplanes. Disadvantages of electrification include the capital investment to install electrical substations and catenaries, together with a lack of flexibility for locomotives to move into other service areas not covered by electrification. In two papers by Marin et al. (2010a, b), a case study of hydrogen train implementation in Ontario is presented for the 115 km "GO Transit Lakeshore" corridor, through Toronto, between Oshawa and Hamilton, Ontario, which currently carries a total of approximately 100,000 passengers each weekday. The Lakeshore corridor operates locomotives on diesel fuel at a rate of approximately 5 L/km pulling/pushing 10 or 12 carriages with up to 1,540 passengers during peak hours.

Passenger use projections for this corridor are expected to grow by about 90 % for Lakeshore East and 65 % for Lakeshore West between 2007 and 2031, respectively. A sensitivity analysis was performed over a range of operational costs for a hydrogen train, with variability of feedstock prices, fuel cell power density, and expected return on capital investment. According to general estimates (see Marin et al. 2010a), the initial capital cost for train electrification in 2007 was in the order of 5.9 billion

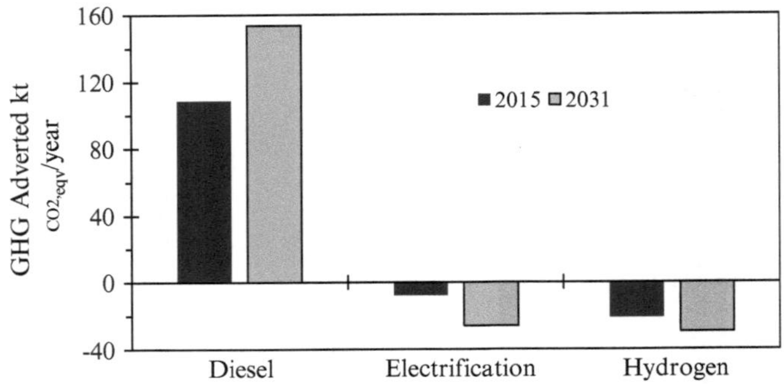

Fig. 2.24 Total equivalent GHGs averted by train transport [data from Marin et al. (2010a)]

Canadian dollars. The operation of the electrified corridor varied between 78 and 145 $/train-km, projected to 2015, while hydrogen operation had a cost varying between 79 and 151$/train-km, with a capital cost varying from 12.5 to 16 %.

The estimated power at the prime mover is on the order of 0.09 GJ/km. This figure offers a basis for modeling of fuel cell trains with an added benefit of higher fuel efficiency between 53 and 58 %. One drawback of fuel cell trains is the required refurbishing at approximately 5,000 h of operation, much less when compared to a diesel engine that requires a complete overhaul every 15,000–25,000 h of operation, although continual advances are being made to improve fuel cell durability.

Figure 2.24 provides a comparison of averted emissions based on different power sources, against distance travelled by trains annually. On the basis of the source energy, whether nuclear, Ontario Power Generation (OPG) mix in Ontario, Canada, or hydrogen from SMR, the utilization of heat rejected from thermal powerplants, combined with hydrogen generated through a thermochemical Cu–Cl cycle and proton exchange membrane fuel cell (PEMFC) trains, results in the lowest level of GHG emissions, on the order of 0.10 kg/GJ.

The higher efficiency of the PEMFC complements the performance of thermochemical hydrogen production, yielding the lowest GHG emissions at the prime mover, on the order of 0.04 kg/km (345 t by 2015, and 488 t by 2031). The actual impact of CO_2 emission reduction is significant when compared to the diesel alternative. Further GHG emission forecasts from GO trains—depending on the propulsion option—are indicated in Fig. 2.25.

When comparing the electrification to hydrogen train alternatives, it is assumed that capital expenditures for infrastructure not related to hydrogen remain unchanged. The cost per kilogram of hydrogen is the combined result of several factors, including the capital cost of building the hydrogen plant, the cost of operating and maintaining the plant, the cost of feedstock (whether it is gas, electricity, or thermal energy), the cost of capital invested, the cost of compression and storage, the cost of transportation and delivery, and other miscellaneous external costs not listed here, all included in this analysis. The economic parameters used to generate the results—for the case study of implementation of hydrogen trains in Ontario—are listed in Table 2.9.

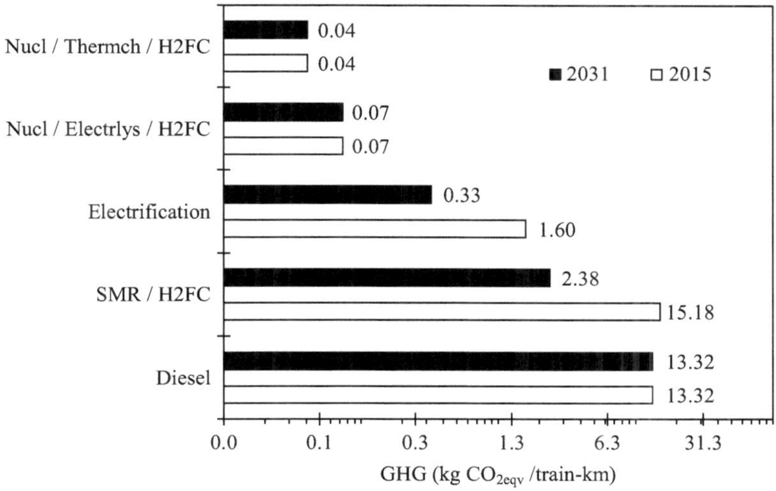

Fig. 2.25 Total GHG emissions from GO trains predicted between 2015 and 2031 [data from Marin et al. (2010a)]

The energy requirement for locomotive propulsion is given by the following equations for thermochemical hydrogen production, electrolysis, and electrification, respectively:

$$\left.\begin{aligned} E_{\mathrm{th}} &= \frac{E_{\mathrm{pm}}}{\eta_{\mathrm{fc}}\,\eta_{\mathrm{c}}\,\eta_{\mathrm{th}}} \\ E_{\mathrm{e}} &= \frac{E_{\mathrm{pm}}}{\eta_{\mathrm{fc}}\,\eta_{\mathrm{c}}\,\eta_{\mathrm{e}}\,\eta_{\mathrm{pp}}} \\ E_{\mathrm{el}} &= \frac{E_{\mathrm{pm}}}{\eta_{\mathrm{t}}\,\eta_{\mathrm{pp}}} \end{aligned}\right\}, \tag{2.12}$$

where E is energy, η is efficiency, and the subscripts signify thermochemical (th), electrolysis (e), electric train (el), prime mover (pm), fuel cell (fc), compression (c), turbine conversion of electrical power (pp), and transmission and distribution of electric power (t).

The results of cost modeling are indicated in Fig. 2.26, showing the cost contributors to the production of hydrogen as a percentage of total cost. The feedstock cost, capital cost, and transportation costs are included. Transportation of hydrogen is typically in the range of 0.3–1.3$/kg. It is compared at operational costs for annual payment of compound interest and capital at the rate of 12.5–16 %, with natural gas costs varying from 5 to 9$/GJ.

The analysis from Fig. 2.27 assumes electricity costs to vary from 32 to 100$/MWh; and thermal power costs vary from 9 to 11.6$/GJ. Figure 2.26 illustrates the capital cost estimates for hydrogen trains, with consideration given to the most probable scenario of fuel cell cost, at the time of introduction into the market. The costs of capital and electricity influence significantly the operational cost of trains when hydrogen is generated via electrolysis. Figure 2.28 shows the expected range

Table 2.9 Economic parameters considered for hydrogen train modeling

Parameter	Equation	Description
Annual payment of compound interest	$A = C \times (1+i)^n \times \frac{i}{(1+i)^n - 1}$	Determines the contribution of capital cost to the cost of hydrogen. It is applied to the construction phase and long-term capital financing of a project
Capital cost contribution	$\mathrm{Hc} = \frac{A}{\mathrm{Dy} \times 24 \times \mathrm{Rh}}$	Represents the contribution of capital costs to hydrogen product cost
Present value of investment	$C = \frac{\mathrm{Ck}}{(\mathrm{Ae}/\mathrm{Ne})^y}$	Predicts the cost of a scalable chemical plant as a function of known capital cost (Ck)
Cost of hydrogen transportation	$\mathrm{Ht} = \mathrm{const} \times \mathrm{Dp}^{1.0146}$, where $\mathrm{const} = 0.0027 - 0.0031$	Represents the contribution of hydrogen transportation towards the cost of hydrogen product via pipeline

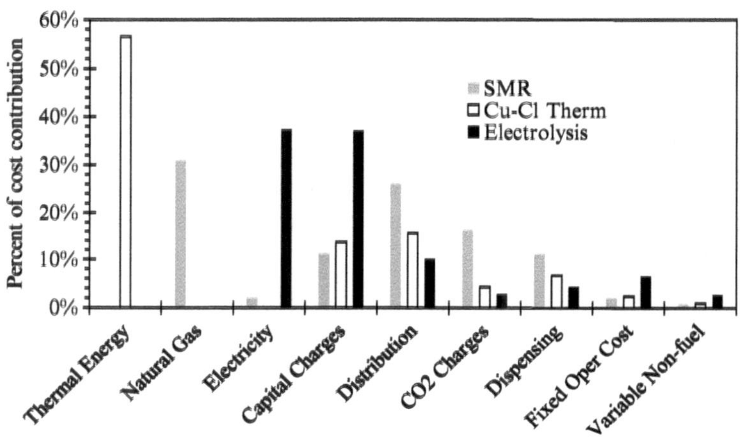

Fig. 2.26 Cost contributors to the production of hydrogen as a percent of total cost [data from Marin et al. (2010a)]

of trains versus onboard hydrogen storage. Under the assumption that GO trains stay at stations an average of 14 % of total traveling time, a 5 % minimum safe hydrogen mass remaining in storage before refuelling, and fuel cell efficiencies at 50 %, the expected range with an initial hydrogen storage of 1,800 kg is approximately 1,000 km.

Another important aspect which influences the economics of either electrification or "hydrogenenation" of GO Lakeshore trains is represented by transportation of the energy carrier from its production source to the locomotive. In the case of electrification, it is necessary to transport the electricity from remote power plants, which may involve up to 9 % transmission losses. These losses are seen as a proportional increase of electricity price (as delivered to the locomotive). Pipeline transportation of hydrogen at high volumes can be as low as $0.1/kg, while liquid hydrogen transportation reaches $0.3/kg. From an economics perspective, the investment required to provide power substations and catenaries for electrification is contrasted

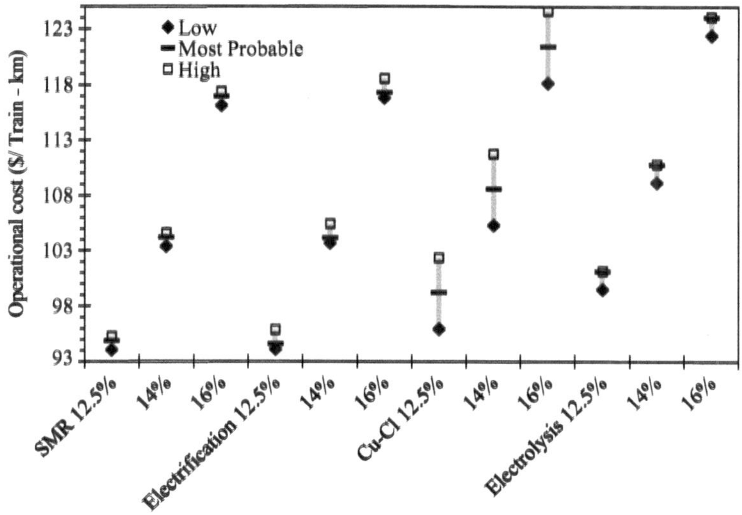

Fig. 2.27 Range of total operational cost of GO train Lakeshore Corridor, based on variable feedstock costs and equal annual payments at 12.5, 14, and 16 % compound interest on capital [data from Marin et al. (2010a)]

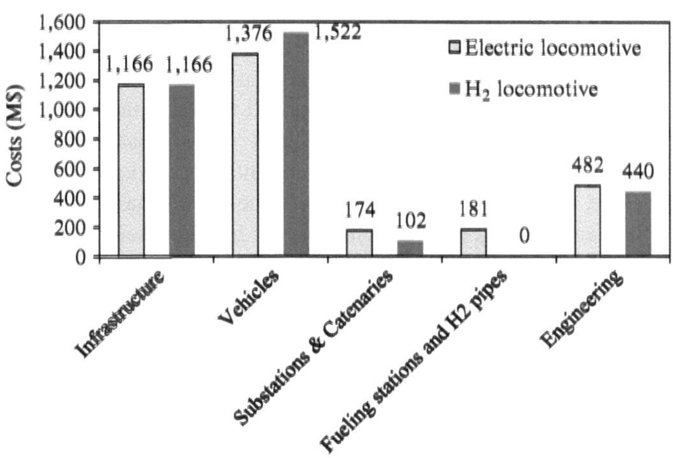

Fig. 2.28 Comparative cost of capital investments via electrification vs. hydrogen fuel cell locomotives [data from Marin et al. (2010a)]

with the investment in refuelling stations and hydrogen delivery via piping, respectively. Electrical transformer substations in the case of electrification have a cost impact that can be compared to filling stations in the case of hydrogen fuel.

Based on the above results, it appears that the GO hydrogen train is promising with major environmental advantages, but also drawbacks such as high cost and relatively short lifetime of PEMFCs. It was found in Marin et al. (2010b) that scaled-up fuel cells can fit within the existing Bombardier ALP-46A locomotives, including hydrogen gas storage. The electrification of train operations in the Lakeshore corridor may also be affected by CO_2 legislation in direct proportion to the electricity mix established by OPG. Since this energy mix is projected to contain a portion based on fossil fuels, there is some uncertainty in the potential increase in the cost of operation per train-km. This uncertainty contrasts with the uncertainty of fuel cell prices in the future. While the cost associated with GHG emissions will inevitably increase in the future, the cost of fuel cells is projected to be lower as the technology develops, matures, and reaches mass production.

2.5.4 Hydrogen-Fuelled Airway Transport

Hydrogen is also a promising alternative for future aircraft fuel because of its high energy content per unit of mass and favorable combustion kinetics, reduced environmental impact, and increased sustainability. The environmental impact of liquid hydrogen fuel with respect to the reference kerosene jet fuel is compared in a study by Nojoumi et al. (2009). According to past Airbus research, for 3.3 m^3 of LH_2 and a wind speed of 4 m/s in the case of an accidental crash, an area of about 1,000 m^2 of fire carpet would arise, in comparison to 13,500 m^2 for propane. In this respect, LH_2 is a safer aviation fuel than kerosene.

During the flight, airplanes form condensation trails that represent artificially induced cirrus clouds. The tank of a proposed hydrogen aircraft can be placed above the payload, above and aft of the payload, or fore and aft of the payload section. The fore tank can be located between the cockpit and forward passenger compartment bulkhead, and the aft tank is located between the aft passenger compartment bulkhead and tail area. The tanks can either be integral or nonintegral parts of the fuselage structure. Storing LH_2 instead of kerosene by volume has a ratio of 4:1. An additional heat exchanger is needed with respect to a kerosene fuelling configuration to heat the liquid hydrogen to a temperature that is suitable for injection into the combustion chamber.

As derived from the combustion equation, $C_{12}H_{23} + 17.75O_2 + 66.77\ N_2O_2 \rightarrow 12CO_2 + 11.5H_2O + 66.77N_2$ for complete stoichiometric combustion of 1 kg of kerosene; 3.4 kg of oxygen is needed, producing 3.16 kg of CO_2, 1.24 kg of water vapor, 1 g of NO_x, and between 1 and 2.5 g of CO. A typical passenger airplane consumes for a trip from Toronto to Calgary about 13,937 kg of kerosene (for a distance of 2,716 km). Table 2.10 shows detailed emission analysis of flight, landing, and takeoff for the Airbus A300 airplane at selected destinations departing from Toronto.

A liquid hydrogen (LH_2) aircraft has quieter operation with respect to conventional aircraft due to a smaller engine, which has a more compact cooling system with liquid hydrogen rather than air. The engine life is increased by about 25 % and maintenance requirements are also reduced. Figure 2.29 compares volumetric specific fuel consumption of an Airbus A300 when operated with conventional fuel or cryogenic hydrogen.

Table 2.10 Total emissions for Airbus A300 airplane at selected destinations from Toronto

City	Montreal	Calgary	London
Distance (km)	504	2716	5710
Fuel consumption (kg)	3,677.3	13,937.3	19,005
Climb/cruise/descent (kg)	2,875	13,751.4	17,553.7
Specific consumption (kg/km)	7.29	5.13	5.55
NO_x (kg) total	56.17	171.4	251.7
Climb/cruise/descent	45.29	171.5	214.0
HC (kg) total	2.43	4.19	5.10
Climb/cruise/descent	0.5	2.30	3.03
CO (kg) total	21.9	32.2	39.2
Climb/cruise/descent	3.1	10.25	12.74
GHG emissions (kg $CO_{2,eqv}$)	22.34	15.71	19.85

Data from Nojoumi et al. (2009)

HC hydrocarbon emissions, *CO* carbon monoxide emissions, *NO_x* nitrogen oxide emissions

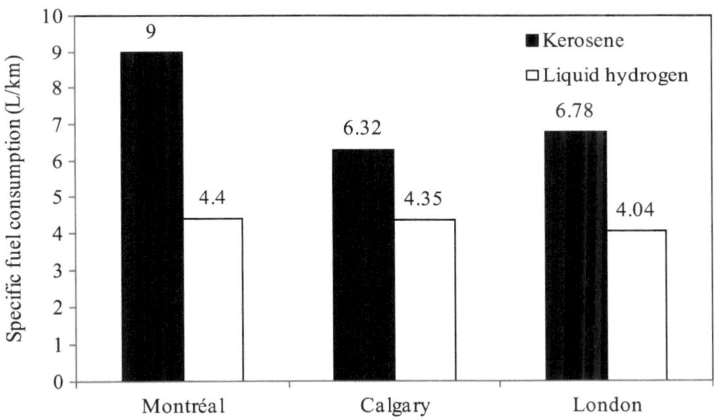

Fig. 2.29 Specific fuel consumption for conventional and hydrogen Airbus A300 departing from Toronto towards three destinations [data from Nojoumi et al. (2009)]

2.5.5 Hydrogen Role for Reducing Environmental Impact in Transportation Sector

If hydrogen is used as fuel in transportation sector, the associated GHG emissions and global warming effect can be fully mitigated. GHGs emitted are inevitable with fossil fuel-propelled transportation vehicles that are majorly in use today, but in hydrogen vehicles there are zero GHGs emitted and therefore no effect on global warming potential (excluding the emissions from the process of generating hydrogen).

The concentration of GHGs in the atmosphere influences the radiative balance on earth. The *radiative forcing* (denoted with ΔF) represents the net change in radiation balance at the tropopause level, produced by a specified cause. It induces an increase of planet temperature when the radiative forcing is positive. Typically the radiative forcing is smaller than 2 W/m^2. The global mean concentration of CO_2 in the atmosphere in 2005 was 379 ppm, leading to a ΔF of +1.66 [±0.17] W/m^2.

The NO_x emissions from transportation vehicles of all kinds tend to increase the ozone level. As an example, for supersonic aircraft at high altitudes above 18 km, ozone is depleted in large extend. An increase in the concentration of ozone in the upper troposphere is more effective in increasing ΔF than lower altitude. Therefore, LH_2 aircraft should be used at lower altitudes below 12 km. Other emissions from a hydrogen airplane is water vapor, but their effect on radiative forcing is minimal because their lifetime in the troposphere is short (around 10 days), and typically removed with precipitation. Condensation trails produced from high-altitude aircraft exhaust cause radiative forcing and thus they may affect climate. They can persist for many hours, depending on their condition and local weather patterns, reflect solar radiation, and absorb and emit thermal infrared radiation. Aviation in general has the potential to change cirrus clouds in a number of ways. Contrails do not always evaporate after a short time. If the local atmosphere is supersaturated with respect to ice, the clouds will form into larger cirrus clouds called contrail cirrus, which cannot be distinguished from contrails, if the concentration is unknown.

The radiative forcing is calculated by using the formula recommended by Intergovernmental Panel on Climate Change (IPCC):

$$\Delta F = 6.3 \times \ln (C/C_0), \tag{2.13}$$

where C_0 is the concentration of carbon dioxide at a reference time, which typically is taken as 280 ppm, corresponding to the beginning of the industrial era; C is the change in carbon dioxide concentration.

With this formula, the radiative forcing can be calculated for specific means of transportation based on an initial concentration of carbon dioxide in the atmosphere. The total mass of the atmosphere is about 5×10^8 kg with an average temperature of 15 °C; therefore a volume estimated with the ideal gas law is 9.2×10^{14} m^3. The mass of carbon dioxide at the beginning of the industrial era is estimated based on C_0 and it is about 920 Gt. Using the data from the Canadian transportation sector which emits in total 180 Mt GHG per annum, the radiative forcing can be determined when produced if this sector is supplied with nonpolluting energy. The result is −12.3 mW/m^2 for a decade. Figure 2.30 presents this analysis for four transportation means in Canada: car, heavy truck, train, and ship. On the same plot, the specific GHG emissions are shown per MJ of energy consumed by the transportation sector. The statistical data from NRCAN (2010), regarding GHG emissions and fuel consumption, are used in this analysis (see Fig. 2.19). It can be observed that vehicles generate the smallest quantity of GHGs while rail transport generates the highest.

Note that the radiative forcing influences global warming through the so-called climate sensitivity factor defined by

$$\gamma = \frac{\Delta T_e}{\Delta F}, \tag{2.14}$$

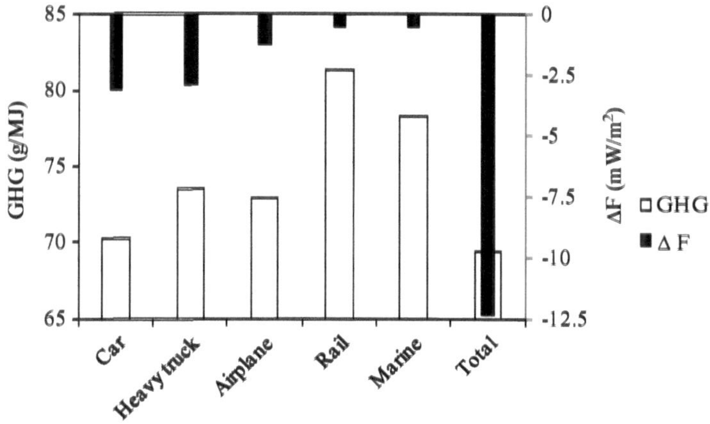

Fig. 2.30 Specific GHG emissions by means of transportation in Canada and the radiative forcing due to GHG mitigation in this sector

where ΔT_e represents the increase of earth's temperature due to the radiative forcing ΔF. The current estimated value of the climate sensitivity factor is 0.6 m^2K/W (see Dincer and Zamfirescu 2011). With a total anthropogenic radiative forcing of 61 mW/m^2 per decade, it shows that the temperature increase within 10 years may be about 0.04 °C. Using an average of -10 mW/m^2 for the radiative forcing created by GHG gases from the entire Canadian transportation sector, it yields a result wherein the total radiative forcing may decrease to 51 mV/m^2; therefore the increase of the earth's temperature may reduce to 0.03 °C.

2.6 Conclusions

In this chapter, the role of nuclear energy in hydrogen production has been examined. It was shown that hydrogen is important for better environmental performance, efficiency, and cost effectiveness in various industrial and other sectors. The roles of hydrogen in three sectors are studied in more detail, namely, electricity generation and distribution, petrochemical operations, and the transportation sector. These sectors are significant in the future hydrogen economy. Nuclear power offers a base load capability and all other power generators operate at fluctuating loads as a function of demand and fuel availability. Operation at part load is not optimal and leads to more fuel consumption and more GHG emissions for the same output (power generation).

The role of nuclear power is crucial in the future for a base load source of power generation without GHG emissions. Generation IV nuclear power plants (e.g., SWCR-CANDU) can be coupled with thermochemical cycles (such as the Cu–Cl cycle) to generate hydrogen at base load conditions. Nuclear hydrogen production can operate in synergy with water electrolysis at delocalized stations which operate intensely during off-peak periods. During the peak periods, hydrogen can be converted back to electricity with fuel cells. When it is advantageous, hydrogen can be sold as a fuel to the transportation sector, or feedstock in various chemical and other industries. Load levelling for the grid through nuclear hydrogen is a useful outcome with better matching between supply and demand.

Another major application is hydrogen use for upgrading of oil sands in petrochemical operations. Both conventional oil and oil sands are envisaged as high future consumers of hydrogen. Nuclear energy can provide heat, power, and hydrogen as outputs for processing of fossil fuels (including extraction, upgrading, distillation, and hydro-treating).

Hydrogen is often cited to become a major fuel for the transportation sector worldwide in the future. It is important that hydrogen is generated from clean sources to reduce GHG pollution and the global warming effect. Various scenarios of hydrogen vehicle implementation in society were analyzed in this chapter. Hydrogen can be readily used for vehicles although further developments are required for fuelling and storage systems, fuel cells, and infrastructure. Sustainable hydrogen production is an essential requirement and it will have a crucial role in facilitating the future hydrogen economy.

Nomenclature

C	Cost, any currency
E	Energy (MJ)
F	Radiative forcing (W/m^2)
G	Gibbs free energy (MJ/kg)
H	Reaction enthalpy (kJ)
$\dot{m}$	Rate of hydrogen production (t/day)
Q	Heat flux (kJ)
S	Reaction entropy (kJ/kg K)
T	Temperature (K)
x	Molar fraction of methane

Greek Letters

γ	Climate sensitivity factor
η	Efficiency

Subscripts

0	Reference value
c	Compression

ΔG	Gibbs free energy
e	Electrolysis or earth
el	Electric power train
fc	Fuel cell
pm	Prime mover
pp	Turbine conversion of electrical power
ref	Reference
SMR	Steam methane reforming
t	Transmission and distribution of electric power
tot	Total
th	Thermochemical, thermal
$T\Delta S$	Heat transfer irreversibilities
w	Per unit of energy

References

Carty RH, Mazumder MM, Schreiber JD, Pangborn JB (1981) Thermo-chemical hydrogen production. Final report IGT Project 30517, Institute of Gas Technology, USA

Dincer I, Rosen MA (2007) Exergy: energy environment and sustainable development. Elsevier, Boston, MA

Dincer I, Rosen MA, Zamfirescu C (2010) Economic and environmental comparison of conventional, hybrid, electric and hydrogen fuel cell vehicles. In: Pistoia G (ed) Electric and hybrid vehicles: power sources, models, sustainability, infrastructure and market. Elsevier, Amsterdam

Dincer I, Zamfirescu C (2011) Sustainable energy systems and applications. Springer, New York

Environment Canada (2010) Electricity intensity tables. http://www.ec.gc.ca/ges-ghg/default.asp?lang=En&n=EAF0E96A-1

Fang Y (2010) Extraction of hydrocarbons from oil sand using supercritical carbon dioxide. MSc thesis, University of Alberta

IESO (2011) Monthly market report: July 2011. Independent Electricity System Operator. www.ieso.ca. Accessed 16 Aug 2011

Marin G, Naterer GF, Gabriel KS (2010a) Rail transportation by hydrogen vs. electrification—case study for Ontario Canada, I: propulsion and storage. Int J Hydrogen Energy 35:6084–6096

Marin G, Naterer GF, Gabriel KS (2010b) Rail transportation by hydrogen vs. electrification—case study for Ontario Canada, II: energy supply and distribution. Int J Hydrogen Energy 35:6097–6107

Naterer GF, Fowler M, Cotton J, Gabriel K (2008) Synergistic roles of off-peak electrolysis and thermochemical production of hydrogen from nuclear energy in Canada. Int J Hydrogen Energy 33:6849–6857

Nojoumi H, Dincer I, Naterer GF (2009) Greenhouse gas emissions assessment of hydrogen and kerosene fueled aircraft propulsion. Int J Hydrogen Energy 34:1363–1369

NRCAN (2010) National energy use database. Natural resources of Canada. http://oee.nrcan.gc.ca/corporate/statistics/neud/dpa/data_e/databases.cfm?attr=0

Orhan MF (2011) Conceptual design, analysis and optimization of nuclear-based hydrogen production via copper-chlorine thermochemical cycles. PhD thesis, University of Ontario Institute of Technology

Wang Z, Naterer GF (2010) Greenhouse gas reduction of oil sands upgrading and extraction operations with thermochemical hydrogen production. Int J Hydrogen Energy 35:11816–11828

Wang Z, Naterer GF, Gabriel K (2010) Clean hydrogen for oil sands upgrading with CANDU and SCWR nuclear reactors. Second Canada-China joint conference on supercritical water-cooled reactors, Toronto, ON, 25–28 April

Chapter 3
Worldwide Nuclear Energy Research Programs

Abstract In this chapter, worldwide research efforts on advanced nuclear energy systems for power, heat, and hydrogen generation are presented. It is shown that the next generation of nuclear reactors will create a new energy paradigm shift by significantly improving the fuel utilization efficiency of nuclear reactors via increased operating temperatures and a reactor's capability to cogenerate high-temperature process heat, power, and hydrogen. In former generations of nuclear plants, the coolant temperature was typically limited to ~300 °C leaving only larger size scale-up as the option for efficiency increases and cost per kWh reductions.

Generation of hydrogen via coal gasification, natural gas reforming, and petroleum naphtha reforming is used by many countries as a transitional phase towards a fully implemented hydrogen infrastructure which will use extraction of hydrogen from water. The Generation IV International Forum (GIF) was formed as an initiative of the US Department of Energy to lead an international cooperation on development of the next generation of nuclear reactors with hydrogen production capabilities. Six reactor concepts were selected by GIF as the most promising for commercial implementation. In addition, there are two major research and development efforts worldwide on thermochemical water splitting processes: the sulfur–iodine cycle and the copper–chlorine cycle.

3.1 Introduction

Peaceful use of nuclear energy emerged after World War II mainly as a technology to generate electricity from nuclear fuel (uranium). A controlled fission reaction is conducted in the nuclear reactor which steadily generates heat, after which the heat is transferred to a Rankine cycle (or other power cycle) to operate an electric generator.

In 1951 at Idaho (USA), the first full-scale nuclear power plant was commissioned to generate 100 kW of electricity. In 1954, Atomic Energy of Canada Limited (AECL) tested its first nuclear reactor, which a few years later became a commercial power plant technology known as CANadian Deuterium

G.F. Naterer et al., *Hydrogen Production from Nuclear Energy*,
DOI 10.1007/978-1-4471-4938-5_3, © Springer-Verlag London 2013

Uranium (CANDU). Although continuous efforts have been made in the last ~70 years to generate controlled heat from a fusion reaction, the only currently available technology remains nuclear fission. There is a large variety of fission-based nuclear reactors around the world.

In the first years after World War II, many countries started the development of nuclear power programs at the national level, for both peaceful and military applications. The geopolitical context evolved and international associations on nuclear energy emerged. The International Atomic Energy Agency (IAEA) was established in 1957 with its headquarters in Vienna (Austria) with a mandate to promote peaceful use of nuclear energy. Much international and regional cooperation was observed after 1970 as the number of countries willing to implement nuclear power plants increased. During the Iron Curtain years, there was a remarkable competition between North Atlantic Treaty Organization (NATO) countries and former communist states in nuclear power research, extended from military to peaceful applications. The states within the former Soviet bloc and partner countries (e.g., CAER countries—Council for Mutual Economic Assistance) acted under the political influence of the Soviet Union and mostly licensed the Russian nuclear power generation technology in their territory. For example, the Russian VVER pressurized water reactor (PWR) was exported to Bulgaria, China, the Czech Republic, East Germany, Hungary, Slovakia, and Ukraine. An exception was Romania, which although part of CAER, opted for CANDU reactors. In the NATO countries, a large variety of commercial power plants was developed mainly due to their open policy which engaged government as well as private funding involvement in nuclear power plant development and operation.

The energy crisis in the 1970s increased the national research programs in nuclear power, for four main goals: improvement of nuclear power plant efficiency by augmenting the temperature of the reactor coolant; expanding the use of nuclear energy as process heat; and improved safety (including a reduced risk of terrorist use of radioactive materials) and sustainability. There are many processes that require heat at high temperatures, and nuclear reactors provide a clean way to obtain high temperatures. Also, there are no direct greenhouse gas (GHG) emissions associated with nuclear plant operation.

The international and regional cooperation flourished after the fall of the Iron Curtain. In the 1990s, the European Union started to expand and launched many policies encouraging collaboration of member countries in nuclear energy, including nuclear process heat. The Frame Programs of the European Commission led to the development of significant nuclear hydrogen research, especially in the last decade when "Networks of Excellence" were established. In pace with the globalization process, the nuclear hydrogen programs expanded at a global level. The Generation IV International Forum (GIF) is a major initiative that commits 13 countries to participate in a number of selected advanced nuclear reactors for power, heat, and hydrogen production.

Commercial nuclear reactor technologies were grouped over four generations which cover a time frame of 70 years (1950–2020). The so-called Generation IV of nuclear reactors is expected to be commercially available by 2030. Due to a significantly higher Carnot efficiency, it becomes economically viable to construct modular reactors which cover a very large power generation range (~20–2,000 MW).

In this chapter, there are a number of research programs described in the direction of nuclear hydrogen production, at local, regional, and international levels. In the first part of the chapter, the nuclear reactor technologies are reviewed. The second part is dedicated to GIF activities and the research and development (R&D) roadmap. Non-GIF research programs and initiatives are described separately. A subsection is dedicated to nuclear hydrogen programs. Detailed information is given on the international research program on the copper–chlorine cycle development for thermochemical water splitting, which is a promising technology for medium-temperature Generation IV reactors.

3.2 Development Programs of Advanced Nuclear Reactors

Nuclear reactors evolved significantly in the last seven decades and they will continue to evolve in order to achieve a higher level of fuel utilization, efficiency, safety, sustainability, and cost-effectiveness. Nuclear energy passed through several stages of development which were influenced by geopolitical situations, technological levels, security, and other issues. In the early development phase, nuclear research was funded mostly at a local level, as it was always considered a topic of military and national defense interest. Regional collaborations started to appear after some decades of technological advancement, and eventually, in the last two decades, international collaboration on nuclear research started to be encouraged by governments. Figure 3.1 presents a brief outlook of advanced nuclear energy research programs around the world. In this section, the design, features, and development of nuclear reactor designs, as well as major international research initiatives on nuclear energy systems, are discussed.

3.2.1 Generations of Nuclear Reactors

At the beginning of the twenty-first century, the US Department of Energy (DOE) started to engage governments, industry, and the global research community in a wide exchange of technologies regarding the next generation of nuclear reactors. Commercial nuclear reactors are classified in four generations, as proposed by DOE (2002), of which Generation IV is in a very early phase of R&D.

First-generation reactors were mostly a breeder type. The process of transforming fertile materials in fissionable fuels is known as breeding. Breeder reactors use either natural uranium or thorium-232 as fuel (called fertile materials). When bombarded with neutrons, the fertile materials transform into fissionable fuel. Uranium-238 can be made fissile in about 2½ days; thus natural uranium-based reactors are called "fast breeding reactors (FBR)." In an FBR, ~2.7 neutrons are generated for one absorbed neutron to produce fission, while 1.7 neutrons produce fuel breeding. Thorium

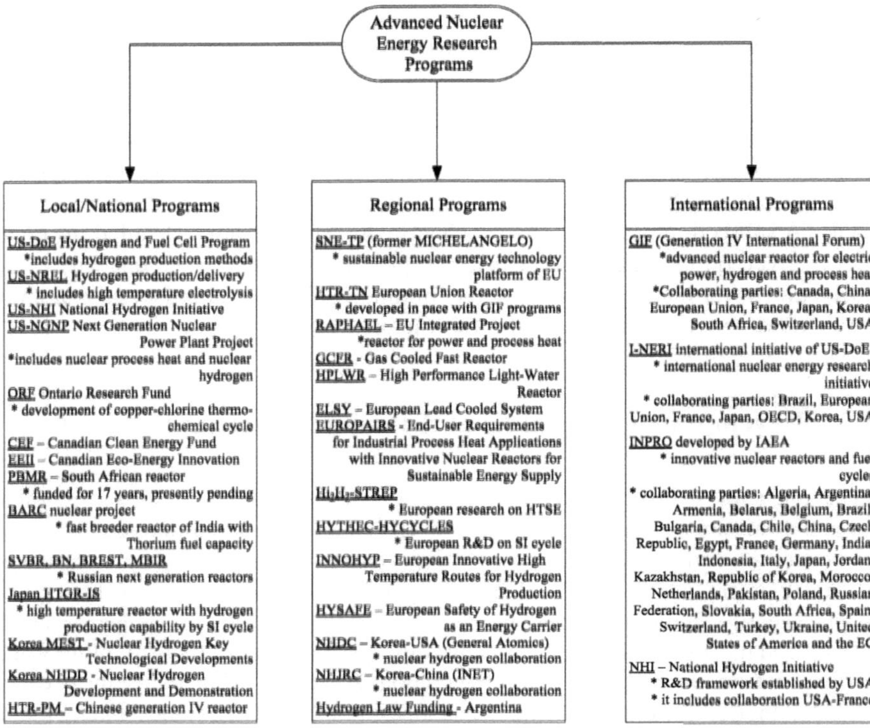

Fig. 3.1 Advanced nuclear energy research programs

breeds slower than natural uranium, i.e., after about 21 days. Thus, thorium-based reactors are called "thermal breeding reactors (TBR)."

Another reactor included in the first generation is the early prototype of boiling water reactors. Any nuclear reactor comprises a "reactor vessel" in which the radioactive material is placed and surrounded by a fluid coolant and a "moderator." The moderator is a material which has two roles: the main role is to decelerate the scattered neutrons generated by the fission reaction (process called thermalization) and the secondary role is to cool the fuel cladding. The nuclear fuel is embedded in a material with a low absorption of neutrons, typically made in zircaloy, magnesium, and steel. This material is called cladding and it has the main role to prevent any radioactive materials to escape into the coolant and contaminate it. Only neutrons can go through the cladding during normal operation. The pace of a nuclear reaction is regulated with the help of "control rods" which are elements made in a neutron-absorbing material and have a role to absorb neutrons, slowing down the fission reaction. Typical materials for control rods include silver, boron, iridium, cadmium, and others.

The first generation of nuclear reactors represent early power plant (experimental) prototypes commissioned between 1954 and 1965. None of the first-generation reactors are still in use today. The following list gives a description of the most significant first-generation reactors.

- Obninsk Nuclear Power Station—Built in the former Soviet Union (FSU), commissioned in 1954 as the first commercial power plant with 5 MWe, and decommissioned in 2002. It is a breeder reactor, graphite moderated, and using light water as a coolant and natural uranium.
- Mangox 1—First-generation nuclear reactor built in the United Kingdom at Calder Hall having 60 MWe and operating between 1956 and 2003. It has a dual commercial and military application. It is a fast breeder reactor which used pressurized carbon dioxide as a coolant (~7 bar working pressure and ~390 °C maximum temperature), magnesium based for fuel cladding, boron-steel control rods, and a graphite moderator.
- Fermi 1—A nuclear power plant prototype developed with a fast breeder reactor at a location close to Monroe (USA, Lake Erie) with a capacity of ~100 MWe, commissioned in 1957 and decommissioned in 1972.
- Shippingport Atomic Power Station—The first full-size nuclear power plant commissioned in 1958 in Pennsylvania (25 km from Pittsburgh), operated until 1982. It is a thermal breeder with a light water moderator with a power output of 60 MWe.
- Dresden Generating Station 1—Commissioned in 1960 in Illinois comprising a boiling water reactor of 210 MWe (based on a design developed at the Idaho National Laboratory and General Electric in the 1950s).
- Nuclear Power Demonstration (NPD) reactor—Prototype of a CANDU reactor commissioned in Rolphton, Ontario, in 1961 with 19.5 MWe. It is a pressurized heavy water reactor (PHWR) operating with natural and slightly enriched uranium. It was decommissioned in 1987.

The second generation of commercial nuclear reactors had its reactors commissioned between ~1975 and ~1990. It includes five kinds of reactors: CANDU, PWR, "water=water energetic reactor" (VVER-440, which is the Russian version of PWR), boiling water reactor (BWR), and the advanced gas reactor (AGR).

The CANDU-6 is the sixth design evolution of AECL reactors and corresponds to the second generation of nuclear reactors, as per the classification of the DOE. It uses natural uranium (UO_2) as fuel and heavy water as a moderator and coolant, while the cladding is made in zircaloy. The heavy water absorbs fewer neutrons than light water, leaving enough of them to initiate reactions in the fuel bundles with natural (unenriched) uranium. The principle of the CANDU reactor is presented in Fig. 3.2. The fuel is placed in tubes through which pressurized heavy water coolant flows. The coolant transfers the nuclear heat to a heat exchanger, which is typically a steam generator; the produced heat flows at about 350 °C. Outside the pressure tubes, there is a moderator consisting of a heavy water bath in which the tubes are submerged. The pressure in the moderator is around 1 bar. The moderator is continuously cooled and heat is rejected at about 80 °C. The fuel rods can be replaced or extracted independently at any time through special mechanisms.

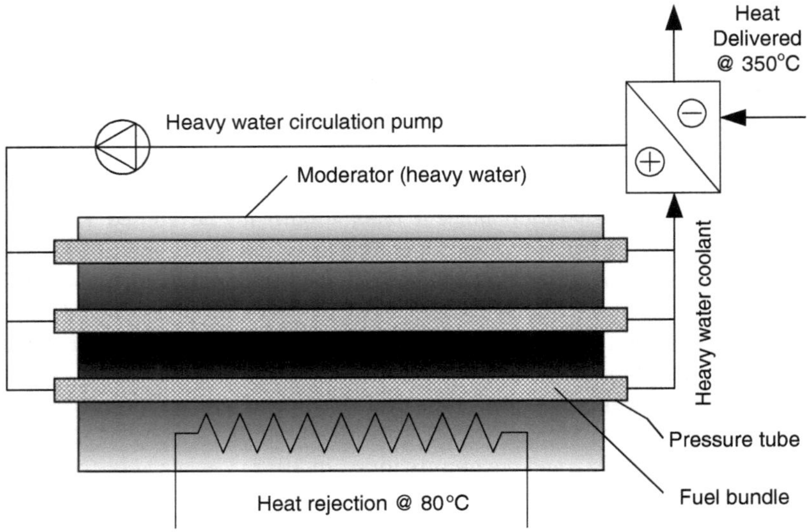

Fig. 3.2 Principles of operation of the CANDU reactor

In the moderator bath, a series of graphite rods (not shown) is placed with the role to attenuate the neutron flux and thus control the reaction rate and criticality. The CANDU-6 generates 500–600 MWe.

The PWR comprises the fuel bundles, moderator vessel, pressurized vessel, heat exchanger (steam generator), and coolant pump. It uses slightly enriched uranium in the form of UO_2 containing about 3 % $^{235}_{92}U$ and placed in tubes of zircaloy and sunk in light water which serves the role of moderator. The fuel and moderator are kept in a pressure vessel designed to operate at about 155 bar, at a pressure at which water boils at 344 °C. However, in the reactor (pressure) vessel, water heated from 275 to 315 °C always remains in a liquid state. In order to enhance the heat transfer between the fuel rods and the water, the flow is set such that subcooled nucleate boiling occurs at the surface of the rods. The small vapor bubbles formed are immediately absorbed in the subcooled liquid water. Overall, the operation is very stable. The light water is an excellent moderator. In order to control the reaction rate, additional moderators are used in the form of movable bars made from boron carbide or Ag-In-Cd.

The pressure is maintained through a pressure vessel equipped with submerged electric heaters to maintain a vapor pressure of 155 bar. The vapor space is located at the top of the pressurizer, while its bottom is always full of subcooled liquid. A heat exchanger is used to deliver the nuclear heat from the primary circuit to a secondary circuit, which may be a steam generator aimed to drive a turbine which turns a generator or propels a marine vehicle. The generated steam is typically 275 °C and 60 bar.

Similar to the PWR, the BWR uses water as a moderator and coolant at the same time, with the difference that the reactor produces saturated steam. There are no primary and secondary circuits like for PWR; instead, subcooled water enters the reactor where it is preheated and boiled to the saturation point. Typically the pressure in the BWR is around 75 bar and the corresponding temperature is 285 °C. The reactor has a special construction which allows for a stable boiling process. When fluid enters the reactor, water is first guided into a down-comer where it is preheated. It rises thereafter and flows around vertical fuel rods through the reactor core region. The resulting two-phase flow of ~15 % vapor quality is directed towards a cyclone separator and steam dryer at the top of the reactor, where saturated steam is collected. In general, the reactor is kept in a containment structure, while the steam turbine, condenser, and pumps are located at the exterior.

The VVER 440 which refers to a second-generation nuclear reactor was first commissioned before 1970 with the most common design version known as V230 with 440 MWe output. It is cooled and moderated with light water; therefore the Russian VVER is a version of PWR. There are some design features in VVER that distinguish it from PWR, namely, its use of a horizontal boiler instead of a vertical arrangement.

A Generation III nuclear reactor is based on Generation II design principles but with improved efficiency, modularity, safety, standardization of components, 1.5 times increased lifetime (~60 years), and reduced capital and maintenance costs. The reactors of Generation III were commissioned between ~1990 and 2005. It includes the enhanced CANDU-6 (EC6) reactor, VVER-1000/392 pressurized water reactor, BARC-AHWR (Indian advanced heavy water reactor), advanced pressurized water reactor (APWR), and the advanced boiling water reactor (ABWR). Note that ABWR was developed by a GE-HNE consortium which comprises General Electric (GE) and Hitachi Nuclear Energy (HNE). The APWR was developed by Mitsubishi Heavy Industries (MHI). The EC6 has an output of 740 MWe and it has the ability to operate at part-load conditions. The VVER-1000 has about 33 % efficiency with a net power generation of 1,000 MWe and a hottest coolant temperature of 321 °C. The APWR capacity is ~1,500 MWe and units exist in operation in Japan. It is expected that some ABWR units will be constructed in the USA in the future.

Generation III+ nuclear reactors will extend from 2010 until 2030. Its designs include the following commercial reactor models: Advanced CANDU (ACR-1000) reactor, the European pressurized reactor (EPR), VVER-1200, APWR-IV, economic simplified boiling water reactor (ESBWR), AP-1000 reactor, APR-1400, and EU-APWR. The ACR-1000 is a hybrid design that uses heavy water as a moderator and light water as a coolant to generate 1,200 MWe. It includes enhanced safety features with respect to former CANDU versions and it is planned to be commissioned by 2016. The AP-1000 is the Westinghouse design of Generation III+ of ~1,150 MWe output. The EPR was designed by an international consortium of Siemens AG, Areva NP, and Electricité de France (EDF). It is a PWR-type reactor with 37 % efficiency generating 1,650 MWe. In Finland, the first EPR power plant in construction started in 2005 and is planned to be commissioned in 2013. The ESBWR is a BWR designed by GE-HNE with a capacity of 1,520 MWe and lifetime of 60 years. APR-1400 is an APWR adopted by South Korea, with 1,455 MWe and first commissioning expected in 2013. The VVER-1200 of Generation III + is an evolution of VVER-1000 with a lifetime expectancy of 50 years and power range of 1,200–1,300 MWe. The EU-APWR is a version of APWR by MHI to be commissioned in Europe at a power generation capacity of ~1,600 MWe.

The fourth generation of nuclear reactors is represented by six reactor concepts selected by the GIF. These concepts will be discussed in more detail in Chap. 7. Here they are briefly introduced with the purpose of facilitating the discussion of the technology roadmap, research issues, and existing research programs. GIF aims at the development of safer, sustainable, more economical, physically secure, and proliferation-resistant nuclear reactors. Generation IV reactors are intended to become commercially available by 2030. According to GIF (2011), the Generation IV nuclear reactors include the following designs.

- Gas-cooled fast reactor (GFR)—It features a fast-neutron-spectrum, helium-cooled reactor and closed fuel cycle (spent fuel is reprocessed and reused); it generates 1,200 MWe and process heat at 850 °C.
- Very-high-temperature reactor (VHTR)—It is a graphite-moderated, helium-cooled reactor with a once-through uranium fuel cycle (spent fuel is not reprocessed); it has a projected core outlet temperature of 1,000 °C and the units have 650 MW output.
- Supercritical-water-cooled reactor (SCWR)—This is a high-temperature, high-pressure water-cooled reactor that operates above the thermodynamic critical point of water; it is planned for 1,700 MWe output and a core outlet temperature in the range of ~500–650 °C.
- Sodium-cooled fast reactor (SFR)—It features a fast-spectrum, sodium-cooled reactor and closed fuel cycle for efficient management of actinides and conversion of fertile uranium; its core outlet temperature is 550 °C and it is planned for three capacities, i.e., low (50–150 MWe), intermediate (300–600 MWe), and large (600–1,500 MWe).
- Lead-cooled fast reactor (LFR)—It features a fast-spectrum lead/bismuth eutectic liquid-metal-cooled reactor and a closed fuel cycle for efficient conversion of fertile uranium and management of actinides. It includes two versions: (1) small size at 19.8 MWe and 567 °C core outlet temperature, and (2) high power at 600 MWe and 480 °C.
- Molten salt reactor (MSR)—It produces fission power in a circulating molten salt fuel mixture with an epithermal-spectrum reactor and a full actinide-recycling fuel cycle. The reference capacity is 1,000 MWe with a coolant exit from the core at 700–800 °C.

Two main parameters define a reactor system: the core outlet temperature and the power generation capacity (in MWe). Figure 3.3 summarizes these parameters for past, present, and future generations of nuclear reactors. The historical development of nuclear reactors can be split into two periods. In the first period, from 1954 to ~2030 the nuclear reactors are used primarily for power generation. Their core outlet temperature is kept around 300–350 °C for safety reasons. The trend in the first period was to increase the power generation capacity. This is the extension of a

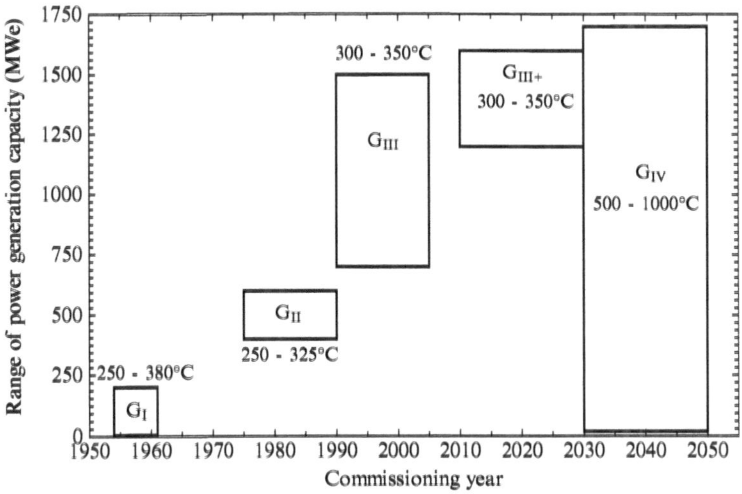

Fig. 3.3 Power generation capacity range vs. period of commissioning of commercial reactors of Generations I–IV and the respective outlet temperature ranges

logical approach: since the hot outlet temperature was limited, incremental enhancements of power generation efficiency can be obtained by increasing the capacity scale. Larger plants have in general better efficiency than smaller plants. Generation IV is a paradigm change. The coolant temperature is much higher; therefore the associated Carnot efficiency of Generation IV reactors is in the range of 60–75 %, much higher than 45–52 % for Generations I–(III+). As a consequence of the higher temperature, Generation IV reactors are capable to deliver process heat for thermochemical production of hydrogen and other applications. Furthermore, the ability to produce heat at higher temperatures makes it possible for the fourth generation of nuclear reactor to cover a very broad spectrum of generation power outputs (from ≈20 to ~1,700 MWe). This facilitates the use of nuclear reactors for a large range of processes and integrated systems, for example:

- Thermochemical hydrogen production
- Desalination at low temperature ranges
- Steam generation
- Process heat for oil refineries
- Coal-to-liquid production
- High-temperature steam for industrial complexes
- Steam-methane reforming for hydrogen production
- High-efficiency electricity generation at medium temperature ranges
- Coal gasification
- Petroleum refining
- Oil recovery
- Coal and natural gas derivatives, and petrochemical production

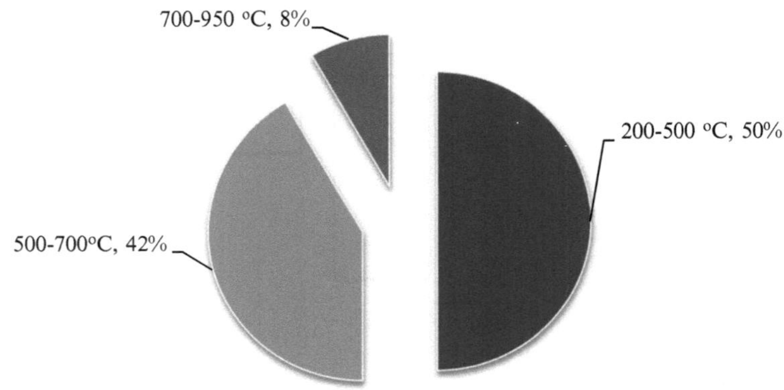

Fig. 3.4 Temperature level and the percentage of total thermal power of required heat by industry in the USA [data from Latour et al. (1982)]

- Synthetic fuel production (e.g., Fischer–Tropsch or methanol)
- Fertilizer production (e.g., ammonia or urea)
- District heating

The industrial process heat has a vast market for nuclear reactors which definitely will benefit the nuclear technology development. Based on a study by Latour et al. (1982), one can conclude that around 50 % of required heat by the US industry is at a level of 200–500 °C. Figure 3.4 illustrates the temperature distribution of the industrial heat demand in the USA.

3.2.2 *GIF and Its Proposed Technology Development Roadmap*

The US DOE promoted an international initiative on the development of advanced nuclear reactors for power and process heat which was established in 2001 as the Generation IV International Forum (GIF). The concept "Generation IV" has been proposed by DOE (as mentioned previously) as a framework to facilitate international cooperation towards the development of advanced nuclear systems by 2030s and beyond. On the DOE (2011) Web site, the department seeks international collaboration for shared R&D costs in the field of nuclear energy. DOE implemented a Generation IV national program staring in 2002, where the terminology of "nuclear energy systems" is used instead of nuclear reactors to indicate that the fourth generation is aimed at power and also process heat production. In January 2000, the Office of Nuclear Energy, Science and Technology of the DOE facilitated discussions of senior representatives from 9 countries (Argentina, Brazil, Canada, France, Japan, Republic of Korea, South Africa, the United Kingdom, and the USA) aimed to establish an international collaboration towards development of Generation IV nuclear energy systems. Three other countries later adhered to the charter: Switzerland (2002), China, and the Russian Federation (2006). The European

Table 3.1 Framework participants within the Generation IV International Forum

Organization	Country or region	Technology[a]			
		VHTR	GFR	SCWR	SFR
Natural Resources Canada (NRC)	Canada	No	No	Yes	No
China Atomic Energy Authority (CAEA) and Ministry of Science and Technology (MST)	China	Yes	No	No	No
Euratom Joint Research Centre (EURATOM-JRC)	European Union	Yes	Yes	Yes	Yes
Commissariat of Atomic Energy (CEA)	France	Yes	Yes	No	Yes
Agency for Natural Resources and Energy (ANRE) and the Japan Atomic Energy Agency (JAEA)	Japan	Yes	Yes	Yes	Yes
Korea Science and Engineering Foundation (KSEF) and Ministry of Education, Science and Technology (MEST)	Republic of Korea	Yes	No	No	Yes
Departments of Minerals and Energy (DM & DE)	South Africa	No	No	No	No
Paul Scherrer Institute (PSI)	Switzerland	Yes	Yes	No	No
Department of Energy (DOE)	USA	Yes	No	No	Yes

Source: DOE (2011)
GFR gas-cooled fast reactor, *SCWR* supercritical water reactor, *SFR* sodium fast reactor, *VHTR* very-high-temperature reactor
[a]Technology R&D as engaged by participants in 2008

Union adhered in 2003 with its atomic energy agency—Euratom. The International Atomic Energy Agency (IAEA) and the Nuclear Energy Agency of the Organisation for Economic Co-operation and Development (OECD) participate in the charter as permanent observers.

Nine of the GIF members signed a Framework Agreement through which they formally participate in the development of at least one of the nuclear energy systems of Generation IV. GIF identified a number of research topics as follows:

- Nuclear fuel and fuel cycle
- Advanced nuclear fuel
- Fuel qualification
- Fuel core, materials, and fuel cycle
- Hydrogen production
- Computational methods (including validation and benchmarking)
- Materials and chemistry
- Global actinide cycle
- Component design for the balance of plant
- Safety and operation
- Thermal hydraulics
- Component design and safety
- System integration and assessment

Among the six nuclear energy systems of GIF, only four will be investigated within the Framework Agreement as indicated in Table 3.1. Note that Europe participates with three jurisdictions in the GIF effort: France, Switzerland, and the European Union. South Africa participates in the research agreement within GIF, but they have their own nuclear energy system technology which is not part of the selected Generation IV reactors. Therefore, they are not listed in Table 3.1 for any specific design. The South African reactor system is the pebble bed modular reactor (PBMR) developed in 1994; this is discussed in a subsequent section. Canada will not produce a VHTR but participates in the research within the GIF framework. Korea participates in R&D on SCWR but does not plan to construct its own system (Table 3.1).

The milestones of GIF project are given in Table 3.2. According to DOE (2002), the GIF together with the US DOE planned a budget for its own Generation IV reactor development, which is about $5,770M with a scope until 2020. The percent-based distribution of this budget for each of the adopted nuclear energy systems is presented in Fig. 3.5.

In the USA, the Energy Policy Act was approved in 2005 by the Congress with a mandate of the DOE to establish a Next Generation of Nuclear Power Plant Project (NGNP) with a planned budget of $164M for FY 2010. The Idaho National Laboratory (INL) was selected by DOE to lead the implementation of NGNP. The high-temperature gas reactor (HTGR) has been selected as a preferred technology to be implemented within NGNP.

3.2.3 Non-GIF Reactor Development Programs and Initiatives

Although the GIF roadmap is the most significant research and development framework in nuclear energy systems at a global level, in addition, there are other relevant research and development programs, at national, regional, and international levels, to develop alternative options of nuclear reactors of the next generation for power and process heat applications. Some of them collaborate directly or indirectly with GIF, while some others are independent. This subsection discusses examples of those nuclear energy system programs which are not coordinated by GIF.

The International Nuclear Energy Research Initiative (I-NERI) is another research program with international cooperation established by DOE in 2004 with the mandate to support R&D in topics within those promoted by its Office of Nuclear Energy (DOE-NE), namely: (1) fuel cycle development, (2) nuclear reactor technologies, and (3) hydrogen production from nuclear energy. I-NERI has eight international participants as indicated in Table 3.3. The USA has matched funds with its laboratories on a 50–50 share with the international project participants. According to I-NERI (2009), until 2009, the USA invested $118.5M in the project while other participants contributed $102.7M. The share of each participant in percentages is shown in Fig. 3.6.

The I-NERI program tries to collaborate synergistically with the other three main programs of DOE in nuclear energy, namely, Generation IV of Nuclear Reactors Program (with 54 % of its funding), Fuel Cycle Research and Develop-

Table 3.2 Milestones of GIF project as established by countries in 2008

Project module	Technology milestone			
	VHTR[a]	GFR[b]	SCWR[b]	SFR[a]
I	2010: Viability, preliminary design and safety	2012: Balance of plant	2016: Balance of plant	2016: Advanced fuel system
II	2015: Performance, final design, and safety analysis	2020: Fuels and materials (core materials testing by 2014)	2020: Fuels and materials (mechanical properties by 2012, radiolysis by 2014, core materials by 2016, water chemistry by 2018)	2015: Component design and balance of plant
III	2020: Demonstration, construction, preliminary testing	2018: Reactor systems	2016: Reactor systems	2012: Global actinide cycle demonstration
IV	N/A	2020: Safety (safety approach by 2011)	2020: Safety (approach by 2007, integral experiments by 2018)	2011: System integration and assessment
V	N/A	2020: Design and evaluation (pre-conceptual design by 2014)	2020: Design and evaluation (pre-conceptual by 2015)	2012: Safety and operation
VI	N/A	2020: Fuel cycle (viability assessment by 2014)	2020: Fuel cycle	N/A

[a]According to 2008 planning (DOE 2011)
[b]According to 2002 planning (DOE 2002)

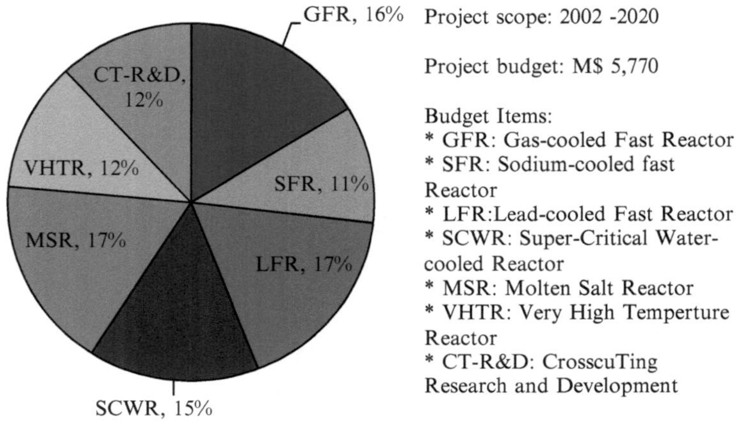

Fig. 3.5 Generation IV International Forum Budget and research items (*source* DOE 2002)

Table 3.3 Participant countries in the I-NERI program of DOE

Participant	Funding	Institutions	Projects
Brazil	$2.3M	3	2
Canada	$19.2M	8	12
European Union	$12.3M	6	15
France	$30.2M	6	21
Japan	$2.7M	7	2
OECD	N/A	1	1
Republic of Korea	$36.0M	11	37
USA	$118.5M	27	90[a]

Source: INERI (2009)

[a]According to I-NERI protocol, the USA is involved in all projects

Fig. 3.6 Budget share among the participants in the I-NERI program

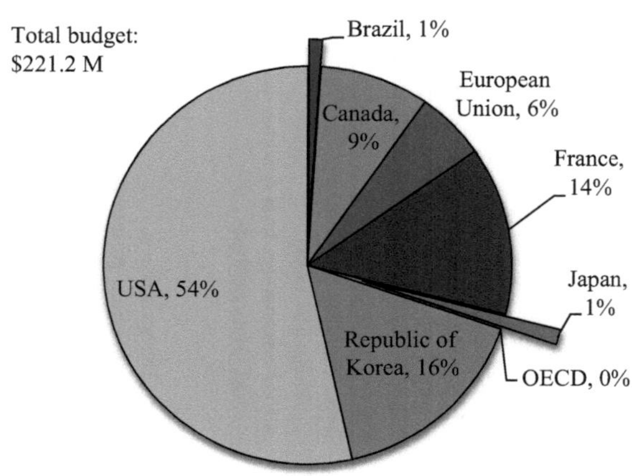

ment Program (with 24 % of its funding), and the Nuclear Hydrogen Initiative Program (with 12 % of its funding). As a consequence, I-NERI has established its R&D priority topics as follows (see INERI 2009):

- Advanced reactors, systems, and components
- Alternative energy conversion cycles
- Design and evaluation methods
- Advanced structural materials
- Thermochemical cycles of hydrogen production
- Advanced fuels
- Transmutation science and engineering
- Fuel cycle systems analysis
- High-temperature electrolysis
- Reactor-hydrogen production process interfaces
- Spent nuclear fuel separation technologies

Another research program is the International Project on Innovative Nuclear Reactors and Fuel Cycles (INPRO) initiated in 2000 by the IAEA. Its mandate is to promote a sustainable nuclear energy technology for the future. INPRO is funded mainly by extra-budgetary resources.

As per 2011, it includes 34 member countries (or regions), namely: Algeria, Argentina, Armenia, Belarus, Belgium, Brazil, Bulgaria, Canada, Chile, China, the Czech Republic, Egypt, France, Germany, India, Indonesia, Italy, Japan, Jordan, Kazakhstan, Republic of Korea, Morocco, the Netherlands, Pakistan, Poland, Russian Federation, Slovakia, South Africa, Spain, Switzerland, Turkey, Ukraine, the USA, and the European Commission. The INPRO developed six programs. Each of the programs funds a number of collaborative projects, as indicated in Table 3.4.

In South Africa, the PBMR was under development for almost 16 years under the funding of Eskom, its partners, and the South African Government (Koster et al. 2003). No equipment failure or human error can produce an accident with PBMR because its design and construction are inherently safe. This greatly simplifies the safety design of the system. Its design evolved from the high-temperature reactor of Siemens to a scheme that uses a Brayton cycle turbine with a closed loop. The hot outlet temperature is 900 °C with around 400 MW thermal output per unit. Currently, the PBMR technology is pending as it has not found the investors needed for the technology.

India is the sole country that has developed a research program on the thorium fuel cycle for nuclear power reactors. The government typically has policy programs formulated for 5-year plans. Five “nuclear energy parks” were set by the Indian Government. The Bhabha Atomic Research Centre (BARC) of India leads the research activities on the country’s nuclear reactors. Two of the Indian nuclear power stations are coupled with desalination plants (WNA 2011). In the vicinity of Madras, a prototype of an FBR of 500 MWe is in construction by an Indian consortium including the government and Bharatiya Nabhikiya Vidyut Nigam Ltd.

Table 3.4 Program and specific projects of INPRO

Program	Projects
Nuclear Energy System Assessment (NESA)	1. PRADA: proliferation resistance: acquisition/diversion pathway analysis 2. ENV: Environmental impact benchmarking applicable to nuclear energy systems under normal operation
Global Vision on Sustainable Nuclear Energy	1. GAINS: Global architecture of innovative nuclear systems based on thermal and fast reactors including closed fuel cycles 2. FINITE: Fuel cycles for innovative nuclear energy systems based on integrated technologies 3. ThFC: Investigation of the ^{233}U/Th fuel cycle 4. RMI: Meeting energy needs in the period of raw material insufficiency during the twenty-first century
Innovations in Nuclear Technology	1. COOL: Investigation of technological challenges related to the removal of heat by liquid metals and molten salt coolants from reactor cores operating at high temperatures 2. DHR: Decay heat removal system for liquid metal cooled reactors 3. AWCR: Advanced water-cooled reactors 4. PGAP: Performance assessment of passive gaseous provisions
Innovations in Institutional Arrangements	1. SMALL: Implementation issues for the use of nuclear power in small countries 2. TNPP: Legal and institutional issues of transportable nuclear power plants
INPRO Dialog Forum	N/A
Policy Coordination, Communication, and INPRO Management	N/A

Source: INPRO (2011)

The FBR reactor is a thorium-based reactor project of India. It will be fuelled with uranium–plutonium fuel and it will have a blanket of uranium/thorium to breed fissile ^{233}U. The thorium-based breeder may become one of the most remarkable non-GIF next generation commercial reactors, as it is able to use large amounts of thorium resources in India. Six more FBRs are planned by 2020. In a third phase, the fissile ^{233}U and plutonium from FBR will be transferred to existing Indian advanced heavy water reactors where it will generate about 1/3 of the power, the rest being supplied by additional thorium fuel.

The Russian Federation is very advanced in FBR research. Russian federal funding in novel nuclear reactors is around two billion USD for a multiyear period (until ~2020, see Table 3.5). The modular SVBR 100 concept is a low-power lead–bismuth cooled FBR for pressurized process steam or power generation. This small reactor is very versatile and it can be configured for many applications. Russia led the IAEA INPRO project in 2001.

Table 3.5 Russian next generation of commercial nuclear reactors

Reactor	Characteristics	Applications	Federal funding	Commissioning
SVBR-100	Lead–bismuth-cooled fast breeder reactor	Modular power or process steam generation	USD 428 M	2017
	265–280 MW thermal or ≈100 MWe	Coastal or floating power generators		
	Steam generation @: 440–482 °C, 4.7–9.5 MPa, 460–580 t/h	Desalination Process heat		
BN 600/800	Sodium-cooled fast neutron reactor	Power generation	USD 173 M	2016
BREST 300	Lead-cooled fast reactor 300 MWe	Power generation	USD 830 M	2020
MBIR 150	150 MW thermal fast neutron reactor of IV generation	Multipurpose research	USD 531 M	2020

Source: WNA (2011)

3.3 R&D Programs on Nuclear Hydrogen Systems

Hydrogen is a key energy carrier for sustainability in the future energy economy. It was discussed in Chap. 2 in regard to the role of nuclear energy in hydrogen production. The next generation of nuclear reactors will be able to produce power and high-temperature process heat, which will then be able to drive water splitting processes for thermochemical hydrogen production.

Possible routes to generate hydrogen from water using nuclear energy were discussed in Chap. 2, and presented in Fig. 2.1. These routes are also depicted in Fig. 3.7 with additional paths which assume other processes than water decomposition, e.g., coal gasification, natural gas reforming, petroleum naphtha reforming, or hydrogen sulfide cracking. In all such processes, high-temperature nuclear heat is used to conduct the required chemical reactions. Only the methods that are based on electrical power and/or high-temperature heat inputs are illustrated in Fig. 3.5. Coal gasification and natural gas reforming methods include the water gas-shift reaction to generate additional hydrogen and reduce the carbon monoxide to CO_2. Nuclear-based reforming of fossil fuels is a promising method of high viability in the transition towards a fully implemented hydrogen economy which will reduce the world dependence on fossil fuel energy.

Although fossil fuel reforming can be used to generate hydrogen, it is attractive to use nuclear heat to synthesize Fischer–Tropsch fuels or methanol. Such fuels can also be used for transportation vehicles. In this respect, nuclear heat is used to

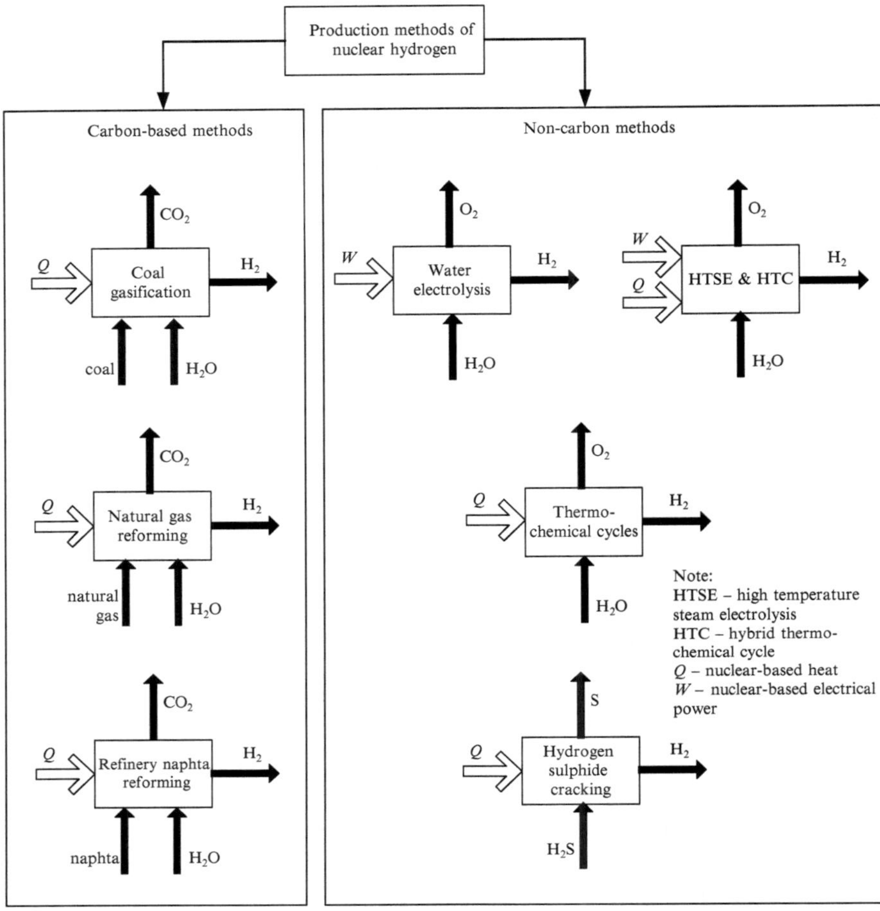

Fig. 3.7 Production methods of nuclear hydrogen which use heat and/or electricity inputs

generate synthesis gas, and then to generate either diesel or methanol. Another process of high interest is fertilizer production such as ammonia and urea with nuclear energy. A nuclear ammonia/urea production facility would comprise the nuclear reactor, a heat transfer system for high-temperature nuclear process heat, a nuclear power plant, a nuclear hydrogen generation unit, an air separation unit, H_2/CO_2 separators, a Haber–Bosch synthesis unit, and a NH_3/CO_2 reactor to generate urea.

Hydrogen sulfide is a potential source of hydrogen. Hydrogen sulfide can be extracted from some geothermal wells, oil wells, volcano sites, and seas. The Black Sea represents the highest reserve of hydrogen sulfide in the world. In the Black Sea, the hydrogen sulfide can be viewed as a renewable resource because it is generated by microorganisms under the influence of solar radiation. Typically, thermo-catalytic cracking of H_2S can be applied to generate hydrogen and sulfur, as two valuable products using nuclear energy.

As mentioned above, the next generation of nuclear reactors will have the coproduction of high-temperature process heat and electricity commodities which can be used for many purposes. Because of its need as a basic chemical in many industrial processes (e.g., aluminum smelting and steel processing, petrochemicals, etc.) and an energy carrier, hydrogen produced with nuclear energy is a key element in the development of the next generation of nuclear reactors. Consequently, many nuclear R&D programs include a significant component focused on nuclear hydrogen production, which roughly represents 20 % of the program financial budget.

This section discusses some of the nuclear-based hydrogen production programs, categorized by countries, regions, and international initiatives.

Japan has a nuclear hydrogen program that is coordinated at the highest level by Japan Atomic Energy Agency (JAEA) and Japan Atomic Energy Commission (JAEC). The nuclear program started in ~1970 with the design and prototyping of the Japanese HTGR with an aim to start its commercialization for nuclear hydrogen production by 2020. The Oarai R&D Center of JAEA developed a test HTGR in operation with 30 MW power output. The sulfur–iodine thermochemical cycle was selected as the thermochemical cycle in conjunction with HTGR. As a consequence of this choice, the program research is denoted as HTGR-IS. The test prototype of the sulfur–iodine (S–I) cycle is demonstrated by hydrogen production for 175 h at a rate of 31 NL/h production and an efficiency of 43 % (Kubo et al. 2004). The S–I thermochemical cycle is examined in detail in Chap. 5. The cycle has been proposed by General Atomics (GA) in the 1970s and GA examined its application to nuclear and solar energy in the 1980s. At a typical commercial scale, a 600 MW thermal reactor can produce around 0.06 Mtons/year of hydrogen.

In the USA, in the National Hydrogen-fuel Initiative (NHI), a national research and development program mandated to establish a GHG-free technology of hydrogen production. The NHI partners were General Atomics, Areva NP, and Westinghouse Electric Company for nuclear development of a commercial nuclear hydrogen production system. Various technologies and plant settings were considered:

- Gas Turbine Modular Helium Reactor (GHTR) developed by GA which can produce electricity at ~50 % efficiency and process steam at 585 °C.
- Placement of a sulfur–iodine thermochemical plant at ~90 m from the nuclear reactor building.
- Development of an S–I cycle with 44.5 % hydrogen production efficiency at a 900 °C heat source temperature.
- Using HTSE (high-temperature electrolysis) as an alternative hydrogen generation method with an envisaged efficiency of 53 % at 860 °C source temperature.

Table 3.6 Research milestones on S–I cycle research at General Atomics

Year	Milestone
1972	GA initiates thermodynamic based investigation into possible thermochemical cycles
1974	Begin active investigation into practical thermodynamic cycles
1975	Apply initial patent for S–I cycle to make sulfuric acid and hydrogen
1977	Follow with multiple patents covering complete S–I cycle
1978	Begin DOE support of S–I cycle development
1979	Start construction of glass loop
1981	Demonstrate Bunsen flow reactor at 60 Lph equivalent hydrogen
1980	S–I tops DOE evaluation of S–I, HyS, and LTE
1981	DOE ends support of nuclear S–I cycle
1981	DOE starts support of solar S–I
1983	Demonstrate solar decomposition of sulfuric acid on GIT power tower, atm operation, limited operation due to lack of sunshine
1986	Complete high-pressure simulated solar sulfuric acid decomposition 5–10 atm operation, Pt and Fe catalysts, 700–900 °C outlet temperature
1999	GA begins NERI evaluation of potential nuclear TC processes, jointly with SNL and UL
2000	Identify top two processes: S–I top candidate, UT-3 second place
2003	Complete NERI final report on S–I flowsheet and begin the first-phase I-NERI
2005	Start the second-phase I-NERI, jointly of SNL, GA CEA
2009	Conclude I-NERI S-I ILS experiments at the GA site

Three main NHI projects are distributed as follows.

- S–I cycle development in an international consortium led by General Atomics and comprising also the Sandia National Laboratory (SNL) and the French Commissariat of Atomic Energy (CEA). The milestones of research and development led by General Atomics on the S–I cycle are listed in Table 3.6.
- High-temperature steam electrolysis development at INL.
- Hybrid sulfur cycle development (which requires power and heat as input, see Chap. 5) by the Savannah River National Lab (SRNL).

A list of research topics was proposed by NHI for the development of an integrated system for nuclear hydrogen production which combines a high-temperature reactor with the sulfur–iodine thermochemical cycle which includes:

- System layout and efficiency assessment
- System configuration
- System performance
- Sulfur–iodine dynamics
- Thermo-fluids and safety
- Fuel and graphite
- Coolant activity
- Mass and energy discharge
- Fission products and containment
- Atmospheric dispersion and public dose
- Graphite corrosion
- Thermo-mechanical graphite seismic effects

- Transport lattice, fuel block, pebble, reflector
- Burnup, diffusion core calculation
- Physics analysis

In the European countries, there is significant research under way on hydrogen energy and its nuclear-based production. In Europe, 170 GWe nuclear power plants are installed which represent 32 % from nuclear electric power generated globally. The executive body of the European Union (EU) is the European Council which elaborates the development policies and strategies and also funds research. Collaborative networks of scientific research are encouraged among European countries.

An important research network sustained by the EU between 2001 and 2005 was the Michelangelo Network with activities in nuclear and hydrogen energy. Michelangelo continued and it is presently active as SNE-TP—Sustainable Nuclear Energy Platform. The SNE-TP has a mandate to develop nuclear energy systems of III, III+, and IV generations for nonelectric applications such as hydrogen production and synthetic fuel production. Other uses of nuclear heat are envisaged in Europe, as follows:

- Cogeneration for industrial use or district heating
- Nuclear heat for crude oil processing with reduced GHG emissions
- Desalination
- Nuclear-based steam reforming
- Nuclear-based coal gasification

Another collaborative research network established in the EU is the European high-temperature reactor network (HTR-TN) with a mandate to develop the European version of a high-temperature nuclear reactor of Generation IV in coordination with GIF. After 2005, the EU established several novel policy instruments for collaborative research funding, including the Networks of Excellence (NoEs), Integrated Projects (IPs), Technological Platforms (TPs), Joint Technology Initiatives (JTIs), and Joint Research Centers (JRCs).

Collaborative research is funded—among other means—via frame programs which are planned for a 5-year period. The FP7 (the EU's seventh frame program) provided research funding of ~50 billion USD. One of the main funds is given to a JTI focused on hydrogen energy and fuel cell development. Another research focus is thermochemical water splitting with the S–I cycle and reforming of fossil fuels to hydrogen in combination with carbon dioxide sequestration.

EURATOM is the acronym for the European Atomic Energy Community which is mandated for nuclear energy research in the European Union. A JTI is active within Euratom for development of Generation IV nuclear reactors for the generation of electricity and process heat. For Generation III+ nuclear reactors, there is a proposed EPR which is under development.

During the Frame Program Six (FP6, 2005–2010) and Integrated Project, the reactor for process heat and electricity Generation (RAPHAEL) has been established with participation of 10 European countries and 33 research partners aiming to design and develop a very-high-temperature reactor for process heat at 800–1,000 °C.

A number of specific targeted research projects (STRPs) were funded in the last 6 years on advanced nuclear reactors, which includes research on subtopics of nuclear hydrogen, namely:

GCFR—"Gas Cooled Fast Reactor" (2005–2009)
HPLWR—"High Performance Light-Water Reactor" (SCWR, 2006–2010)
ELSY—"European Lead Cooled System"—LFR (2006–2010)
EISOFAR—Road map for a "European Innovative Sodium-cooled FAst Reactor"—SFR (2007–2008)
ALSIA—"Assessment of Liquid Salts for Innovative Applications"—MSR (2007)
EUROPAIRS—"End-User Requirements for Industrial Process Heat Applications with Innovative Nuclear Reactors for Sustainable Energy Supply" (started in 2009 and funded by FP7)
CHIRISGAS—IP, which is an integrated project on coal-biomass gasification with an important research pilot plant at Delft University in the Netherlands
Hi_2H_2-STREP—which is a research project to develop solid oxide electrolysis cells (SOEC) for high-temperature electrolysis, which started in 2004 and ended in 2007 with a target cost of ~700 per kWe
HYTHEC-HYCYCLES which is a research project led by CEA (France) on sulfur–iodine cycle development with a maximum process heat limited at 950 °C (2004–2008)
INNOHYP—"Innovative High Temperature Routes for Hydrogen Production" (2004–2008)
HYSAFE—"Safety of Hydrogen as an Energy Carrier," a 5-year project within FP6 comprising an NoE of 25 partners

The projected hydrogen economy in the Republic of Korea requires that 25 % of total hydrogen is supplied by advanced nuclear reactors by 2040. This amount of hydrogen is around 3 Mt/year and it is expected to be produced in 50 nuclear hydrogen units. The nuclear policy in Korea is led by its Atomic Energy Commission (AEC) which collaborates with the Korean Institute of Energy Research (KIER), Korean Atomic Energy Research Institute (KAERI), and the Korean Institute of Science and Technology (KIST).

Korea launched its nuclear hydrogen program in 2004 with two targets: (1) generation of hydrogen for fuel cell applications such as electricity generation, passenger vehicles, and residential power and heating, and (2) lowering hydrogen costs and improving efficiency of the related processes.

The following nuclear hydrogen programs were approved by AEC.

MEST—"Nuclear Hydrogen Key Technological Developments" launched by KAERI and meant to run in phase with GIF projects (2006–2017).
NHDD—"Nuclear Hydrogen Development and Demonstration" program which will start after 2011 and go up to 2030 (milestones: 2022—prototype construction, 2026—technology demonstration, 2030—technology commercialization).
Hydrogen production program with two phases:

- I: Hydrogen production from natural gas, petroleum naphtha, and electricity (ending in 2025)
- II: Hydrogen production from coal, nuclear energy, and renewable energy (ending in 2040)

VHTR-H_2 program—"Very-High-Temperature Reactor for Hydrogen Production" with three applications:

- Hydrogen production for use as fuel
- Hydrogen production for iron ore processing
- Nuclear steam methane reforming with CO_2 capture and recycling to produce synthetic methanol

For the reference case design of the VHTR-H_2 system, an underground VHTR reactor of 200 MW thermal output will be coupled with an S–I cycle to generate H_2 from water. An alternative design is considered also with an advanced high-temperature reactor generating heat at 850 °C for hydrogen reforming from natural gas and methanol production.

- SI program—as a collaboration of KIER and KIST with an option of pressurized S–I cycle development. It runs in parallel with NHDD and GIF programs and it has two phases:
  - I: 2006–2011 for development of key technologies
  - II: 2012–2017 for performance improvement and validation
- I-NERI—Korea participates in the I-NERI program of DOE with joint projects with the Idaho Nuclear Laboratory and Argonne National Laboratory

Korea also established two major joint research agreements, namely: (1) Nuclear Hydrogen Joint Development Centre (NHDC) with General Atomics, and (2) Nuclear Hydrogen Joint Research Center with China (via INET).

In China, the Institute of Nuclear and New Energy Technology (INET) was established in 1960 at Tsinghua University with a role as an advisory body for government policy in energy technology. In 2007, the Chinese State Medium–Long-Term Development Program of Nuclear Power aimed to increase the nuclear energy generation capacity of the country up to 9 GWe (three times more than the 2005 level). R&D of high-temperature nuclear reactors (HTR) has been conducted. Some key nuclear/hydrogen Chinese projects are listed as follows.

- HTR-10—a test reactor project of 10 MWe constructed with spherical fuel elements since 2000, presently in operation.
- HTR-PM—a Generation IV reactor project pursued in coordination with GIF projects which include a nuclear hydrogen development component. The nuclear reactor is planned for completion by 2013 with 200 MWe capacity.
- Nuclear Hydrogen Production Project—includes the production system for nuclear-produced hydrogen, to be coupled with the Chinese high-temperature reactor. Two technologies were selected for R&D funding: the S–I cycle and

HTSE. The research on high-temperature steam electrolysis focuses on the development of solid oxide electrolysis cells (SOEC) with high efficiency and advanced electrode materials. The nuclear hydrogen production project with the S–I cycle includes three phases.

- Phase I (2008–2014)—process verification of nuclear hydrogen process and bench-scale test of the S–I cycle. In 2009, a test sulfur–iodine plat called IS10 was demonstrated at a lab-scale production of 10 NL/h (normal liters per hour). The bench-scale production is planned to be 100 NL/h.
- Phase II (2015–2020)—includes R&D on coupling technology with a reactor, nuclear hydrogen production safety, and pilot-scale test at 100,000 NL/h.
- Phase III (after 2020) is the technology commercialization stage.

Canada aligned its federal R&D policy so as to contribute to GIF-led nuclear hydrogen development efforts and at the same time achieve a national strategic goal to develop its SCWR-CANDU reactor of Generation IV. In addition, Canada contributes to the GIF effort of VHTR development. The Office of Energy Research and Development (OERD) coordinates and leads Generation IV R&D support and it has the mandate to develop a nuclear program for power generation and process heat. One specific target is to use nuclear reactors for oil sand applications. The R&D activities are focused in the following directions:

- Advanced materials for SWCR and VHTR
- Chemistry research support for SWCR and VHTR
- Integration systems of SWCR with nuclear hydrogen production
- Advanced fuels and fuel cycles
- Reactor safety

The following federal Canadian programs managed by OERD are conducted:

- Clean Energy Fund with CND 795M invested for a 5-year period
- Eco-Energy Innovation Initiative with CND 97M budget in 2011 for demonstration projects

Argentina is also committed to nuclear hydrogen development. Its National Atomic Energy Commission (CNEA) and its governmental R&D Agency for the Promotion of Science and Technology (APST) committed a research budget of USD 206M in 2011. Since 2006, the Argentinean Congress outlined a "Hydrogen Law 26123" which promotes technology development on production, use, and applications of hydrogen as a fuel and energy carrier. The main research focus areas include:

- Nuclear-assisted steam reforming
- High-temperature electrolysis
- Development of thermochemical cycles based on metallic chlorides (copper–chlorine cycle, iron–chloride cycle, manganese–chlorine cycle)
- Nuclear coal gasification at 700–800 °C
- Nuclear coal pyrolysis at 300–500 °C
- Development of ceramic ion transport membranes for high-temperature electrolysis

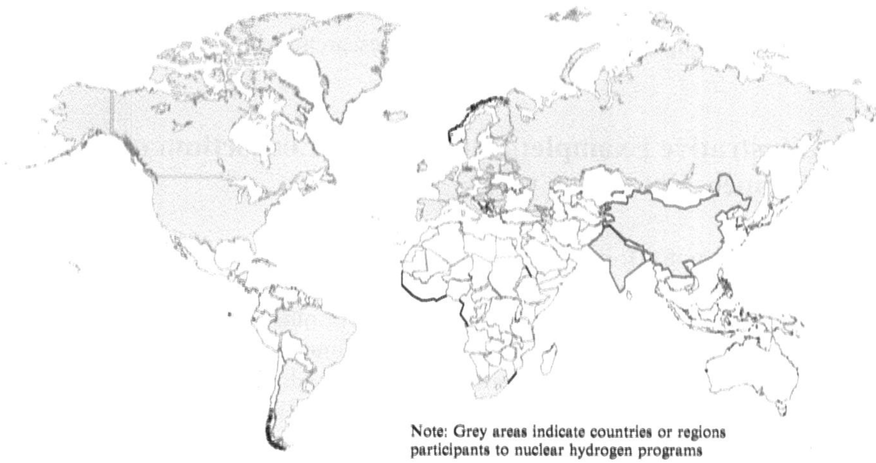

Fig. 3.8 World map showing countries and regions active in nuclear hydrogen research

Based on the above review of worldwide research and development programs, nuclear hydrogen plants are expected to be commercially available worldwide by 2030–2040. The majority of nuclear hydrogen projects in course aim to develop large-scale demonstration pilot plants by 2020. The USA, Japan, and the European Union have significant investors in nuclear hydrogen technology as well as the development of next-generation nuclear reactors. Canada, China, Russia, and India are other major developers of specific nuclear programs. India has been less active in the past with international nuclear trade and exchange programs due to its military activity; however, in recent years, this situation changed and India is now actively developing at full pace its own next-generation program of nuclear reactors. It is the most advanced on thorium-based fuels. Russia—with vast experience in nuclear science—is the leader in FBR technology and participates actively in GIF efforts. Russia will launch within a few years its modular, high-temperature reactor for power and process heat, SVBR 100, which is expected to become a commercial breakthrough in the nuclear market as it has flexibility in implementation and customization capability for specific applications.

Canada's efforts are focused on its next-generation CANDU reactor technology for process heat and power with a capability to be coupled to the copper–chlorine thermochemical cycle for hydrogen production. A particular strategic goal of Canada is to apply its nuclear hydrogen technology to Alberta's oil sand extraction and upgrading processes. Many other countries are developing the sulfur–iodine cycle for hydrogen production (with various versions), including the USA, China, Japan, France, and Europe. This cycle requires VHTR which will be challenging to construct due to material requirements to operate at high temperatures. South Africa is involved in GIF; however, its next-generation nuclear reactor, although a very promising concept, is pending further funding. The map in Fig. 3.8 illustrates the geographical coverage of nuclear hydrogen programs worldwide. It shows that around ¾ of the globe is actively engaged in nuclear hydrogen research and development.

3.4 Illustrative Example: International Consortium on the Copper–Chlorine Cycle

A particular research effort led by Canada is the development of the copper–chlorine thermochemical water splitting cycle with a particular goal to be coupled with the SWCR, although it can also be readily linked with other Generation IV reactors. A first-phase 5-year project ended in 2010 with comprehensive results and a lab-scale demonstration of all unit operations and material development. The first phase was supported primarily by Ontario, Canada, through its Ontario Research Fund (ORF) of the Ministry of Research and Innovation. A "Clean Energy Research Laboratory" was built at the University of Ontario Institute of Technology (UOIT) to house the large-scale unit operations of the Cu–Cl cycle. Several funding sources were made available to support the first phase of copper–chlorine cycle development, including $3M in a provincial budget awarded in 2007 and $6M funding by ORF and the Canadian federal research funding agency NSERC (Natural Sciences and Engineering Research Council of Canada).

A second phase of 5-year funding was awarded in 2011 for a total project valued at about $10M on the integration and further scale-up of the copper–chlorine cycle at UOIT, as well as solar photocatalytic hydrogen production. Funding sources included ORF—Research Excellence Program—which includes funding from the Ontario Ministry of Research and Innovation, AECL, and University Network of Excellence in Nuclear Engineering (UNENE, 5 %), among several others. The international team includes 33 members from eight collaborating institutions and universities from five countries, six industrial partners, and two local organizations in Durham Region (Ontario). The institutional participants in the consortium are listed in Table 3.7. The international consortium on clean hydrogen production led by UOIT comprises the following team members:

- *UOIT team*: Dr. Greg Naterer (principal investigator); Edward Secnik (project manager); Dr. Zhaolin Wang; Dr. Ibrahim Dincer; Dr. Kamiel Gabriel; Dr. Marc Rosen; Dr. Brad Easton; Dr. Igor Pioro; Dr. Brian Ikeda; Dr. Ghaus Rizvi; Dr. Lixuan Lu; Dr. Matthew Kaye; Dr. Dan Zhang; and Dr. Franco Gaspari.
- *Industry and community partners*: AECL; UNENE; Ontario Power Generation; Phoenix Canada Oil Company Limited; Gerdau; St. Mary's Cement; Marnoch Thermal Power; Durham Strategic Energy Alliance; and the Regional Municipality of Durham.
- *Ontario partners*: Dr. Jin Jiang (University of Western Ontario); Dr. Javad Mostaghimi, Dr. Vladimiros Papangelakis (University of Toronto); Dr. Michael Fowler (University of Waterloo); and Dr. Gillian Goward (McMaster University).

Table 3.7 Participants in the international consortium of the Cu–Cl cycle development

Canadian institutions and partners	International partnerships
1. University of Ontario Institute of Technology (UOIT)	1. Generation IV International Forum
2. Memorial University of Newfoundland	2. Argonne National Laboratory (ANL)
3. University of Toronto	3. Pennsylvania State University
4. University of Guelph	4. Gas Technology Institute (GTI)
5. Western University	5. Generation IV International Forum
6. University of Waterloo	6. University of Maribor
7. McMaster University	7. Czech Academy of Science
8. University Network of Excellence in Nuclear Engineering (UNENE)	8. Atomic Energy National Commission of Argentina
9. Atomic Energy of Canada Limited (AECL)	9. National Hydrogen and Fuel Cell Centre (Romania)
10. Ontario Power Generation (OPG)	
11. Regional Municipality of Durham	
12. Marnoch Thermal Power Inc.	
13. Gerdau Ameristeel	
14. St. Mary's Cement	

- *International partners*: Dr. Michele Lewis, Dr. Shabbir Ahmed, and Dr. Magali Ferrandon (Argonne National Laboratory, Argonne, Illinois); Dr. Serguei Lvov (Pennsylvania State University, University Park, Pennsylvania); Dr. Jurij Avsec (University of Maribor, Slovenia); Dr. Daniel Pasquevich (Atomic Energy National Commission, Argentina); and Dr. Iordache Ioan (National Hydrogen and Fuel Cell Centre, Romania).

Some of the key current research activities include the development and scale-up of each unit operation reactor and electrochemical cell, system integration studies, exergo-economic modeling and life cycle analysis, heat exchanger analysis for heat recovery, safety issues, thermochemical data development, and advanced materials and applications. A comprehensive list of the research topics of the UOIT-led team on the Cu–Cl cycle and related issues is presented in Table 3.8. The R&D milestones and working groups of the project for the first phase of funding are shown in Table 3.9. Numerous newspaper and magazine articles including the Globe and Mail, Toronto Star, and Maclean's magazine have referred to the consortium's R&D progress and successful achievements.

The UOIT in Oshawa, Ontario, has a unique location in center of the nuclear and auto industries in Ontario. It is anticipated that the nuclear industry will become a major producer of hydrogen in the future, while the auto industry will become a major user of hydrogen. In this way, Oshawa and UOIT are strategically located as development centers for nuclear hydrogen production research and development.

Many synergistic advances in science and engineering are expected as a result of the collaborative efforts within the international consortium with respect to fundamental processes and systems related to the copper–chlorine cycle. In Canada, the focus of technology development is the coupling of the CANDU-SWCR reactor with the copper–chlorine cycle and its strategic applications like hydrogen fuel production, synthetic fuel production, oil sand upgrading for Alberta's oil sands,

Table 3.8 Research and development topics of Cu–Cl technology

Research topic	Description
Electrochemical cell	Required development of cost-effective electrodes and separation technology
	Selection/development of a cost-effective membrane without significant crossover issues
	Membrane characterization: Permeability and conductivity determination
Drying aqueous $CuCl_2$	Spray drying system development
	Characterization of $CuCl_2$ particle morphology
	Development of crystallization process for drying
Hydrolysis reaction	Experiments to characterize the reaction, $H_2O + 2CuCl_2 \rightarrow Cu_2OCl_2 + 2HCl$
	Fluidized bed reactor design and analysis
Copper oxychloride decomposition	Thermochemical characterization of Cu_2OCl_2
	Performing decomposition experiments and improving the reactor
	Heat recovery system design from molten CuCl
System integration	All independent experimental units for each reaction step must be integrated
Aspen-Plus simulation	Simulation aiming to accurately predict the system performance and efficiency
	Design of heat exchanger network for internal heat recovery
Exergo-economic modeling	Application of design tools such as exergy-cost-energy-mass
Life cycle assessment	Determination of environmental impact and life cycle cost of the technology
Waste heat recovery	Investigating ways to use waste heat and to upgrade the temperature level
Linkage to nuclear plant	Investigation of possible ways to couple Cu–Cl cycle and nuclear reactors
Safety and reliability	Development of dynamic flowgraph methodology to model system behavior
Thermochemical data	Chemical potentials and solubility data
	Formation of Cu(I) and Cu(II) complexes
	Synthesis of copper oxychloride
	Properties of HCl and copper–chlorine compounds
	Determination of reaction rate constants
Advanced materials	Anode electrodes of electrochemical cell
	Development of thermal spray coatings
	Characterization and development of corrosion resistance coating
Economics and applications	Economics of nuclear hydrogen production
	Applications to oil sand upgrading
	Hydrogen for transportation sector

and other process heat usage. Other countries or regions may be interested in other technological applications like coupling of the copper–chlorine cycle with intermediate-temperature solar concentrators, or with heat recovery systems from specific industries that release waste heat at high temperatures (e.g., metallurgy).

Table 3.9 Milestones of copper–chlorine project at UOIT for phase I (2005–2011)

Phase	Research item	Year
Group 1: Thermal efficiency and economics		
1 (a) Cu–Cl cycle simulation and modeling studies		
1	Simulation of system processes with Aspen Plus software	2008
2	Completed models of transport phenomena for system and components	2009
3	Determination of optimum operating conditions in the Cu–Cl cycle	2010
4	Design of new system configuration for performance improvement	2010
5	Recommendations for implementation of the above new configuration	2011
1 (b) Energy/exergy/exergo-economic analysis of Cu–Cl cycle		
1	Energy, exergy, and exergo-economic models of system and components	2008
2	Sensitivity results from efficiency analysis for performance improvement	2009
3	Development of better design schemes for Cu–Cl cycle components	2010
4	Recommendations for implementation of the above new design schemes	2011
1 (c) Life cycle assessment		
1	Scope definition of hydrogen plant to identify all system components	2008
2	Inventory model of the plant and its components and impact analysis	2009
3	Impact analysis of environmental and ecological aspects	2010
4	Improvement method for the plant and its components	2011
5	Recommendations for usage and disposal of Cu–Cl plant materials	2012
1 (d) Studies of combined and hybrid systems with nuclear power		
1	Model of hybrid system and component configurations	2008
2	Model of a top-up cycle for a CANDU reactor	2009
3	Completion of economic model for combined and hybrid systems	2010
4	Recommendations for implementation of hybrid system	2011
Group 2: Process optimization, thermochemistry		
2 (a) Experimental measurements on $CuCl/CuCl_2/H_2O/HCl$ system		
1	Measure equilibrium constants for copper (II) complexation with chloride ions as a function of temperature	2009
2	Combine (1) with literature values of copper (I) complexation constants to yield a thermodynamic database for copper–chloride–water system	2009
2 (b) Electrochemical cell fundamentals		
1	Model of electrode kinetics in the electrolysis cell under conditions relevant to the thermochemical cycle	2008
2	Model for membrane processes (steps 1–2 combined)	2008
3	Model for copper deposition and removal (steps 1–2 not combined)	2009
4	Laboratory-scale demonstration of electrochemical apparatus	2009
5	Optimization of electrode materials and electrolytic process conditions	2010
2 (c) Theoretical predictions of $CuCl/CuCl_2/H_2O/HCl$ system properties		
1	Develop and verify molecular simulation methodology for property predictions of HCl/H_2O mixtures, including azeotrope formation	2008
2	Develop and test molecular simulation methodology for solubility predictions of $CuCl/CuCl_2$ in H_2O/HCl mixtures	2009
3	Incorporate predictions into Aspen Plus simulation software	2010
Group 3: Heat exchangers and fluid equipment		
3 (a) Design of heat exchanger network and heat management		
1	Model of piping network and heat matching for subsystem components	2008
2	Model of intermediate heat exchanger between nuclear and hydrogen plants	2009

(continued)

Table 3.9 (continued)

Phase	Research item	Year
3	New methods of heat recovery to optimize heat management	2009–2011
	3 (b) Heat transfer apparatus for high-temperature conditions	
1	Design of experimental apparatus for multiphase test cell	2008
2	Experimental test cell completed for heat exchange apparatus	2009
3	Temperature and heat flow measurements in multiphase test cell	2010
4	Methods for reducing flow blockage in multiphase flow conditions	2011
5	Methods of handling high temperature and corrosive working fluids	2012
	3 (c) Energy-efficient methods of separation	
1	Separating water from HCl solutions	2008
2	Laboratory-scale demonstration of spray dryer for $CuCl_2$ separation	2009
2	Separating hydrogen from HCl if Cu(s) is produced electrochemically	2009
3	Separating chlorine from oxygen in Cu_2OCl_2 decomposition	2010
4	Separating solid particles from fluid stream in multiphase flows	2011
5	Implementation of separation technique in fluid engineering equipment	2012
	3 (d) Selection of heat exchangers	
1	Determine transport properties (thermal conductivity and density)	2008
2	Establish operating and flow conditions for heat exchangers	2009
3	Model of flow configurations and selection of heat exchangers	2010
4	Modeling of thermal performance for printed circuit heat exchanger	2011
	3 (e) Selection of fluid engineering equipment	
1	Experimental flow apparatus for corrosive working fluids	2008
2	Laboratory-scale demonstration of fluidized bed and hydrogen units	2009
3	Establish operating and flow conditions for fluid equipment	2010
4	Design and selection of pumps and gas blowers	2011
5	Design and selection of valves, filters, and fluid separation units	2012
Group 4: Mechanical design and materials		
	4 (a) Fabrication with advanced materials	
1	Evaluation and design of strategies for ceramic coating of metal alloys	2008
2	Model of wear and corrosion resistance of various materials/substrates	2009
3	Model of densification of coating materials	2010
4	Minimization of effects of bond deterioration due to thermal cycles	2011
5	Minimization of film thickness for heat exchanger materials	2012
6	Method for increasing thermal conductivity of heat exchanger materials	2012
	4 (b) Multidisciplinary optimization algorithm for optimal design	
1	Formulation of multidisciplinary objective and constraint functions	2008
2	Development of MDO algorithm for optimal design	2009
3	Application of MDO algorithm to optimization of coating process	2010
4	Application of MDO algorithm to other systems, such as heat exchangers	2011
5	Application of MDO algorithm to the overall system	2012
	4 (c) Corrosion of equipment surfaces	
1	Build apparatus for electrochemical testing in molten CuCl	2008
2	Establish effects of metal surface on oxygen reduction kinetics in CuCl	2009
3	Demonstrate metal dissolution reaction of model metal systems in CuCl	2010
4	Conference paper on corrosion of candidate metal alloys in molten CuCl	2011
Group 5: Controls, instrumentation, and safety		
	5 (a) Monitoring and safety of a hydrogen plant	

(continued)

Table 3.9 (continued)

Phase	Research item	Year
1	Investigate strategies for online monitoring of equipment performance	2005–2006
2	Perform calculations to determine the optimal measurement locations	2008
3	Develop new techniques to improve measurement accuracy	2009
4	Integrate control and measurement systems and test their performance	2010
5	Verify performance of integrated sensors and control systems	2011
6	Perform a reliability and risk analysis of the hydrogen plant	2011
5 (b) Control systems to manage plant operations		
1	Perform calculations to determine optimal control variables	2008
2	Select and place orders for control systems	2008
3	Design and implement control systems for heat exchangers	2009
4	Implement control systems for other critical subsystems	2009
5	Establish the operating performance of control systems	2010
6	Install control systems in lab-scale demonstration of the Cu–Cl cycle	2011
NSERC program: Thermochemical equipment design		
Group 1: Experimental demonstration of the oxygen production cell		
1	Design and construction of experimental unit	2007–2008
2	Measurements of particle descent, shape, and decomposition	2008–2009
3	Lumped capacitance model of particle melting and decomposition	2009–2010
4	Investigation of oxygen bubble formation and free surface interaction	2010–2011
5	Effects of agitation and free surface level on liquid outlet stream	2011–2012
6	Empirical correlation of liquid removal at varying particle supply rates	2011–2012
2. Corrosion and advanced materials for CuCl pipe flow		
1	Predictive model of corrosion lifetime	2007–2008
2	Experimental unit for corrosion testing of CuCl pipe flow	2008–2009
3	Experimental testing of surface metal ion release	2009–2010
4	Strategy and development of suitable coating process	2010–2011
3. Experimental unit for CuCl(l) heat exchanger		
1	Heat matching within Cu–Cl cycle for heat recovery	2007–2008
2	Exergy modeling of heat exchangers to optimize thermal effectiveness	2008–2009
3	Design and construction of experimental spray heat exchanger	2009–2010
4	Effects of inert gas crossflow to enhance heat transfer	2010–2011
5	Correlation of droplet breakup and solidification	2011–2012
4. Electrowinning process for electrochemical cell design		
1	Correlation of electrowinning to predict the rate of copper deposition	2007–2008
2	Experimental benchtop demonstration of electrowinning process	2008–2009
3	Different configurations and materials for electrode	2009–2010
4	Experimental benchtop demonstration of CuCl–HCl electrolysis	2010–2011

3.5 Conclusions

In this chapter, the main research and development programs on nuclear energy around the world have been presented. There are two major research directions: (1) development of nuclear reactor technologies for electric power and process heat, and (2) development of nuclear hydrogen production systems. The next generation of nuclear reactors (called Generation IV) is expected to become commercially available by 2030. A paradigm change in nuclear technology development is envisaged—as illustrated by Fig. 3.2. After around 70 years (1950–2020) of nuclear power reactors with gradually improved efficiency via scale increases (from a few MWe to ~1,700 MWe), the planned new generation will have a remarkable efficiency improvement (from ~30 % to ~50 %) by allowing higher outlet temperatures (in the range of 500–1,000 °C). The economics of the next generation of nuclear reactors will make them more attractive for a range of capacities covering between 20 and ~2,000 MWe. In addition, generation of process heat and hydrogen will become major applications of nuclear plants.

Hydrogen generation will become one of the most important applications of nuclear process heat. Nuclear-produced hydrogen can be derived from water, as well as from fossil fuels such as coal, natural gas, or petroleum naphtha, and from hydrogen sulfide. Although water splitting for hydrogen generation is attractive, the policy of some countries may encourage nuclear processing of fossil fuels to generate hydrogen and synthetic fuels (such as methanol or diesel) for transportation fuel. This is another potential transition path towards a fully implemented hydrogen economy.

For water splitting, three main technologies are attracting the attention of the research community. These are high-temperature electrolysis, sulfur–iodine water splitting processes, and copper–chlorine water splitting process. The research programs developed worldwide have a wide range of challenges related to nuclear technology implementation. Most of the nuclear reactor programs include the development of high-temperature electrolysis or a selected thermochemical water splitting cycle. The nuclear hydrogen element of nuclear reactor programs extends to about ~20 % of the budget. There are independent programs for nuclear hydrogen programs around the world, such as the Nuclear Hydrogen Initiative in the USA, Nuclear Hydrogen Programs in Japan and Korea, among several others. In addition, there are many programs worldwide on sustainable hydrogen production relevant to nuclear hydrogen research, which include components such as sustainable high-temperature electrolysis and thermochemical water splitting with the copper–chlorine cycle (Canada).

References

DOE (2002) A technology roadmap for generation IV nuclear energy systems. US DOE Nuclear Energy Advisory Committee and the Generation IV International Forum

DOE (2011) Department of Energy—Generation IV Nuclear Energy Systems. http://www.ne.doe.gov/genIV/neGenIV3.html. Accessed October 2011

GIF (2011) Generation IV International Forum. http://www.gen-4.org. Accessed October 2011

INERI (2009) International Nuclear Energy Research Initiative. Annual report. Department of Energy

INPRO (2011) International Project on Innovative Nuclear Reactors and Fuel Cycles. http://www.iaea.org/INPRO/. Accessed October 2011

Koster A, Matzner HD, Nicholsi DR (2003) PMBR design for the future. Nucl Eng Des 222:231–245

Kubo S, Nakajima H, Kasahara S, Higashi S, Masaki T, Abe H, Onuki K (2004) A demonstration study on a closed-cycle hydrogen production by the thermochemical water-splitting iodine–sulfur process. Nucl Eng Des 233:347–354

Latour SR, Menningmann JG, Blaney BL (1982) Waste heat recovery potential in selected industries. US Environmental Protection Agency—Industrial Environmental Research Laboratory Report # EPA-600/S7-82-030

WNA (2011) World Nuclear Association. http://www.world-nuclear.org. Accessed October 2011

Chapter 4
Water Electrolysis

Abstract In this chapter, water electrolysis technology and its applications for nuclear hydrogen production are discussed. In the first part of the chapter, a general classification of water electrolysis systems is given, the fundamentals of water electrolysis are explained, and the relevant notions are introduced. Calculations of reversible potentials and over-potentials have a major importance in electrolyzers' design and modeling. Some technological aspects of four types of electrolyzers are discussed: alkaline, proton exchange membrane, solid oxide electrolyzers with oxygen-ion and with proton conduction. Some special types of electrolyzers—hybrid electrolysis systems—are presented. The chapter ends with a discussion of possible nuclear-electrolytic hydrogen plant configurations as stand-alone systems installed at future nuclear reactor sites dedicated to large-scale hydrogen production.

4.1 Introduction

Nuclear based water electrolysis is a method to generate hydrogen using electricity derived from a nuclear power plant. Water electrolysis represents the process of water decomposition with separation of hydrogen and oxygen under the influence of a direct electric current which induces and sustains redox reactions.

As mentioned in Chap. 2, the process of water electrolysis was discovered in around the year 1800 by Nicholson—a prolific chemist and philosopher—who made a series of experimental tests regarding the effect of Volta's-pile direct electric current passage through water. In his water electrolysis experiments, Nicholson was assisted by a surgeon—Carlisle. The discovery of electrolysis was one of the major landmarks in electrochemistry. Another fundamental landmark was the formulation of electrolysis laws by Faraday in 1834 which states that the mass of reacted matter at an electrode is proportional to the circulated (consumed) electric charge. The law gives a quantitative relationship between the mass of products and reactants of an electrochemical reaction and the consumed electric current. It represents the most basic tool for the design of electrolysis systems.

G.F. Naterer et al., *Hydrogen Production from Nuclear Energy*,
DOI 10.1007/978-1-4471-4938-5_4, © Springer-Verlag London 2013

Its importance is crucial as it was demonstrated by the substantial technological progress on electrochemistry and water electrolysis during the next fifty years, immediately after its formulation.

Electrolysis achieved an industrial scale by the late 1800s being stimulated by a high demand of hydrogen for catalytic ammonia synthesis in a newly born fertilizers industry (Guerrini and Trasatti 2009). The following information—regarding production scales of electrolytic hydrogen by the end of the 1800s—is obtained from Hale (1919). One of the first commercial systems generating approximately 300 standard liters of hydrogen per day was devised in 1885 at Royal College of France which operated an alkaline electrolyte comprising 30 % KOH. By 1890, the reported scale of production of hydrogen was 250 standard liters per hour in bipolar alkaline electrolyzers. Schmidt constructed an industrial plant in Zurich by 1899 with bipolar alkaline electrolyzers including diaphragms made in asbestos which achieved 54 % energy efficiency with a production rate of ~1.4 m^3 hydrogen per hour and a capital cost of ~ GBP 2,000 (or equivalent to ~ $200,000 in US$ 2011). Note that 1.4 standard m^3 of hydrogen per hour represents about 115 g/h which requires ~7 kW electricity at 54 % efficiency.

The interest of developing electrolysis at a larger production scale diminished sharply in the first half of twentieth century because coal gasification and steam methane reforming became the most cost-competitive technologies. Water electrolysis was used in niche areas like the pharmaceutical or food industry where high-purity hydrogen is needed or the chlorine–alkali industry where alkaline electrolysis is typically applied to extract chlorine from saline water. In the military marine sector, nuclear powered water electrolyzers were used to generate oxygen from water for deep-sea subsistence. For space applications, proton exchange membrane electrolyzers were developed in the 1960s for space missions.

The energy crisis from the 1970s brought again more interest in electrolysis by scientists—this time as a means to produce hydrogen from renewable resources. A 10-year research program on electrolysis for clean hydrogen production from renewable sources was funded by the European Community. In the same period, the US Department of Energy continued work and in Japan, there were research programs in alkaline water electrolyzers with operating temperatures over 120 °C (Guerrini and Trasatti 2009). In the year 1980, seven worldwide major manufacturers were mentioned by Tilak et al. (1981) which produce ammonia with alkaline electrolysis plants operating at capacities from 535 to 30,000 standard cubic meters of hydrogen per hour (with an equivalent electric power consumption in the range 2–100 MW).

At capacities of 10–20 standard cubic meters per hour, it was found that alkaline electrolyzers are the most economic hydrogen production technologies, better than coal gasification and natural gas reforming. At a capacity higher than 100,000 m^3/h, the economically most competitive technology is coal gasification (Bocris et al. 2000). Advanced alkaline electrolysis systems with "zero-gap" electrolytes were developed in recent years. With a zero-gap configuration, the ohmic losses through the electrolyte are minimized and better purity of hydrogen is obtained (Guerrini and Trasatti 2009). By the end of the twentieth century, proton exchange membrane electrolyzers became commercially available up to production capacities of about 100 kW (on a lower heating value basis of produced hydrogen). More recently, solid oxide electrolysis cells made much progress in research and development as one of the most promising technologies to generate electrolytic hydrogen at a large scale.

In the first part of this chapter, the water electrolysis systems are classified and the fundamentals of water electrolysis are introduced. All major types of electrolysis systems are discussed in detail with respect to their construction, specific processes, performance parameters, and applications. Some hybrid electrolysis systems are treated in a separate section. In the last part, there are a number of integrated systems presented to generate electrolytic hydrogen at nuclear reactor sites.

4.2 Fundamentals of Water Electrolysis

In this section, the fundamentals of water electrolysis are briefly introduced and the main concepts are explained. The water electrolysis process can be viewed as a superposition of concurrent or sequential electrochemical reactions occurring in the vicinity of electrodes (half-reactions) with the overall effect of splitting the water molecule and separating the gaseous products (hydrogen and oxygen). The overall water-splitting reaction—according to Guerrini and Trasatti (2009)—is written as

$$H_2O + 2F \rightarrow H_2(g) + 1/2\,O_2(g), \tag{4.1}$$

where F represents 1 mol of electric charge defined by the Faraday constant $F = N_A e$, where $N_A = 6.022 \times 10^{23}$ is Avogadro's number and $e = 1.602 \times 10^{-19}$ C is the elementary charge; thus $F = 96{,}490$ C/mol. Equation (4.1) states that for electrolyzing 1 mol of water, an electric charge of $2F$ Coulombs must be provided so that one mole of hydrogen and half mole of oxygen are formed.

There are several electrochemical reactions—such as reduction and oxidation—which may occur during the electrolysis process alongside with other physical and chemical processes like adsorption, electron transfer, formation of molecular hydrogen from atomic hydrogen, etc. The types of electrochemical reactions involved in the process depend upon the specific kind of electrolyzer. The general layout of an electrolyzer—regardless its specific type—is presented in Fig. 4.1 and consists of two electrodes (anode and cathode) and an electrolyte, with the whole assembly being connected to a direct current (DC) power supply. The electrode connected at the positive pole of the power source is denoted as an anode, while the negative pole is the cathode. Oxygen evolves at the anode while hydrogen is produced at the cathode.

Electrolytes are media which allow conduction of ions. Under the influence of the electric field, positive ions (or cations) move from the anode to cathode. The anions (or negatively charged ions) move through the electrolyte from the cathode to anode. Depending on the nature of the electrolyte, the ions that are most

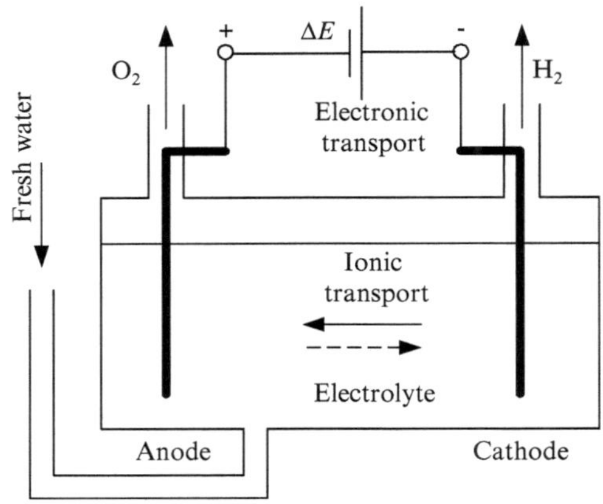

Fig. 4.1 General layout of an electrolysis system

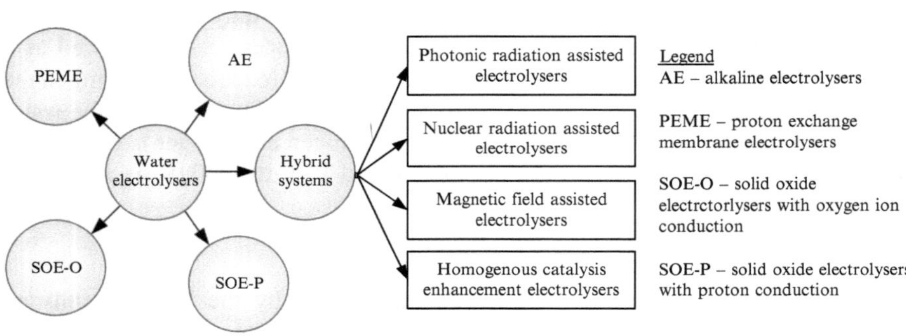

Fig. 4.2 Classification of water electrolyzers

mobile may be anions or cations. In acidic electrolytes, cations are those that are transported while in alkaline electrolytes, anions are transferred.

A classification of water electrolyzers which considers the nature of the electrolyte and the types of ions transferred is presented in Fig. 4.2. Each kind of electrolyzer system indicated in the figure is analyzed in more detail in subsequent sections of this chapter. Hybrid electrolysis systems are those that use other forms of energy in addition to electricity as input. Examples of additional forms of energy which may assist electrolysis are photonic radiation, nuclear radiation, and magnetic fields. Furthermore, hybrid electrolysis can be implemented—as indicated by recent advances in the field—by superimposing heterogeneous electrocatalysis at electrode surfaces with homogeneous catalysis in the whole volume.

The analysis of an electrolyzer system must consider various aspects. The most fundamental analysis is the thermodynamic part which determines the driving forces of the process based on an application of conservation laws such as energy conservation (first law of thermodynamics), thermodynamic equilibrium principles, and other specific methods. The thermodynamic analysis must be accompanied by a kinetic analysis which studies the reaction rates. Moreover, the study of transport phenomena within electrolyzers is important and cannot be neglected. Additional aspects can be derived from material analysis, economic analysis, and environmental impact analysis. This section focuses on electrolysis modeling through thermodynamics, kinetics, mass transport, and electrical circuit equivalency analyses.

4.2.1 *Thermodynamic Analysis*

Faraday stated in 1834 a law of electrochemistry—known as Faraday's law of electrolysis—which represents a form of conservation of electric charge. The Faraday law states that the electric charge consumed by an ionic species reduced (or oxidized) in an electrolyzer equals to the electric charge of electrons flowing through the exterior circuit. Mathematically, this law is expressed as

$$\int_0^t I(t)\,\mathrm{d}t = z \times F \times n, \tag{4.2}$$

where z represents the valence of ionic species of interest (the number of electrons per ion), F is the Faraday number, n is the number of moles of released product, $I(t)$ is the momentarily intensity of electric current through the exterior circuit (in Amperes), and t is time. For a hydrogen-evolving reaction, $z = 2$, while for an oxygen-evolving reaction, $z = 4$.

Any electrochemical reaction is defined thermodynamically by three parameters, such as temperature, pressure, and electrochemical potential (where electrochemical potential is a superposition of chemical potential and any potential energy caused by the presence of the electric field). The electrochemical potential for reactions given in (4.1) is defined based on chemical potential and electrical potential. The chemical potential is based on the Gibbs free energy of the reaction, $\Delta G = 2 \times F \times E$, where E represents the potential difference between the two electrodes.

Table 4.1 lists electrochemical reactions relevant to water electrolysis and their associated values of reaction entropy, enthalpy, and Gibbs free energy. According to the definition of the Gibbs energy, this is calculated with $G = H - T \times S$, where H and S represent the molar enthalpy and entropy, respectively, while T is the temperature in K. The reaction enthalpy and entropy are calculated as the difference between the product and reactant enthalpy and entropy, respectively. Then, the Gibbs energy of the reaction is calculated with $\Delta G = \Delta H - T \times \Delta S$. In Table 4.1, the quantities are expressed in electron-Volt (eV) units, where an electron-Volt is defined as the energy needed to displace an elementary electrical charge in an electrical field, across a potential difference of 1 V. The following notations are used for quantities expressed in eV, namely, $\mathtt{G}$, $\mathtt{H}$, $\mathtt{S}$. An Engineering Equation Solver (EES) code was used to calculate data for Table 4.1 (see Klein 2011).

Table 4.1 Thermodynamic properties of some relevant electrochemical reactions[a]

Reaction type	#	Chemical equation (half-cell)	ΔH (eV)	ΔS (meV/K)	ΔG (eV)	E (V)	E_{th} (V)
H_2 evolving ($z = 2$)	1	$H_2O(g) + 2e^- \rightarrow H_2(g) + O^{(2-)}$ at 1,000 K	1.901	0.949	0.952	0.476	0.951
	2	$2H_2O(l) + 2e^- \rightarrow H_2(g) + 2OH^-(aq)$	2.914	3.413	1.896	0.948	1.457
	3	$2H^+ + 2e^- \rightarrow H_2(g)$	−31.840	−1.339	−31.44	−15.72[b]	−15.92
O_2 evolving ($z = 4$)	4	$2H_2O(l) \rightarrow O_2(g) + 4H^+ + 4e^-$	69.610	6.062	67.800	16.95[b]	17.40
	5	$2H_2O(g) \rightarrow O_2(g) + 4H^+ + 4e^-$ at 775 K	69.320	4.852	65.560	16.39	17.33
	6	$2O^{2-} \rightarrow O_2(g) + 4e^-$ at 1,000 K	1.335	−0.753	2.087	0.522	0.334
	7	$4OH^-(aq) \rightarrow 2H_2O(l) + O_2(g) + 4e^-$	0.097	−2.707	0.904	0.226	0.024
Overall reaction	8	$H_2O(l) + 2F \rightarrow H_2(g) + 0.5O_2(g)$	2.962	1.691	2.458	1.229	1.481
	9	$H_2O(g) + 2F \rightarrow H_2(g) + 0.5O_2(g)$ at 1,000 K	2.569	0.571	1.998	0.999	1.284

[a] If not otherwise specified, the reaction temperature is 298.15 K and the partial pressure for each species is assumed to be $P = P_0 = 101.325$ kPa

[b] Here the Gibbs free energy of protons is calculated as per vacuum conditions

The differences of potential E and E_{th} are calculated with

$$\Delta G = -z \times E \text{ and } \Delta H = -z \times E_{th}, \tag{4.3}$$

where E represents the potential of an electric field across which z electrical charges are displaced to transfer an energy equivalent to ΔG (in eV for one molecule of hydrogen). This potential is denoted as the absolute equilibrium potential for the electrochemical reaction. In an electrochemical reaction, the energy that is transferred in the form of electricity from an external supply must compensate at least for the Gibbs energy of the reaction. After supplying the Gibbs energy by electrical means, the remaining part of the energy required by the reaction is $T\Delta S$ which can be provided either in the form of electricity or thermal energy. Depending on the operating parameters, in electrolysis, one can have either $\Delta H = \Delta G + T\Delta S \geq \Delta G$ (case in which the process may be endothermic) or $\Delta H \leq \Delta G$ (case in which the system rejects heat—exothermic). In an endothermic reaction, the required heat may be provided via a Joule effect (electric heating of the electrolyte due to the ionic current passing through it). In such a situation, the operation is called thermo-neutral, meaning that the reaction is adiabatic (no heat transfer is required). The electric thermo-neutral potential is defined in absolute terms by (4.3) as E_{th}, where ΔH is given in units of eV.

In practical situations, the species involved in electrochemical reaction (see Table 4.1) may be at a different partial pressure than P_0 (standard pressure) or at a thermodynamic activity different than 1 (see subsequently a definition of the "activity") and this influences the value of the Gibbs free energy of the reaction. In a mixture of chemically reacting species (e.g., ions such as OH^-, H^+, O^{2-} or molecules such as H_2O, H_2, O_2), the Gibbs energy variation can be calculated using chemical potentials μ, defined for the component "i" of the system according to the equation

$$dG = \sum \mu_i \times dn_i, \text{ where } \mu_i = \frac{\partial G}{\partial n_i}. \tag{4.4}$$

If a mixture of gases obeys the ideal gas law, the following equation expresses the variation of chemical potential with pressure: $\mu_i(P_i) - \mu_i(P_0) = R \times T \times \ln(P_i/P_0)$, where P_0 is a reference pressure while P_i is the partial pressure of component "i." A similar equation does exist for ideal solutions, where partial pressure is replaced by molar concentration, namely, $\mu_i(c_i) - \mu_i(c_0) = R \times T \times \ln(c_i/c_0)$, where c_0 represents the concentration at reference conditions. For real chemical systems, in order to account for nonideal conditions and still keep the same form for the relationship describing the variation of chemical potential, the thermodynamic activity "a_i" is defined by

$$\mu_i(a_i) - \mu_i(a_i = 1) = R \times T \times \ln(a_i), \tag{4.5}$$

where the chemical potential for $a_i = 1$ must be defined and it is known as the standard state thermodynamic potential, denoted by $\mu_i^0 = \mu_i(a_i = 1)$.

The thermodynamic activity has a measure of concentration and is often expressed through a dimensionless activity coefficient f which quantifies the abatement from ideal conditions of a component of a chemical system through the equation $a_i = c_i \times f_i$. According to the definitions for chemical potential and thermodynamic activity given in (4.4) and (4.5), the Gibbs energy of a chemical reaction has the general expression as follows:

$$\Delta G = \sum \left(n_i \times \mu_i^0 \right) + R \times T \times \ln\left(\Pi (c \times f)_i^{n_i} \right), \tag{4.6}$$

where the term $\sum \left(n_i \times \mu_i^0 \right)$ can be expressed as the sum of products (P) minus the sum of reactants (R)—where reactants are the chemical species that are consumed, and products that are produced; thus, $\sum \left(n_i \times \mu_i^0 \right) = \sum_P \left(n_i \times \mu_i^0 \right) - \sum_R \left(n_i \times \mu_i^0 \right) = \Delta G^0$ or

$$\Delta G = \Delta G^0 + R \times T \times \ln(K_{eq}), \tag{4.7}$$

where $K_{eq} = \Pi (c \times f)_i^{n_i}$ is known as the equilibrium constant of the reaction. An electrochemical reaction of major importance is the reaction of proton reduction,

$$2H^+(aq) + 2e^- \rightarrow H_2(g). \tag{4.8}$$

This equation is also noted in Table 4.1 where it was assumed that the proton evolves from vacuum. The Gibbs energy of $2H^+ + 2e^- \rightarrow H_2(g)$ when a proton evolves from vacuum is $-3,033.79$ kJ/mol (or -31.44 eV/molecule, see Table 4.1). According to Trasatti (1986), the removal into vacuum of protons from water according to the equation $2H^+(aq) \rightarrow 2H^+$ requires a Gibbs free energy of $2,176\,\mathrm{k/mol}$. Therefore, the Gibbs energy of the reaction (4.8) is $\Delta G = -858.24$ k/mol. Consequently, the absolute difference of potential for reaction (4.8) becomes $E = -4.44$ V.

The reverse of reaction (4.8), namely, $H_2(g) \rightarrow 2H^+(aq) + 2e^-$, defines by convention the potential of the hydrogen electrode, which is adopted as the standard for electrochemical potentials. The absolute value of a standard hydrogen electrode is $+4.44$ V; however, by convention, the standard hydrogen electrode is set to zero and it is denoted with $E^0(H/H^+) = 0$ V. The reversible potentials of all other electrochemical reactions can be expressed with reference to the standard hydrogen electrode.

The standard hydrogen electrode is defined for 25 °C at pH = 0. In chemistry, pH is defined as $\mathrm{pH} = -\log_{10}(a_{H^+})$. Water with pH = 0 has a protonic activity of $a_{H^+} = 1$ mol/L. From (4.3) and (4.7), the well-known Nernst equation can be derived, which expresses the variation of electrode potential with operating conditions, namely,

$$E_{rev} = E^0 + \frac{RT}{zF} \ln(K_{eq}), \tag{4.9}$$

where the subscript rev stands for reversible reaction. For the hydrogen electrode, the activity of hydrogen gas—modeled as an ideal gas—is equal to the concentration; therefore the Nernst equation takes the following particular form:

$$E_{\mathrm{rev,H^+}} = \frac{RT}{2F}\ln\left(\frac{(f \times c)^2_{\mathrm{H^+}}}{c_{\mathrm{H_2}}}\right). \tag{4.10}$$

Here, it can be observed that at 25 °C for pH $= 0$ when proton activity is $a_{\mathrm{H^+}} = (f \times c)_{\mathrm{H^+}} = 1$, the electrode potential is zero; for pH $= 7$, it is -0.414 V, and for pH $= 14$, it is -0.828 V. However, the hydrogen electrode potential is a half-cell potential in acidic electrolyte—specific to PEM electrolyzers (pH < 7). If the electrolyte is basic (pH > 7), hydroxyl ions are formed instead of protons, according to reaction #2 in Table 4.1. In this case, one has a hydroxyl half-cell—specific to alkaline electrolyzers with the Nernst equation taking the following specific form:

$$E_{\mathrm{rev,OH^-}} = E^0_{\mathrm{OH^-}} + \frac{RT}{2F}\ln\left(\frac{a^2_{\mathrm{H_2O}}}{c_{\mathrm{H_2}} \times a^2_{\mathrm{OH^-}}}\right), \tag{4.11}$$

where in standard conditions the activity of water is 1, the activity of hydroxyls is $a_{\mathrm{OH^-}} = 1$, and the hydrogen concentration is $c_{\mathrm{H_2}} = 1$. Since (4.10) and (4.11) are equivalent, imposing their equality, one obtains the standard potential in basic solutions:

$$E^0_{\mathrm{OH^-}} = \frac{RT}{2F}\ln\left(a^2_{\mathrm{H^+}}\right). \tag{4.12}$$

The equilibrium constant of water dissociation in solution $H_2O(l) \rightarrow H^+(aq) + OH^-(aq)$ is given by the constant $K_{\mathrm{eq,w}} = a_{\mathrm{H^+}} \times a_{\mathrm{OH^-}}$ which has a value of $1.27 \times 10^{-14}(\mathrm{mol/L})^2$ at 25 °C. For $a_{\mathrm{OH^-}} = 1$, it results in $a_{\mathrm{H^+}} = 1.27 \times 10^{-14}$ and from (4.11), one has $E^0_{\mathrm{OH^-}} = -0.822$ V. Determining the reversible potential of half-cell reactions is essential in calculations for reaction kinetics at each electrode (see Sect. 4.2.2). For the overall cell, the reversible potential can then be calculated with $E = E_a - E_c$, where the subscript "a" stands for anode, and subscript "c" for cathode. The Nernst equation for the overall reaction is

$$E_{\mathrm{rev}} = E^0 + \frac{RT}{2F}\ln\left(\frac{c_{\mathrm{H_2}} \times c^{0.5}_{\mathrm{O_2}}}{a_{\mathrm{H_2O}}}\right), \tag{4.13}$$

where the standard potential at 25 °C is $E^0 = 1.229$ V. In many practical situations, in (4.13), the activity of water is 1, the hydrogen concentration is $c_{\mathrm{H_2}} = P_{\mathrm{H_2}}/P_0$, and the oxygen concentration is $c_{\mathrm{O_2}} = P_{\mathrm{O_2}}/P_0$.

4.2.2 Kinetic Analysis

In order to generate hydrogen and oxygen from water, a direct current of higher voltage than the reversible potential must be supplied to the electrolysis cell. Denote the actual voltage with E_{tot} and the current consumed with I. The energy rate balance on the electrolysis cell is $E_{tot} \times I = E \times I + \dot{W}_{loss}$, where E is the reversible cell voltage given by (4.13) and $\dot{W}_{loss}$ represents the energy dissipated due to various losses. An expression for current (I) can be obtained by differentiating (4.2) according to $I \times dt = z \times F \times dn$; thus

$$I = z \times F \times \dot{n}, \tag{4.14}$$

where $\dot{n}(t) = dn/dt$ is the reaction rate.

A kinetic analysis aims to determine the relationship between reaction rate and operating conditions as well as between the current and the voltage consumed by the electrolysis cell. Moreover, the kinetic analysis determines the relationship between the rate of reaction at the electrodes and the electrode over-potential and the relation between the current of ionic species across the electrolyte and the electric potential over it. When the cell assembly is more complex, comprising membranes or diaphragms, a kinetic analysis accounts for the over-potentials across them and the currents of matter and electric charges through them.

The presence of the electric field in the vicinity of electrodes influences the reactions. Reduction reactions (viz. (4.8)) can occur only at the electrode in the vicinity of which the potential is more negative than the reversible potential of the reaction, whereas oxidation reactions require an electrode with a potential more positive than it. Reaction kinetics can be described with the help of a potential surface diagram as shown in Fig. 4.3 for an "oxidation ⇆ reduction" process. There are two potential surfaces, one corresponding to reactants and another for products. Moving along the reaction coordinate, the free energy of the system decreases and reaches a minimum in #2. An energy barrier $2 - 3$ of magnitude ΔG_{act} has to be overcome in order for the system to move on the product's potential surface, toward the right. If an electric field is applied by polarization of the electrode (anode for the case shown in Fig. 4.3), the whole potential surface characterizing the thermodynamic system of the reaction products displaces at lower energy with a magnitude proportional to the polarization potential, namely, $F \times \Delta E_{act}$, where ΔE_{act} represents the activation over-potential. Consequently, the activation energy of the reaction under the influence of polarization decreases as it can evolve along the path $1 - 2 - \overline{3} - \overline{4}$.

It can be observed from the figure that the activation energy of the forward reaction in the presence of the electric field $\overline{\Delta G}_{act}$ is reduced with respect to the activation energy in the absence of polarization ΔG_{act} with a quantity which is proportional to the intensity of the electric field. Mathematically, one can write

$$\overline{(\Delta G)}_{act} = \Delta G_{act} - \alpha \times F \times \Delta E_{act}, \tag{4.15}$$

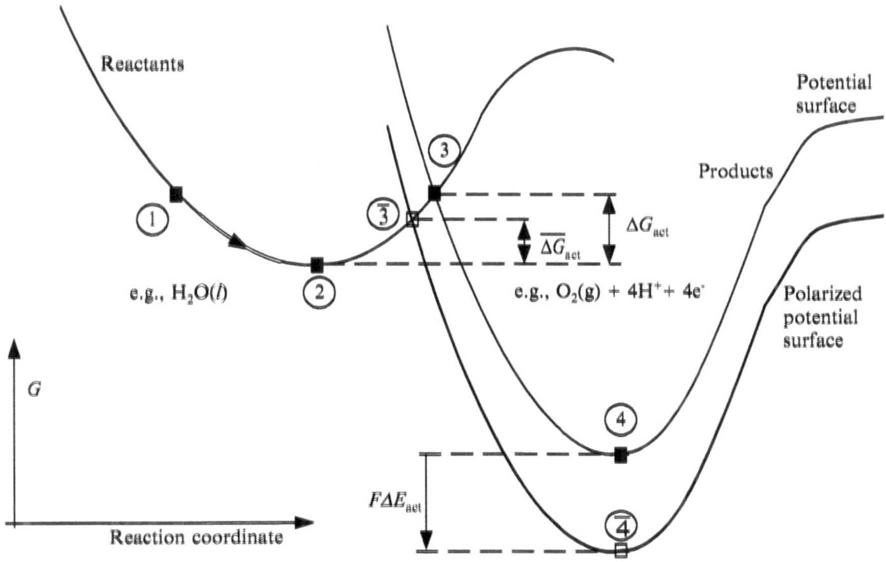

Fig. 4.3 Potential surfaces and activation energy for electrochemical oxidation reaction

where $\alpha \in [0, 1]$ is a parameter known as "transfer coefficient." Note that the backward reaction can also occur (along path $4 - 3 - 2$ in the absence of the electric field, or $\overline{4} - \overline{3} - 2$ in its presence). For the situation depicted in the figure, the backward reaction is disfavored by the presence of the electric field, because its activation energy (corresponding to the path $\overline{4} - \overline{3}$) is increased. The activation energy for the backward reaction can be written with the help of the transfer coefficient according to

$$\overline{(\Delta G)}_{b,act} = \Delta G_{b,act} + (1 - \alpha) \times F \times \Delta E_{act}. \tag{4.16}$$

The rate of reaction (either forward or backward) is proportional to the rate constant and the concentration of consumed and/or generated species (depending on the reaction order). The rate constant can be written based on the Arrhenius equation, $k = A \times \exp(-\Delta G_{act}/RT)$, where A is known as a pre-exponential factor. The net rate of reaction is the difference between the rate of forward reaction and backward reaction as $\dot{n} = \dot{n}_f - \dot{n}_b = k_f \times c_f - k_b \times c_b$, where c refers to the molar concentration of reactants (f) and products (b). Further, using (4.13), it results in $I = z F \times (k_f c_f - k_b c_b)$. Instead of net current for electrochemical reactions occurring at the surface of an electrode, employing current density is more useful. The current density is defined by $J = I/A$ where A is the effective area of the electrode. Using the Arrhenius equation and activation energy as expressed in (4.16) and (4.17), one obtains an expression for the current density at the electrode which is known as the Butler–Volmer equation

$$J = zFAk^0 \times \left[\frac{c_R}{c_{R,0}} \exp(\alpha F \Delta E_{act}/RT) - \frac{c_O}{c_{O,0}} \exp(-(1-\alpha)F\Delta E_{act}/RT)\right], \quad (4.17)$$

where k^0 is the rate constant for a reversible operation, c_O is the concentration of oxidized species, and c_R is the concentration of the reduced species, both at the surface of the electrode; the subscripts "R,0" and "O,0" refer to concentrations in bulk solution. For the simplified case when mass transport influence can be neglected at the electrodes, then $c_R/c_{R,0} = 1$ and the Butler–Volmer equation takes the form (see Rubinstein 1995)

$$J = J_0 \times [\exp(\alpha F \Delta E_{act}/RT) - \exp(-(1-\alpha)F\Delta E_{act}/RT)], \quad (4.18)$$

where J_0 represents the current density specific to the reversible reaction, also known as exchange current density; its dependence on temperature is related to the Arrhenius equation, namely, $J_0 = zFk = zFA \times \exp(-\Delta G_{act}/RT)$. There are two limiting situations for (4.18), the first occurring when the electrode polarization is low (i.e., ΔE_{act} is relatively small such that $\Delta E_{act} \leq 0.2\ RT/\alpha F$); and (4.18) can be linearized according to (see Rubinstein 1995)

$$\Delta E_{act} = \frac{RT}{J_0 F} \times J. \quad (4.19)$$

The current is proportional with an overvoltage near the equilibrium. If the polarization is high, the system is far from equilibrium ($\alpha \cong 1 \to \alpha \Delta E_{act}/RT \gg (1-\alpha)\Delta E_{act}/RT$), so it results in

$$\Delta E_{act} = \frac{RT}{\alpha F} \times \ln(J/J_0). \quad (4.20)$$

Julius Tafel established experimentally in 1905 the well-known Tafel equation (see Burstein 2005) which relates a thermodynamic parameter (the activation over-potential of a polarized electrode) to a kinetic parameter (the current density). The Tafel equation has its theoretical basis on relationship (4.20) and it is written as

$$\Delta E_{act} = a + b \times \ln(J). \quad (4.21)$$

The parameter a in Tafel equation represents the value of polarization potential at a unit current density. Typically, the parameter a has values in the range of $0.03-3$ V. The parameter b is known as "Tafel slope" and its range of values is $\sim 0.04-0.150$ V at ambient temperature. From (4.20) and (4.21), $b = RT/\alpha F$ and $a = b \times \ln(J_0)$.

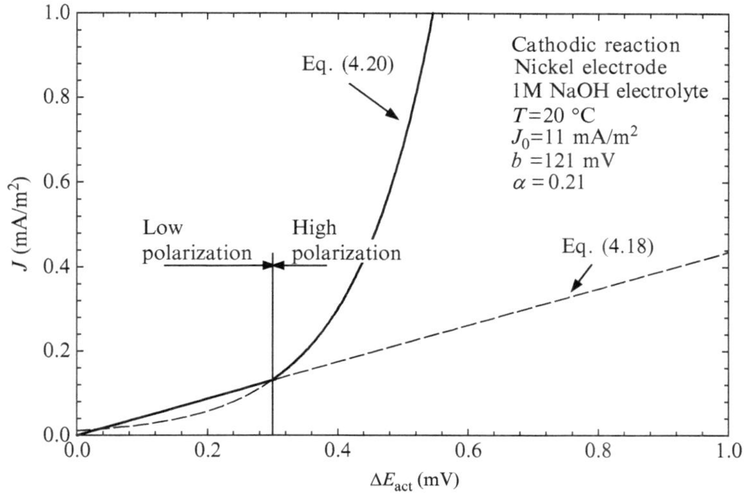

Fig. 4.4 Activation over-potential of a kinetically controlled cathodic reaction

In Fig. 4.4, a typical current vs. activation over-potential diagram is shown for an alkaline electrolyte with a Ni electrolyte; the reaction conditions and the Tafel slope and exchange current are indicated. It is observed that at low polarization, the curve is linear (according to (4.19)), whereas at higher polarizations, the behavior becomes exponential (according to (4.21)). Note that this behavior is valid when the mechanism that controls the reaction is kinetically controlled.

4.2.3 *Transport Process Analysis*

Equations (4.18)–(4.21) are valid only if mass transport processes within the electrolyte can be neglected, a case in which the kinetic mechanism controls the reaction (i.e., activation energy influences the process). For much higher polarization, however, mass transport processes may become dominant (depending on the geometrical configuration of the electrolysis cell). The mass transport of ionic species within an electrolyte is influenced by the electric field (migration), hydrodynamics (convection), and concentration gradients (diffusion).

Mass transport processes at a polarized electrode are illustrated in a simplified diagram in Fig. 4.5. The diagram depicts the case of a diffusion-controlled hydrogen evolution reaction. At the interface of any two material media, a Helmholtz layer is formed with a typical thickness on the order of 1 Å. The electrons must be transferred across the Helmholtz layer in order to reach the reaction sites at the very vicinity of the reactant molecules. Depending on the polarization and materials involved, a diffusion layer is formed at the electrode surface with a thickness directly influenced by the thickness of the Helmholtz layer. As shown in the figure, a control

Fig. 4.5 Helmholtz layer and mass diffusion process at a polarized electrode

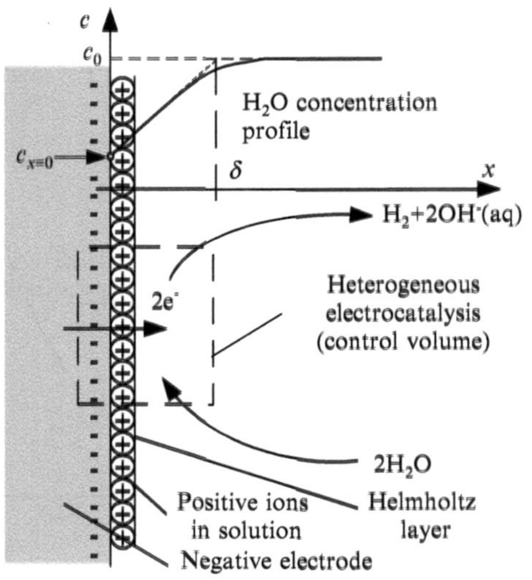

volume can be delimited in the vicinity of the reaction site. The concentration of reactant molecule (e.g., water) decreases toward the electrode surface from a value specific to the bulk electrolyte c_0 to the surface value, $c_{x=0} < c_0$. Based on Fick's law of diffusion, the current density can be written as

$$J_R = zFAD_R \times \left(\frac{c_0 - c_{x=0}}{\delta}\right)_R \text{ and } J_O = zFAD_O \times \left(\frac{c_{x=0} - c_0}{\delta}\right)_O, \tag{4.22}$$

where D is the diffusion coefficient, δ is the thickness of the diffusion layer, and the index R refers to reduced species while the index O refers to the oxidized species. It is observed that the current density in (4.22) displays a maximum limiting value which is obtained when the reactant concentration at electrode surface is zero. The diffusion-limited current density is therefore

$$J_{\text{lim,R}} = zFAD_R \frac{c_{0,R}}{\delta_R} \text{ and } J_{\text{lim,O}} = zFAD_R \frac{c_{0,O}}{\delta_O}, \tag{4.23}$$

where it is taken into account that both the forward and backward reactions can occur.

The practical range of diffusion layer thickness in water electrolysis is 50–500 μm. The limiting current density is in the range of 8,000–10,000 A/m^2; it is about 10^6 times higher than the exchange current (Bocris et al. 2000). If the limiting current is known, then the ratio between bulk concentrations and surface concentrations can be obtained by dividing (4.22) into (4.23), yielding

$$\left(\frac{c_{x=0}}{c_0}\right)_R = 1 - \frac{J}{J_{\text{lim,R}}} \text{ and } \left(\frac{c_{x=0}}{c_0}\right)_O = 1 + \frac{J}{J_{\text{lim,O}}}. \tag{4.24}$$

For the diffusion processes such as $c_{0,R} \rightarrow c_{x=0,R}$ and $c_{x=0,O} \rightarrow c_{0,O}$, the relationship (4.7) derived in Thermodynamic Analysis section is applicable. If one divides (4.7) with zF and one assumes that in the vicinity of the electrode, the solution has an ideal behavior, i.e., $K_{eq} = c_{x=0}/c_0$, one obtains

$$\Delta E_{conc,R} = \frac{RT}{zF}\ln\left(\frac{c_{x=0}}{c_0}\right) = \frac{RT}{zF}\ln\left(1-\frac{J}{J_{lim,R}}\right) \text{ and}$$
$$\Delta E_{conc,O} = \frac{RT}{zF}\ln\left(1+\frac{J}{J_{lim,O}}\right). \tag{4.25}$$

The net concentration over-potential is therefore $\Delta E_{conc} = \Delta E_{conc,R} - \Delta E_{conc,O}$, i.e.,

$$\Delta E_{conc} = \frac{RT}{zF} \times \ln\left(\frac{1+\dfrac{J}{J_{lim,R}}}{1-\dfrac{J}{J_{lim,O}}}\right), \tag{4.26}$$

which is valid for the case when the reaction kinetics is very fast such that all reactants that diffusively reach the reaction site are converted to products; this is the diffusion-limited process. This hypothetical situation can be a valid assumption for some special cases. In many practical problems, the kinetic process is not fast enough to reach 100 % yield. Equations (4.19) and (4.21) describe the other extreme situation, when kinetics is relatively slow and it controls the conversion process. For such a case, it is assumed that due to the slow rate of the kinetics, there is no concentration gradient at the electrode surface; therefore, there is no diffusion process. Actual water electrolysis systems may operate such that none of the processes are dominant. For a combined kinetic-diffusion controlled regime, a superposition of (4.20) and (4.25) may be obtained. According to Bagotski (2006), the following equation describes a combined kinetic-diffusional polarization process:

$$J = J_0 \times \frac{\exp(\alpha F/RT \times \Delta E_{pol}) - \exp(-(1-\alpha)F/RT \times \Delta E_{pol})}{1+\exp(\alpha F/RT \times \Delta E_{pol}) \times J_0/J_{lim,R} + \exp(-(1-\alpha)F/RT \times \Delta E_{pol}) \times J_0/J_{lim,O}}, \tag{4.27}$$

where ΔE_{pol} is the electrode polarization over-potential due to combined effects of activation energy for the reaction to occur and due to concentration gradients that need to be generated for reactants to be supplied at the reaction sites. Figure 4.6 depicts the current vs. polarization over-potential diagram for a hydrogen-evolving platinum electrode and a cell configuration in which the limiting forward direct is 5,000 A/m^2, while the backward current is 4,000 A/m^2. Equations (4.21), (4.26), and (4.27) are used to construct the plot.

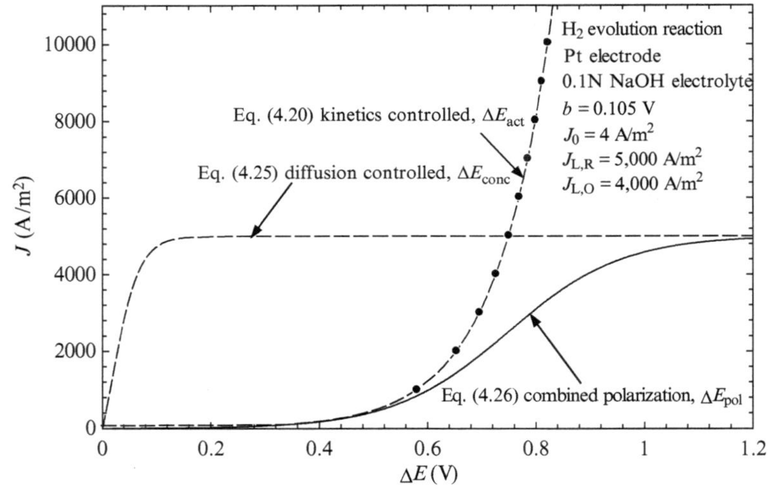

Fig. 4.6 Polarization potential at a hydrogen-evolving platinum electrode

4.2.4 Equivalent Electric Diagram-Based Analysis

In the previous sections, two main obstacles must be overcome in order to obtain a net evolution of hydrogen and oxygen from water. These are the reversible potential and the polarization over-potential. In water electrolysis, two half-reactions must be conducted concurrently: hydrogen and oxygen evolutions. An electrical charge circuit must be closed by forcing the movement of ionic species between the anode and cathode, and that of the electrons through the external circuit. Therefore, additional electric power is required to overcome these losses (overpotentials), which are ohmic in nature, namely, $\dot{W}_{\Omega} = I \times \sum \Delta E_{\Omega}$, where the sum of all ohmic over-potentials is denoted with $\sum \Delta E_{\Omega}$. The ohmic losses have different components depending on the actual type of electrolyzers (in subsequent sections, these are detailed for some specific cases). The total overpotential for an electrolysis cell becomes

$$\Delta E_{\text{tot}} = \Delta E_{\text{pol,a}} + \sum \Delta E_{\Omega} + \Delta E_{\text{pol,c}}, \tag{4.28}$$

which can be divided by the net current I to obtain an equivalent circuit representation of the electrolysis cell as a series of electrical resistances, $R_{\text{eq}} = R_{\text{a}} + \sum R_{\Omega} + R_{\text{c}}$. This circuit is illustrated in Fig. 4.7. The circuit includes two diodes in direct polarization which suggest the reversible voltages required by the anode and cathode. Note that for an ideal diode in direct polarization, the voltage does not vary with the current; the same is the situation with reversible potentials of half-reactions. Kirchhoff's law for the equivalent circuit is

$$E_{\text{tot}} = E_{\text{rev,a}} + E_{\text{rev,c}} + I \times R_{\text{eq}}. \tag{4.29}$$

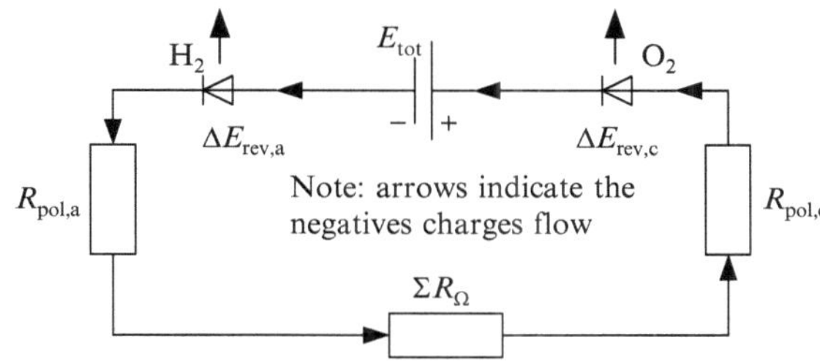

Fig. 4.7 Equivalent electrical circuit of a water electrolysis cell

Equation (4.30) can be solved for E_{tot} if the current is specified and the equivalent resistance is known. All components of electrical resistance ($R_a, \sum R_{\Omega}$, R_c) can be calculated provided that the current density is known. The cathodic and anodic resistances can be determined based on (4.27) noting that $(\Delta E_{pol})_{a,c} = J \times (A \times R_{pol})_{a,c}$.

The ohmic resistances $\sum R_{\Omega}$ are those responsible for heat generation within the electrolyte and in the exterior electrical circuit due to the Joule effect. Several ohmic resistances must be taken into account. One major type of ohmic loss occurs when the current passes across the electrolyte. To same extent, the ohmic losses can be modeled, in general using the electrical conductivity (σ) and the thickness of the conductive layer (δ), according to

$$R_{\Omega} = \delta/\sigma. \tag{4.30}$$

The dissipated power in the form of heat (due to the Joule effect) can be calculated with

$$\dot{W}_{loss} = R_{eq} \times I. \tag{4.31}$$

The design of the electrolysis system must be made such that the dissipation is minimized. If $\dot{W}_{loss} > \Delta H - \Delta G$ where ΔH and ΔG are reaction enthalpy and Gibbs free energy of water splitting at operating conditions, then the electrolyzer rejects heat to the exterior (it must be cooled) so that $\dot{Q}_{out} = \dot{W}_{loss} - (\Delta H - \Delta G)$. If $\dot{W}_{loss} < \Delta H - \Delta G$ then heat must be heated at a rate of $\dot{Q}_{in} = \Delta H - \Delta G - \dot{W}_{loss}$. If $\dot{W}_{loss} = \Delta H - \Delta G$ then the electrolyzer operates at thermo-neutral conditions.

The energy efficiency of an electrolyzer can be expressed as the energy retrieved in the produced hydrogen by the energy input. The energy carried by a molar flow rate of hydrogen $\dot{n}$ is calculated based on the molar higher heating value HHV = 283.6 kJ/mol, namely, as the product $\dot{n} \times$ HHV. The total electric power supplied to the electrolyzer is $E_{tot} \times I$. Accounting for (4.13), where $z = 2$, it results in

$$\eta_{el} = \frac{\text{HHV}}{2FE_{tot}} = \frac{1.4696}{E_{tot}}. \tag{4.32}$$

If heat is supplied to the electrolyzer (such as in high-temperature electrolysis cells) then this is used as an input in the efficiency formulation; in this case, (4.32) is modified as follows: $\eta_{el} = 1.4696/(E_{tot} + Q/2F)$. The exergy efficiency of electrolysis systems is calculated using the molar chemical exergy of hydrogen ($ex^{ch} = 236.12\,\mathrm{kJ/mol}$):

$$\psi_{el} = \frac{ex^{ch}}{zFE_{tot}} = CF \times \eta = \frac{1.224}{E_{tot}}, \tag{4.33}$$

where $CF = 0.83$ is a quality factor expressing the ratio between chemical exergy and the higher heating value of hydrogen. If the cell operation is assisted by heat input, then (4.33) is modified to $\psi_{el} = 1.224/(E_{tot} + ex^{Q}/2F)$, where ex^{Q} is the exergy associated with the heat input. In many practical situations, it is reasonable to determine an equivalent temperature (T^{Q}) at which the heat is supplied to the electrolyzer. Then the associated exergy is $ex^{Q} = (1 - T_0/T^{Q}) \times Q$, where T_0 is the reference temperature of the environment.

When compared to other hydrogen production technologies, the efficiency of the water electrolysis process must account for the electrical power generation efficiency. In such a case, (4.32) and (4.33) must be written as $\eta = \eta_{el} \times \eta_{pg}$ and $\psi = \psi_{el} \times \psi_{pg}$, where the subscript "pg" refers to power generation. In addition to (4.32) and (4.33), the efficiency of water electrolysis can be expressed by the Faradaic efficiency, which represents the ratio between the reversible cell voltage and the actual cell voltage, according to

$$\eta_{Faradaic} = \frac{E_{rev}}{E_{tot}}, \tag{4.34}$$

where E_{rev} is the reversible cell voltage at given operating conditions.

Optimizing the design of water electrolyzers implies calculations of polarization and ohmic resistances. The modeling can follow a systematic approach such as: (1) selection of the electrodes and electrolyte and defining the operating conditions; (2) calculation of the reversible potentials at the cathode and anode; (3) determination of polarization resistances for an assumed current density; (4) determination of all ohmic resistances for the same current density; (5) variation of the current density to generate a current voltage diagram; (6) finding the optimum current density at which the electrolyzer efficiency is maximum; and (7) repetition of all calculations for different design parameters such as electrode types, electrolyte, operating pressure, and temperature in order to find the best design.

4.3 Alkaline Electrolyzers

As the name suggests, alkaline electrolyzers (AE) use an alkaline electrolyte (pH > 7). In an alkaline electrolyte, the mobile ionic species are anions (negative ions) of the hydroxyl group, OH^-. Typical electrolytes are liquid solutions of bases such as KOH or NaOH. The electrochemical half-cell reactions in an alkaline electrolyzer cell are the following:

$$\begin{cases} 2H_2O + 2e^- \rightarrow H_2(g) + 2OH^- & \text{at cathode} \\ 2OH^- \rightarrow H_2O + 0.5O_2(g) + 2e^- & \text{at anode} \end{cases}. \tag{4.35}$$

The concentration of the electrolyte must be high enough such that it assures a good mobility of ions; typically for KOH this is 30 %. The electrodes are made of a planar geometry to enhance the contact area with the electrolyte and, most importantly, they do not require expensive catalysts (normally carbon steel coated with nickel is used). Due to these characteristics, alkaline electrolyzers can be constructed in a cost-effective manner for larger production capacities. Commercial alkaline electrolyzers operate at temperatures of 80–200 °C with production capacities in the range of $500-30,000\,\mathrm{Nm^3/h}$ (or equivalent $0.5-40\,\mathrm{MW}$ per unit with respect to the LHV of produced hydrogen). The purity of hydrogen is higher than 99.7 % and the energy efficiency (electricity to hydrogen) of commercial electrolyzers is in the range of $55-90\,\%$. When operating at higher temperatures (e.g., over 120 °C), a part of the energy needed to split water can be transmitted by heat transfer from an external/sustainable heat source.

Alkaline electrolyzers are constructed in two configurations: monopolar and bipolar. In a monopolar configuration, several electrolysis cells are packed in series such that the voltage applied to the stack is the sum of voltages for each cell, $E_{AE} = \sum E_{cell} \cong (N_{cell} - 1) \times E_{cell}$, while the current through each cell is the same. The monopolar configuration is sketched in Fig. 4.8a. In a bipolar configuration—Fig. 4.8b—the cells are connected in parallel such that the voltage is the same for all cells while the total current is $I_{AE} = \sum I_{cell} \cong N_{cell} \times I_{cell}$. Due to constructive reasons, the ohmic losses in monopolar electrolyzers are higher with respect to a bipolar configuration. The gap between electrodes must be reduced as much as possible, but in a manner that avoids any surge of current between the two opposite electrodes. Cells with a zero-gap configuration were developed to minimize the ohmic losses through an electrolyte while increasing the purity of hydrogen. In a zero-gap configuration, an electrode–diaphragm assembly is devised as a sandwich-like geometry having in the middle the diaphragm, permeable only to ionic species, while on both sides are attached the cathode and anode electrodes which have pierced holes of 0.5 mm (in average) to allow for ionic transport. The electrolyte must be recirculated in order to minimize the concentration overpotentials.

Some typical electrode materials for cathode in alkaline electrolyzers are Ni, Fe, Co, Zn, Pb, Pd, Pt, and Au while for anode, Ni, Pt, Ir, Ru, Rh, TiO_2, Co, and others are used. Table 4.2 lists the main kinetic parameters of some electrodes commonly used in alkaline water electrolyzers. Typically, the range of the operating current density of commercial alkaline electrolyzers is $1-3\,\mathrm{kA/m^2}$, while for advanced alkaline electrolyzers, it covers $2-15\,\mathrm{kA/m^2}$. One of the main problems with the electrode is due to the alteration in time by various chemical reactions that may occur

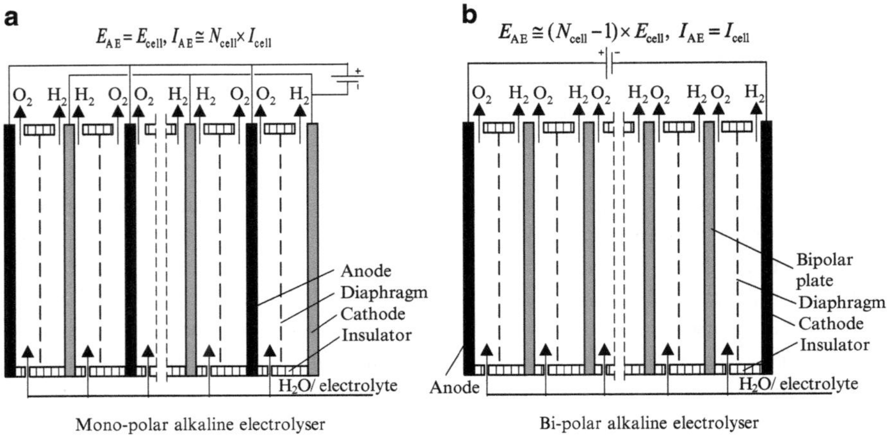

Fig. 4.8 Constructive configurations of alkaline electrolyzers—(**a**) monopolar, (**b**) bipolar

at their surface; this leads to deactivation of electro-catalysts (if any are used) and a sensible decrease of current density while the over-potentials increase. With Ni-based electrodes, deactivation is due to the formation of a nickel hydride (Zeng and Zhang 2010) which can be "removed" by maintenance operations requiring circulation of dissolved vanadium species in the electrolyte. The enhancement of electrode performance is typically obtained by doping its surface with components that may act as very active electro-catalysts. For example, electrodes in alkaline electrolyzers can be doped with Pt, Pd, Mo, Ru, and Ir. The structure of the electrode surface plays a major role. Nano-structured surfaces were recently engineered to obtain higher surface area and better current densities.

Regarding the electrolyte in alkaline electrolyzers, there are two specific issues: (1) mass transport must be controlled to reduce the diffusion layer, and (2) the unavoidable gas bubbles formed at an electrode surface must be removed as best as possible to avoid an excessive increase of ohmic losses. Both these problems can be reasonably solved with appropriate additives dissolved in the electrolyte. The additives aim to reduce surface tension and thus to increase diffusion mechanisms at the electrode surface, and to accelerate the bubble detachment. Note that the bubble detachment requires an activation energy which translates to a slight increase of over-potential. Based on liquid–vapor surface tension (γ) and the contact angle (θ), the overpotential due to bubble detachment can be approximated with $zF\Delta E_{\text{bubble}} = \gamma \times (\cos\theta - 1)$; see Zeng and Zhang (2010). The ohmic losses within the electrolyte are usually determined by the molar conductivity (Λ) in $\text{m}^2\,\Omega^{-1}\,\text{mol}^{-1}$ which is defined as electrical conductivity (σ) in $\Omega^{-1}\,\text{m}^{-1}$ divided to molar concentration (c) in mol/l. Ohmic losses are relatively insensitive to current density but may be sensitive to temperature variations.

In Fig. 4.9, a typical variation of over-potentials in commercial electrolyzers is shown for industrial applications. The ohmic losses in electrodes are the smallest while the over-potential at the anode is the highest. Furthermore, it can be observed that at the higher range of current densities, the ohmic losses in the electrolyte

Table 4.2 Activity parameters for some electrodes specific to alkaline electrolyzers

Cathode material	Electrolyte	T (°C)	J_0 (A/m^2)	b (mV)	Anode material	Electrolyte	T (°C)	J_0 (A/m^2)	b (mV)
Ni[a]	50 % KOH	80	1.10	140	Ni	50 % KOH	80	4.2×10^{-2}	95
		150	13.1	167			150	1.8	125
		208	40.0	195			208	20	110
		264	43.0	133			264	10	32
Pt	0.1 N NaOH	25	0.676	114	Pt	30 % KOH	80	1.2×10^{-5}	46
Zn	9 N KOH	25	1.5×10^{-5}	124	Co	30 % KOH	80	3.3×10^{-2}	126
Zn + 2 % Hg	9 N KOH	25	2.7×10^{-6}	116	TiO_2 + 0.3 NiO	30 % KOH	80	1.4	259
Fe + 75 % Ni	0.1 N NaOH	25	3.5×10^{-2}	115	Fe	30 % KOH	80	0.17	191
Carbon	40 % NaOH	40	0.29	148	0.75 Ni + 0.25 Ru	30 % KOH	80	8.8×10^{-3}	67

[a]High-surface-area electrode (0.1–30 m^2/g)

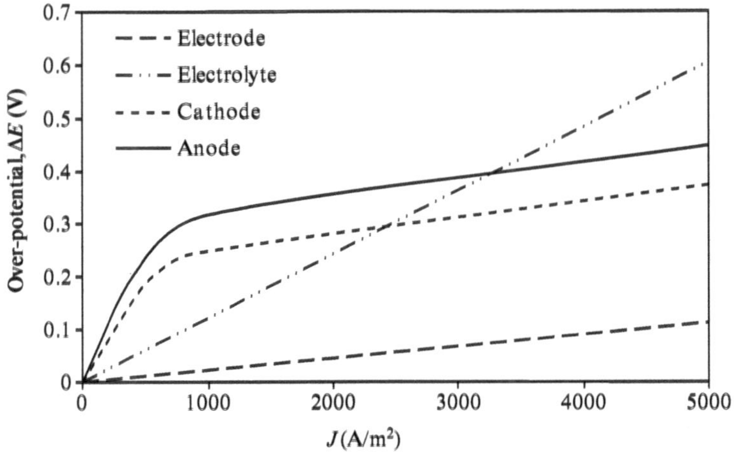

Fig. 4.9 Typical over-potentials for commercial alkaline electrolyzers [data from Zeng and Zhang (2010)]

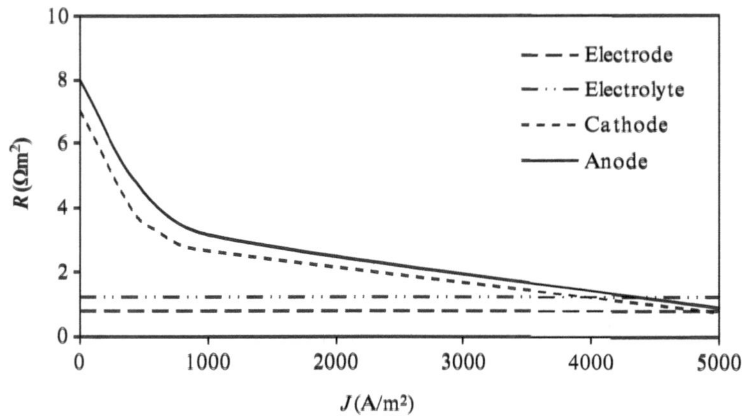

Fig. 4.10 Equivalent resistances within an alkaline electrolyzer for a range of current densities [data from Zeng and Zhang (2010)]

become significant. The equivalent resistances for each of the losses are presented in Fig. 4.10. The equivalent electrical resistance corresponding to polarization decreases with the current density. Figure 4.11 depicts the range of cell voltage for alkaline electrolyzers as a function of current density. Commercial alkaline electrolyzers have an efficiency in the range of 60–70 %, while advanced systems—currently in development—can surpass 90 %.

The purity of water can be a problem of much concern in water electrolysis. The presence of metal atoms in water such as calcium or magnesium can cause occurrence of side reactions at the electrode surface that eventually leads to scale formation with consequence of a reduction of active surface area and enhanced clogging of the diaphragm. The salinity of water is important; if water contains

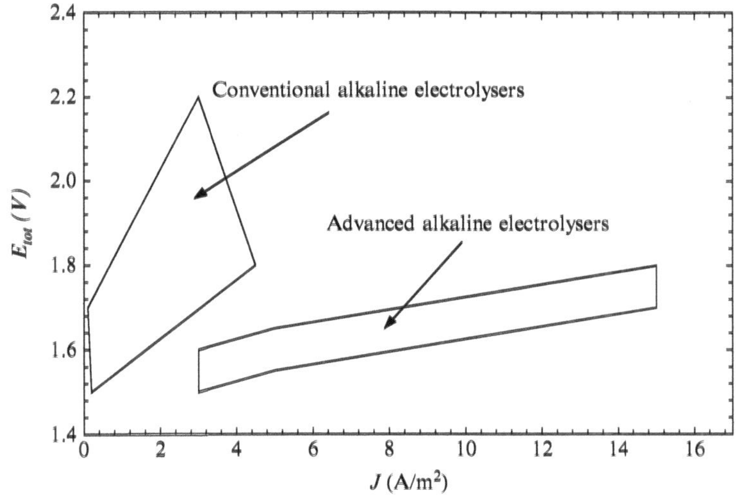

Fig. 4.11 Range of cell voltages for alkaline electrolyzers as a function of current density [data from Guerrini and Trasatti (2009)]

NaCl, or other sources of chlorine ions, chlorine gas will emanate at the cathode with a highly corrosive effect. Therefore, all industrial electrolysis plants are equipped with water treatment stations. Figure 4.12 depicts the general layout of a hydrogen production plant with alkaline electrolyzers.

4.4 Proton Exchange Membrane Electrolyzers

A proton exchange membrane electrolyzer (PEME) is a device which splits water by electrolysis using an acidic electrolyte. Across acidic electrolytes, the species that are transported are positive ions (or cations). In the case of water electrolysis, the ionic transport results in a net transfer of protons from anode to cathode. The overall electrochemical reactions for acid electrolyte electrolyzers are

$$\begin{cases} H_2O \rightarrow 2H^+ + 1/2O_2(g) + 2e^- & \text{at anode} \\ 2H^+ + 2e^- \rightarrow H_2(g) & \text{at cathode} \end{cases}. \tag{4.36}$$

The most common acid electrolyte is a perfluorosulfonic acid polymer of which the most well-known trade name is NAFION®. The sulfonate groups in a polymeric chain make the electrolyte behave as a strong acid. NAFION® presents a thin, solid membrane. The ionic species which permeates through NAFION® is the hydronium (H_3O^+) in the aqueous solution. The membrane must be always wet to allow for ionic transport. Moreover, one needs to compensate for a backward permeation process of molecular oxygen through the membrane by increasing the electrical current density by ~5 %.

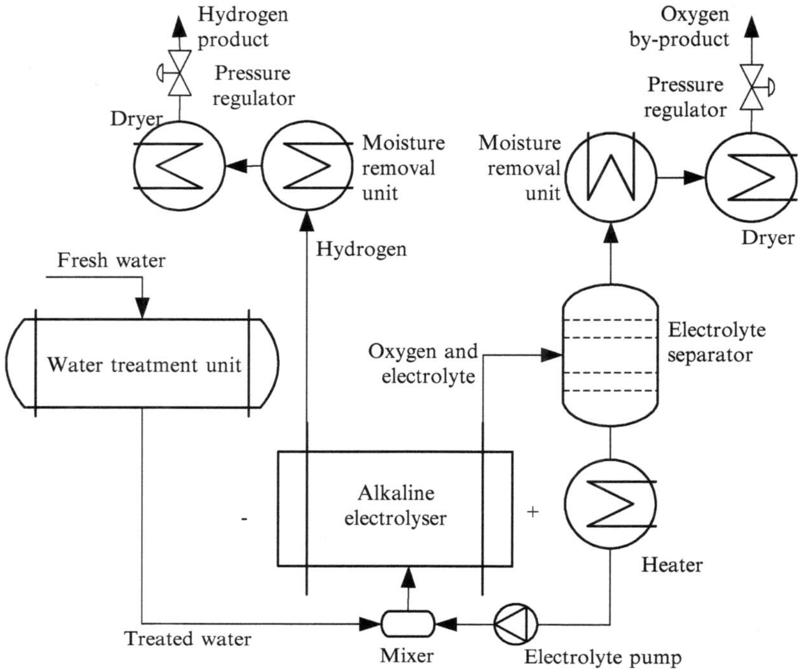

Fig. 4.12 General layout of a water alkaline electrolysis plant

The operating temperature range of a PEME is restricted to 25–80 °C. Typical electricity consumption of a commercial PEME is 23–26 MJ electric per normal cubic meter (Nm^3) of hydrogen produced or approx. 540–580 MJ per kg of hydrogen produced. The energy efficiency of PEME drops quasi-exponentially with the current density and reaches 54 % at ~10 kA/m^2 with 80 – 85 % conversion. The generated hydrogen has a purity higher than 99.999 %. The capacity range of an industrial PEME is ~0.2–60 Nm^3/h or from about 10 g/h to approx. 2.5 kg/h. The associated electrical energy consumption of a commercially available PEME can go up to 400 kW while the water consumption reaches 25 l/h.

The capital cost of PEME is relatively high due to the cost of the NAFION® proprietary membrane (approximately \$600/m^2) and the expensive metals required at the electrodes and bipolar plates. From the total electrolytic hydrogen cost, approximately 58 % represents the operation cost and the remaining 32 % reflects the capital cost. A PEME requires ultrapure distilled water for proper operation.

The common stack arrangement for a PEME uses bipolar plates. Bipolar plates comprise extruded channels which facilitate a two-phase water-gas flow on both sides. Figure 4.13 depicts the geometrical arrangement of a PEME stack. It comprises several bipolar plates with channels on both sides (only one is illustrated in the figure) and two end plates; each plate is firmly attached to a planar, porous,

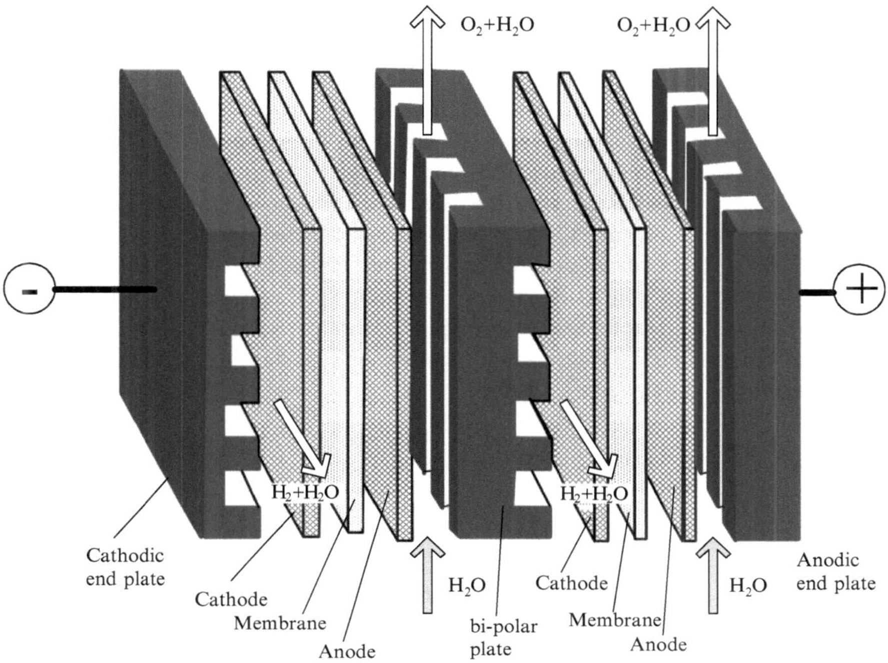

Fig. 4.13 Geometrical arrangement of a PEM electrolyzer stack

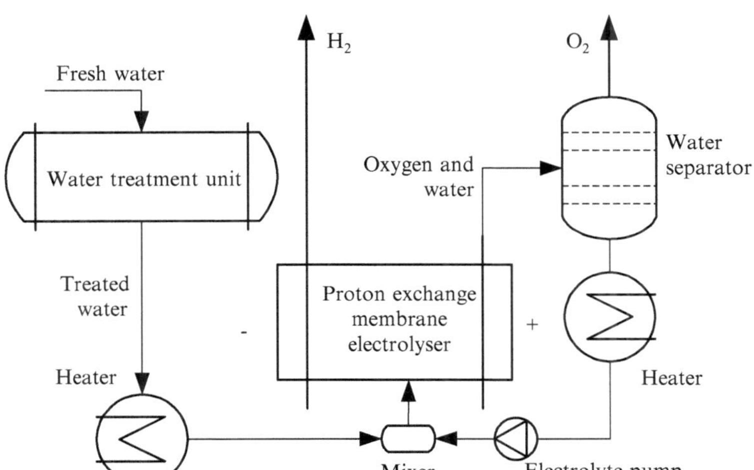

Fig. 4.14 General layout of a PEM electrolyzer system

electrically conductive layer which has the role of the electrode (cathode or anode). Between the electrodes, a proton exchange membrane is placed. The general layout of a PEME system is presented in Fig. 4.14.

The electrodes and bipolar plates of a PEME must have high resistance to corrosion because the membrane creates an acidic medium. Therefore, expensive platinum group metal must be used for the electrodes to coat the bipolar plates. At the cathode, the higher over-potential occurs; for its minimization, electrocatalysts such as oxides of Ir and Ru are commonly used. Thus, the cost of PEME electrodes is perceived to be relatively high.

The resistance associated with the Nafion membrane is another factor which influences the efficiency of PEME. It was found that increasing the membrane thickness from 50 to 200 μm decreased the efficiency of a PEME from 60 to 56 % at a current density of 9,800 A/m^2 due to the increase in required cell voltage (Meng et al. 2008).

Calculations of ohmic losses through PEM make use of specific equations for electrical conductivity across the membrane. This can be calculated with the following formula (Meng et al. 2008):

$$\sigma(c,T) = (0.5139 \times c - 0.326) \times \exp\left(1{,}268 \times \left(\frac{1}{303} - \frac{1}{T}\right)\right), \tag{4.37}$$

where c is the local molar concentration of water inside the membrane and T is the membrane temperature. The molar concentration varies across the membrane thickness from a high concentration Λ_a at the anode (where water is supplied) to a lower concentration Λ_c at the cathode. At any distance across the membrane, $x \in [0, \delta]$, the molar concentration is $c(x) \in [c_c, c_a]$, where $x = 0$ is the anode side and $x = \delta$ is the cathode side of the membrane. It is reasonable to assume that the molar water content varies linearly across the membrane (Nieminen et al. 2010):

$$c(x) = c_c + \frac{c_a - c_c}{\delta} \times x. \tag{4.38}$$

The electrical conductivity is defined by $\sigma \equiv \mathrm{d}x/\mathrm{d}R$, where R is the electrical resistance; thus, the differential equation $\mathrm{d}R = \sigma^{-1}\mathrm{d}x$ can be integrated from 0 to δ:

$$R_{\Omega,\mathrm{PEM}} = \int_0^{\delta} \sigma^{-1}(c(x),T)\,\mathrm{d}x. \tag{4.39}$$

The concentration over-potentials for PEME must take into account the variation of water vapor concentration across the membrane. In a recent analysis, Nieminen et al. (2010) assumed a limiting current density $J_{\mathrm{lim}} = 20{,}000\ \mathrm{A/m^2}$ and solved the following equation for concentration over-potentials:

$$\Delta E_{\mathrm{conc}} = J^2 \times \left[\beta\left(\frac{J}{J_{\mathrm{lim}}}\right)^2\right], \tag{4.40}$$

where the factor β is defined by

$$\beta = \begin{cases} (7.16 \times 10^{-4}T - 0.622) \times P + (-1.45 \times 10^{-3}T + 1.68) \text{ if } P < 2\,\mathrm{atm} \\ (8.66 \times 10^{-5}T - 0.068) \times P + (-1.6 \times 10^{-4}T + 0.54) \text{ if } P \geq 2\,\mathrm{atm} \end{cases}, \tag{4.41}$$

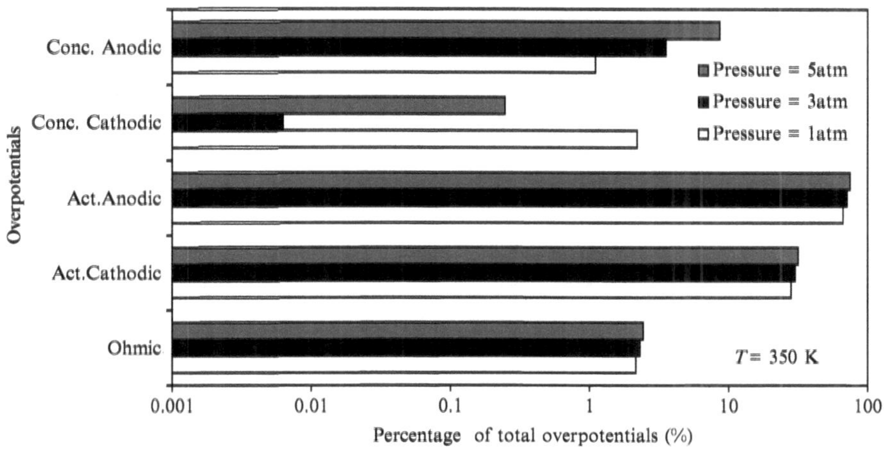

Fig. 4.15 Over-potential distribution in PEM electrolyzer at various pressures [data from Nieminen et al. (2010)]

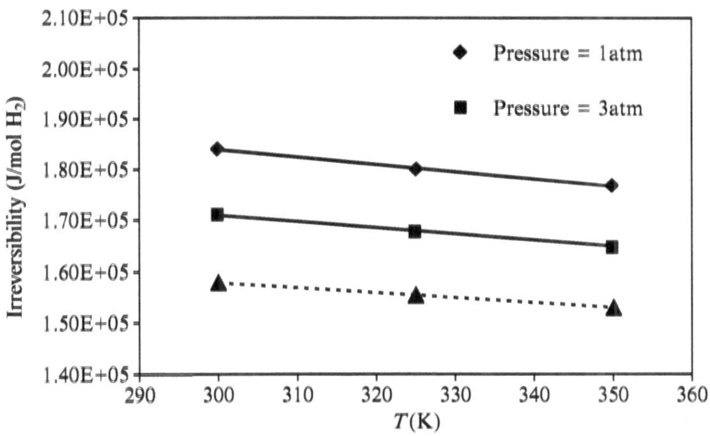

Fig. 4.16 Irreversibility within a PEM electrolyzer operating at various pressures [data from Nieminen et al. (2010)]

where $P = P_i/0.1173 + P_{sat}$ is the local pressure at the anode or cathode, P_i is the partial pressure at the anode or cathode, and P_{sat} is the saturation pressure of water at the operating temperature; the index i becomes "a" for anode and "c" for cathode.

In Fig. 4.15, it can be seen that the ohmic over-potentials in a PEME account for less than 5 % of the total over-potentials while the anodic activation over-potentials in a PEME occupy approximately 60–80 % of the total over-potential. This suggests that hydrogen transport through a NAFION® membrane is highly efficient; however, there are significant losses due to polarization. Figure 4.16 illustrates the magnitude of total irreversibilities of a PEME and their trend with operating pressure and temperature. The irreversibilities are decreasing with higher operating temperature

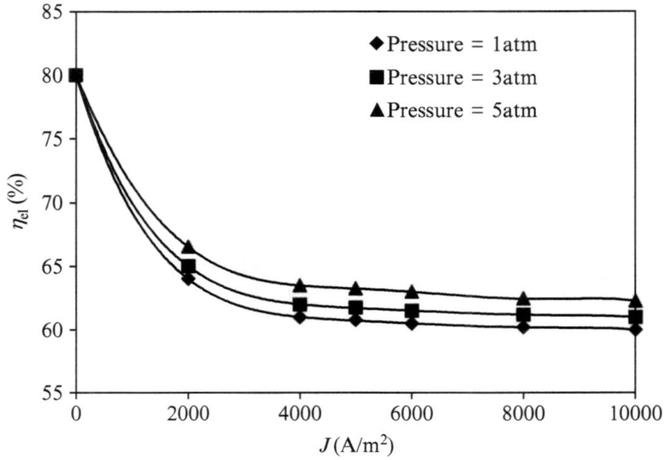

Fig. 4.17 Energy efficiency of PEM electrolyzer operating at 350 K at various pressures [data from Nieminen et al. (2010)]

Fig. 4.18 Cathodic activation over-potentials of a PEM electrolyzer for various current densities [data from Nieminen et al. (2010)]

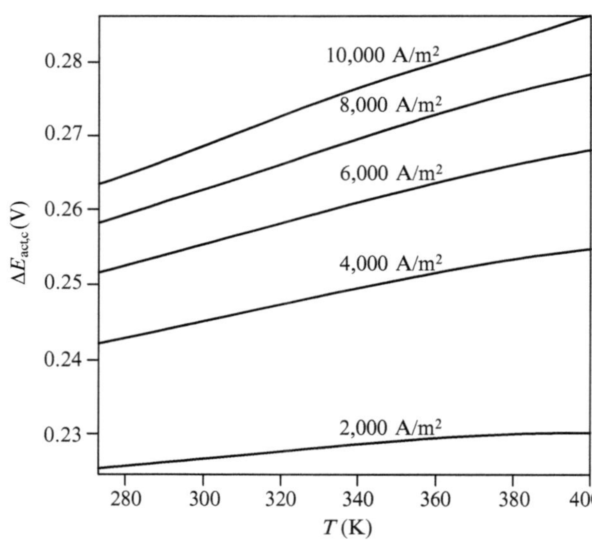

and pressure. As a result, a minor increase in efficiency is occurring with higher temperature and pressure. This observation is confirmed by the efficiency profiles shown in Fig. 4.17 where the influence of current density on electrolyzer performance is also studied.

The variation of over-potentials in PEME with a current density is shown in Figs. 4.18, 4.19, 4.20, 4.21, and 4.22. The current density varied between 2,000 A/m^2 and 10,000 A/m^2 in increments of 2,000 A/m^2. The activation over-potential ranges from 0.22 V to 0.28 V at the cathode and 0.55 to 0.75 V

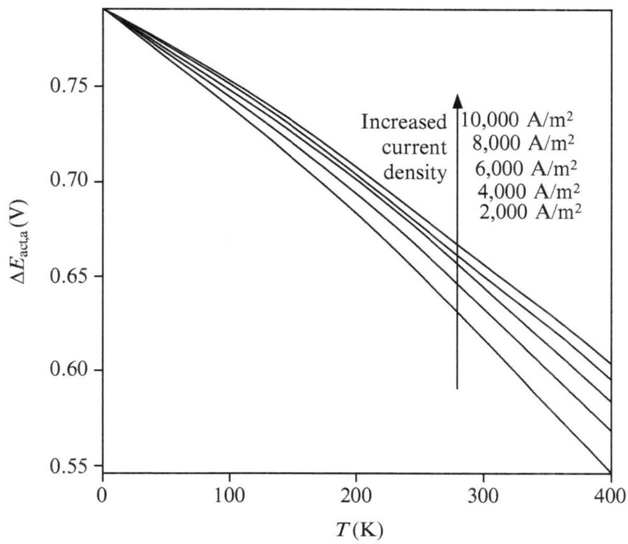

Fig. 4.19 Anodic activation over-potentials in a PEM electrolyzer for various current densities [data from Nieminen et al. (2010)]

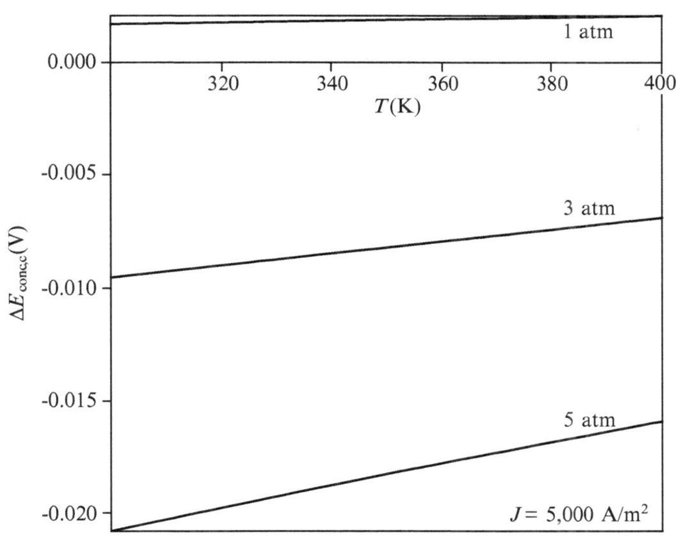

Fig. 4.20 Cathodic concentration overpotentials in a PEM electrolyzer for various pressures [data from Nieminen et al. (2010)]

at the anode. The concentration over-potentials are negligible for current densities below 10 kA/m^2; in Figs. 4.20 and 4.21, it is shown that for 5 kA/m^2, the absolute value of concentration over-potential is below 20 mV for the cathode and below 5 mV for the anode.

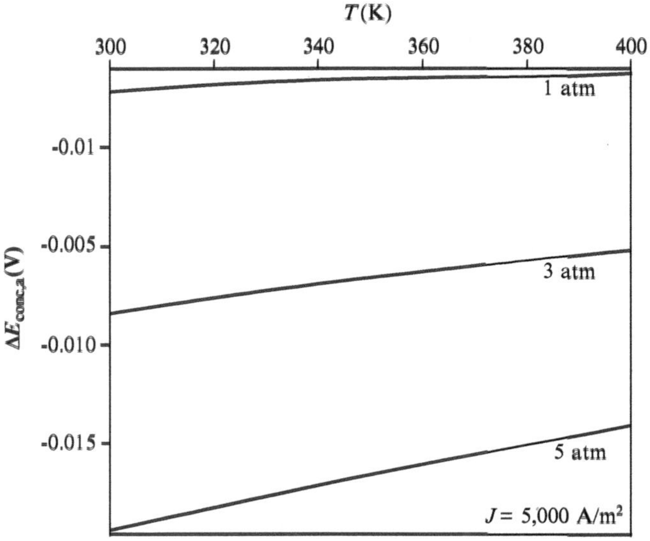

Fig. 4.21 Anodic concentration over-potentials in a PEM electrolyzer for various pressures [data from Nieminen et al. (2010)]

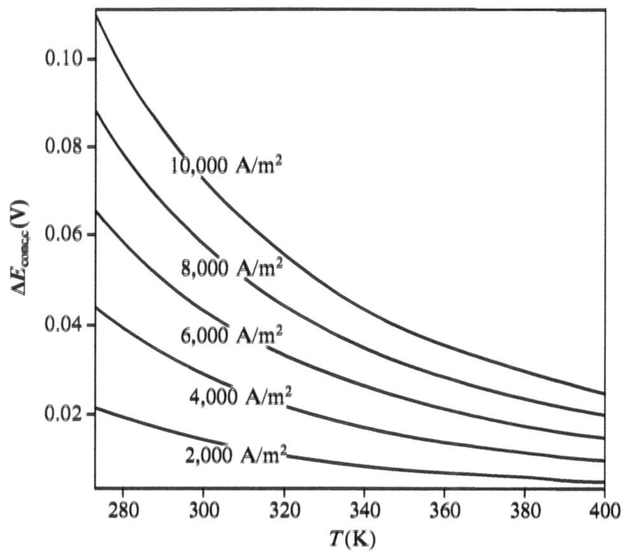

Fig. 4.22 Ohmic over-potentials in a PEM electrolyzer for various current densities [data from Nieminen et al. (2010)]

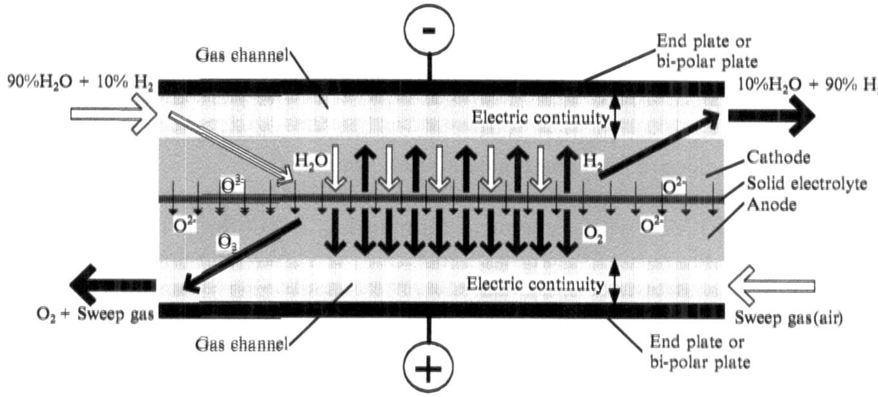

Fig. 4.23 Operating principle and geometrical configuration of an SOE-O cell

4.5 Solid Oxide Electrolyzers with O^{2-} Conduction

Solid oxide electrolyzers with oxygen ion conduction (SOE-O) emerged in recent years as a promising technique to split water in the form of steam at high temperatures. They use a solid, nonporous metal oxide electrolyte, which is conductive to oxygen ions, O^{2-}. The two half-reactions in an SOE-O are

$$\begin{cases} 2O^{2-} \rightarrow O_2(g) + 4e^- \text{ at anode} \\ H_2O(g) + 2e^- \rightarrow H_2(g) + O^{2-} \text{ at cathode} \end{cases} \tag{4.42}$$

Ceramic-based electrolytes comprising yttrium-stabilized zirconium (YSZ) or scandium-stabilized zirconium (ScSZ) are commonly used in SOE-O construction. At temperatures of 1,000–1,300 K, these electrolytes allow the passage of oxygen ions O^{2-} under a driving force created by an electrical field and/or concentration gradient. Typically, zirconium dioxide (ZrO_2) is doped with 8 % yttrium trioxide (Y_2O_3) or with scandium trioxide (Sc_2O_3) in order to stabilize its crystalline structure for the operating temperature range and to produce crystallographic defects which eventually allow for the passage of oxygen ions.

A cross-sectional view through an SOE-O cell is illustrated in Fig. 4.23, which shows the geometrical arrangement of its components (bipolar or end plates, gas channels, anode, cathode, and electrolyte) and the flowing paths of chemical species. The electrolyte–electrode assembly consists of a thin layer of solid oxide packed in a sandwich-like structure between a porous cathode and a porous anode. The cathode is permeable to hydrogen and steam.

The cathode has a composite structure comprising a porous matrix made in YSZ combined with nickel and/or cerium. The role of the electro-catalyst is served by nickel atoms, which are dispersed over the porous matrix. The hydrogen-evolving reaction occurs within the porous structure of the cathode in the vicinity of active centers. Once formed, molecular hydrogen diffuses toward the collecting channel under the influence of a concentration gradient. Excess steam must be present in the cathodic channel to reduce the partial pressure of hydrogen and to facilitate its diffusion. Oxygen ions diffuse toward the positively charged anode under the influence of an electric field and concentration gradient.

The anode is normally composed of lanthanum and manganese oxides doped with strontium (LMS) with a chemical formula $La_{0.8}MnO_3Sr_{0.2}$. The oxygen ion oxidizes within the porous structure of an anode which is maintained at a positive potential by the electrical power supply. The manufacturing of the electrode–electrolyte assembly is made using the anode (~1.5 mm thickness) as a mechanical support on which is deposited first the electrolyte layer (~140 μm) and then, over the electrolyte, the cathode is deposited (~13 μm thickness); over the cathode, a layer of pure nickel of ~10 μm thickness is applied. The bipolar plates are made in ferritic stainless steel which is resistant to high temperatures and presents good electric conductivity. In order to facilitate the diffusion of molecular hydrogen through the anode, a sweep gas is in general used, e.g., air, which decreases the partial pressure of oxygen.

The electrical conductivity across the solid oxide electrolyte (σ in (4.30)) varies exponentially with temperature according to (Meng et al. 2007)

$$\sigma = 33,000 \times \exp(-10,300/T). \tag{4.43}$$

The calculation of reversible half-cell potentials, and anodic and cathodic polarization over-potential, takes into account that the reactions occur in a gas phase rather than in solution. Therefore, the activity can be replaced with fugacity and molar concentrations with a partial pressure when the reversible half-cell potentials are calculated (see Sect. 4.2.1). Some specific ways to calculate the polarization potentials are given below from Meng et al. (2007).

The activation over-potentials associated with the operation of an SOE-O are formulated as

$$\Delta E_{\mathrm{act,i}} = \frac{RT}{F} \ln\left(\frac{J}{2J_{o,i}} + \sqrt{\left(\frac{J}{2J_{\mathrm{o,i}}}\right)^2 + 1}\right), \tag{4.44}$$

where i = a,c and the exchange current density have an order of magnitude of ~2,000 A/m^2 for the anode and ~5,000 A/m^2 for the cathode. Its dependence on temperature is described by the Arrhenius equation (see Sect. 4.2.2). For SOE-O, the order of magnitude of activation energy is around 100 kJ/mol, while the pre-exponential factors are around $\sim 5,000\,\mathrm{s}^{(-1)}$ for the anode and $\sim 50,000\,\mathrm{s}^{(-1)}$ for the cathode.

Table 4.3 Practical parameters for SOE-O modeling

Parameter	Symbol	Value
Activation energy (cathode and anode, about the same)	ΔG_{act}	100 kJ/mol
Pre-exponential factor for anode reaction	A_a	5,000 s^{-1}
Pre-exponential factor for cathode reaction	A_c	50,000 s^{-1}
Electrode's average pore radius	r	0.5 μm
Electrode porosity	ε	0.4
Electrode tortuosity	ξ	5.0
Cathode thickness	δ_c	50 μm
Electrolyte thickness	δ	50 μm
Anode thickness	δ_a	500 μm
Characteristic length H_2	λ_{H_2}	2.827 *A*
Characteristic length H_2O	λ_{H_2O}	2.641 *A*
Lenard-Jones potential H_2	ϵ_{H_2}	5.144 meV
Lenard-Jones potential H_2O	ϵ_{H_2O}	69.72 meV
Coefficients for collision integral Ω_D, (4.47)	a_0	1.06036
	a_1	0.15610
	a_2	0.19300
	a_3	0.47635
	a_4	1.03587
	a_5	1.52996
	a_6	1.76474
	a_7	3.89411

Source: Meng et al. (2007)

For SOE-O, two distinct equations can be formulated for concentration over-potentials for both the anode and cathode. For the anode:

$$\Delta E_{conc,a} = \frac{RT}{4F} \ln\left(\sqrt{\left(P_{O_2}^{in}\right)^2 + JRT\mu\delta_a/(2FB_g)} \Big/ P_{O_2}^{in}\right). \tag{4.45}$$

In (4.45), P_{O2}^{in} is partial pressure of oxygen in the gas stream at the inlet port on the anode side, μ is the dynamic viscosity of molecular oxygen at an operating temperature, δ_a is the thickness of the anode, and B_g is the flow permeability which is determined by the Carman–Kozeny relationship as follows:

$$B_g = 2r^2\varepsilon^3 \,/\, [72 \times \xi \times (1 - \varepsilon)^2]. \tag{4.46}$$

At the cathode, the concentration over-potential is a function of many parameters such as effective diffusion coefficient of water $\left(D_{H_2O}^{eff}\right)$, Knudsen diffusion coefficient of water $\left(D_{H_2O,k}\right)$, molecular binary diffusion coefficient of the hydrogen—water mixture $\left(D_{H_2O,H_2}\right)$, molecular mass ($M$) of water and hydrogen, electrode porosity (ε), electrode tortuosity (ξ), mean electrode pore radius (r), temperature (T), pressure (P), collision–diffusion integral based on Lenard-Jones potential (Ω_D), and mean characteristic length for species H_2 and H_2O ($\bar{\lambda}$). Table 4.3 lists some practical values of

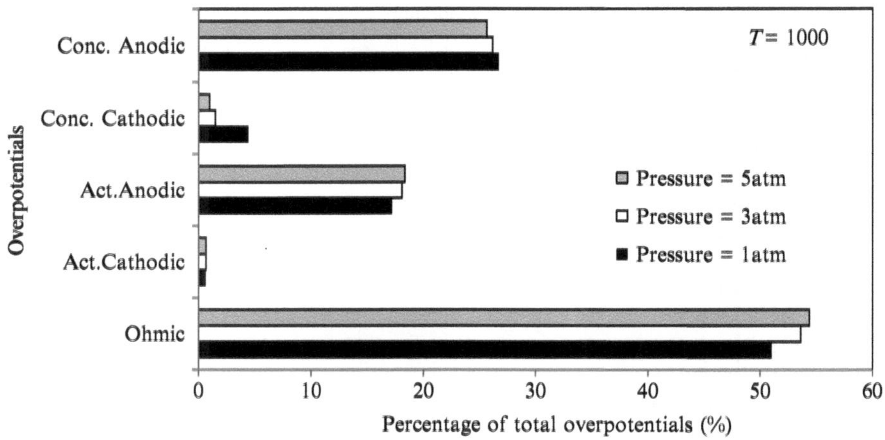

Fig. 4.24 Over-potential distribution in an SOE-O at various pressures [data from Nieminen et al. (2010)]

these parameters for SOE-O modeling. The concentration over-potential at the cathode is calculated with the following set of equations:

$$\begin{cases} \Delta E_{\text{conc,c}} = \dfrac{RT}{2F} \ln\left(\left[1 + \left(\dfrac{JRT\delta_c}{2FD^{\text{eff}}_{\text{H}_2\text{O}} P^{\text{in}}_{\text{H}_2}}\right)\right] \Big/ \left[1 - \left(\dfrac{JRTd_c}{2FD^{\text{eff}}_{\text{H}_2\text{O}} P^{\text{in}}_{\text{H}_2\text{O}}}\right)\right]\right) \\ D^{\text{eff}}_{\text{H}_2\text{O}} = \varepsilon / \left\{\xi \times \left[(D_{\text{H}_2\text{O,H}_2})^{-1} + (D_{\text{H}_2\text{O,K}})^{-1}\right]\right\} \\ D_{\text{H}_2\text{O,K}} = \dfrac{4r}{3}\sqrt{8RT/(\pi M_{\text{H}_2\text{O}})} \\ D_{\text{H}_2\text{O,H}_2} = 1.33 \times 10^{-3} T^{1.5} \sqrt{M_{\text{H}_2} + M_{\text{H}_2\text{O}}} / (P\bar{\lambda}\Omega_{\text{D}}) \\ \Omega_{\text{D}} = a_0/\tau^{a_1} + a_2/\exp(a_3\tau) + a_4/\exp(a_5\tau) + a_6/a_7\tau, \end{cases} \tag{4.47}$$

where $\bar{\lambda} = 0.5 \times (\lambda_{H_2} + \lambda_{H_2O})$ and the collision–diffusion integral use a mean collision time calculated with $\tau = k_b T/\bar{\epsilon}$, where $k_b = 1.38066 \times 10^{-23}$ J/K is the Boltzmann constant, and $\bar{\epsilon}$ is the characteristic Lenard-Jones potential for the binary system (H_2, H_2O) given by the geometrical average of the Lenard-Jones potentials of the components, $\bar{\epsilon} = (\epsilon_{H_2}\epsilon_{H_2O})^{0.5}$.

In Fig. 4.24, it can be observed that the ohmic over-potentials account for 50–70 % of all over-potentials. This suggests that the hydrogen transport through the nonporous solid oxide electrolyte is not efficient. However, the evolution of hydrogen at the cathode is a relatively efficient process due to the cathodic concentration and activation over-potentials being less than 10 % of the total of over-potentials.

Figures 4.25, 4.26, 4.27, 4.28, 4.29, 4.30, 4.31, and 4.32 exemplify the effects of operating temperature and pressure on SOE-O operation. Overall, there is a decrease in over-potentials with higher temperature and pressure. The main cause

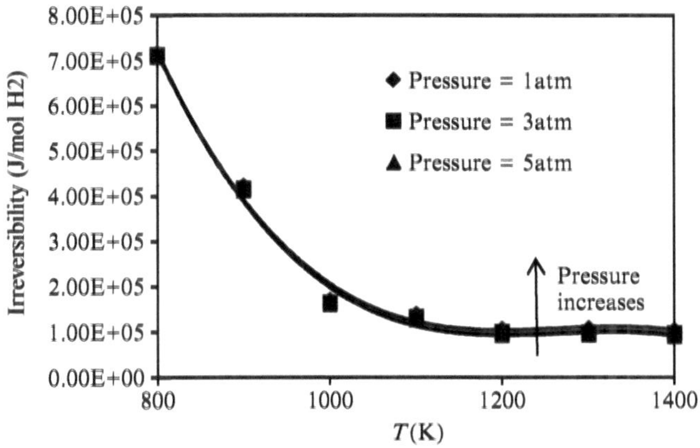

Fig. 4.25 Irreversibility of an SOE-O for various pressures [data from Nieminen et al. (2010)]

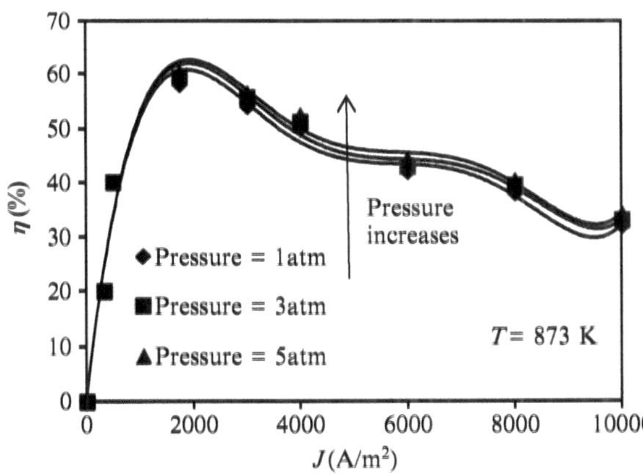

Fig. 4.26 Energy efficiency of SOE-O for various pressures [data from Nieminen et al. (2010)]

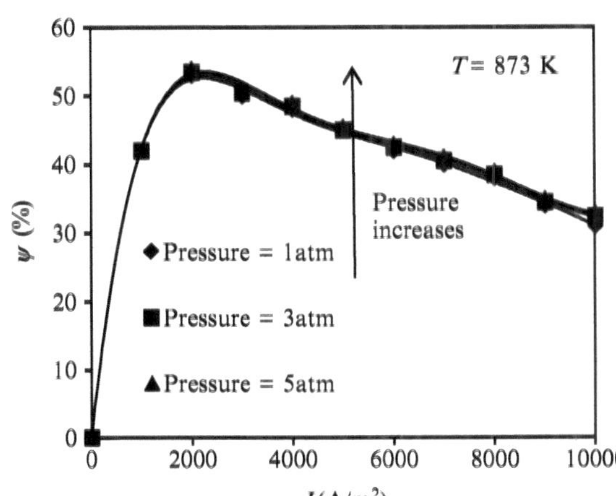

Fig. 4.27 Exergy efficiency of an SOE-O for various pressures [data from Nieminen et al. (2010)]

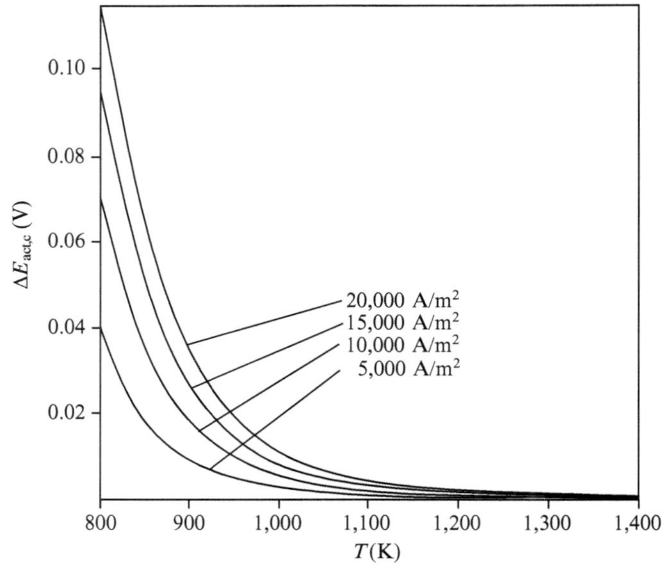

Fig. 4.28 Cathodic activation over-potentials in an SOE-O at various current densities [data from Nieminen et al. (2010)]

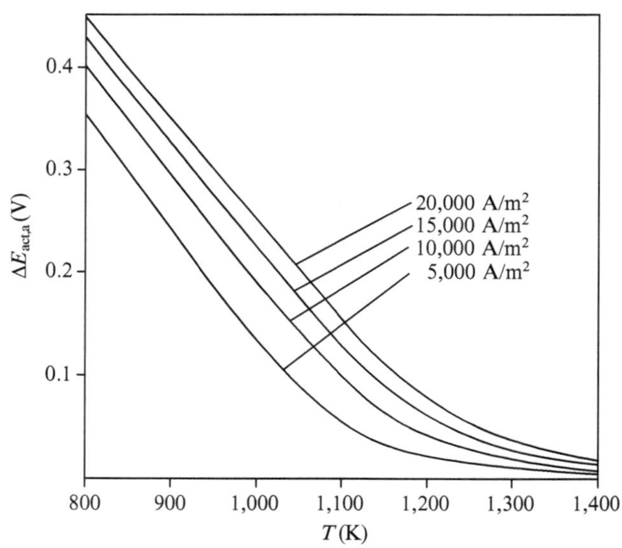

Fig. 4.29 Anodic activation over-potentials in an SOE-O at various current densities [data from Nieminen et al. (2010)]

of this trend was found to be the major decrease in ohmic over-potential due to the increased ionic conductivity of the SOE-O electrolyte at higher temperatures. This suggests that H_2 production via an SOE-O plant is more efficient at higher

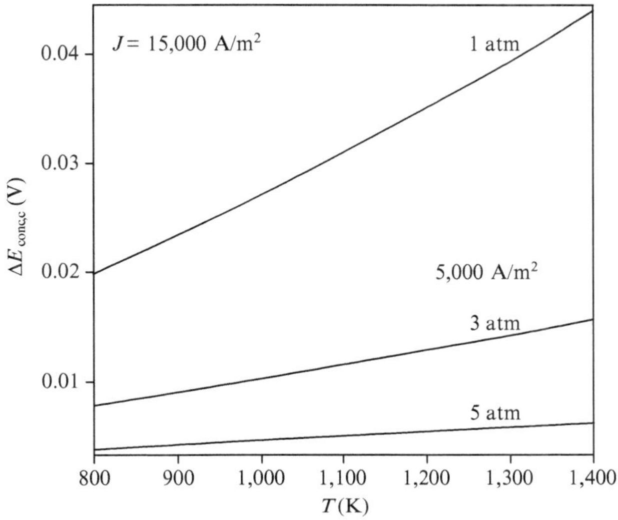

Fig. 4.30 Cathodic concentration over-potentials in an SOE-O at various pressures [data from Nieminen et al. (2010)]

Fig. 4.31 Anodic concentration over-potentials in an SOE-O at various pressures [data from Nieminen et al. (2010)]

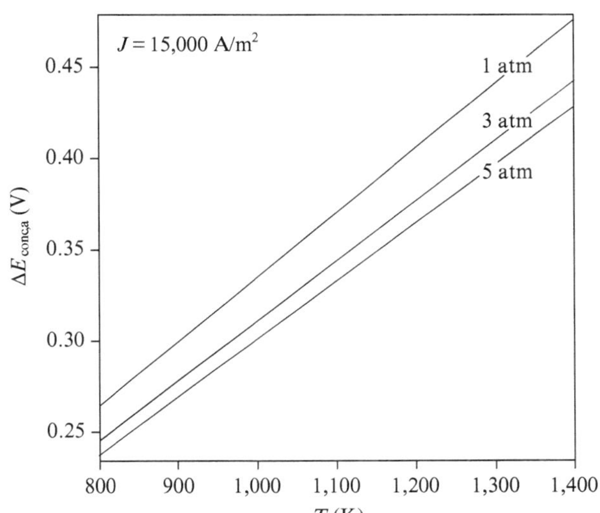

temperatures and pressures. This makes the SOE-O an option that depends on whether or not there is an economical and reliable source of high-temperature heat available to meet the large thermal energy demand.

In SOE-O operation, there is a large thermal energy requirement and therefore different behaviors for energy and exergy efficiencies. In general, the thermal energy input associated with higher temperature operation reduces the energy and

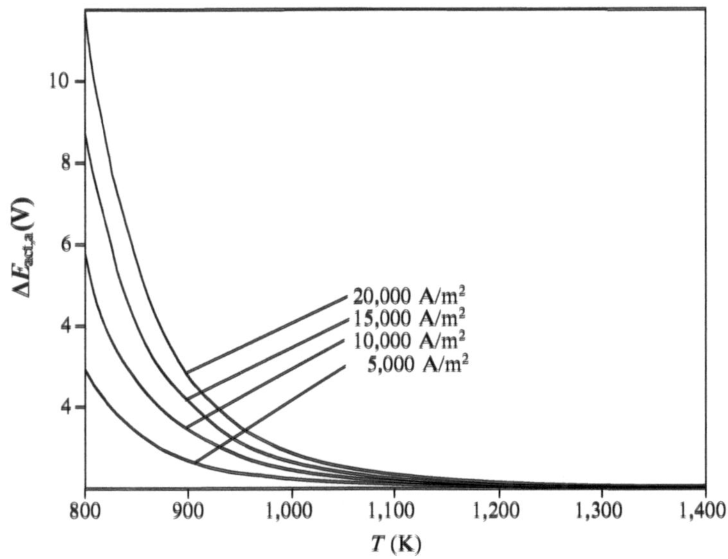

Fig. 4.32 Ohmic over-potentials in an SOE-O at various current densities [data from Nieminen et al. (2010)]

exergy efficiencies. As an order of magnitude, an SOE-O plant operating at ~875 K shows a maximum energy efficiency at approximately 52, 53, and 54 % for 1.0, 3.0, and 5.0 atm, respectively, at a current density of ~2,000 A/m^2. Also, the maximum exergy efficiency is approximately 61 % for 1.0 atm with a 1 – 2% increase if the pressure is raised up to 5.0 atm.

The increase in both energy and exergy efficiency is primarily due to the increase in pressure, which caused a decrease in concentration over-potentials. This further causes a decrease in thermodynamic irreversibility, which results in an increase in both efficiencies.

A simplified layout of a water electrolysis system with SOE-O is presented in Fig. 4.33. The plant includes a water treatment station (not shown), three-flow heat exchanger network, heat exchangers for product cooling, preheating heat exchanger, water pump for water recycling, water separator, and product gas dryers (not shown). The fresh water at 25 °C is mixed with recycled water at 100 °C (which results from condensing steam), and partially boiled in a heat exchanger network. Further water is heated from an external heat source to 800 °C and supplied to the SOE-O unit. Power and high-temperature heat is also supplied to the SOE-O unit. The product gases—at 800 °C temperature—are diverted to the heat exchanger network for heat recovery purposes; there they are cooled to 100 °C and water is fully condensed. Finally, the product hydrogen and oxygen gases are cooled to 25 °C and the heat is rejected to the ambient air.

For 100% hydrogen as equivalent energy flow, the system requires ~100.3% electric power supply and 32.6% input energy in the form of heat and 0.5% in the form of water enthalpy; the total energy input is 133.4%. The energy efficiency

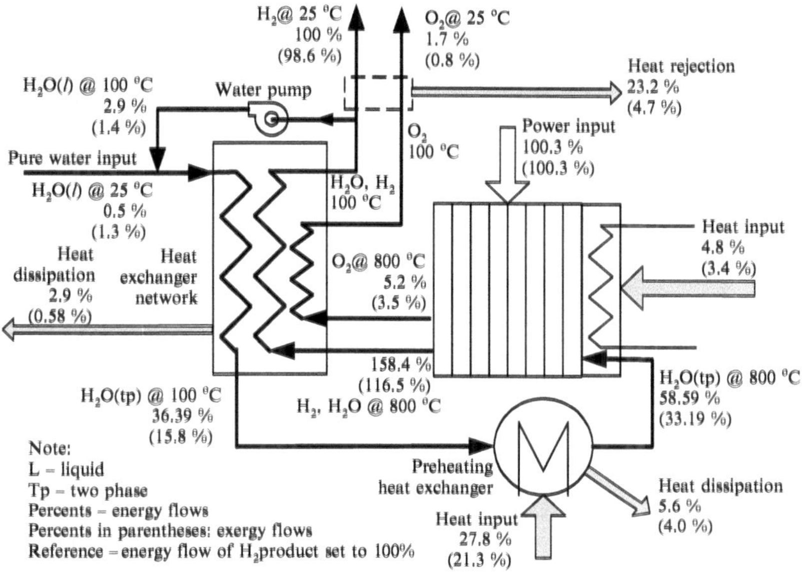

Fig. 4.33 Simplified layout of an SOE-O plant showing the energy and exergy flows [adapted from Meng et al. (2007)]

of the system is $100/133.4 = 75\%$. If one assumes that electricity is generated with an efficiency of 30 %, then the system efficiency becomes 27%. The exergy input in the system is 126.3%. Therefore, the exergy efficiency of hydrogen production is 78%. The exergy of the output streams is 108.7%; it yields exergy destroyed within the system as $126.3 - 108.7 = 17.6\%$.

The data from Fig. 4.33 can be used to determine the energy and exergy portions for the SOE-O system. The energy charts from Fig. 4.34 represent percentages of output flow from the total input flow. In order to obtain these charts, first the total energy and energy inputs in the system are calculated. Then, each output flow is expressed as a percentage from the total input flow of energy and exergy, respectively. In the case of the exergy diagram—Fig. 4.34b—the exergy destroyed (Ex_{d}) for the overall system is indicated. Exergy destroyed results from the exergy balance for the overall system: $(\sum Ex)_{\mathrm{in}} = (\sum Ex)_{\mathrm{out}} + Ex_{\mathrm{d}}$.

4.6 Solid Oxide Electrolyzers with H+ Conduction

Proton-conducting solid oxides can be used as electrolyte in intermediate-temperature electrolysis cells for steam. The electrolyte becomes conductive to ions in a range of temperatures of $\approx 400 - 750\,^{\circ}\mathrm{C}$. Research work on intermediate-temperature electrolyses cells with proton conduction (SOE-P) is in its early stage. The electrochemical half-cell reactions are identical to those of a PEME but they are conducted at

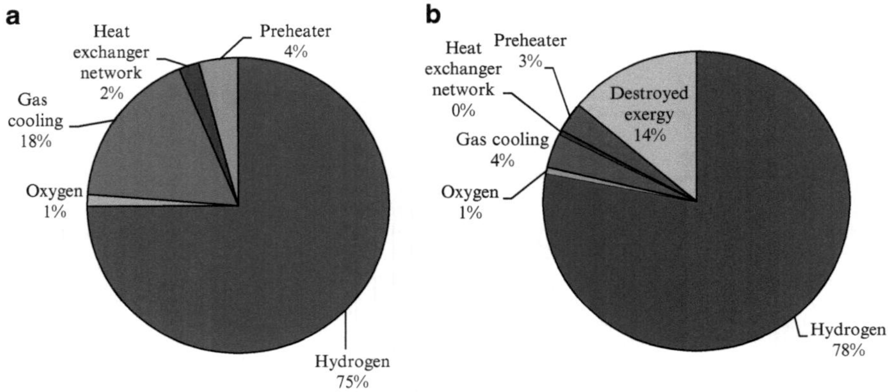

Fig. 4.34 (**a**) Energy and (**b**) exergy portions for the SOE-O system based on the data presented in Fig. 4.33

much higher temperature (see, for example, reaction #5 in Table 4.1). Also, the construction of the SOE-P is very similar to that of PEME, consisting of bipolar plates with channels for the gas flow, porous electrodes, and electrolyte.

Materials based on cerium (Ce) were discovered as good conductors of protons. One of the most cited proton-conducting electrolytes is $BaCeO_3$; the other is $SrCeO_3$. The governing mechanism of proton conduction is based on the movement of protons between adjacent oxygen ions that are bounded to the atomic structure of the solid oxide electrolytic membrane. The protons, which are extremely small, necessitate reduced activation energy as compared to the case of oxygen ion transport through the most advanced solid electrolytes for SOE-O, e.g., electrolytes based on YSZ or doped ceria systems (see Sect. 4.5).

The main challenge with barium cerate results from the difficulty to sinter it in the form of a solid membrane of reduced thickness. One possibility to obtain membranes of 50 μm or smaller is by doping barium cerate with samarium (Sm) and a conductivity up to 0.01 S/cm can be obtained. Sakai et al. (2009) studied strontium cerate-based membranes doped with ytterbium (Yb) which is packed with a layer based on strontium–zirconium–yttrium. One advantage of the proton-conducting electrolyte is that it allows for the absence of steam at the cathode; therefore there is no need for hydrogen to be separated.

The fundamental electrochemical theory of electrolysis presented in Sect. 4.2 can be applied to SOF-P modeling. Reversible potentials are calculated by (4.9) where the equilibrium constant is estimated based on partial pressures of the gases. The activation overvoltage is calculated with (4.20). In proton-conducting solid oxide electrolysis cells, the concentration over-potential at the cathode side is (see Sakai et al. 2009)

$$\Delta E_{\mathrm{conc,c}} = \frac{RT}{2F} \times \ln\left(\frac{P_{\mathrm{H_2}}(J)}{P^0_{\mathrm{H_2}}}\right), \tag{4.48}$$

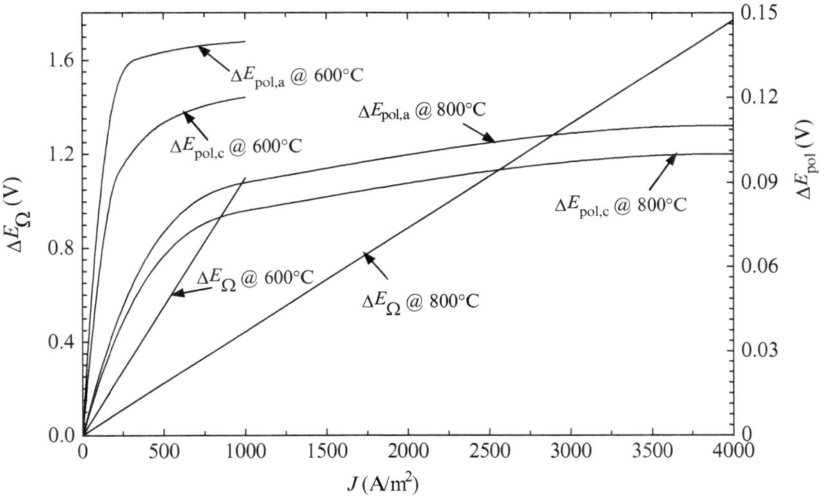

Fig. 4.35 Over-potentials in an SOE-P with SSC-55|SZCY-541|SCYb|Ni configuration [experimental data from Sakai et al. (2009)]

where $P_{H_2}(J)$ is the partial pressure of hydrogen at a current density J, while the exponent 0 indicates the partial pressure of hydrogen if no current flows through the circuit. The pressure $P_{H_2}(J)$ can be measured directly, or (4.26) can be applied provided that the limiting current density is known. The anode side concentration over-potential is (see Sakai et al. 2009)

$$\Delta E_{\mathrm{conc,a}} = -\frac{RT}{2F} \times \ln\left(\frac{P_{H_2O}(J)}{P^0_{H_2O}} \times \left(\frac{P^0_{O_2}}{P_{O_2}(J)}\right)^{0.5}\right). \tag{4.49}$$

In Fig. 4.35, over-potentials of an SOE-P cell are illustrated based on Sakai et al. (2009). The cell has an anode made of strontium–samarium and cobalt trioxide $Sr_{0.5}Sm_{0.5}CoO_3$ characterized by a conductivity of 1,000 S/cm; the anode is denoted with the short notation SSC-55. The anode is in contact with a cerium-based proton-conducting strontium zirconate electrolyte SZCY-541 with a chemical formula $SrZr_{0.5}Ce_{0.4}Y_{0.1}O_{3-\alpha}$, where α denotes the oxygen vacancy in the crystalline structure. The cell has a double-layer cathode SCYb|Ni formed by a strontium–cerate–ytterbium (SCYB) interlayer with a chemical formula $SrCe_{0.95}Yb_{0.05}O_{3-a}$ which is in contact with the electrolyte and a Ni substrate deposited on the SCYB.

The over-potentials were measured at 600 and 800 °C operating cell temperatures. It is observed in Fig. 4.35 that polarization over-potentials are negligible with respect to the ohmic values. Furthermore, at 800 °C, over-potentials are significantly lower than at 600 °C. Based on concentration over-potentials that result from Fig. 4.35, one can calculate backwards the output concentrations with (4.48) and (4.49). The partial

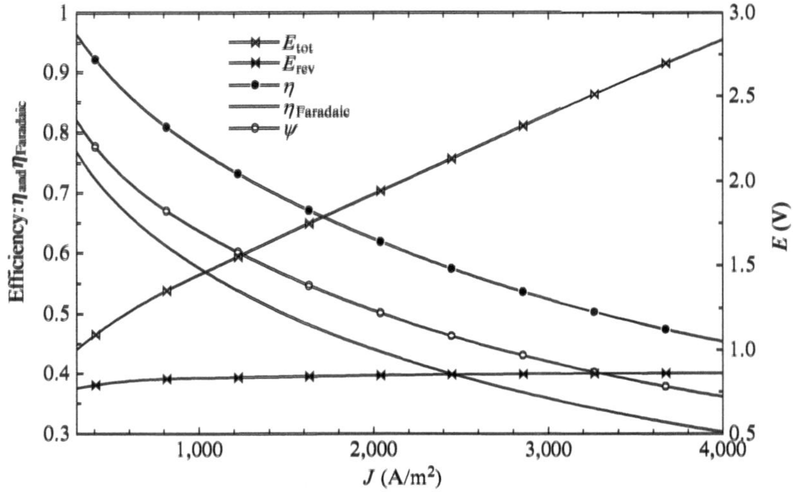

Fig. 4.36 Efficiency and cell potentials for the SSC-55|SZCY-541|SCYb|Ni at 800 °C

pressure of water, oxygen, and hydrogen must be known for the condition when no current is applied to the cell. Sakai et al. (2009) used argon as sweep gas and set the idle concentrations to ~50 % for water, 0.05 % for oxygen, and 1 % for hydrogen.

With concentrations determined for any current density, the reversible cell potential can be calculated with (4.9); further, based on the total cell potential, (4.29), one can determine the efficiencies η_{Faradaic}, η_{el}, and ψ_{el} according to (4.32)–(4.34). In the equations for η_{el} and ψ_{el}, it must be taken into account that the system requires heat input in order to operate. The required heat input is given by $Q_{\text{req}} = \Delta H - \Delta G$. A part of this thermal energy is supplied by the Joule effect within the electrolyte, i.e., $Q_{\Omega} = \Delta E_{\Omega} \times J \times A$, where A is the area of the electrode. The heat input provided to the electrolysis cell is therefore $Q_{\text{inp}} = Q_{\text{req}} - Q_{\Omega}$. Based on these considerations, the results from Fig. 4.35 were further expanded as indicated in Fig. 4.36 where it is shown that the cell efficiency depends on the current density. The same figure also presents the dependence of total and reversible potential with current density. The Faradaic efficiency drops from ~75 % at low current density to 30 % at 4,000 A/m^2. The energy efficiency of the cell varies in the range of 45–95 %, and the exergy efficiency varies in the range of 35–80 %.

4.7 Hybrid Electrolysis Systems

Several systems exist that perform water electrolysis in a hybrid mode, by using electricity and other forms of energy (see Fig. 4.1). The most common example of this kind of system is the photo-electrochemical cell. In this section, some relevant hybrid electrolysis systems are presented.

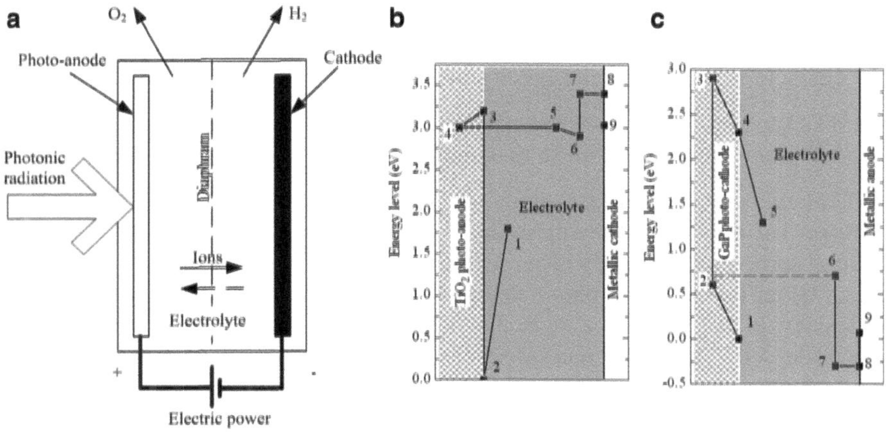

Fig. 4.37 Schematic of photo-electrochemical cell and its operation: (a) general cell layout, (b) cell with TiO_2 photo-anode and external 0.5 V source bias, (c) GaS photo-cathode and 1 V source bias

A photo-electrochemical cell is a device that uses heterogeneous photocatalysis to drive water electrolysis and to generate hydrogen from water. The system requires a source of photons and a source of electrical power (direct current). The source of photons can be derived from solar radiation, but other sources of photons may exist, e.g., Cherenkov radiation (blue light) in spent fuel pools of nuclear reactors. A prototypical photo-electrolysis cell is illustrated schematically in Fig. 4.37a. The electrodes are made commonly from various kinds of semiconductors, possibly doped or coated with other compounds that have a role of photo-catalysts or photosensitizers. The cell comprises a photo-electrode immersed in an electrolyte that is exposed to light through a transparent window. In addition, there is connected a power supply to the electrodes which generates an electric bias.

The system from Fig. 4.37b uses titanium dioxide as a photo-anode. The process 1–2 represents water oxidation, which is followed by photon absorption (3.2 eV energy gap between valence and conduction band of TiO_2), a fall of energy level corresponding to the difference between the semiconductor surface and the bulk (process 3–4), electron transport to the cathode (4–5), energy fall on the Fermi level (5–6), voltage bias generated by the external power source (6–7), and ohmic losses at the metallic cathode. The energy efficiency of system is

$$\eta_{el} = \frac{\dot{n} \times HHV}{j_{ph} \times \sum A}, \tag{4.50}$$

where $\sum A$ represents the total area exposed to incident photonic radiation and j_{ph} is the intensity of the photonic current (in W/m^2).

The system in Fig. 4.37c is analogous to (b) except that it uses a photo-cathode instead of a photo-anode. This system uses a GaP semiconductor of p-type to evolve hydrogen at its surface, while oxygen evolves at a metallic anode. The system requires at least 1 V bias generated by an external power source. Due to the characteristics of the solar spectrum:

- $SrTiO_3$ photo-anode absorbs a maximum 2.62 % of incident solar light energy.
- GaP photo-cathode absorbs a maximum 21.9 % of incident light energy.

Consequently, the surface of the photo-anode must be 92 % from the total incident surface, while the surface of the photo-cathode is only 8 %. The total light absorbed by the cell will represent 7.4 % from incident radiation, which is much better than a single photo-electrode cell ($SrTiO_3$ with 2.62 %). The energy efficiency of the coupled cell is 2.8 times higher than that of a single photo-electrode cell. There are many possible combinations of photo-electrodes for a coupled photo-electrochemical cell.

Various types of radiation are derived from radioactive material that may be used in conjunction with water electrolysis. For example, strontium-90 generates beta radiation consisting of accelerated electrons. Electrons injected in water will thermalize and in some conditions induce redox reactions. It was found that gamma and alpha particles have the property to ionize water. With catalysts such as BeO, SiO_2, TiO_2, and ZrO_2, water radiolysis may be induced under exposure to gamma radiation (see Cecal et al. 2001). If one denotes the radio-catalyst with X, the following types of reactions are believed to occur:

$$\begin{cases} X \xrightarrow{\gamma} \mathrm{X}^*, \text{ activated catalyst} \\ \mathrm{X}^* + \mathrm{H_2O} \rightarrow [\mathrm{X} \cdot \mathrm{H_2O}], \text{ activated complex} \\ [\mathrm{X} \cdot \mathrm{H_2O}] \rightarrow \mathrm{H}^* + \mathrm{OH}^* + \mathrm{X} + hv, \text{ excited species} \end{cases} \tag{4.51}$$

It is known that any excited species becomes very reactive (viz H^* and OH^* in above equation). If an electric field is present in the vicinity of a reaction site, a redox reaction may occur with lower activation energy. This may be a pathway to devise high-performance gamma radiation-assisted electrolysis systems for water splitting. Nuclear power generation sites are an excellent location for development of hybrid radiation-assisted electrolysis. Electricity is available as well as facilities to handle radioactive materials and their radiation.

Magnetic fields are known to influence electric charges through the Lorentz force which increases the convection mechanisms in the electrolyte solution through magneto-hydrodynamic processes. If properly applied, the magnetic field may reduce the concentration potentials substantially as well as reduce the ohmic losses of an electrolyte due to a better transport of charges—see Lin et al. (2011). Three classes of electrodes were analyzed by Lin et al. (2011) to investigate the influence of a magnetic field in electrolysis processes: ferromagnetic (e.g., nickel), paramagnetic (e.g., platinum), and diamagnetic (e.g., graphite). The Lorentz force is given by

$$\vec{\mathbb{F}} = q\vec{\mathbb{E}} + q\vec{\mathbb{V}} \times \vec{\mathbb{B}}, \tag{4.52}$$

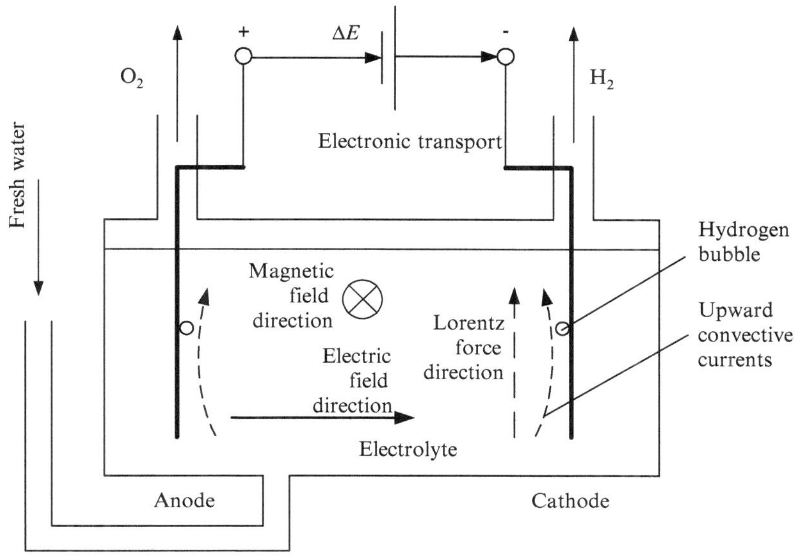

Fig. 4.38 Effect of magnetic field on gas bubble convection in an alkaline electrolyzer

where q is the electrical charge of mobile ions, $\mathbb{E}$ is the intensity vector of the electric field (in N/C), $\vec{v}$ is velocity vector of electric charges (in m/s), and $\mathbb{B}$ is the vector magnetic flux density (in T, Tesla).

In a magnetically assisted electrolysis cell, the magnetic field is perpendicular to the electric field. Figure 4.38 illustrates the effect of a magnetic field on convective currents in an alkaline electrolyzer. The magnetic field is applied by using permanent magnets placed on two opposite sides of the electrolysis bath such that an upward Lorentz force is created. The effect of the Lorentz force is a net convective current with upward direction at the electrode surface which facilitates the removal of gas bubbles. Consequently, the concentration over-potentials decreases and ohmic resistances due to bubble presence also decrease. It was found by Lin et al. (2011) that with platinum electrodes, the current density increased by 1,222 A/m^2 for the same cell voltage in the presence of the magnetic field. If nickel electrodes are used instead of platinum, the increase is more significant, namely, 2,467 A/m^2 in similar conditions. The best electrodes with respect to magnetic field-assisted alkaline electrolysis appear to be those based on nickel due to its good electrochemical activity together with ferromagnetic behavior.

An interesting hybrid technique to conduct water electrolysis is to couple the heterogeneous electro-catalysis process occurring at the electrode surface with a homogeneous catalytic process that takes place in a solution volume. This technique is applicable to systems that use a liquid electrolyte. A net enhancement of hydrogen production is obtained due to the action of homogeneous catalysts that work in parallel with the heterogeneous process (see Fig. 4.2 for general classification). This type of hybrid electrolysis process was applied mostly for organic compounds rather than water. An example of a process is formaldehyde production in solution,

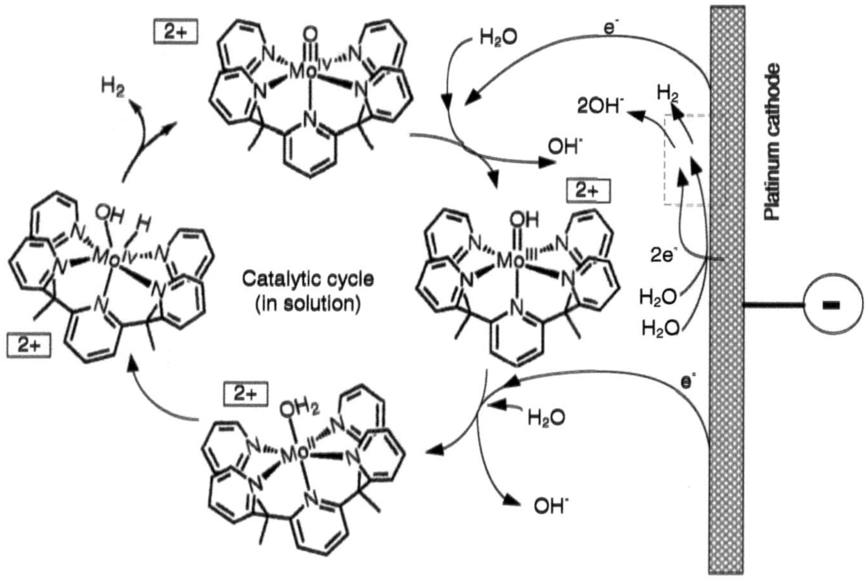

Fig. 4.39 Hybrid catalysis (heterogeneous/homogeneous) for hydrogen-evolving reaction [modified from Karunadasa et al. (2010)]

$H_2C(OH)_2 \leftrightarrows H_2CO$ in solution, together with $CH_3OH + ne^- \rightarrow H_2CO$ at the electrode surface. A generic scheme of this reaction is (see Bret and Bret 1993)

$$\begin{cases} A + ne^- \rightarrow B, \text{ at the electrode (heterogeneous electrocatalysis)} \\ C \leftrightarrows B, \text{ in the solution(homogeneus catalysis),} \end{cases} \tag{4.53}$$

where A, B, C are chemical species.

Recently, Karunadasa et al. (2010) developed a molybdenum-based oxo-catalyst for a water reduction half-reaction at a platinum electrode. The heterogeneous and homogeneous reaction mechanisms compete. The general reaction scheme can be expressed as

$$\begin{cases} 2H_2O + 2e^- \rightarrow H_2(g) + 2OH^-(aq), \text{ at the electrode (heterogeneous catalysis)} \\ \left.\begin{array}{l} CAT + 2e^- \rightarrow CAT^{2-}, \text{ at electrode surface} \\ 2H_2O + CAT^{2-} \rightarrow H_2(g) + 2OH^-(aq), \text{ in solution} \end{array}\right\} \text{(homogeneus catalysis cycle)} \end{cases} \tag{4.54}$$

The reaction mechanism from Karunadasa et al. (2010) is illustrated in Fig. 4.39. According to (4.54), it consists of a main catalysis cycle which requires electron transfer from the electrode to the macromolecular catalyst, which is dissolved in solution. The turnover frequency obtained with this catalyst is very high and

Table 4.4 Some operational parameters of CANDU ACR-1000 power plant

Parameter	Values
High-pressure steam	278 °C, 6.3 MPa, 2,777 kJ/kg, $x = 70\,\%$, 1,664 kg/s
Generated energy	3,070 MW thermal, 1,087 MW electrical, $\eta = 35\,\%$
Moisture separator/reheater	Steam entering: 186.4 °C, 2,522.6 kJ/kg, 1,166 kg/s
	Condensate entering: 186.4 °C, 788.1 kJ/kg, 169.7 kg/s
	First reheater: 240.6 °C, 2,681.1 kJ/kg, 82.53 kg/s
	Second reheater: 278.8 °C, 2,777 kJ/kg, 62.79 kg/s

x vapor quality
Source: Ryland et al. (2007)

reaches a limiting value of 1,700 mol of hydrogen per mole of catalyst per hour. Moreover, this is considered as a non-expensive catalyst as it does not comprise noble metals.

4.8 Nuclear-Electrolytic Hydrogen Production Systems from Water

Water electrolysis technology is relatively mature, although new advances in the field aim to increase the efficiency, relax the stringent requirements of water quality, increase the durability of materials, and increase the operating temperature and pressure of the systems. Nuclear energy is an ideal supply of electricity, heat, and radiation which can be used to split water and generate hydrogen. This section discusses some possible schemes of nuclear hydrogen production via water electrolysis. The information is based on past studies in the open literature.

Ryland et al. (2007) analyzed electrolytic hydrogen production with CANDU reactors. It was shown that an advanced CANDU reactor ACR-1000 coupled with an electrolysis facility can reach 33–34 % as compared to 27 % for conventional electrolysis (these figures are estimated considering the power generation efficiency of conventional nuclear reactors). The parameters of an ACR-1000 nuclear power plant are listed in Table 4.4.

The first option to consider is to use the electrical output of CANDU reactor to drive an electrolysis plant made with multiple conventional alkaline electrolyzer units. The block diagram of a CANDU ACR-1000 nuclear power plant for hydrogen production is shown in Fig. 4.40. It will generate hydrogen continuously at a scale of production comparable to steam methane reforming in industrial chemical plants with an overall thermal-to-hydrogen efficiency of 27 %. The electrolyzer can operate under pressure at 30 atm; further compression of gases for storage limits the compression factor to reasonable values of 10–20.

The second option considers alkaline electrolyzers operating at 200 °C assisted by thermal energy from the nuclear power plant. At this temperature, the reversible cell potential is 88 % from the value at standard conditions. The system block diagram and performance parameters in this case are presented in Fig. 4.41.

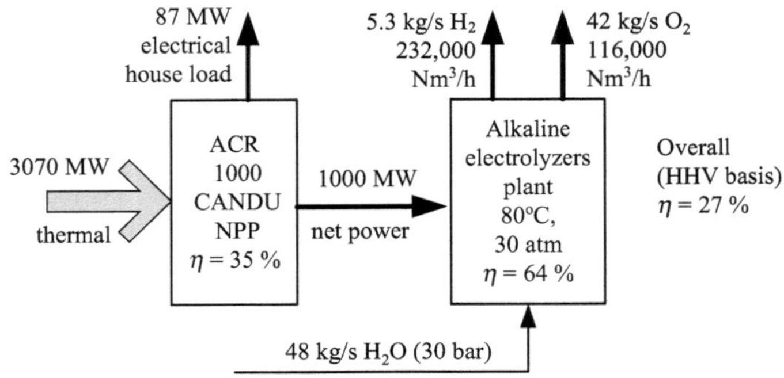

Fig. 4.40 CANDU ACR-1000 nuclear power plant (NPP) for hydrogen production with conventional alkaline electrolyzers [data from Ryland et al. (2007)]

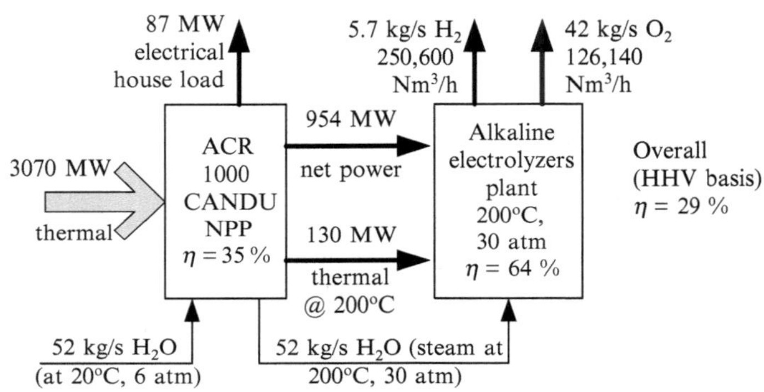

Fig. 4.41 CANDU ACR-1000 nuclear power plant (NPP) for hydrogen production with high-temperature alkaline electrolyzers

An amount of ~45–50 kg/s intermediate-pressure steam at 200 °C is extracted from the plant (while the feed pumps will supplement the loop with fresh water) to be used for electrolysis. Some modifications are necessary for the balance of plant such that this steam could be transferred to the electrolysis plant and, furthermore, at least 130 MW of thermal energy at 200 °C is also transferred. In these conditions, the power plant will generate only 954 MW net power which is fully used for the electrolyzer. The overall hydrogen production efficiency increases to 29 % while the production rate increases by ~ 7.5 %.

The third option proposes to use an SOE-O plant operated in a thermo-neutral regime. For an operating temperature of 827 °C of the cell and assuming at cell inlet 50 % steam concentration which is diverted from the steam generator of the nuclear power plant coming at 278 °C and fed to the electrolyzer at 50 atm (see Ryland et al. 2007), the thermo-neutral potential is 1.4 V. This assumes that there is 75 % hydrogen in the outlet stream and 50 % oxygen. Figure 4.42 depicts the SOE-O-based system. Its overall hydrogen production efficiency reaches 34 %. The medium temperature level heat at 278 °C is extracted from steam at constant temperature (steam is condensed at high pressure); the return high-pressure water

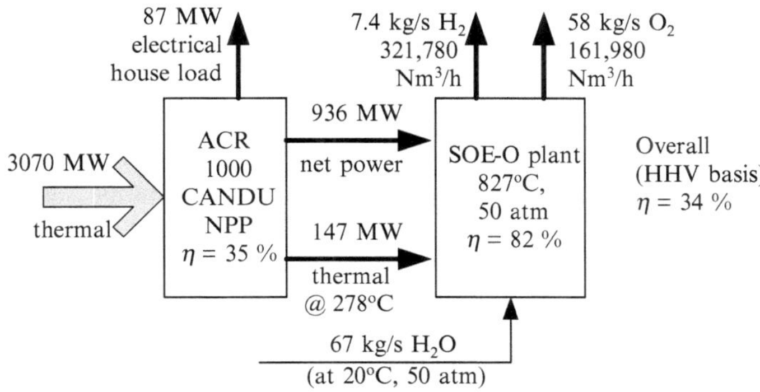

Fig. 4.42 CANDU ACR-1000 nuclear power plant (NPP) for hydrogen production with SOE-O plant [data from Ryland et al. (2007)]

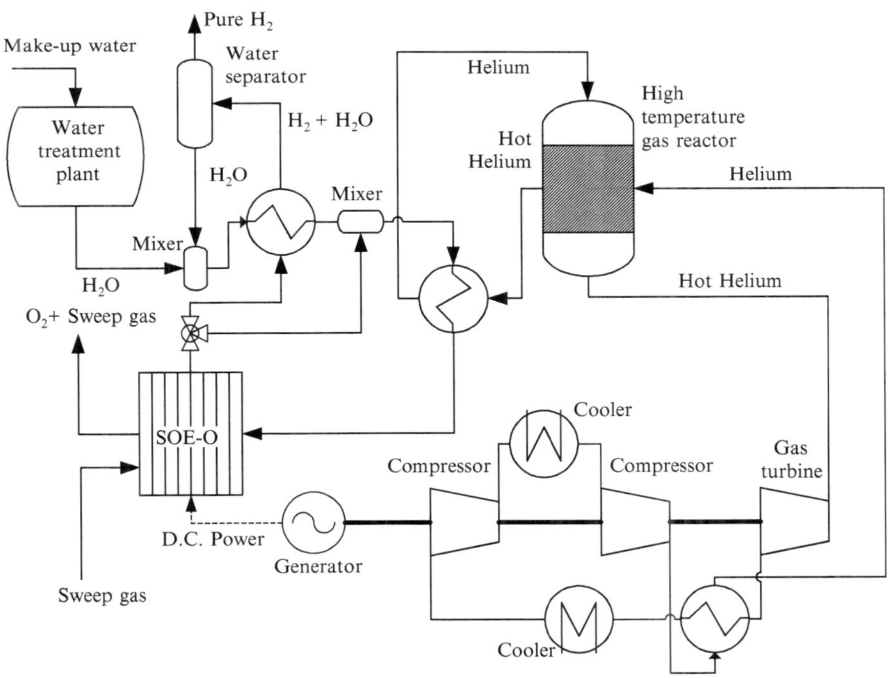

Fig. 4.43 General layout of a high-temperature gas-cooled reactor coupled with SOE-O

is reinjected into the Rankine plant and recycled. The heat is needed to preheat water prior to its feed into the SOE-O plant.

High-temperature reactors have the ability to generate heat at a temperature level which is compatible with SOE-O operation (about 1,000 K). Figure 4.43 shows the general layout of a high-temperature gas-cooled reactor dedicated to hydrogen production. There are two helium-heating loops. The primary loop flows through a Brayton cycle operating with helium as a working fluid. It comprises a turbine, reheaters, and two-stage compression station with intercoolers. The second heating loop runs through an intermediate heat exchanger to transfer heat to an SOE-O-based plant. This heats the reactants consisting of steam and hydrogen which are fed to the SOE-O cathode. A sweep gas is used (air or steam) to enhance the reaction rate. The resulting product gas comprising hydrogen and steam passes through a cooling system with water condensation, separation, and recycling. According to Ryland et al. (2007), a system such as in Fig. 4.43 can generate about 85,000 Nm^3/h hydrogen with a 600 MW thermal output reactor.

Although an alkaline electrolyzer and solid oxide fuel cell with oxygen ion conduction appear to be the most preferred water electrolysis technologies in conjunction with nuclear reactors, other systems such as PEME or intermediate-temperature solid oxide electrolyzers with proton conduction can find their niche application in nuclear hydrogen production. For example, SOE-P appears to be a suitable option in conjunction with a generation IV supercritical water reactor that operates in the range of temperatures of 450–650 °C. Regarding PEME, due to their specific construction, they are adapted for lower production capacities; therefore, most likely they can be applied as distributed systems, connected to the electrical grid. Such systems are a promising solution for grid regulation and load levelling (see Chap. 2).

4.9 Conclusions

In this chapter, water electrolysis systems for nuclear hydrogen production were discussed. Fundamental concepts of electrochemical process analysis are essential in understanding the principles of water electrolysis and the essential parameters for design and technology assessment. In the first part of the chapter, the fundamentals of water electrolysis were reviewed. Focus is given to defining the principal concepts such as reversible potentials, standard potentials, reaction rates, etc. Analysis, modeling, and design of water electrolysis systems must consider at least three aspects—thermodynamics, kinetics, and transport phenomena. Modeling of the overall system by an electric circuit analogy is also introduced as a useful technique for analysis and design.

Most established industrial systems for electrolytic hydrogen production are alkaline electrolyzers which can be found commercially up to capacities of a few hundred MW equivalent hydrogen production based on lower heating values. Advanced alkaline electrolyzers can operate at high pressure and temperature up to 200 °C. At higher temperatures, water electrolysis can be performed in the gaseous phase. Electrochemical steam decomposition is performed using either oxygen-ion or proton-conducting electrolytes. Solid oxide electrolyzers with oxygen-ion conduction appear to be the most established as well as alkaline electrolysis technology. PEME are relatively well developed and adapted to operate at low temperatures and lower production capacities. Some special hybrid systems which relate at least two sources of energy (from which one is electrical) are also discussed. A typical example of hybrid electrolysis systems is the photo-electrochemical cell which uses photonic radiation to assist electrolysis. Other systems were identified, such as nuclear radiation-assisted electrolysis or magnetic field-assisted electrolysis. Hybrid electrolysis can be implemented also based on two parallel reaction techniques: heterogeneous electro-catalysis at an electrode surface and homogeneous catalysis in liquid solutions.

Although the hybrid method presents interest for niche applications, for nuclear hydrogen production, alkaline electrolyzers and solid oxide electrolyzers are the most promising technologies. Several examples of nuclear-integrated electrolysis systems are given at the end of the chapter. With present technology, high-temperature alkaline electrolyzers can be adopted in the short term for nuclear hydrogen production at scales comparable to natural gas reforming and an overall efficiency of nearly 30 %. Generation IV reactors are suitable for high-temperature electrolysis with SOE-O or SOE-P with a projected efficiency of 34 %.

Nomenclature

a	Thermodynamic activity, mol/dm^3
a	Tafel polarization potential, V
A	Area, m^2
A	Pre-exponential factor, s^{-1}
b	Tafel slope, V
$\vec{\mathbb{B}}$	Magnetic flux density, T (Tesla)
B_g	Flow permeability
c	Molar concentration, mol/dm^3
CF	Quality factor
D	Diffusion coefficient, m^2/s
e	Elementary electrical charge, C
ex	Molar exergy kJ/mol
Ex	Exergy, kJ
E	Electric potential, V
f	Activity coefficient
F	Faraday constant, C/mol
$\vec{\mathbb{F}}$	Lorentz force, N
G	Gibbs free energy, eV
G	Molar Gibbs free energy, kJ/mol
H	Enthalpy, eV
H	Molar enthalpy, kJ/mol
HHV	Higher heating value, kJ/mol
J	Intensity of photonic current, W/m^2
k	Reaction rate constant, s^{-1}
k_b	Boltzmann constant, J/K
K_{eq}	Equilibrium constant, s^{-1}
M	Molecular mass, kg/kmol
n	Number of moles, mol

$\dot{n}$	Molar flow rate, mol/s
N_A	Avogadro's number, molecules/kmol
I	Intensity of electric current, A
J	Current density, A/m^2
P	Pressure, Pa
q	Electrical charge, C
r	Mean electrode pore radius, m
R	Universal gas constant, kJ/mol K
R	Electrical resistance, Ω
S	Entropy, eV/K
S	Molar entropy, J/mol K
t	Time, s
T	Temperature, K
v	Velocity, m/s
$\dot{W}$	Work rate, W
x	Length coordinate, m
x	Vapor quality
z	Valence of ionic species

Greek letters

α	Transfer coefficient
β	Factor defined by (4.41)
γ	Surface tension, N/m
δ	Thickness, m
ϵ	Lennard-Jones potential, meV
ε	Electrode porosity
η	Energy efficiency
λ	Characteristic length, A
μ	Chemical potential, kJ/mol
μ	Dynamic viscosity, Ns/m^2
ψ	Exergy efficiency
σ	Electric conductivity, S/cm
τ	Collision time, s
ξ	Electrode tortuosity
Ω_D	Collision–diffusion integral

Subscripts

0	Reference (standard) state
0	Exchange current density
a	Anode
act	Activation

b	Backward reaction
c	Cathode
conc	Concentration
d	Destroyed
el	Electrolyzer
eq	Equilibrium or equivalent
f	Forward reaction
Faradaic	Faradaic efficiency
i	General index
in	Input
k	Knudsen
lim	Limiting
loss	Dissipated
O	Oxidized species
out	Output
P	Products
PEM	Proton exchange membrane
pg	Power generation
ph	Photonic
pol	Polarization
R	Reactants or reduced species
rev	Reversible
sat	Saturation
th	Thermo-neutral
tot	Total
w	Water
Ω	Ohmic

Superscripts

$\overrightarrow{()}$	Vectorial quantity
0	Reference (standard) state or reversible reaction
ch	Chemical
eff	Effective
in	Inlet
Q	Heat

References

Bagotski VS (2006) Fundamentals of electrochemistry, 2nd edn. Hoboken, New Jersey, Wiley Interscience

Bocris JO'M, Reddy AKN, Gamboa-Aldeco M (2000). Modern electrochemistry, vol. 2A Fundamentals of electrodics, 2nd edn. Kluwer Academic, New York

Bret CMA, Bret AMO (1993) Electrochemistry—principles, methods, applications. Oxford University Press, New York

Burstein GT (2005) A hundred years of Tafel's equation: 1905–2005. Corrosion Sci 46:2858–2870

Cecal A, Goanta M, Palamaru M, Stoicescu T, Popa K, Paraschivescu A, Anita V (2001) Use of some oxides in radiolytical decomposition of water. Radiat Phys Chem 62:333–336

Guerrini E, Trasatti S (2009) Electrocatalysts in water electrolysis. In: Bardaro P, Bianchini C (eds) Catalysis for sustainable energy production. Wiley-VCH, Weinheim

Hale AJ (1919) The manufacture of chemicals by electrolysis. Constable & Co, London

Karunadasa HI, Chang CJ, Long JR (2010) A molecular molybdenum oxo-catalyst for generating hydrogen from water. Nature 464:1329–1333

Klein SA (2011) Engineering Equation Solver. http://www.fchart.com/assets/downloads/ees_manual.pdf accessed in March 2012

Lin M-Y, Hourng L-W, Kuo C-W (2011) The effect of magnetic force on hydrogen production efficiency in water electrolysis. Int J Hydrogen Energy 37:1311–1320

Meng N, Leung MK, Leung DY (2007) Energy and exergy analysis of hydrogen production by solid oxide steam electrolyser plant. Int J Hydrogen Energy 32:4648–4660

Meng N, Leung MK, Leung DY (2008) Energy and exergyanalysis of hydrogen production by a PEM electrolyser plant. J Energy Convers Manag 49:2748–2756

Nieminen J, Dincer I, Naterer G (2010) Comparative performance analysis of PEM and solid oxide steam electrolysers. Int J Hydrogen Energy 35:10842–10850

Rubinstein I (1995) Fundamentals of physical electrochemistry. In: Rubinstein I (ed) Physical electrochemistry—principles, methods and applications. Marcel Dekker, New York

Ryland DK, Li H, Sadhankar RR (2007) Electrolytic hydrogen generation using CANDUnuclear reactors. Int J Energy Res 31:1142–1155

Sakai T, Matsushita S, Matsumoto H, Okada S, Hashimoto S, Ishihara T (2009) Intermediate temperature steam electrolysis using strontium zirconate-based protonic conductors. Int J Hydrogen Energy 34:56–63

Tilak BV, Lu PWT, Colman JE, Srinivasan S (1981) Electrolytic production of hydrogen. In: Rubinstein I, Bocris JO'M, Conway BE, Yeager E, White RE (eds) Comprehensive treatise of electrochemistry, vol. 2, Electrochemical Processing. Plenum, New York

Zeng K, Zhang D (2010) Recent progress in alkaline water electrolysis for hydrogen production and applications. Prog Energy Combustion Sci 36:307–326.s

Chapter 5
Thermochemical Water-Splitting Cycles

Abstract This chapter presents and analyzes thermochemical cycles, which are promising methods of nuclear produced hydrogen at a large scale. The introduction presents the origins of concepts and a historical perspective on the technology development. In the first part, the most important aspects and fundamental concepts for cycle modeling and synthesis are introduced and detailed. The discussion proceeds from single-step thermochemical water-splitting processes, to two-step and multi-step processes, followed by a presentation of hybrid cycles. Relevant analysis methods are introduced in the context of each type of cycle presentation. These concepts include chemical equilibrium, chemical kinetics, reaction rate and yield, and others. Analysis of the practicality of chemical reactions is established based on their yield. A large number of reactions and thermochemical cycles are compiled, categorized, and discussed. In total, the chapter presents 122 thermochemical cycles, 25 hybrid cycles, and six special cycles (assisted with photonic or nuclear radiation).

The most important reactions, encountered in pure and hybrid cycles, are analyzed in detail. For example, both the Deacon reaction and H_2SO_4 decomposition methods are the most encountered oxygen-evolving reactions. Hydrogen iodide decomposition has a major role as a hydrogen-evolving reaction. The Bunsen reaction is also significant. In thermochemical cycle synthesis and assessment, it is important to account for the energy associated with chemical separation, chemical recycling, and material transport; this is explained and exemplified in the chapter. Another discussion involves cycle synthesis and a down selection process, a methodology that systematically leads to identification of the most promising cycles. A comparative assessment of cycles is presented and the use of exergy as a potential analysis tool is introduced. The final part of the chapter refers to three thermochemical plants which are considered as the most promising. These are plants based on the sulfur–iodine cycle, the hybrid sulfur cycle, and the hybrid copper–chlorine cycle. Some bench-scale or pilot plants exist for the sulfur–iodine and hybrid sulfur plants and they are in the course of development for the copper–chlorine cycle.

G.F. Naterer et al., *Hydrogen Production from Nuclear Energy*,
DOI 10.1007/978-1-4471-4938-5_5,

5.1 Introduction

Thermochemical water-splitting cycles represent technological processes that decompose the water molecule while separate streams of hydrogen and oxygen gases are released via a closed sequence of chemical reactions. Besides the chemical elements constituting the water molecule (H and O), the chemical compounds in multi-step thermochemical water-splitting cycles comprise other elements (e.g., the copper–chlorine thermochemical cycle involves compounds of Cu and Cl; the sulfur–iodine thermochemical cycle involves chemical compounds of S and I). In a thermochemical cycle (1) only water is consumed; (2) the only products generated are hydrogen and oxygen as separated streams; and (3) all other chemicals involved in particular reaction steps are completely recycled.

Thermochemical plants are supplied only with heat and water to operate. Their overall process is $H_2O + \text{heat} \rightarrow [\text{cyclic chemical reactions}] \rightarrow H_2 + O_2$. In addition to pure thermochemical cycles, there are hybrid thermo-electro-chemical cycles (or shortly named "hybrid cycles") which use, besides heat, a relatively small amount of electricity to drive some electrochemical reactions. A hybrid cycle process is described by the following equation: $H_2O + \text{heat} + \text{electricity} \rightarrow [\text{cyclic chemical and electrochemcical reactions}] \rightarrow H_2 + O_2$.

Thermochemical cycles are a promising choice for large-scale hydrogen production which can be achieved at nuclear reactor facilities (i.e., $\gtrsim$ 100,000 m^3/h or $\gtrsim$ 10 t/h or $\gtrsim$ 325 MW on the HHV basis of hydrogen). Although large alkaline electrolyzer plants formed of multiple commercial units can be installed to generate hydrogen in the capacity range of hundreds of thousands of standard cubic meters per hour, hybrid and thermochemical cycles maintain the advantage of a reduced requirement of electricity. In pure thermochemical cycles, there are requirements of electricity for pumping, compressing, transport of chemicals, and other operations such as stirring, separation of chemicals, or other loads (in power plants, the electrical load—i.e., auxiliary power needed to maintain the plant operation, including the office building etc.—represents about 2–3 % of the generated power; around the same percentage is expected to be valid for large-scale industrial plants, viz. thermochemical cycles).

Regarding the electricity requirement in hybrid cycles, it is much lower than what is normally required for any water electrolysis technology, for the same hydrogen output. The reduced electricity requirement makes thermochemical and hybrid cycles dependent mainly on high-temperature thermal energy sources. In contrast, electrolyzers linked to nuclear reactors require two energy conversion steps, namely, nuclear-derived thermal energy $\rightarrow$ electrical energy $\rightarrow$ chemical energy (i.e., hydrogen). At a large production scale, the overall efficiency becomes closely dependent on any additional energy conversion step. On this basis, a direct process such as nuclear-derived thermal energy to hydrogen-based chemical energy may be more cost competitive even if it is assisted with a relatively small amount of electrical power input for its sub-processes.

A brief review of thermochemical cycles was presented in Chap. 2 of this book. This chapter provides further details on the origins of thermochemical water-splitting cycles and the subsequent technological developments. According to Funk (2001), the first worldwide systematic funding on nuclear hydrogen energy was initiated in 1963 in the framework of the Energy Depot program of the US Department of Defense. This project focused on the development of a nuclear military combat reactor (MCR) facility to generate high-temperature heat in a combat zone, whereby the heat is used by appropriate technologies to fabricate synthetic fuels from air and water (see Grimes 1966, and Funk and Reinstrom 1964, 1966). The envisaged fuels were hydrogen, ammonia, and hydrazine as they were believed to be applicable to the existing military equipment propelled by internal combustion engines, without major modifications. The envisaged technology in the Energy Depot Project was nuclear power-to-electricity-to-electrolyzer-to-hydrogen with subsequent ammonia or hydrazine synthesis. This approach was found to be inadequately justified, economically, for a large-scale operation.

The results were published in 1964–1966 (see Grimes 1966, Funk and Reinstrom 1964, 1966, and the review by Funk 2001) with the recommendation that water electrolysis should be replaced with a direct thermochemical process in order to avoid the electricity generation and its associated energy losses. Funk (2001) mentioned at least 20 publications worldwide after the communication of the Energy Depot Project conclusions—involving the development of thermo-chemical water-splitting processes, suitable for nuclear reactor linkage. A first roundtable on nuclear hydrogen production by thermochemical cycles was held at Ispra, Italy, in 1969, where a process—called Mark 1—was proposed as a promising thermochemical cycle. Mark 1 operates with chemical elements based on Ca, Br, and Hg and requires high-temperature heat at ~1,000 K.

At The Hydrogen Economy Miami Energy (THEME) conference held in Miami (FL) in March 1974, around 30 established thermochemical water-splitting cycles were described, which were proposed and developed at seven leading laboratories, namely, ISPRA (Euratom), Jülich Center (University of Aachen), Argonne National Laboratory (ANL), Los Alamos Scientific Laboratory, General Atomics, General Electric, and Allison Laboratory of General Motors. The European Commission initiated a Joint Research Centre at Ispra in Italy in conjunction with the University of Aachen in Germany—with an annual budget for the period 1973–1983, dedicated to Mark cycles development. After a first phase of research—with mercury-based compounds—better subsequent cycles were developed within the Ispra-Mark program extending from iron–chlorine cycles, to hybrid sulfuric acid cycles, to sulfur–iodine cycles. In total, 24 versions of Mark cycles were reported by the end of the program (Beghi 1986).

The Ispra Mark 13 cycle—a hybrid type with Br and S compounds—was also studied at the Los Alamos National Laboratory. Sulfur and iodine compounds were used in the Mark 11 cycle. Mark 11 uses sulfuric acid decomposition to evolve oxygen at 1,120 K, according to the reaction $H_2SO_4 \rightarrow H_2O + SO_2 + 0.5O_2$. In Ispra's Mark 11 version, there is an electrochemical H_2-evolving reaction as follows: $H_2O + SO_2 \rightarrow H_2SO_4 + H_2$. Exactly the same cycle as Ispra's Mark 11 was pursued by Westinghouse for further development (Funk 2001). The Ispra cycle Mark 16 is a variant of Mark 11 which uses another hydrogen-evolving scheme—which is purely thermochemical—consisting of two reaction steps,

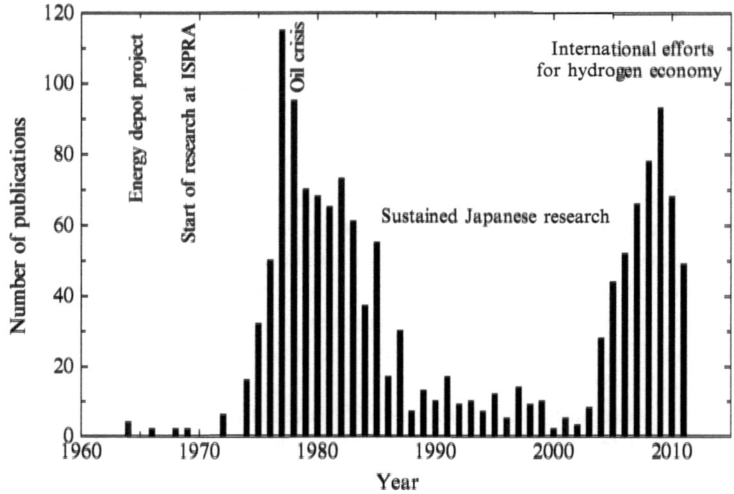

Fig. 5.1 Publications on thermochemical water-splitting cycle per year [data from Brown et al. (2002) and Engineering Vilage (2012)]

namely, $2H_2O + I_2 + SO_2 \rightarrow H_2SO_4 + 2HI$ and $2HI \rightarrow I_2 + H_2$. This cycle (Mark 16) was adopted and pursued for development by General Atomics in the USA, which contributed significantly to process engineering and scaling-up.

By the end of 1976, the University of Tokyo proposed a cycle based on Ca, Br, and Fe compounds called UT-3. This cycle has been extensively developed, as it shows a promising energy efficiency of 49 % (Sakurai et al. 1996). Another cycle is the copper–chlorine cycle which has recently been proven at a laboratory scale, and several commercially appealing variants are being developed for heat sources of around 350–550 °C. For this temperature range, the Cu–Cl cycle appears to be one of the most promising cycles among other options. This cycle has an estimated efficiency of 40–50 % (see Naterer et al. 2011a, b).

Several recent reviews on thermochemical cycles show that there is intensive research worldwide in this direction. Three types of heat sources for hydrogen production with thermochemical cycles are envisaged: concentrated solar heat, nuclear heat, or geothermal heat. In general, solar radiation can operate at lower production scales and higher temperatures, whereas nuclear systems are more suitable for intermediate temperatures and higher production scales. Abanades et al. (2006) screened 280 cycles. Rosen (2010) presented a review of the most promising thermochemical cycles accompanied by an exergy analysis of Mark 11 and Cu–Cl cycles. Six low-temperature cycles (450–700 °C) and their sustainability were assessed by Dincer and Balta (2011), especially for nuclear applications. Six cycles were analyzed in Balta et al. (2010) for geothermal-based hydrogen production. Ten hybrid thermochemical cycles with operating temperatures of 800–1,200 K were assessed by Lewis and Masin (2009) from multiple points of view. Two edited books on nuclear hydrogen production include chapters on thermochemical water-splitting cycles (Verfonden 2007, Yan and Hino 2011).

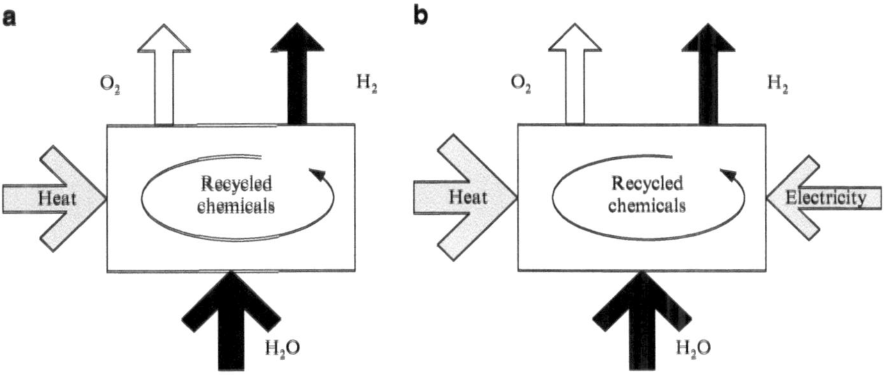

Fig. 5.2 Thermochemical water-splitting cycles: (**a**) thermochemical cycle, (**b**) hybrid cycle

Figure 5.1 shows the number of publications on thermochemical water-splitting for each year since 1964 until the present time.

In this chapter, thermochemical cycles are introduced and their analysis and assessment are presented. The most relevant cycles are categorized as hybrid and pure thermochemical. They are analyzed in a grouped manner with respect to the number of reaction steps. Significant parameters such as temperature level, energy requirements, and efficiency are provided. Full sub-sections are devoted for the most promising thermochemical plants based on sulfur–iodine and hybrid sulfur cycles. The copper–chlorine cycle is examined extensively in Chap. 6.

5.2 General Aspects

The simplified black-block representation of a thermochemical water-splitting cycle is shown in Fig. 5.2. In general, the cycle requires heat and water and generates two output streams, hydrogen and oxygen (Fig. 5.2a). The simplest process of this kind consists of a direct water thermolysis reactor coupled to a separation system for hydrogen and oxygen extraction. In essence, water thermolysis is the basic process for single-step thermochemical water splitting via high-temperature heat. Water thermolysis is technically possible at temperatures over 2,500 K, but it is rather difficult to be obtained.

Several multi-step processes involving recycled chemicals were developed such as two-step, three-step, four-step, five-step, six-step, etc., where "steps" refers to the number of chemical reactions (or sometimes to important physical processes such as drying, extraction, crystallization, etc.). Figure 5.2b illustrates the black-box representation of hybrid cycles, showing that besides heat, they also require electricity. The scope of a hybrid cycle is for medium and intermediate temperature applications,

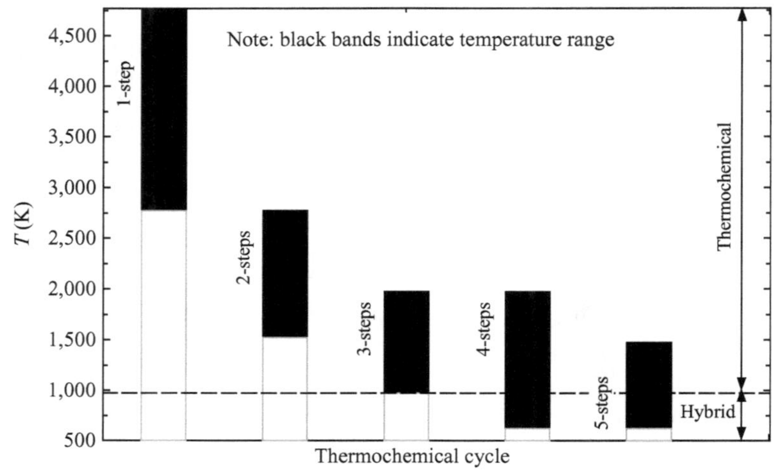

Fig. 5.3 Range of required temperature level required for thermochemical and hybrid cycles

while pure thermochemical cycles are designed for the highest range of source temperatures.

Figure 5.3 shows the range of required temperatures correlated to the number of chemical reaction steps. This figure was obtained based on a literature screening of cycle parameters and it corresponds with other results such as in Dincer and Balta (2011). As mentioned, direct thermolysis requires the highest temperatures. The level of temperature decreases with the number of chemical steps involved in the cycle. There are both pure-thermochemical and hybrid processes with 3 and 4 steps. However, all cycles with a temperature level below ~1,000 K generally must be hybrid.

The development of thermochemical cycles implies an identification of certain intermediate chemical compounds and chemical or electro-chemical reactions which facilitate the overall water-splitting and product separation processes. This involves a trial-and-error search of the best feasible cycle which should be conducted and searched in a systematic way.

5.3 Thermochemical Water Splitting in a Single Reaction Step

In this section, thermodynamic and kinetic analyses of the single-step thermochemical decomposition of water (thermolysis) are presented. Such analyses are useful because they offer a comparison basis for assessment of multi-step thermochemical cycles in regard to the required temperature level, specific hydrogen production rates, required energy input, and efficiency. When a direct thermolysis reactor is coupled to chemical separation devices capable of extracting pure hydrogen and pure oxygen from a reaction mixture, a complete, single-reaction-step water-splitting system is obtained. In this section, after the general theoretical treatment of water thermolysis, additional information and performance parameters of single-reaction-step thermochemical water-splitting systems are given.

A thermodynamic system for direct thermolysis of water can be represented as depicted in Fig. 5.2a: water crosses the system boundary (water is assumed in a vapor phase at 100 % molar concentration); heat is applied to the system; and the following chemical reaction occurs:

$$H_2O \rightarrow \xi \times (y_{H_2}H_2 + y_{O_2}O_2 + y_{H_2O}H_2O), \tag{5.1}$$

where ξ represents the number of moles in the product stream per mole of reactant and takes values between $\xi = 1$ when no reaction occurs and $\xi = 1.5$ when the reaction goes to completion ($H_2O \rightarrow H_2 + 0.5O_2$); remark that ξ quantifies the extent of reaction. Initially, it is useful to assume that there are no other chemicals involved in the process except H_2, O_2, and H_2O. Unreacted water from the product stream can be recycled. In (5.1), the product stream consists of H_2 at a molar fraction y_{H_2} and O_2 at molar fraction y_{O_2}; in addition, some unreacted steam leaves the system (not shown in Fig. 5.2a) at a molar fraction of y_{H_2O}. The molar fraction, y, of a species is defined as the number of moles of the respective species divided by the total number of moles; therefore one has

$$1 = y_{H_2} + y_{O_2} + y_{H_2O}, \tag{5.2}$$

From chemical thermodynamics, it is known that for a general reacting system in the gas phase described by $\sum_i rR \rightarrow \sum_i pP$ where R is for reacting species and P is the produced species, the Gibbs free energy of the reaction is expressed by

$$\Delta G = \Delta G^0(T) + RT \times \ln\left(\Pi_j y_P^p \times \Pi_i y_R^{-r} \times (P/P_0)^{\sum_j p - \sum_i r}\right), \tag{5.3}$$

where R is the universal gas constant, T is the reaction temperature, P is the reaction pressure and the superscript 0 indicates reaction conditions at standard pressure ($P_0 = 101.325$ kPa) for a maximum extent of reaction (the reaction goes to completion). For reaction (5.1), the following particular form is taken by the general (5.3):

$$\Delta G = \Delta G^0(T) + RT \times \ln\left(y_{H_2}^{(\xi y_{H_2})} \times y_{O_2}^{(\xi y_{O_2})} \times y_{H_2O}^{(\xi y_{H_2O} - 1)} \times (P/P_0)^{\xi - 1}\right). \tag{5.4}$$

The stoichiometric balance in (5.1) requires that the number of atoms of chemical elements in the process is conserved. Therefore,

$$\begin{cases} 2 = 2\xi(y_{H_2} + y_{H_2O}) \\ 1 = \xi(y_{O_2} + y_{H_2O}). \end{cases} \tag{5.5}$$

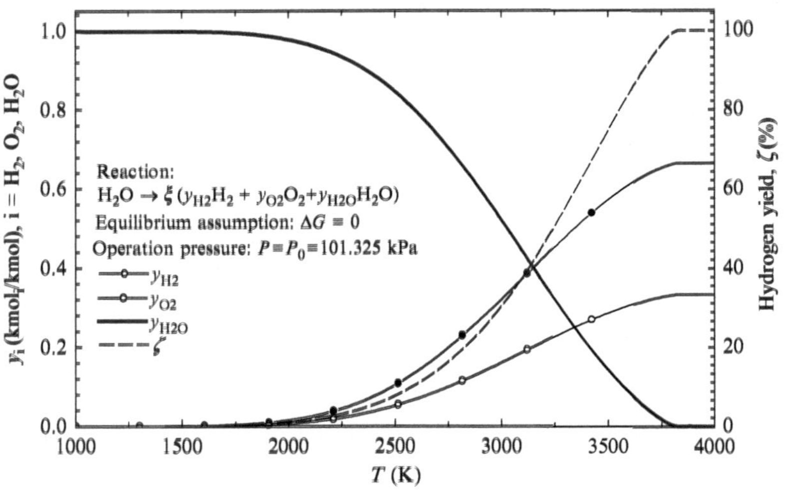

Fig. 5.4 Thermodynamic equilibrium for water thermolysis based on reaction system (5.1)

The equilibrium conditions of the thermodynamic system represented in (5.1) require that the Gibbs free energy of the reaction is zero.

$$\Delta G = 0 \tag{5.6}$$

Equations (5.2, and 5.4–5.6) are solved simultaneously to determine ξ (or the extent of reaction) and the molar concentration of product species at equilibrium. The system was solved with Engineering Equation Solver (EES, Klein 2011) and the results are shown in Fig. 5.4. At around 3,820 K, the reaction completion is nearly achieved with 0 % unreacted steam and 66 % hydrogen in the output stream. At 2,500 K, $\xi = 1.05$ and the equilibrium concentrations are 10 % H_2, 6 % O_2, and 84 % H_2O.

The reaction yield—denoted with ζ—is defined as the number of moles of hydrogen in the product stream per number of moles of reacted water. Ideally for one mole of water, there is one mole of hydrogen; therefore the ideal yield is one. The yield for reaction (5.1) at equilibrium conditions is calculated with $\zeta = \xi \times y_{H_2}/\xi_{max}$ (where $\xi_{max} = 1.5$) and it is shown in Fig. 5.3. It reaches the maximum of 100 % at 3,830 K when the hydrogen concentration in the product stream is 66 %. A high level of temperature is necessary to generate the maximum hydrogen yield from single-step thermochemical water dissociation. Below 1,500 K, the yield becomes zero.

At equilibrium conditions, there is no net production of any chemical because the reaction rate of the forward reaction ($\dot{n}_f$) is equal to the reaction rate of the backward reaction ($\dot{n}_b$). In other words, the produced chemical species are consumed to generate reactants without any net change. In (5.4), if $\Delta G = 0$ at standard pressure $P = P_0$ then:

$$K_{eq} = \frac{y_{H_2}^{(\xi y_{H_2})} \times y_{O_2}^{(\xi y_{O_2})} \times y_{H_2O}^{(\xi y_{H_2O})}}{y_{H_2O}} = \exp\left(-\frac{\Delta G^0}{RT}\right). \tag{5.7}$$

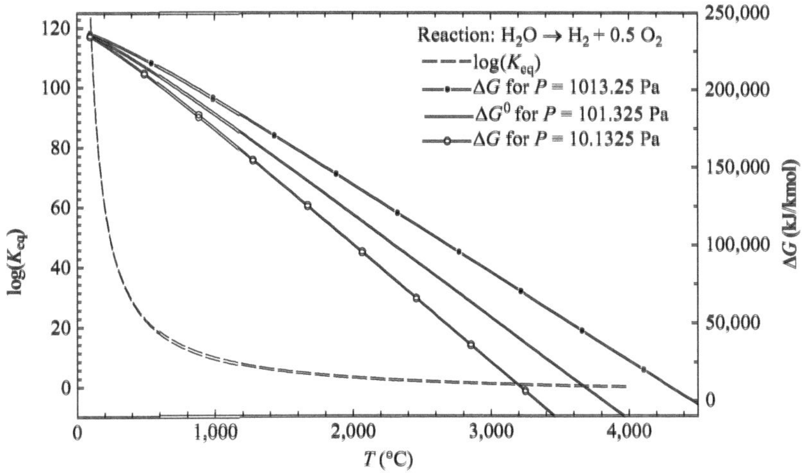

Fig. 5.5 Equilibrium constant and molar Gibbs free energy for water-splitting reaction

Here, (5.7) introduces the equilibrium constant (K_{eq}) of the reaction which is an important parameter for the thermodynamic and kinetic analyses. Figure 5.5 shows the variation of equilibrium constant with temperature as well as the variation of molar Gibbs energy of the reaction. The reactant and products are assumed in the gas phase and the products exist at the stoichiometric molar fraction $y_{H_2} = 2/3$ and $y_{O_2} = 1/3$. The equilibrium constant—according to (5.7)—indicates that the concentrations of products and reactants occur in a determined ratio at equilibrium. The forward reaction rate is increased if the concentration of reactant (y_{H_2O}) is higher. The backward reaction rate increases if the concentration of products is higher. The following proportionality must be established between the reaction rates and concentrations in the vicinity of equilibrium:

$$\begin{cases} \dot{n}_f = k_f \times y_{H_2O} \\ \dot{n}_b = k_b \times y_{H_2}^{(\xi y_{H_2})} \times y_{O_2}^{(\xi y_{O_2})} \times y_{H_2O}^{(\xi y_{H_2O})}. \end{cases} \tag{5.8}$$

In (5.8), the proportionality factors k_f and k_b are denoted as reaction rate constants for the forward and backward reactions, respectively. In practical situations, the molar fractions are replaced with molar concentrations in (5.8). A simple linear relationship can be used between the molar fraction and molar concentration, namely, $c \times v = y$, where v is the molar volume. At equilibrium, in (5.8), the forward and backward rates must be equal; thus

$$k_f \times y_{H_2O} = k_b \times y_{H_2}^{(\xi y_{H_2})} \times y_{O_2}^{(\xi y_{O_2})} \times y_{H_2O}^{(\xi y_{H_2O})}. \tag{5.9}$$

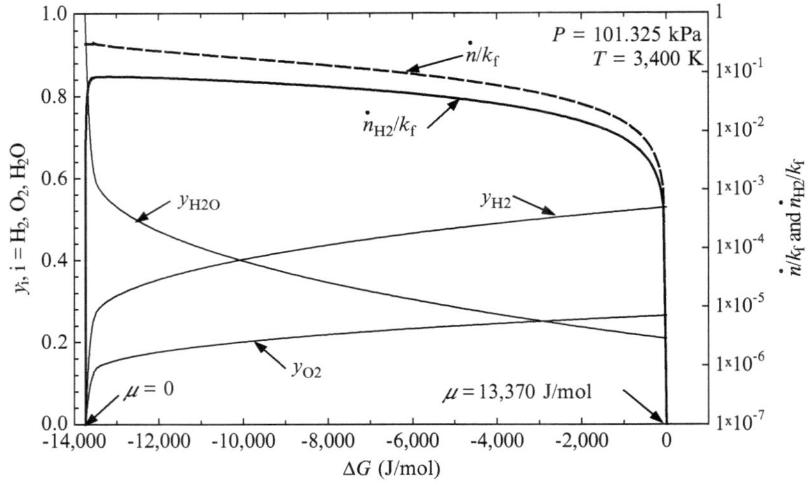

Fig. 5.6 Rate-equilibrium diagram for single-step water dissociation

From (5.8) and (5.9),

$$K_{eq} = k_f / k_b. \tag{5.10}$$

Equation (5.10) is important in kinetic calculations of thermochemical reactions. It allows to express the net rate of forward reaction $\dot{n} = \dot{n}_f - \dot{n}_b$ as a function of rate constant of the forward reaction, equilibrium constant, and concentrations (or molar fractions). For water thermolysis, one has

$$\dot{n} = k_f \times \left[y_{H_2O} - y_{H_2}^{(\xi y_{H_2})} \times y_{O_2}^{(\xi y_{O_2})} \times y_{H_2O}^{(\xi y_{H_2O})} / K_{eq} \right]. \tag{5.11}$$

In (5.11), it can be observed that the net rate achieves a maximum when there are no molecules of the product present in the vicinity of the reaction site (molar fraction of hydrogen and/or oxygen equals zero, $y_{H_2} = 0$, $y_{O_2} = 0$). The maximum rate is thus $\dot{n}_{max} = k_f \times y_{H_2O}$.

Figure 5.6 illustrates what happens with the chemical system described in (5.1) if it is moved away from equilibrium in the forward direction at a constant temperature of 3,400 K. This graph is obtained by solving (5.2), (5.4), and (5.5) which forms a closed system with unknowns being ξ and molar fractions (note that ΔG must be specified). In order for the reaction to move forward spontaneously, a negative value for ΔG is needed. It can be observed in the figure, away from equilibrium, that the reactant's (steam) molar fraction increases while the product molar fraction decreases (until it becomes zero).

With known concentrations and reaction extents, the dimensionless quantity $\dot{n}/k_f$ is calculated with (5.11). This quantity expresses the reaction rate. Furthermore, the dimensionless quantity $\dot{n}_{H_2}/k_f = y_{H_2} \times \dot{n}/k_f$ is calculated. This quantity represents a measure of the hydrogen production. In a steady-state operation, the actual molar flow rate of hydrogen that leaves the unit of the control volume will be $\dot{n}_{H_2} = y_{H_2} \times \dot{n}$. There are some important conclusions that can be observed from a kinetics analysis illustrated in Fig. 5.6:

- At a point away from equilibrium in the forward direction, the reaction rate is maximum.
- Where the reaction rate is the maximum, there is no reaction product at the reaction site; the molar fraction of products must be zero.
- Because the molar fraction is zero, there is no net production of species when the system operates at a maximum rate; this is a "short-circuit" type of operation.
- Once products appear at the reaction site, the system moves toward equilibrium and the reaction rate decreases.
- When the system is at chemical equilibrium, there is no net reaction rate because the forward and backward rates are equal; there is no net production of species although products are present at the reaction site at their maximum possible concentration; this is an "open-circuit" type of operation.
- Once products are moved out from a reaction site at equilibrium, the system will be moved out of equilibrium toward higher reaction rates in the forward direction.
- In between the two extremes (maximum production rate and equilibrium condition) there is an operation condition where product generation is maximized; this condition indicates the optimal product concentration in the system.
- The Gibbs energy controls the operation of the system in the sense that it indicates the system's departure from the equilibrium and allows for a unique determination of the molar fractions (or concentration) of the species as well as determination of the reaction rate and product generation rate.

The chemical potential is a more convenient quantity to indicate the reaction departure from equilibrium. In Chap. 4, the chemical potentials were introduced in (4.4) and (4.5). For a gas-phase reaction, the chemical potential of species "i" is expressed by

$$\mu(y_i) - \mu(y_i^0) = RT\ln(y_i),$$

where the chemical potential for the reference state $\mu(y_i^0)$ must be defined. For any reaction system at constant temperature, $\mu - \mu^0 = \Delta G - \Delta G^0$, as can be made clear from (5.3) and (5.4). By convention, $\mu = 0$ at the "short-circuit" condition.

For the case studied in Fig. 5.5 at a short-circuit condition, $\Delta G = 13,370$ J/mol and $\Delta G^0 = -5,828$ J/mol (for 3,400 K); thus set $\mu^0 = 7,542$ J/mol. At equilibrium, where $\Delta G = 0$, $\mu = 13,370$ J/mol. When there is no chemical potential applied to the system, the reaction rate is maximum but there is no flow exiting the system (short-circuit situation). When the chemical potential is maintained at the maximum, there is no possibility to extract any chemicals out of the system without perturbing its equilibrium (open-circuit situation).

Table 5.1 Representative elementary reactions for water thermolysis

Reaction[a]	ΔH^0 (kJ/mol)	ΔS^0 (J/mol K)	ΔG^0 (kJ/mol)	Direction	ΔE_{act}[a] (kJ/mol)	A[a] (m^3/mol s)
$2H_2O \rightleftarrows H + OH + H_2O$	231.8	120.2	196.0	F	439.8	2.2×10^{10}
				B	$2RT\ln(T)$	1.4×10^{11}[b]
$H + H_2O \rightleftarrows H_2 + OH$	−200.6	−51.5	−185.2	F	85.2	9.0×10^7
				B	21.5	2.2×10^7
$H + OH \rightleftarrows H_2 + O$	−202.9	−52.62	−187.2	F	29.1	8.3×10^3
				B	37.2	1.8×10^4
$O + OH \rightleftarrows O_2 + H$	198.5	104.5	167.3	F	0.0	2.0×10^7
				B	70.0	2.2×10^8

[a]*Source*: Lede et al. (1982); data valid for temperature range of 2,000–4,000 K
[b]Unit is m^6/mol s; *F* forward; *B* backward

The above theoretical analysis indicates that the operating temperature of single-step thermochemical water-splitting systems must be over 2,500 K to obtain a reasonable yield. The kinetic analysis shows that the system should work far from equilibrium. This suggests that the output stream for a maximum production rate will be relatively diluted of hydrogen. Consequently, one may expect bulky designs characterized by low production per unit of volume. In the theoretical diagram from Fig. 5.6, it is also observed that the maximum hydrogen production is relatively flat. Therefore, the selection range of parameters for operation at a maximized production rate appears to be wide.

Many technical aspects must be considered for a system's practical design: material selection, heat recovery systems, heat sources, possibility of using catalysts, product separation and purification, etc. One primary issue is the influence of side reactions. It was found that when water is heated at temperatures over 2,000 K, it dissociates thermally into many species, with significant molecular compounds as follows: H_2O, H_2, O_2, O, H, and OH (Lede et al. 1982). Table 5.1 indicates the representative elementary reactions for water dissociation and their thermodynamic and kinetic parameters at temperatures between 2,000 and 4,000 K. The reaction rate for each elementary reaction is calculated with the Arrhenius equation at process temperatures as follows:

$$k = A \times \exp(-\Delta E_{act}/RT), \tag{5.12}$$

where A is the pre-exponential factor and ΔE_{act} is the activation energy. Chemical equilibrium of water thermal dissociation based on representative elementary reactions as indicated in Table 5.1 can be calculated by solving the equilibrium equations in a manner similar to that illustrated above for the simplified case that neglected the side reactions according to (5.1).

Figure 5.7 presents the equilibrium concentration and hydrogen yield based on the reaction system described in Table 5.1. It can be observed that the hydrogen concentration reaches a maximum at ~3,400 K while the maximum hydrogen yield occurs at around 3,600 K. The yield is slightly over 26 %; this means that maximum

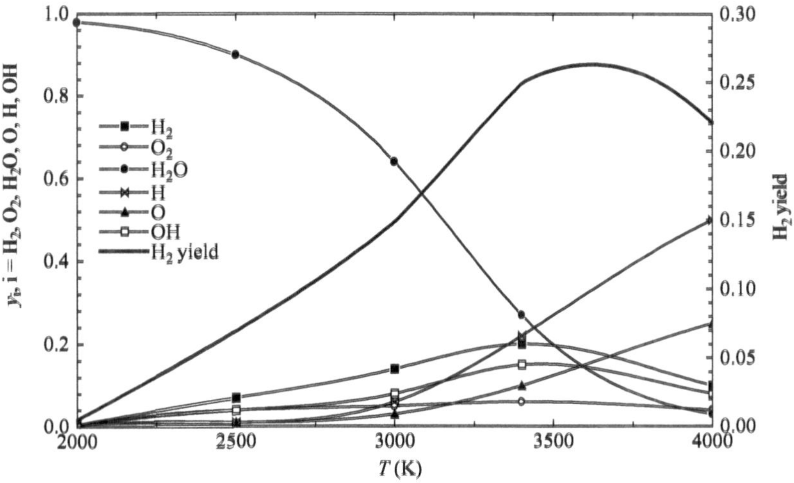

Fig. 5.7 Equilibrium concentrations at H_2O thermolysis for reaction system in Table 5.1 [data from Lede et al. (1982)]

0.26 mol of hydrogen can be ideally obtained from 1 mol of water in a thermolysis reactor.

The efficiency of water single-step thermochemical splitting by thermolysis can be calculated based on energy (η) or exergy (ψ) input as follows:

$$\eta = \dot{n}_{H_2} \times \frac{\mathrm{HHV}}{\dot{Q}_{\mathrm{in}}} \text{ and } \psi = \dot{n}_{H_2} \times \frac{ex^{\mathrm{ch}}}{\dot{Q}_{\mathrm{in}} \times (1 - T_0/T)}, \tag{5.13}$$

where $\dot{n}_{H_2}$ is the molar rate of hydrogen production. Also, HHV and ex^{ch} are the molar higher heating value and chemical exergy of hydrogen, respectively, and $\dot{Q}_{\mathrm{in}}$ is the heat input rate provided to the reaction, T is the reaction temperature and T_0 is the reference temperature.

The upper-bound limit of efficiencies can be calculated, by assuming (5.13) that hydrogen is extracted at an equilibrium yield and the heat input required by the process is equal to the enthalpy of reaction (there are no heat losses). The results of these calculations are shown in Fig. 5.8. It appears that the ideal maximum efficiencies are around 60 % at 2,500 K. The exergy efficiency is slightly higher than the energy efficiency. Note that at 2,500 K, the Carnot factor is 88 %. The only irreversibilities in the process model are due to the side reactions and because the reaction does not go to completion. Nevertheless, there are other irreversibilities in the process due to heat transfer, departure from equilibrium, hydrogen separation from the output stream, and efficiency losses due to gas separation (Nakamura 1977).

Gas separation technology at high temperatures necessitates high-pressure gradients to generate a sufficient driving force across the separation membranes. Another issue is quenching the product gases, which is required as a measure to

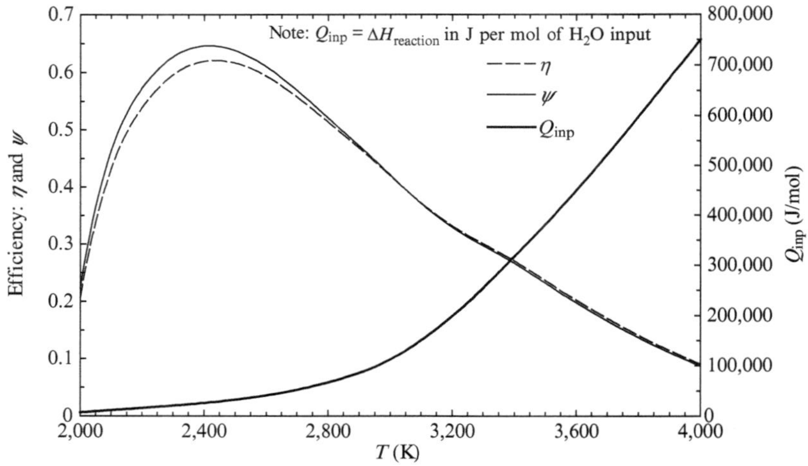

Fig. 5.8 Reaction heat and ideal maximum efficiencies for direct water thermolysis in the reaction system in Table 5.1

avoid a recombination of hydrogen and oxygen. It is estimated that the energy efficiency of practical water-splitting systems based on direct water thermolysis is in the range of 25–55 % (Nakamura 1977, Sato 1979).

5.4 Two-Step Thermochemical Cycles

In the seminal paper by Funk and Reinstrom (1966), two-step thermochemical water-splitting cycles were first proposed as a means to improve the overall energy efficiency of hydrogen production and to widen the applications with respect to other heat-driven technologies, namely, water thermolysis and water electrolysis (it is assumed that electrical energy for the electrolyzer is derived from a high-temperature heat source). Two types of two-step cycles were proposed originally in Funk and Reinstrom (1966): oxide type and hydride type. Nakamura (1977) advanced an additional type of two-step process called a hydroxide type. Bilgen and Bilgen (1982) suggested an oxide-sulfate scheme for two-step thermochemical water-splitting cycles. A comprehensive list of two-step thermochemical water-splitting cycles is compiled in McQuillan et al. (2010); they indicate an additional type of multi-component cycle which is based on mixed-metal oxides. Figure 5.9 describes the types of chemical reactions of two-step thermochemical water-splitting cycles.

When compared to electrolysis, thermochemical cycles eliminate or substantially reduce the irreversibility of power generation. Two-step water-splitting cycles offer the following advantages with respect to the single-reaction-step process based on water thermolysis:

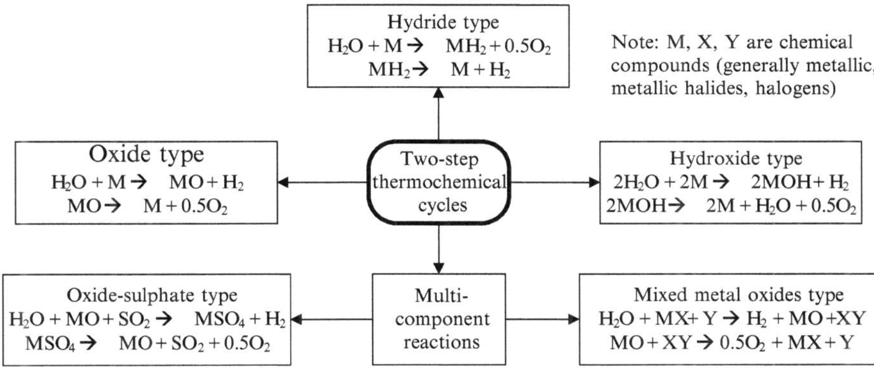

Fig. 5.9 Classification of two-step thermochemical water-splitting cycles

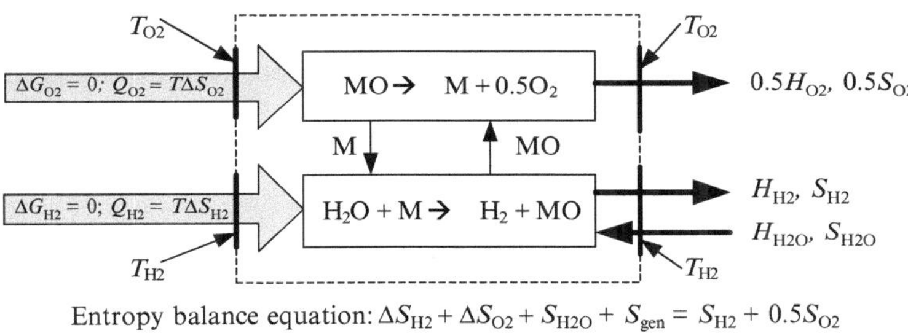

Fig. 5.10 Energy flows and entropy balance for a generic two-step water-splitting process assuming both reactions at equilibrium

- The product gases (hydrogen and oxygen) are chemically separated as they are produced in distinct chemical reactors; therefore the risk of hydrogen combustion is reduced.
- The temperature level of the reaction is significantly lower (below ~2,500 K) which is better for material selection, while at the same time, a larger range of sustainable heat sources may be identified at lower temperatures.

A basic thermodynamic analysis of a two-step thermochemical water-splitting, using the first and second laws of thermodynamics, can be formulated based on the schematic in Fig. 5.10. A process consisting of two reaction steps, from which one evolves hydrogen at a temperature T_{H2} and the other evolves oxygen at temperature T_{O2}, is illustrated. The temperature for each reaction is set such that the forward reaction is spontaneous; thus $\Delta G_{H2} \lesssim 0$ and $\Delta G_{O2} \lesssim 0$. It is desired to have as low as possible reaction temperatures. Therefore assume that both reactions operate at equilibrium. It is also assumed that water (steam), hydrogen, and oxygen are at the reaction temperature. Since ΔG is zero for both reactions, the heat input in the process is given by $Q_{H_2} = T_{H_2}\Delta S_{H_2}$ at T_{H_2} and $Q_{O_2} = T_{O_2}\Delta S_{O_2}$ at T_{O_2}. The first law of thermodynamics (energy balance) for this system yields

$$Q_{H_2} + Q_{O_2} + Q_{reheat} + H_{H_2O} = H_{H_2} + 0.5H_{O_2}, \quad (5.14)$$

where H is the molar enthalpy and Q_{reheat} is the net heat input necessary to reheat the chemicals circulated between the lower temperature reaction to the high-temperature reaction. For preliminary analyses, it is reasonable to assume that $Q_{reheat} \cong 0$ because the heat rejected by the chemicals cooled from a high temperature to low temperature level is approximately equal to the heat required to augment the temperature of reactants of the high-temperature reaction. The second law of thermodynamics is written as an entropy balance equation using molar entropies (S), as follows:

$$\Delta S_{H_2} + \Delta S_{O_2} + S_{H_2O} + S_{gen} = S_{H_2} + 0.5S_{O_2}. \quad (5.15)$$

In (5.14) and (5.15), the molar entropies and enthalpies are evaluated at the corresponding reaction temperatures and ΔS_{H_2} and ΔS_{O_2} are the reaction entropies for the hydrogen- and oxygen-evolving reaction, respectively. Combining (5.14) and (5.15) and stipulating that the entropy generation is positive ($S_{gen} \geq 0$), it results in the following conditions that must be satisfied by chemical reactions that form a two-step thermochemical water-splitting process:

$$\begin{cases} \Delta S_{H_2} + \Delta S_{O_2} \leq S_{H_2} + 0.5S_{O_2} - S_{H_2O} \\ T_{H_2}\Delta S_{H_2} + T_L\Delta S_L = H_{H_2} + 0.5H_{O_2} - H_{H_2O} \\ \Delta G_{H_2} \cong 0 \\ \Delta G_{O_2} \cong 0 \\ T_{H_2} < 3,815\,\text{K} \\ T_{O_2} < 3,815\,\text{K}. \end{cases} \quad (5.16)$$

The last two conditions in (5.16) state that the reaction temperatures must be smaller than the temperature of 3,815 K at which the reaction $H_2O \rightarrow H_2 + 0.5O_2$ has $\Delta G = 0$; this condition relates to one of the technological advantages of two-step thermochemical cycles, namely, by reducing the process temperature with respect to water thermolysis. Equation (5.16) provides a useful set of criteria for screening chemical reactions for two-step water-splitting cycles. These criteria are useful only for preliminary screening. In some cases, reactions that do not fit the 3,815 K restriction can be selected. For example, reduction of WO_3 shows $\Delta G = 0$ at temperatures higher than 3,815 K (namely, 4,183 K for $1/3WO_3 \rightarrow 0.5O_2 + 1/3W$); however this reaction can be conducted with a reasonable efficiency and yield at 2,000 K (see Bilgen and Bilgen 1982). Another example is MoO_2 with $0.5MoO_2 \rightarrow 0.5O_2 + 0.5Mo$ at equilibrium at 3,896 K; the reaction is feasible at 2,000 K (see Bilgen and Bilgen 1982).

Funk and Reinstrom (1966) identified 22 oxide-type and 16 hydride-type reactions that satisfy (5.16) and therefore possible for two-step thermochemical water-splitting cycle development. However, very few of these cycles have been

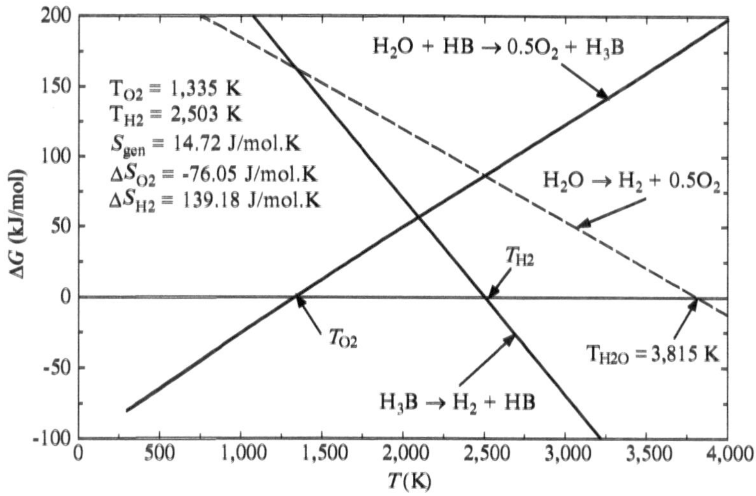

Fig. 5.11 Free Gibbs energy vs. temperature for a hydride-type cycle with boron-hydride (hydrogen-evolving reaction is at the highest temperature)

attractive for practical implementation. An example of an early-proposed hydride-type cycle that may be practically attractive is based on boron hydride (HB) and borane (H_3B), both reacting in a gas phase at high temperature (Funk and Reinstrom 1966). Boron hydride is reacted with steam at 1,335 K to generate oxygen and borane via a chemical reaction at equilibrium. Direct borane thermolysis is performed at 2,503 K with production of hydrogen and boron hydride. The cycle is described by the equations

$$\begin{cases} H_2O + HB \rightarrow 0.5O_2 + H_3B, \quad \text{at } 1,335\,\text{K} \\ H_3B \rightarrow H_2 + HB, \quad \text{at } 2,503\,\text{K}. \end{cases} \tag{5.16}$$

Figure 5.11 presents Gibbs free energy vs. temperature for chemical reactions involved in the two-step boron hydride thermochemical cycle, calculated with EES software (see Klein 2011). The diagram shows that the Gibbs energy for the oxygen-evolving reaction has a positive slope with respect to temperature. The reaction is spontaneous at low temperatures. At equilibrium, it shows a negative reaction entropy, which is an indication that the reaction is exothermic. The hydrogen-evolving reaction reaches equilibrium at a higher temperature, but still substantially lower than the equilibrium temperature for direct water thermolysis (2,503 K vs. 3,815 K). The slope of its Gibbs free energy has a negative slope with respect to temperature, so the reaction may become spontaneous at higher temperatures. The reaction entropy is positive and high, which indicates that the reaction is endothermic. The generated entropy by the two-step water-splitting cycle is positive (14.72 J/mol K), implying that the process is feasible according to the second law of thermodynamics. In the figure, for comparison purposes, the

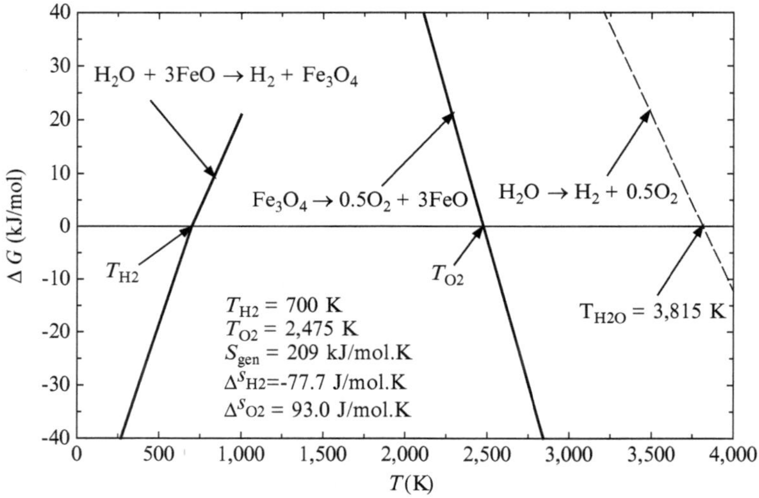

Fig. 5.12 Free Gibbs energy vs. temperature of an oxide-type cycle with ferrous oxide (hydrogen-evolving reaction is at the lowest temperature) [data from Nakamura (1977)]

Gibbs free energy curve is superimposed, characterizing the water decomposition reaction (dashed line).

A process similar to that described in (5.16) is mentioned in Chao (1974). This includes a first reaction of the hydrolysis of chlorine according to the reverse Deacon reaction, namely, $H_2O + Cl_2 \rightarrow 0.5O_2 + 2HCl$. This reaction can be conducted at 873 K. It is a reaction that is encountered in many multistep thermochemical water-splitting cycles. Its importance is described in the next section of this chapter where further details are given. The hydrogen evolution reaction is the thermolysis of hydrochloric acid, $2HCl \rightarrow H_2 + Cl_2$, which can be conducted at 973 K (see Chao 1974). This process is very difficult due to separation of hydrogen and chlorine at high temperature, although membrane reactors for hydrogen extraction can be devised with existing technology.

In Fig. 5.12, an oxide-type two-step cycle for thermochemical water splitting that operates with ferrous oxide is presented in a $\Delta G - T$ diagram. This cycle was discussed by Nakamura (1977). Steam reacts with ferrous oxide and generates hydrogen at a lower temperature, 700 K, while ferrous ferric oxide (Fe_3O_4) is formed. At a higher temperature of 2,475 K, the direct thermal decomposition of Fe_3O_4 occurs with a release of oxygen. The reactions for this cycle are

$$\begin{cases} H_2O + 3FeO \rightarrow H_2 + Fe_3O_4, & \text{at}\, 700\ \text{K} \\ Fe_3O_4 \rightarrow 0.5O_2 + 3FeO, & \text{at}\ 2,475\ \text{K}. \end{cases} \quad (5.17)$$

The hydrogen-evolving reaction is exothermic at equilibrium (with a negative reaction entropy), while the oxygen-evolving reaction is endothermic. The generated entropy of the overall two-step process is positive and high (209 kJ/mol K) which is an indication of the process feasibility, but with significant irreversibilities.

Table 5.2 shows a number of two-step thermochemical water-splitting cycles which are grouped with respect to reaction type. Only cycles that are believed to be the most promising for practical implementation are listed in the table. The cycles are sorted in descending order with respect to the highest process temperature.

The temperatures for which the reaction reaches equilibrium are also indicated as $T_{\Delta G=0}$. The cycle energy efficiency is given from literature sources and it is calculated as a function of HHV of hydrogen, heat input (Q_{inp}) into the process, and the electrical energy (W_{aux}) needed for the auxiliary systems to generate one mole of hydrogen.

It is assumed that the electricity is generated from a high-temperature heat source with an efficiency of $\eta_{el} = 50\%$ which corresponds to future generations of nuclear reactors. This energy efficiency definition is consistent with that used by McQuillan et al. (2010). The energy and exergy efficiencies are given by

$$\eta = \mathrm{HHV}/(Q_{inp} + W_{aux}/\eta_{el}) \text{ and } \psi = \mathrm{ex}^{ch}/[(Q_{inp} + W_{aux}/\eta_{el}) \times (1 - T_0/T_H)], \quad (5.18)$$

where T_H is the highest reaction temperature.

The two-step cycles have a relatively simple construction because they involve processing of a very limited number of chemicals and only two chemical reactors. When metallic oxides in a solid phase are involved in a chemical reaction, the solid–gas separation is facilitated. Despite these advantages, there is a major barrier with coupling two-step thermochemical cycles to nuclear reactors. This is due to the high level of temperature required for the oxygen-evolving reaction. According to Table 5.2, the cycle with a minimum temperature requirement is based on nickel–manganese–ferrite ($NiMnFe_4O_6$) at 1,093 K (see the cycle in the table). This temperature is at the compatibility limit with very-high-temperature nuclear reactors of the next generation.

This cycle is attractive also because it involves simpler solid–gas separation of reaction products with the possibility of improved heat recovery. Fluidized bed reactors can be used for both reactions with solid–gas separation; water is separated by condensation.

Figure 5.13 illustrates a simplified diagram of the cycle, useful for preliminary thermodynamic analysis. Liquid water enters in #1 at T_0 and it is heated and partially boiled (#1–#2) using heat recovered from product gases, hydrogen and oxygen (#4–#5 and #12–#13). Water is then boiled and superheated using heat recovered from hydrogen-evolving reactor (#2–#3). Nickel–manganese–wustite ($NiMnFe_4O_8$) is heated (#7–#8) using heat recovered from the oxygen stream (#11–#12) and nickel–manganese–ferrite (#9–#10) to reach the reaction temperature (~1,100 K) of the oxygen-evolving reactor. The product streams (hydrogen and oxygen), still hot, are cooled (without heat recovery) to reach a temperature close to T_0 prior to further compression (#5–#6 and #13–#14).

Figure 5.14 shows a $T - H$ diagram that represents heating and cooling processes within the cycle. The diagram is useful to determine the heat recovery potential.

Table 5.2 Two-step thermochemical water-splitting cycles

Type	ID: Elements	Reactions	T (K)	$T_{\Delta G=0}$ (K)[a]	η (%)	ψ (%)
Oxide-type	1: Si	$H_2O + SiO \rightarrow H_2 + SiO_2$	2,929	2,929	39	35
		$SiO_2 \rightarrow 0.5O_2 + SiO$	3,250	3,250		
	2: Ce	$H_2O + Ce_2O_3 \rightarrow H_2 + 2CeO_2$	700	700	30	28
		$2CeO_2 \rightarrow 0.5O_2 + Ce_2O_3$	2,573	2,273		
	3: Fe	$H_2O + 3FeO \rightarrow H_2 + Fe_3O_4$	450	700	42	39
		$Fe_3O_4 \rightarrow 0.5O_2 + 3FeO$	2,475	2,475		
	4: C	$H_2O + CO \rightarrow H_2 + CO_2$	423	($\Delta G < 0$)	37	35
		$CO_2 \rightarrow 0.5O_2 + CO$	2,473	($\Delta G > 0$)		
	5: Zn	$H_2O + Zn \rightarrow H_2 + ZnO$	1,489	1,489	53	50
		$ZnO \rightarrow Zn + 0.5O_2$	2,337	2,337		
	6: Mo	$H_2O + 0.5Mo \rightarrow H_2 + 0.5MoO_2$	1,816	1,816	35	34
		$0.5MoO_2 \rightarrow 0.5O_2 + 0.5Mo$	2,000	3,896		
	7: W	$H_2O + 1/3W \rightarrow H_2 + 1/3WO_3$	1,157	1,157	55	53
		$1/3WO_3 \rightarrow 0.5O_2 + 1/3W$	2,000	4,183		
	8: In	$H_2O + 2/3In \rightarrow H_2 + 1/3In_2O_3$	1,116	1,116	35	34
		$1/3In_2O_3 \rightarrow 0.5O_2 + 2/3In$	2,000	2,814		
	9: Sn	$H_2O + 0.5Sn \rightarrow H_2 + 0.5SnO_2$	873	873	59	57
		$0.5SnO_2 \rightarrow 0.5O_2 + 0.5Sn$	2,000	3,094		
	10: Fe, Mn, Ni	$H_2O + 0.5NiMnFe_4O_6 \rightarrow H_2 + 0.5NiMnFe_4O_8$	873	n/a	48	55
		$0.5NiMnFe_4O_8 \rightarrow 0.5O_2 + 0.5NiMnFe_4O_6$	1,093	1,650		
Oxide/sulphate-type	11: Sr, S	$H_2O + SrO + SO_2 \rightarrow H_2 + SrSO_4$	1,360	2,597	60	58
		$SrSO_4 \rightarrow 0.5O_2 + SrO + SO_2$	2,038	2,711		
	12: Ca, S	$H_2O + CaO + SO_2 \rightarrow H_2 + CaSO_4$	1,150	2,633	60	59
		$CaSO_4 \rightarrow 0.5O_2 + CaO + SO_2$	1,880	2,735		
	13: Fe, S	$H_2O + FeO + SO_2 \rightarrow H_2 + FeSO_4$	522	1,647	17	18
		$FeSO_4 \rightarrow 0.5O_2 + FeO + SO_2$	1,373	1,926		
	14: Co, S	$H_2O + CoO + SO_2 \rightarrow H_2 + CoSO_4$	473	n/a	35	37
		$CoSO_4 \rightarrow 0.5O_2 + CoO + SO_2$	1,373			

	15: Mn, S	$H_2O + MnO + SO_2 \rightarrow H_2 + MnSO_4$	563	n/a	42	44
		$MnSO_4 \rightarrow 0.5O_2 + MnO + SO_2$	1,373			
	16: Cd, S	$H_2O + CdO + SO_2 \rightarrow H_2 + CdSO_4$	473	n/a	55	59
		$CdSO_4 \rightarrow 0.5O_2 + CdO + SO_2$	1,273			
Mixed metal oxides-type	17: Fe, Zn	$H_2O + 2FeO + ZnO \rightarrow H_2 + Fe_2ZnO_4$	694	n/a	43	41
		$Fe_2ZnO_4 \rightarrow 0.5O_2 + 2FeO + ZnO$	2,358			
	18: C, Na, Fe	$H_2O + 2Na_2FeO_2 + CO_2 \rightarrow H_2 + Na_2CO_3 + 2NaFeO_2$	1,073	n/a	48	46
		$Na_2CO_3 + 2NaFeO_2 \rightarrow 0.5O_2 + 2Na_2FeO_2 + CO_2$	2,273			
	19: Cl, Fe	$4H_2O + 3FeCl_2 \rightarrow H_2 + Fe_3O_4 + 6HCl$	973	1,235	48	48
		$Fe_3O_4 + 6HCl \rightarrow 0.5O_2 + 3FeCl_2 + 3H_2O$	1,773	($\Delta G < 0$)		
	20: Fe, Zn	$H_2O + 1/2Fe_3O_4 + 3/4Zn \rightarrow H_2 + 3/4Fe_2ZnO_4$	873	n/a	23	23
		$3/4Fe_2ZnO_4 \rightarrow 0.5O_2 + 1/2Fe_3O_4 + 3/4Zn$	1,573			
	21: Fe, Mn, Zn	$H_2O + MnFe_2O_4 + 3ZnO \rightarrow H_2 + Zn_3MnFe_2O_8$	1,273	n/a	52	54
		$Zn_3MnFe_2O_8 \rightarrow 0.5O_2 + MnFe_2O_4 + 3ZnO$	1,473			
Hydride type	22: B	$H_2O + HB \rightarrow 0.5O_2 + H_3B$	1,335	1,335	N/A	N/A
		$H_3B \rightarrow H_2 + HB$	2,500	2,503		
	24: Cl	$H_2O + Cl_2 \rightarrow 0.5O_2 + 2HCl$	873	751	N/A	N/A
		$2HCl \rightarrow H_2 + Cl_2$	1,073	($\Delta G > 0$)		

Note: $\Delta G = 0$ at $T = 3{,}815$ K for $H_2O \rightarrow H_2 + 0.5O_2$

Sources: McQuillan et al. (2010), Bilgen and Bilgen (1982), Ewan and Allen (2005), Chao (1974)

[a]Data calculated with EES (Klein 2011)

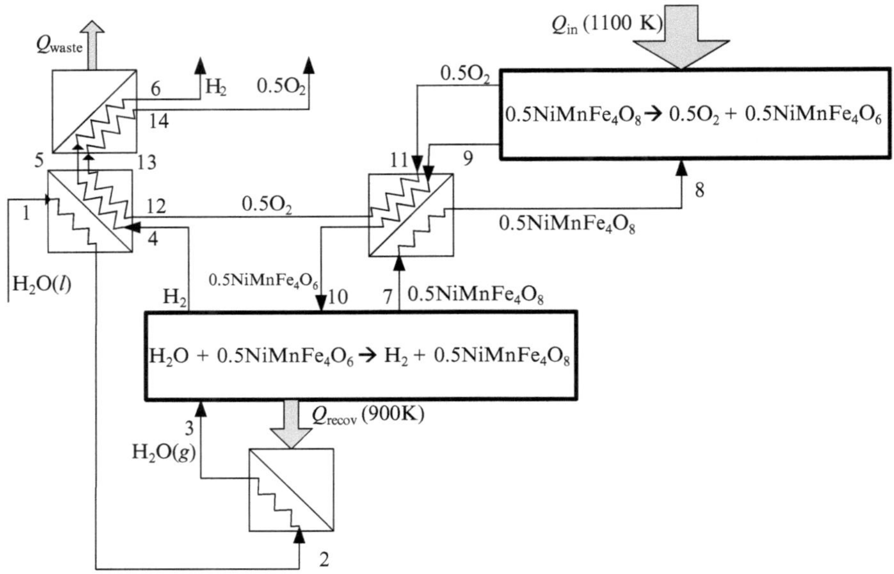

Fig. 5.13 Simplified diagram of two-step nickel–manganese–ferrite cycle

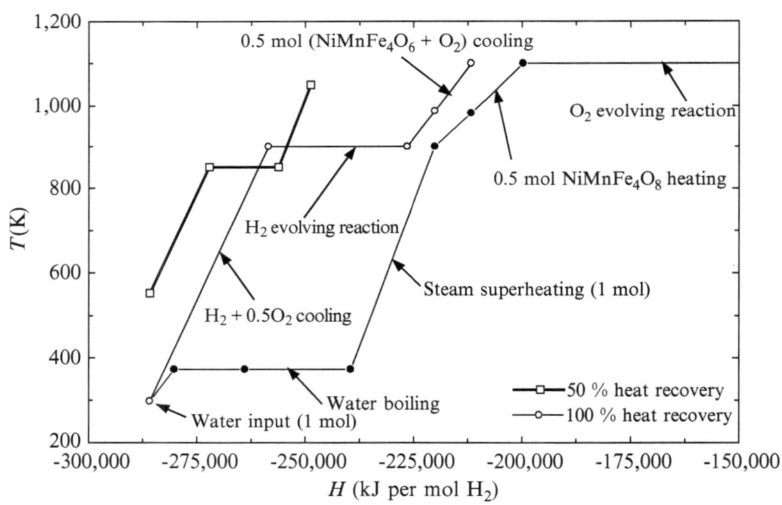

Fig. 5.14 Temperature–enthalpy diagram for internal heat recovery with a two-step nickel–manganese–ferrite cycle

Water is supplied as a liquid at T_0, P_0 and heated to the boiling point, boiled, and superheated to 900 K. An amount of 0.5 mol of $NiMnFe_4O_8$ per mole of generated hydrogen must be heated from 900 K to 1,100 K (see Figs. 5.11 and 5.12 and Table 5.2).

If 100 % heat recovery and re-use within the cycle could be obtained, then the cycle's energy efficiency is 100 % because 1 mol of liquid water generates 1 mol of hydrogen and a half mole of oxygen at standard conditions; therefore the heat required for the process is equal to higher heating value of hydrogen (no heat is lost and other irreversibilities are neglected). Figure 5.14 shows the temperature–enthalpy curves of processes that reject heat. These are cooling of a half mole of O_2 and half mole of $NiMnFe_4O_6$ from 1,100 to 900 K, heat rejected by the hydrogen-evolving reaction at 900 K, and cooling of hydrogen and oxygen streams from 900 K down to 298.15 K.

Assume that only 50 % of the heat rejected can be recovered and reused internally within the cycle (see Bilgen and Bilgen 1982). For this assumption, the heat recovered and reused is represented with a bold line in the diagram. It can be observed that the recovered heat cannot completely boil the water. The estimated heat input requirement in this case is 323 kJ/mol; therefore the HHV-based efficiency calculated with (5.18) under the additional assumption that $W_{aux} = 0$ is 87 %. This figure is high for practical systems that must use additional power input (or additional heat) for species separation and chemical circulation. A way to estimate the energy requirement for chemical separation is based on the Gibbs free energy for component separation of gas mixtures, namely,

$$\Delta G_{sep} = -RT \ln(\Pi_i y_i^{n y_i}), \tag{5.19}$$

where y_i is the molar fraction of component i and n is the number of moles of the gas mixture.

For example, a stoichiometric mixture of one mole of hydrogen and a half mole oxygen will have a molar fraction of hydrogen of 2/3 and that of oxygen of 1/3, while the total number of moles is $n = 1.5$. From (5.19), oxygen and hydrogen separation at 1,000 K requires a Gibbs free energy of 5,292 J/mol. In many practical designs, reactants must be supplied in excess in order to shift forward the reaction equilibrium. As with the nickel–manganese–ferrite cycle, water must be supplied in excess to the hydrogen-evolving reaction. Therefore water–hydrogen separation must be applied. In addition, the oxygen-evolving reaction is very slow. In order to facilitate the reaction, a carrier gas must be used, either steam or other gas; this involves a further oxygen separation step. A more realistic value for the nickel–manganese–ferrite cycle efficiency is 48 %, which accounts for chemical separation in an actual system, power required to circulate the chemicals (W_{circ}), and heat losses (McQuillan et al. 2010). The efficiency is calculated with (5.18) where

$$W_{aux} = W_{circ} + \Delta G_{sep}. \tag{5.20}$$

Figure 5.15 presents a comparison of the energy efficiencies of two-step cycles listed in Table 5.2 with direct water thermolysis (DT). A conservative upper bound of 55 % of direct water thermolysis was adopted according to the estimates discussed in the previous section. This comparison shows that some two-step

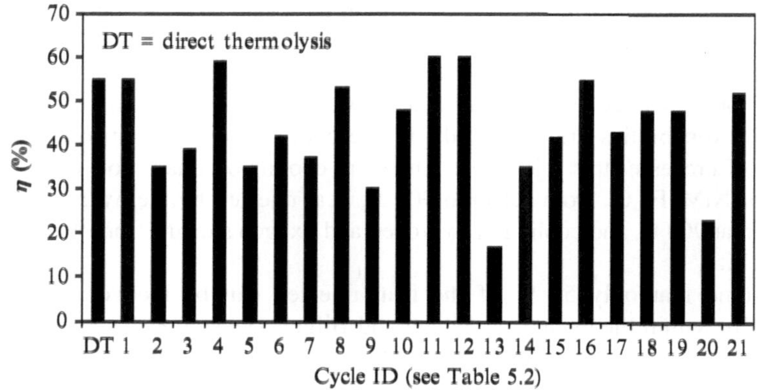

Fig. 5.15 Efficiency comparison of two-step cycles with direct water thermolysis

thermochemical water-splitting cycles can achieve the efficiency level of a single-step thermolysis process at a lower temperature level.

5.5 Multistep Thermochemical Cycles

Although two-step cycles are an attractive option for water splitting due to their unique advantage of simplicity of having only two chemical reactors and intrinsic hydrogen–oxygen separation features, the required working temperature is too high for the linkage to nuclear reactors. If an additional reaction step is added, the maximum temperature level requirement can decrease (see Fig. 5.2) at a temperature level that is easier to be attained by nuclear reactor technologies (e.g., generation IV reactors). In this section, thermochemical cycles with three and more steps are discussed.

Consider the ferrous oxide cycle #6 listed in Table 5.2. This cycle has an oxygen-evolving reaction at 2,475 K, according to $Fe_3O_4 \rightarrow 0.5O_2 + 3FeO$. In order to reduce the temperature, it is desired to find an intermediate compound to replace the direct Fe_3O_4 thermolysis with a two-step reaction process. It was found that one mole of Fe_3O_4 can be combined with three moles of $FeSO_4$ at 1,073 K to release a half mole of oxygen. In a subsequent reaction step at 473 K, three moles of FeO and three moles of $FeSO_4$ are generated (see Williams 1980). Table 5.3 illustrates the modification of the two-step ferrous oxide cycle into a three-step cycle with an oxygen-evolving process performed as two successive reactions. The maximum temperature with a three-step cycle is reduced to 1,073 K as compared to 2,475 K for the two-step cycle.

The second example given in Table 5.3 refers to a transformation of the two-step cycle #19 (Table 5.2) in a multistep cycle, again by replacing its oxygen-evolving reaction (originally conducted at 1,773 K) with a sequence of reactions. In the Ispra Mark 7A cycle, it is proposed to use a four-step reaction process of which one is the

Table 5.3 Multistep cycles obtained by replacing the O_2-evolving reaction of two-step cycles

H_2-evolving reaction	Cycle-closing reactions: For two-step cycle (one reaction)	Cycle-closing reactions: For multistep cycles (multiple reactions)
$H_2O + 3FeO \rightarrow H_2 + Fe_3O_4$ at 450 K (cycle #6, Table 5.2)	Oxygen-evolving reaction at 2,475 K: $Fe_3O_4 \rightarrow 0.5O_2 + 3FeO$	*Cycle name*: Jülich Center EOS $T_{max} = 1,073$ K Oxygen-evolving reaction at 2,073 K: $Fe_3O_4 + 3FeSO_4 \rightarrow 0.5O_2 + 3Fe_2O_3 + 3SO_2$ Intermediates recycling reaction at 473 K: $3Fe_2O_3 + 3SO_2 \rightarrow 3FeSO_4 + 3FeO$
$4H_2O + 3FeCl_2 \rightarrow H_2 + Fe_3O_4 + 6HCl$ at 923 K (cycle #19, Table 5.2)	Oxygen-evolving reaction at 1,773 K: $Fe_3O_4 + 6HCl \rightarrow 0.5O_2 + 3FeCl_2 + 3H_2O$	*Cycle name*: Mark 7A, $T_{max} = 1,273$ K Oxygen-evolving reaction at 1,273 K: $0.5Fe_2O_3 + 1.5Cl_2 \rightarrow 0.5O_2 + FeCl_3 + 0.25O_2$ Intermediates recycling reactions: $Fe_3O_4 + 0.25O_2 \rightarrow 1.5Fe_2O_3$ at 623 K $Fe_2O_3 + 6HCl \rightarrow 2FeCl_3 + 3H_2O$ at 423 K $3FeCl_3 \rightarrow 3FeCl_2 + 1.5Cl_2$ at 693 K *Cycle name*: Mark 9, $T_{max} = 923$ K Oxygen-evolving reaction at 473 K: $Fe_3O_4 + 6HCl + 1.5Cl_2 \rightarrow 0.5O_2 + 3FeCl_3 + 3H_2O$ Intermediates recycling reaction at 693 K: $3FeCl_3 \rightarrow 3FeCl_2 + 1.5Cl_2$

oxygen-evolving reaction, and the other three are reactions necessary for recycling all intermediate chemical compounds.

With Mark 7A, the oxygen evolution reaction is conducted at 1,273 K and this is the maximum temperature required by the cycle; the total number of steps of the Mark 7A cycle is five. A second option for the oxygen-evolving reaction does exist as proposed in the Mark 9 cycle. In this process, chlorine gas is added to hydrochloric acid to facilitate the reaction with magnetite at 473 K, a reaction which evolves oxygen. There is a subsequent chemical reaction at 693 K that recycles the intermediates: $FeCl_3$ and chlorine. The total number of steps of the Mark 9 cycle is therefore three.

Table 5.4 gives the complete list of Mark thermochemical cycles sorted by the number of steps in ascending order and with a maximum temperature in decreasing order. This set of cycles represents an important reference in the field, developed in the 1970s.

The first cycle established at Ispra was defined by De Beni in 1970 and called Mark 1 (see Savage et al. 1973). It successfully demonstrated that thermochemical water splitting is a feasible concept. This is a four-step cycle with a maximum temperature requirement of ~1,000 K. The additional chemical elements involved in the cycle (except H and O) are Br, Ca, and Hg.

The thermodynamic processes in the Mark 1 cycle can be represented in an *H–T* diagram such as the one indicated in Fig. 5.16. This diagram indicates two main enthalpy lines, namely, that of the reactant (water or steam) and that of the products (namely, the stoichiometric mixture of $H_2 + 0.5O_2$). State #1 in the diagram represents liquid water at $T_0 = 298.15$ K, while state #16 represents the final product state at T_0 and P_0. In order to drive water to evolve from state #1 to state #16, an enthalpy change of $\Delta H^0 = 285.813$ kJ must occur for 1 mol of liquid water. The dashed line splits the zone between the enthalpy line of reactant and that of the products in two parts, indicating the amount that must be given in the form of the Gibbs free energy (work) and the amount representing $T\Delta S$ (heat). If the process is conducted in an ideal electrolyzer operating at T_0, the amount of necessary Gibbs energy is $\Delta G^0 = 237,172$ kJ/mol, representing ~83 % of the total enthalpy required for the process.

The line 3–4 represents the enthalpy line for reactants of the highest temperature reaction in the Mark 1 cycle, which is $2H_2O(g) + CaBr_2(s) \rightarrow Ca(OH)_2(s) +$ $2HBr(g)$ at 1,003 K. The enthalpy line of the products (line passing through state point #5) and the line representing the Gibbs energy requirement of this reaction (dashed line passing through state #4) are also indicated in the diagram. As observed, the highest temperature reaction is spontaneous at 1,003 K because its associated Gibbs energy is slightly negative. The processes and state points in the diagram from Fig. 5.16 are described as follows:

- State #1: One mole of liquid water at T_0 and P_0.
- Process 1–2: Water heating, boiling, and steam superheating up to 473 K.
- State #3: Mixing of 1 mol H_2O with 1 mol of recycled H_2O and 1 mol $CaBr_2$.
- Process 3–4: Heating $2H_2O + CaBr_2$ from 473 K up to 1,003 K.

Table 5.4 List of Mark cycles sorted with the maximum temperature in decreasing order

N_s	ID	T_{max} (K)	Reactions	Elements/efficiency
3 Steps	8	1,173	$4H_2O(g) + 3MnCl_2(s) \rightarrow H_2(g) + Mn_3O_4(s) + 6HCl(g)$ at 973 K $1.5MnO_2(s) \rightarrow 0.5O_2(g) + 0.5Mn_3O_4(s)$ at 1,173 K $1.5Mn_3O_4(s) + 6HCl(g) \rightarrow 3MnCl_2(aq) + 1.5MnO_2(s) + 3H_2O(l)$ at 373 K	Cl, Mn/45 %
	16	1,123	$2HI(g) \rightarrow H_2(g) + I_2(g)$ at 773 K $H_2SO_4(g) \rightarrow 0.5O_2(g) + SO_2(g) + H_2O(g)$ at 1,123 K $2H_2O(l) + SO_2(g) + I_2(s) \rightarrow H_2SO_4(aq) + 2HI(aq)$ at 373 K	I, S/45 % (GA cycle)
	1S	1,073	$2HBr(g) + Hg(l) \rightarrow H_2(g) + HgBr_2(s)$ at 473 K $SrO(s) + HgBr_2(s) \rightarrow 0.5O_2 + Hg(g) + SrBr_2(s)$ at 773 K $H_2O(g) + SrBr_2(s) \rightarrow SrO(s) + 2HBr(g)$ at 1,073 K	Br, Hg, Sr/33 %
	2	1,073	$Mn_2O_3(g) + 4NaOH(l) \rightarrow H_2(g) + H_2O(g) + 2Na_2O \cdot MNO_2(s)$ at 1,073 K $2MnO_2(s) \rightarrow 0.5O_2(g) + Mn_2O_3(s)$ at 873 K $2H_2O(g) + 2Na_2O \cdot MnO_2(s) \rightarrow 4NaOH(aq) + 2MnO_2(s)$ at 373 K	Mn, Na/57 %
	9	923	$4H_2O(g) + 3FeCl_2(s) \rightarrow H_2(g) + Fe_3O_4(s) + 6HCl(g)$ at 923 K $Fe_3O_4(s) + 1.5Cl_2(g) + 6HCl(g) \rightarrow 0.5O_2(g) + 3FeCl_3(g) + 3H_2O(g)$ at 473 K $3FeCl_3(g) \rightarrow 3FeCl_2(s) + 1.5Cl_2(g)$ at 693 K	Cl, Fe/44 %
4 Steps	1C	1,173	$4HBr(aq) + Cu_2O(s) \rightarrow H_2(g) + 2CuBr_2(aq) + H_2O(l)$ at 373 K $CuO(s) \rightarrow 0.5O_2(g) + Cu_2O(s)$ at 1,173 K $4H_2O(g) + 2CaBr_2(s) \rightarrow 2Ca(OH)_2(s) + 4HBr(g)$ at 730 K $2CuBr_2(aq) + 2Ca(OH)_2(aq) \rightarrow 2CuO(s) + 2CaBr_2(aq) + 2H_2O(l)$ at 373 K	Br, Ca, Cu/36 %
	2C	1,123	$H_2O(g) + CO(g) \rightarrow H_2(g) + CO_2(g)$ at 773 K $2MnO_2(s) \rightarrow 0.5O_2(g) + Mn_2O_3(s)$ at 873 K $Mn_2O_3(s) + 2Na_2CO_3(s) \rightarrow 2NaO \cdot MnO_2(s) + CO_2(g) + CO(g)$ at 1,123 K $nH_2O(l) + 2Na_2O \cdot MnO_2(s) + 2CO_2(g) \rightarrow 2Na_2CO_2 \cdot nH_2O(aq) + 2MnO_2(s)$ at 373 K	C, Mn, Na/57 %
	12	1,123	$2NH_4I(s) \rightarrow H_{2(g)} + 2NH_3(g) + I_2(g)$ at 973 K $ZnSO_4(s) \rightarrow 0.5O_2(g) + ZnO(s) + SO_2(g)$ at 1,123 K $2H_2O(l) + I_2(s) + SO_2(g) + 4NH_3(g) \rightarrow 2NH_4I(s) + (NH_4)_2SO_4(s)$ at 323 K $(NH_4)_2SO_4(s) + ZnO(s) \rightarrow ZnSO_4(s) + 2NH_3(g) + H_2O(g)$ at 773 K	I, N, S, Zn N/A

(continued)

Table 5.4 (continued)

N_s	ID	T_{max} (K)	Reactions	Elements/efficiency
	3	1,073	$2HCl(g) + 2VOCl(s) \rightarrow H_2(g) + 2VOCl_2(s)$ at 443 K	Cl, V/48 %
			$H_2O(g) + Cl_2(g) \rightarrow 0.5O_2(g) + 2HCl(g)$ at 1,073 K	
			$4VOCl_2(s) \rightarrow 2VOCl(s) + 2VOCl_3(g)$ at 873 K	
			$2VOCl_3(g) \rightarrow 2VOCl_2(s) + Cl_2(g)$ at 473 K	
	4	1,073	$H_2S(g) \rightarrow H_2(g) + 0.5S_2(g)$ at 1,073 K	Cl, Fe, S/35 %
			$H_2O(g) + Cl_2(g) \rightarrow 0.5O_2(g) + 2HCl(g)$ at 1,073 K	
			$2HCl(g) + S(s) + 2FeCl_2(s) \rightarrow H_2S(g) + 2FeCl_3(s)$ at 373 K	
			$2FeCl_3(g) \rightarrow 2FeCl_2(s) + Cl_2(g)$ at 693 K	
	6	1,073	$2HCl(g) + 2CrCl_2(s) \rightarrow H_2(g) + 2CrCl_3(s)$ at 433 K	Cl, Cr, Fe/46 %
			$H_2O(g) + Cl_2(g) \rightarrow 0.5O_2(g) + 2HCl(g)$ at 1,073 K	
			$2CrCl_3(s) + 2FeCl_2(s) \rightarrow 2CrCl_2(s) + 2FeCl_3(g)$ at 973 K	
			$2FeCl_3(g) \rightarrow 2FeCl_2(s) + Cl_2(g)$ at 623 K	
	1	1,003	$2HBr(aq) + Hg(l) \rightarrow H_2(g) + HgBr_2(aq)$ at 523 K	Br, Ca, Hg/26 %
			$HgO(s) \rightarrow 0.5O_2(g) + Hg(g)$ at 873 K	
			$2H_2O(g) + CaBr_2(s) \rightarrow Ca(OH)_2(s) + 2HBr(g)$ at 1,003 K	
			$HgBr_2(s) + Ca(OH)_2(s) \rightarrow CaBr_2(s) + HgO(s) + H_2O(g)$ at 473 K	
	15	923	$4H_2O(g) + 3FeCl_2(s) \rightarrow H_2(g) + Fe_3O_4(s) + 6HCl(g)$ at 923 K	Cl, Fe/46 %
			$H_2O(g) + Cl_2(g) \rightarrow 0.5O_2(g) + 2HCl(g)$ at 873 K	
			$Fe_3O_4(s) + 8HCl(g) \rightarrow FeCl_2(s) + 2FeCl_3(s) + 4H_2O(g)$ at 573 K	
			$2FeCl_3(s) \rightarrow 2FeCl_2(s) + Cl_2(g)$ at 673 K	
5 Steps	7A	1,273	$4H_2O(g) + 3FeCl_2(s) \rightarrow H_2(g) + Fe_3O_4(s) + 6HCl(g)$ at 923 K	Cl, Fe/29 %
			$0.5Fe_2O_3(s) + 1.5Cl_2(g) \rightarrow 0.5O_2(g) + FeCl_3(g) + 0.25O_2(g)$ at 1,273 K	
			$Fe_3O_4(s) + 0.25O_2(g) \rightarrow 1.5Fe_2O_3(s)$ at 623 K	
			$Fe_2O_3(s) + 6HCl(g) \rightarrow 2FeCl_3(s) + 3H_2O(g)$ at 423 K	
			$3FeCl_2(g) \rightarrow 3FeCl_2(s) + 1.5Cl_2(g)$ at 693 K	
	7B	1,273	$4H_2O(g) + 3FeCl_2(s) \rightarrow H_2(g) + Fe_3O_4(s) + 6HCl(g)$ at 923 K	Cl, Fe/48 %
			$1.5Fe_2O_3(s) + 4.5Cl_2(g) \rightarrow 0.5O_2(g) + 3FeCl_2(s) + 1.75O_2(g)$ at 1,273 K	
			$Fe_3O_4(s) + 0.25O_2(g) \rightarrow 1.5Fe_2O_3(s)$ at 623 K	
			$6HCl(g) + 1.5O_2(g) \rightarrow 3Cl_2(g) + 3H_2O(g)$ at 673 K	
			$3FeCl_2(g) \rightarrow 3FeCl_2(s) + 1.5Cl_2(g)$ at 693 K	

	5	1,173	$2HBr(g) + Hg(l) \rightarrow H_2(g) + HgBr_2(s)$ at 873 K $HgO(s) \rightarrow 0.5O_2(g) + Hg(g)$ at 873 K $H_2O(g) + CaBr_2(s) + CO_2(g) \rightarrow CaCO_3(s) + 2HBr(g)$ at 873 K $CaCO_3(s) \rightarrow CaO(s) + CO_2(g)$ at 1,173 K $HgBr_2(s) + CaO(s) + nH_2O(l) \rightarrow CaBr_2 \cdot nH_2O(aq) + HgO(s)$ at 473 K	Br, C, Ca, Hg/23 %
	6C	1,073	$2HCl(g) + 2CrCl_2(s) \rightarrow H_2(g) + 2CrCl_3(s)$ at 443 K $H_2O(g) + Cl_2(g) \rightarrow 0.5O_2(g) + 2HCl(g)$ at 1,073 K $2CrCl_3(s) + 2FeCl_2(s) \rightarrow 2CrCl_2(s) + 2FeCl_3(g)$ at 973 K $2FeCl_3(s) + 2CuCl(s) \rightarrow 2FeCl_2(s) + 2CuCl_2(s)$ at 423 K $2CuCl_2(s) \rightarrow 2CuCl(l) + Cl_2(g)$ at 773 K	Cl, Cr, Cu, Fe/29 %
	7	1,073	$4H_2O(g) + 3FeCl_2(s) \rightarrow H_2(g) + Fe_3O_4(s) + 6HCl(g)$ at 923 K $1.5H_2O(g) + 1.5Cl_2(g) \rightarrow 0.5O_2(g) + 3HCl(g) + 0.25O_2(g)$ at 1,073 K $Fe_3O_4(s) + 0.25O_2(g) \rightarrow 1.5Fe_2O_3(s)$ at 623 K $1.5Fe_2O_3(s) + 9HCl(g) \rightarrow 3FeCl_3(s) + 4.5H_2O(g)$ at 423 K $3FeCl_2(g) \rightarrow 3FeCl_2(s) + 1.5Cl_2(g)$ at 693 K	Cl, Fe/40 %
	14	1,003	$4H_2O(g) + 3FeCl_2(s) \rightarrow H_2(g) + Fe_3O_4(s) + 6HCl(g)$ at 923 K $H_2O(g) + Cl_2(g) \rightarrow 0.5O_2(g) + 2HCl(g)$ at 1,003 K $Fe_3O_4(s) + 0.5Cl_2(g) \rightarrow 1/3\,FeCl_2(g) + 4/3\,Fe_2O_3(s)$ at 573 $4/3\,Fe_2O_3(s) + 8HCl(g) \rightarrow 8/3\,FeCl_2(s) + 4H_2O(g)$ at 443 K $3FeCl_2(s) \rightarrow 3FeCl_2(s) + 1.5Cl_2(g)$ at 553 K	Cl, Fe/45 %
	1B	1,003	$2HBr(aq) + Hg_2Br_2(s) \rightarrow H_2(g) + 2HgBr_2(s)$ at 393 K $HgO(s) \rightarrow 0.5O_2(g) + Hg(g)$ at 873 K $2H_2O(g) + CaBr_2(s) \rightarrow Ca(OH)_2(s) + 2HBr(g)$ at 1,003 K $HgBr_2(s) + Hg(l) \rightarrow Hg_2Br_2(s)$ at 393 K $HgBr_2(s) + Ca(OH)_2(s) \rightarrow CaBr_2(s) + HgO(s) + H_2O(g)$ at 473 K	Br, Ca, Hg/26 %
6 Steps	10	1,123	$2NH_4I(s) \rightarrow H_2(g) + 2NH_3(g) + I_2(g)$ at 900 K $SO_3(g) \rightarrow 0.5O_2(g) + SO_2(g)$ at 1,123 K $2H_2O(l) + I_2(s) + SO_2(g) + 4NH_3 \rightarrow 2NH_4I(s) + (NH_4)_2SO_4(s)$ at 323 K $(NH_4)_2SO_4(s) + Na_2SO_4(s) \rightarrow 2NaHSO_4(s) + 2NH_3(g)$ at 673 K $2NaHSO_4(s) \rightarrow Na_2S_2O_7(s) + H_2O(g)$ at 673 K $Na_2S_2O_7(s) \rightarrow Na_2SO_4(s) + SO_3(g)$ at 825 K	I, N, S/32 %

N_s number of steps, *GA* General Atomics, *N/A* efficiency not available, N_s number of steps

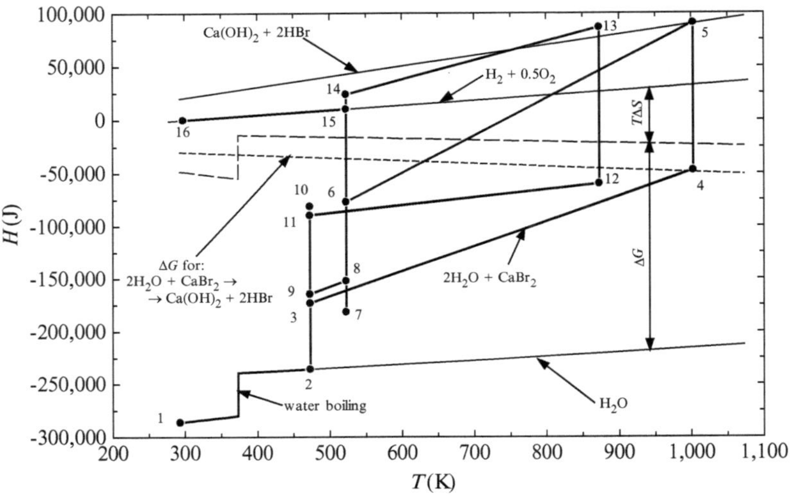

Fig. 5.16 Representation of Mark 1 process in an *H-T* diagram [data from Savage et al. (1973)]

- Process 4–5: Reaction $2H_2O(g) + CaBr_2(s) \rightarrow Ca(OH)_2(s) + 2HBr(g)$ at 1,003 K.
- Process 5–6: Cooling reaction products $Ca(OH)_2(s) + 2HBr(g)$ to 523 K.
- State #6: Separation of 2HBr from $Ca(OH)_2$ at 523 K.
- Process 6–7: Reacting 2HBr from state #6 with Hg from state #14, at 523 K, according to $2HBr(aq) + Hg(l) \rightarrow H_2(g) + HgBr_2(aq)$.
- State #7: Separation of H_2 from $HgBr_2$ from aqueous solution.
- State #8: Mixing $HgBr_2(s)$ from state #7 with $Ca(OH)_2$ from state #6, at 523 K.
- Process 8–9: Cooling $HgBr_2(s)$ and $Ca(OH)_2$ from 523 K to 473 K.
- Process 9–10: Reaction $HgBr_2(s) + Ca(OH)_2(s) \rightarrow CaBr_2(s) + HgO(s) + H_2O(g)$ at 473 K.
- State #11: Extraction of HgO(s) from product stream at 473 K.
- Process 11–12: Heating HgO from 473 to 873 K.
- Process 12–13: Reaction $HgO(s) \rightarrow 0.5O_2(g) + Hg(g)$ at 873 K.
- Process 13–14: Cooling reaction products $0.5O_2$ and Hg from 873 K to 523 K.
- State #14: Separation of $0.5O_2$ from Hg at 523 K.
- Process 15–16: Cooling 1 mol H_2 and 0.5 mol O_2 from 523 K to T_0.

According to Sato (1979) the reported efficiency of the Mark 1 lab-scale prototype was 55 %, although scale-up studies reported 35 % (McQuillan et al. 2010). However, this cycle was abandoned because several side reactions occur and there is a very large quantity of mercury that must be handled for an industrial scale plant to safely operate. Beside its basic version, the Mark 1 cycle has three additional variants: 1B, 1C, and 1S. In version 1B, the hydrogen evolution reaction occurs at 393 K and one additional chemical reaction is required to recycle the chemicals. In the Mark 1S cycle, the number of steps is reduced to three. The need of calcium compounds is eliminated; however, the maximum temperature required is slightly increased to 1,073 K. The cycle Mark 1C eliminates the use of mercury and reduces significantly the ecological risks in case of accidents in a large-scale chemical plant. This cycle uses copper compounds instead of mercury compounds. The main drawback of the Mark 1C variant with respect to the basic Mark 1 cycle is the higher temperature level requirement, which is increased to 1,173 K. The Mark 2 cycles are based on manganese and sodium elements. Mark 3 uses chlorine and vanadium and requires a maximum temperature of 1,073 K. This cycle uses a prototypical oxygen-evolving reaction, namely, $H_2O(g) + Cl_2(g) \rightarrow 0.5O_2(g) + 2HCl(g)$, which is also employed in many other cycles. This reaction is the reverse Deacon reaction, also mentioned in the previous section. Sulfur, iron, and chlorine are the elements used in the Mark 4 cycle where the same oxygen-evolving reaction as with Mark 3 is employed. Mark 5 was a 5-step trial at Ispra, but not pursued because it uses mercury compounds. Mark 6 cycles with two versions are based on chromium, chlorine, and iron.

Three iron–chlorine cycles were also developed as five-step reaction processes: Mark 7, 7A, and 7B. Iron–chlorine cycles refer to a second phase of research and development on thermochemical water splitting at Ispra (Beghi 1986). In this phase, a manganese–chlorine cycle, denoted Mark 8, was studied with three reaction steps. This research phase culminated with the development of a Mark 9 cycle with three reaction steps. This cycle, together with Mark 15 (also based on iron-chloride), has the lowest temperature requirement of 920 K. A hydrogen yield of more than 80 % was obtained in a continuous bench-scale unit (Beghi 1986). For the Mark 15 cycle, a study of a semi-continuous reactor for the hydrogen-evolving reaction was developed and tests were conducted for a production rate of 25 standard liters of hydrogen per hour with 90 % yield. The reverse Deacon reaction was successfully demonstrated at 873 K. The Mark 14 cycle is one of the same iron–chlorine series with a maximum temperature of 1,003 K. However, the researchers at Ispra abandoned Mark 14 and Mark 15 cycles because they found that the thermal efficiency is not satisfactory (~20 % for Mark 14 and ~35 % for Mark 15).

The third phase of research at Ispra focused on sulfur family cycles starting with Mark 10, which is a six-reaction-step process. The development of Mark 12 followed, which is similar to Mark 10, but it includes an additional chemical element (Zn) in a four-step process operating at a similar temperature. The last cycle in the Mark series is the 16thcycle, which is the well-known, three-reaction-step, sulfur–iodine cycle. This was further developed by General Atomics. The efficiency of the Mark 16 cycle is in the order of ~44 %.

The pioneering research at Ispra (Mark cycles) and other research centers in the 1970s led to the establishment of a number of reaction schemes for thermochemical water splitting. Table 5.5 lists the main research centers active in the 1970s in the field of thermochemical water-splitting and the most relevant processes proposed at each center. Sato (1979) classified the reactions for thermochemical water splitting cycles in four categories:

Table 5.5 Main institutions that developed thermochemical cycles in the 1970s

Country	Institution	Developed process	Elements
France	Gas de France	K-process	K
		Souriau process	Sn
Germany	Rheinisch-Westfaelische Technische Hochschule Aachen	"Jülich Fe-S"	Fe, S
Italy	Joint Research Centre, Ispra	Mark series (e.g., #16)	I, S
Japan	Japan Atomic Energy Research Institute (JAERI)	NiIS-Process	I, Ni, S
	Mitsubishi Heavy Industry	Fe–Cu–Cl	Cl, Cu, Fe
	University of Osaka	Osaka-75	Ba, C, I, N
	University of Tokyo	UT-3	Br, Ca, Fe
USA	Argonne National Laboratory	Ag-process	Ag, Br, Na
	General Atomic Company	GA (Mark 16)	I, S
	General Electric	Agnes process	Cl, Fe, Mg
		Catherine process	I, K, Li, Ni
	Institute of Gas Technology (IGT)	A-2	Cl, Fe
	Lawrence Livermore Laboratory, University of California (LLLC)	$MgSeO_4$-Process	Mg, Se
	Los Alamos Scientific Laboratory (LASL)	Li–Mn-process	Li, Mn
	Oak Ridge National Laboratory (ORNL)	Fe-process	Ba, Cr
	University of Kentucky (UoK)	UoK-8	Ba, Fe, S
	Allison Laboratory of General Motors	Funk, Ta-process	Cl, Ta
		Funk, V-process	Cl, V

- Hydrolysis reaction, $H_2O + \sum R \rightarrow \sum P$;
- Hydrogen evolution reactions, $HX + \sum R \rightarrow H_2 + \sum P$;
- Oxygen evolution reactions, $XO + \sum R \rightarrow 0.5O_2 + \sum P$;
- Reactions of regeneration of intermediate reactants $\sum P \rightarrow \sum R$;

where R and P refer to reactants and products, respectively, HX is a hydrogen-containing compound, and XO is an oxygen-containing compound.

Table 5.6 shows some prototypical hydrolysis reactions. In hydrolysis, water is combined with a reactant and it forms intermediate reactants and/or hydrogen or oxygen depending on the reaction type. The reverse Deacon reaction, for example, generates oxygen while hydrolyzing chlorine; the iron-steam reaction produces hydrogen; and the hydrolysis reaction of a halogen such as iodine in the presence of an oxygen acceptor such as SO_2 generates neither hydrogen nor oxygen but rather compounds that must be recycled. Table 5.7 presents some prototypical hydrogen-evolving reactions. Thermolysis of a hydrogen halide is a typical example of a gas-phase reaction of hydrogen evolution. Table 5.8 presents some oxygen evolution reactions such as a metallic oxide or oxyacid thermolysis, or a reaction of a metallic oxide with a halogen or the reverse Deacon reaction. Table 5.9 exemplifies some reactions for regeneration of intermediate reactants in thermochemical cycles.

The practicality of any thermochemical reaction can be determined based on the reaction yield (or the reactant conversion ratio). As a general criterion, one may consider that (see Sato 1979): (1) if the yield (or conversion) is smaller than ~0.001,

Table 5.6 Prototypical water hydrolysis reactions

Reaction	Examples
$nH_2O + MO_m \rightarrow nH_2 + MO_{m+n}$, reaction of a metallic oxide MO_m with water to form a higher oxide MO_{m+n}	$4H_2O + 3Fe \rightarrow 4H_2 + Fe_3O_4$ at 773 K $H_2O + 3FeO \rightarrow H_2 + Fe_3O_4$ at 450 K $2H_2O + 2Cs \rightarrow H_2 + 2CsOH$ at 450 K
$nH_2O + MX_m \rightarrow (n - m/2)H_2 + mHX + MO_n$, $2n \geq m$, reaction of a metallic halide with water to form hydrogen, hydrogen halide, and metallic oxide	$4H_2O + 3FeCl_2 \rightarrow H_2 + 6HCl + Fe_3O_4$ at 923 K $4H_2O + 3MnCl_2 \rightarrow H_2 + 6HCl + Mn_3O_4$ at 923 K $4H_2O + SrBr_2 \rightarrow 2HBr + SrO$ at 1,073 K
$H_2O + X_2 \rightarrow 0.5O_2 + 2HX$, hydrolysis with a halogen	$H_2O + Cl_2 \rightarrow 0.5O_2 + 2HCl$ at 873 K $H_2O + Br_2 \rightarrow 0.5O_2 + 2HBr$ at 873 K
$H_2O + X_2 + A \rightarrow 2HX + AO$, reaction of steam with a halogen in the presence of an oxygen acceptor	$H_2O + 0.5Br_2 + 0.5SO_2 \rightarrow HBr + 0.5H_2SO_4$ at 373 K $H_2O + 0.5I_2 + 0.5SO_2 \rightarrow HI + 0.5H_2SO_4$ at 373 K

Table 5.7 Some types of thermochemical reactions for hydrogen generation

Reaction	Examples
$MX_m + HX_n \rightarrow H_2 + MX_{m+n}$, reaction of hydrogen halide with a metallic salt or a metal to form metallic halide	$2Cu + 2HCl \rightarrow H_2 + 2CuCl$ at 703 K $2VCl_2 + 2HCl \rightarrow H_2 + 2VCl_3$ at 298 K $Hg_2Br_2 + 2HBr \rightarrow H_2 + 2HgBr_2$ at 393 K
$2HX \rightarrow H_2 + X_2$, thermolysis of a hydrogen halide to form hydrogen and halogen	$2HI \rightarrow H_2 + I_2$ at 773 K $2HBr \rightarrow H_2 + Br_2$ at 1,600 K
$nH_2O + MO_m \rightarrow nH_2 + MO_{m+n}$, hydrolysis of a metallic oxide to generate hydrogen and a higher oxide MO_{m+n} (see Table 5.6).	$H_2O + 3MnO \rightarrow H_2 + Mn_3O_4$ at 1,123 K $H_2O + 3FeO \rightarrow H_2 + Fe_3O_4$ at 450 K $H_2O + 3CoO \rightarrow H_2 + Co_3O_4$ at 1,273 K
$nH_2O + MX_m \rightarrow (n - m/2)H_2 + mHX + MO_n$, $2n > m$, hydrolysis of a metallic halide to form H_2, hydrogen halide and metallic oxide (see Table 5.6)	$4H_2O + 3FeCl_2 \rightarrow H_2 + 6HCl + Fe_3O_4$ at 923 K $4H_2O + 3MnCl_2 \rightarrow H_2 + 6HCl + Mn_3O_4$ at 923 K

Table 5.8 Some types of thermochemical reactions for oxygen generation

Reaction	Examples
$MO_{m+1} \rightarrow 0.5O_2 + MO_m$, completed decomposition of an oxide or partial decomposition of a higher oxide	$HgO \rightarrow 0.5O_2 + Hg$ at 873 K $2CuO \rightarrow 0.5O_2 + Cu_2O$ at 1,173 K $SO_3 \rightarrow 0.5O_2 + SO_2$ at 1,073 K
$AO \rightarrow 0.5O_2 + A$, thermolysis of oxyacid (or its salt e.g., chlorate, iodate, sulfate, nitrate) to form a halide (e.g., chloride, iodide, bromide) or oxide	$MgSO_4 \rightarrow 0.5O_2 + MgO + SO_2$ at 1,400 K $H_2SO_4 \rightarrow 0.5O_2 + H_2O + SO_2$ at 1,123 K $Fe_3O_4 \rightarrow 0.5O_2 + 3FeO$ at 2,473 K
$MO + X_2 \rightarrow 0.5O_2 + MX_2$, reaction of a metallic oxide with a halogen to form oxygen and halide	$CuO + 0.5I_2 \rightarrow 0.5O_2 + CuI$, at 973 K $0.25Fe_3O_4 + 2.25Cl_2 \rightarrow 0.5O_2 + 0.75\,FeCl_3$, at 1,273 K
$H_2O + Cl_2 \rightarrow 0.5O_2 + 2HCl$, reverse Deacon reaction of chlorine hydrolysis	Hydrogen evolution in Mark 3, 4, 6, and 6C at 1,073 K Mark 14 at 1,003 K, Mark 15 cycle at 873 K

the reaction is infeasible; (2) the reaction has a marginal yield if the reactant conversion is between ~0.001 and ~0.1; and (3) if the conversion is higher than ~0.1 the reaction has a good yield and it can be accomplished with ease. Consider—as an

Table 5.9 Some types of reactions to recycle intermediary reagents in thermochemical cycles

Reaction	Examples
$M_3O_4 + 8HX \rightarrow 4H_2O + 3MX_2 + X_2$, (or $\rightarrow MX_2 + 2MX_3$) reaction of a metal oxide with a hydrogen halide	$Fe_3O_4 + 8HCl \rightarrow 4H_2O + FeCl_2 + 2FeCl_3$ at 573 K $Fe_3O_4 + 8HBr \rightarrow 4H_2O + 3FeBr_2 + Br_2$ at 573 K
$MX_{n+1} \rightarrow MX_n + 0.5X_2$, thermolysis of a metallic halide to form a lower halide and a halogen	$FeCl_3 \rightarrow FeCl_2 + 0.5Cl_2$ at 693 K $CuCl_2 \rightarrow CuCl + 0.5Cl_2$ at 773 K
$MX_{n+1} + NX_n \rightarrow MX_n + NX_{n+1}$, reaction of two metallic halides that exchange one halogen atom	$CrCl_3(s) + FeCl_2(s) \rightarrow CrCl_2(s) + FeCl_3(g)$ at 973 K $FeCl_3(s) + CuCl(s) \rightarrow FeCl_2(s) + CuCl_2(s)$ at 423 K
$M_{n+1}O_{n+2} + 1/2n\,O_2 \rightarrow (1 + 1/n)M_nO_{n+1}$, oxidation of a metal oxide to form an oxide or oxidation of a metal sulfide to form a sulfate $M_2S + SO_2 + 3O_2 \rightarrow 2MSO_4$	$Fe_3O_4(s) + 0.25O_2(g) \rightarrow 1.5Fe_2O_3(s)$ at 623 K $2Fe_3O_4 + CO \rightarrow 3Fe_2O_3 + C$ at 523 K $Cu_2S + SO_2 + 3O_2 \rightarrow 2CuSO_4$ at 573 K

example—a thermochemical reaction $A(s) \rightarrow B(s) + 0.5C(g)$ which is the thermal decomposition of a single solid reactant $A(s)$ to form a solid product $B(s)$ and evolve a gas $C(g)$. Such reactions are listed in Table 5.9; an example is the reaction of cuprous chlorine decomposition: $CuCl(s) \rightarrow Cu(s) + 0.5Cl_2(g)$.

The Gibbs free energy for the reactions of this type is given by $\Delta G = \Delta G^0 + 0.5RT\ln(y)$, where $y = P/P_0$ quantifies the product yield, P is the pressure of the gaseous product evolved from the reaction, and $P_0 = 1$ bar. At equilibrium ($\Delta G = 0$) it results in $\Delta G^0 = -0.5RT\ln(y)$; for $y \leq 0.001$, one has $\Delta G^0_{0.001} \geq 28.7T$ and for $y \geq 0.1$, one has $\Delta G^0_{0.1} \leq 9.6T$, both in kJ/reaction. Figure 5.17 shows $\Delta G^0(T)$ curves, namely, the Gibbs free energy variation with temperature at standard pressure for some decomposition reactions used in thermochemical cycles. The graph space is divided into three regions by the lines $\Delta G^0_{0.001}$ and $\Delta G^0_{0.1}$: impractical yield, marginal yield, and good yield. Among reaction types analyzed in Fig. 5.17, only reactions with $\Delta G^0 < 10$ kJ/reaction give a good yield at temperatures lower than 1,200 K. Based on this criterion, reactions #6, #8, and #1 may be selected as the most promising reactions. Reaction #2 is impractical for temperatures below 1,100 K, while the other reactions show a marginal yield.

The same procedure can be applied to other types of gas-evolving reactions to determine the magnitude of the yield or the conversion ratio. Several examples of chemical reactions are illustrated in the graph from Fig. 5.18 in terms of the ΔG^0 variation with temperature. Reactions #1 and #2 evolve oxygen by thermolysis of a single gaseous reactant. From 1 mol of reactant, it yields1.5 mol total of gaseous products. If one denotes with y the yield (moles of converted reactant per mole of reactant fed) then at equilibrium, one has $\Delta G^0 = -RT\ln(y^{1.5}/(1-y))$, where $1-y$ represents the unconverted fraction of gaseous reactant.

The reactions #3 to #7 generate 1 mol of gas in the total products from 1 mol of feed; therefore at equilibrium, $\Delta G^0 = -RT\ln(y/(1-y))$; for #8 one has $\Delta G^0 = -RT$

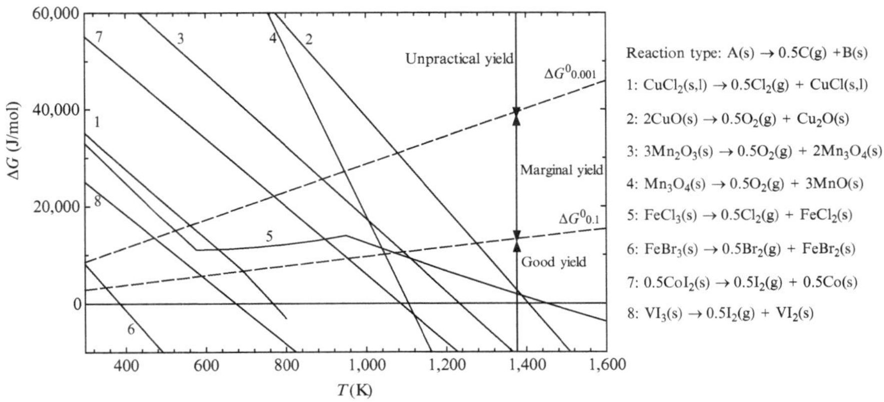

Fig. 5.17 Analysis of decomposition reactions of type $A(s) \rightarrow 0.5C(g) + B(s)$ based on the Gibbs free energy magnitude at various temperatures

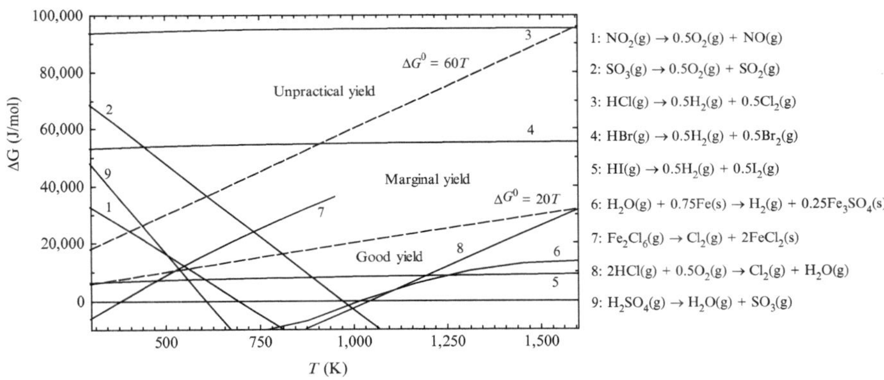

Fig. 5.18 Gibbs energy variation and practical temperature range of some chemical reactions used in multistep thermochemical cycles

$\ln(y^2/(1-y)^{2.5})$ and for reaction #9, one has $\Delta G^0 = -RT\ln(y^2/(1-y))$. Above the line, $\Delta G^0 = 60T$ represented in the graph from Fig. 5.18 (observe the dashed line), the yield for reactions #1 and #2 is therefore 0.008, while for reaction group #3–#7, the yield is 0.0007, and for #8 and #9 it is 0.026 and 0.027, respectively. Therefore, above this line, the reactions #1–#7 are infeasible, while reactions #8 and #9 are difficult. Below line $\Delta G^0 = 20T$ (dashed line), the reaction yields can be characterized as good because the yield is higher than 0.18 for reactions #1 and #2, higher than 0.083 for reactions #3–#7, higher than 0.22 for reaction #8, and higher than 0.23 for reaction #9. From this plot, one learns that the decomposition of HCl and HBr in the gaseous phase are impractical processes, but decomposition of HI (reaction #5) gives a good yield; thus it is practical, even for temperatures well below 1,000 K. Note that the reactions in Fig. 5.18 include oxygen-evolving processes #1 and #2, hydrogen-evolving processes #3–#6, hydrolysis #6, and reactions to recycle intermediate reagents #7–#9.

The reaction #5 is used in the sulfur–iodine cycle and Mark 16 cycle where it is conducted at 773 K. Reaction #8 is the Deacon reaction which was used industrially in the past to generate chlorine gas. This reaction requires a $CuCl_2$ catalyst and operates at a typical temperature of 600–723 K. This reaction is used in Mark 7 and 7B cycles to recycle chlorine.

The reverse Deacon reaction is considered the most important step in many thermochemical cycles, as an oxygen evolution reaction. It can be carried out in quartz reactors without catalysts at temperatures over 1,273 K with a feed ratio of steam/chlorine of ~5, residence time of 8–24 s, and a departure from equilibrium of 13–25 % (see Sato 1979). This reaction is a basic chlorination of water, namely,

$$H_2O(g) + Cl_2(g) \rightarrow 0.5O_2(g) + 2HCl(g). \quad (5.21)$$

In order to improve the yield at lower temperatures, a catalyst (e.g., $CuCl_2$) must be used for the reverse Deacon reaction. The temperature can be lowered to roughly 1,000 K when the yield is ~60 %; at lower temperatures, the reaction rate decreases as the backward reaction is more favored. At 900 K, the yield becomes ~50 %; in the Mark 15 cycle, this reaction is performed at 873 K. An alternative way to conduct the reverse Deacon reaction is via a three-step reaction process with magnesium-based compounds as follows (see Sato 1979):

$$\begin{cases} Cl_2 + Mg(OH)_2 \rightarrow 0.5O_2 + MgCl_2 + H_2O & \text{at } 623\,K \\ MgCl_2 + H_2O \rightarrow Mg(OH)Cl + HCl & \text{at } 570\,K \\ Mg(OH)Cl + H_2O \rightarrow MgCl_2 + HCl & \text{at } 623\,K. \end{cases} \quad (5.22)$$

The chlorination reactions are also important in iron- or copper-based thermochemical processes. Chlorination of ferrites or magnetite is used as an oxygen-evolving process or intermediate ferrite-based reagent recycling. The chlorination can be performed directly with chlorine or with hydrochloric acid. Figure 5.19 presents chlorination results of Fe_3O_4 with HCl at various temperatures. The conversion of Fe_3O_4 to $FeCl_2$ increases with temperature, reaching about 60 % at 923 K. Depending on the temperature and on how the reaction is conducted, this chlorination can evolve according to one of the following sequences of reactions:

$$\begin{aligned} &Fe_3O_4(s) + 8HCl \rightarrow 3FeCl_2(s) + Cl_2(g) + 4H_2O(g) \\ &Fe_3O_4(s) + 8HCl(g) \rightarrow FeCl_2(s) + Fe_2Cl_6(g) + 4H_2O(g) \\ &Fe_3O_4(s) + 6HCl(g) \rightarrow Fe_2Cl_6(g) + 3H_2O(g). \end{aligned}$$

Chlorination of ferrous ferric oxide can also be made with gaseous chlorine, a case in which higher temperatures are required (700–1,300 K). Another class of chlorination reactions of interest is that of copper oxide to release oxygen.

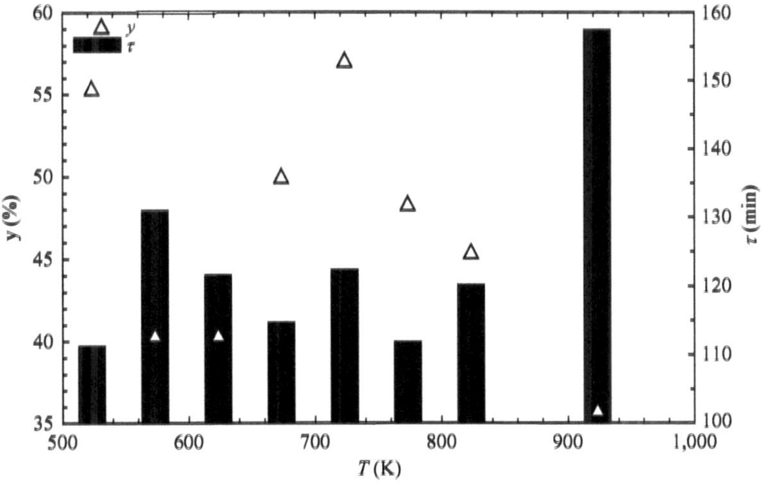

Fig. 5.19 Chlorination of Fe_3O_4 with hydrochloric acid [data from Knoche et al. (1978), y = conversion, τ = residence time]

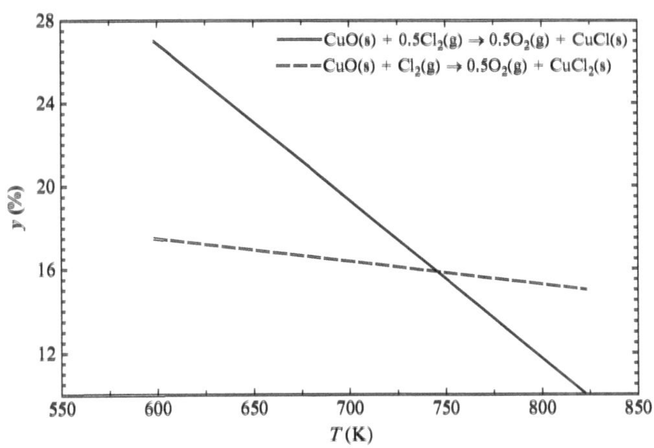

Fig. 5.20 Chlorination of copper oxide [data from Marin et al. (2011), Figure 10, y = conversion of reactants]

Chlorination of copper compounds is important for some copper–chlorine cycles. Two chlorination paths for CuO are indicated in Fig. 5.20.

Apart from chlorination, other halogenation reactions (with bromine or iodine instead of chlorine) are important in some thermochemical cycles, when after hydrolysis, a halide must be recycled. The reaction can be conducted either in the gas phase or in an aqueous solution. Some reactions are shown as follows:

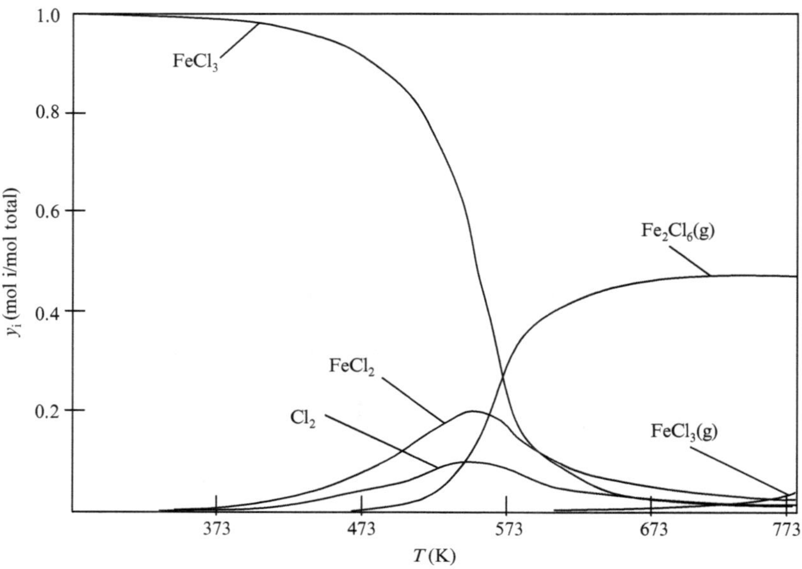

Fig. 5.21 Chemical equilibrium concentrations with ferric chloride decomposition [data from Abanades et al. (2006)]

$$3Fe_3O_4 + I_2 \rightarrow FeI_2 + 4Fe_2O_3$$
$$3Fe_3O_4 + Br_2 \rightarrow FeBr_2 + 4Fe_2O_3$$
$$Fe_3O_4 + 8HI \rightarrow 3FeI_2 + 4H_2O + I_2.$$

The decomposition of a metallic chloride may be used within a thermochemical cycle as a source of chlorine (e.g., for a chlorination reaction). Two examples are the decomposition of copper di-chloride and ferric chloride as shown below:

$$CuCl_2 \rightarrow 0.5Cl_2 + CuCl$$
$$FeCl_3 \rightarrow 0.5Cl_2 + FeCl_2.$$

Figure 5.21 presents the equilibrium concentrations for the decomposition of ferric chloride at standard pressure. It can be observed that this reaction shows a maximum yield at around 550 K. At higher temperatures than 600 K, vapors of ferric chloride start to evolve and the concentration of Fe_2Cl_6 in the gas phase increases sharply, thereby leading to a decrease of $FeCl_2$ yield.

Other reactions of interest are those that lead to mixed acid formation. For example, hydrolysis of iodine in the presence of SO_2 can lead to a mixture of two acids, namely, sulfuric acid and hydroiodic acid, according to the reaction $2H_2O + I_2 + SO_2 \rightarrow H_2SO_4 + 2HI$. This reaction poses problems of acid separation in the liquid phase. The same type of reaction is possible with bromine, namely, $2H_2O + Br_2 + SO_2 \rightarrow H_2SO_4 + 2HBr$.

One of the most stringent conditions of a thermochemical cycle is that the yield of each reaction step must be high. For example, if the cycle has three steps and each reaction in the sequence has 90 % yield, then the overall yield is 73 %; if the cycle has four steps, then the overall yield is 66 %; if the yield of each individual reaction in the sequence is 80 % for a four-step process, the overall yield is 41 %. This low yield poses problems with chemical separation and recycling in a manner such that for smaller yields the process may become either infeasible or not competitive with alternative technologies (such as high-temperature electrolysis (HTE)). Therefore, it is important to reduce the number of steps as much as possible for an enhanced competitiveness of the process.

For the temperature levels compatible with future nuclear reactors for hydrogen production, it appears that the minimum number of reaction steps of thermochemical cycles can be three. A three-step cycle can comprise, for example, hydrolysis with oxygen or hydrogen evolution, a hydrogen or oxygen evolution reaction at lower temperatures, and a reagent-recycling reaction, normally an endothermic decomposition reaction at higher temperature:

$$\begin{cases} H_2O+A \rightarrow 0.5O_2+AH_2 \\ AH_2+B \rightarrow H_2+AB \\ AB \rightarrow A+B. \end{cases} \quad \text{or} \quad \begin{cases} H_2O+A \rightarrow 0.5H_2+AO \\ AO+B \rightarrow 0.5O_2+AB \\ AB \rightarrow A+B. \end{cases} \tag{5.23}$$

The first thermochemical cycle of three steps, based on this sequence of steps—with an oxygen-evolving hydrolysis process—was proposed by Funk and Reinstrom (1964). See also Sato (1979); this is the well-known Tantalum cycle (see Table 5.10). Other three-step reaction schemes are possible although the cycle from (5.23) is most often used. Table 5.10 lists a number of three-step thermochemical water-splitting cycles organized in descending order of maximum reaction temperature requirement.

The efficiencies of thermochemical cycles indicated in Table 5.10 are estimated from theoretical analyses reported by McQuillan et al. (2010) and based on (5.18)–(5.20). Thermochemical cycles with four steps comprise in general a hydrolysis reaction, a hydrogen-evolving reaction, an oxygen-evolving reaction, and a reagent-recycling reaction. The general scheme of a four-step thermochemical cycle is described by the following equation:

$$\begin{cases} H_2O + AB \rightarrow AH_2 + BO \\ AH_2 \rightarrow H_2 + A \\ BO \rightarrow 0.5O_2 + B \\ A + B \rightarrow AB. \end{cases} \tag{5.24}$$

Table 5.10 Significant three-step thermochemical cycles

Cycle: Elements	T_{max} (K)	Reactions	η (%)
GDF, Sourinau process: Sn	1,973	$H_2O + 0.5Sn \rightarrow H_2 + 0.5SnO_2$, 673 K $SnO_2 \rightarrow 0.5O_2 + SnO$, 1,973 K $SnO \rightarrow 0.5SnO_2 + 0.5Sn$, 973 K	37
GA: Fe, Na	1,743	$3H_2O + 3Na_2O \cdot Fe_2O_3 \rightarrow 6NaOH + 3Fe_2O_3$, 803 K $6NaOH + 2Fe_3O_4 \rightarrow H_2 + 3Na_2O \cdot Fe_2O_3 + 2H_2O$, 1,413 K $3Fe_2O_3 \rightarrow 0.5O_2 + 2Fe_3O_4$, 1,743 K	27
Euratom: C, Fe	1,573	$H_2O + C \rightarrow H_2 + CO$, 973 K $3Fe_2O_3 \rightarrow 0.5O_2 + 2Fe_3O_4$, 1,573 K $CO + 2Fe_3O_4 \rightarrow C + 3Fe_2O_3$, 523 K	44
ORNL: Fe, (M = K, Li, Na)	1,573	$3H_2O + 3MFeO_2 \rightarrow 6MOH + 3Fe_2O_3$, 373 K $6MOH + 2Fe_3O_4 \rightarrow H_2 + 3MFeO_2 + 2H_2O$, 773 K $3Fe_2O_3 \rightarrow 0.5O_2 + 2Fe_3O_4$, 1,573 K	40
CNRS, UNLV-133: Fe, S	1,373	$H_2O + 3FeO \rightarrow H_2 + Fe_3O_4$, 823 K $Fe_3O_4 + 3SO_3 \rightarrow 0.5O_2 + 3FeSO_4$, 1,073 K $3FeSO_4 \rightarrow 3FeO + 3SO_3$, 1,373 K	52
GM (Funk): Cl, Ta	1,366	$H_2O + Cl_2 \rightarrow 0.5O_2 + 2HCl$, 1,000 K $2HCl + 2TaCl_2 \rightarrow H_2 + 2TaCl_3$, 298 K $2TaCl_3 \rightarrow 2TaCl_2 + Cl_2$, 1,366 K	20
LASL: Li, Mn	1,273	$3H_2O + 3Li_2O \cdot Mn_2O_3 \rightarrow 6LiOH + 3Mn_2O_3$, 355 K $6LiOH + 2Mn_3O_4 \rightarrow H_2 + 3Li_2O \cdot Mn_2O_3 + 2H_2O$, 973 K $3Mn_2O_3 \rightarrow 0.5O_2 + 2Mn_3O_4$, 1,273 K	56
Miura: Sb, I	1,273	$H_2O + I_2 + 0.5Sb_2O_3 \rightarrow 0.5Sb_2O_5 + 2HI$, 278 K $2HI \rightarrow H_2 + I_2$, 823 K $0.5Sb_2O_5 \rightarrow 0.5O_2 + 0.5Sb_2O_3$, 1,273 K	37
IGT-C7: Fe, S	1,273	$H_2O + Fe_2O_3 + 2SO_2 \rightarrow H_2 + 2FeSO_4$, 400 K $SO_3 \rightarrow 0.5O_2 + SO_2$, 1,273 K $FeSO_4 \rightarrow Fe_2SO_3 + SO_2 + SO_3$, 973 K	51
GA: Cl, Cs	1,273	$H_2O + Cl_2 \rightarrow 0.5O_2 + 2HCl$, 1,273 K $2CeCl_3 + 4H_2O \rightarrow H_2 + 2CeO_2 + 6HCl$, 1,073 K $2CeO_2 + 8HCl \rightarrow 2CeCl_3 + 4H_2O + Cl_2$, 523 K	21
Jülich: C, S	1,223	$H_2O + CO \rightarrow H_2 + CO_2$, 773 K $H_2SO_4 \rightarrow 0.5O_2 + SO_2 + H_2O$, 1,223 K $CO_2 + SO_2 + H_2O \rightarrow CO + H_2SO_4$, 623 K	56
IGT: Ce, Cl	1,200	$H_2O + Cl_2 \rightarrow 0.5O_2 + HCl$, 1,123 K $4H_2O + 2CeCl_3 \rightarrow H_2 + 6HCl + 2CeO_2$, 1,200 K $2CeO_2 + 8HCl \rightarrow 2CeCl_3 + Cl_2$, 383 K	21
GIRIO: Br, Fe	1,123	$H_2O + Br_2 \rightarrow 0.5O_2 + 2HBr$, 923 K $4H_2O + 3FeBr_2 \rightarrow H_2 + 6HBr + Fe_3O_4$, 1,123 K $8HBr + Fe_3O_4 \rightarrow 3FeBr_2 + 4H_2O + Br_2$, 523 K	32
Williams, US-chlorine: Cl	1,123	$H_2O + Cl_2 \rightarrow 0.5O2 + 2HCl$, 1,123 K $2HCl + 2CuCl \rightarrow H_2 + 2CuCl_2$, 473 K $2CuCl_2 \rightarrow 2CuCl + Cl_2$, 773 K	49
GDF, K-process: K	1,100	$H_2O + K_2O_2 \rightarrow 0.5O_2 + 2KOH$, 400 K $2KOH + 2K \rightarrow H_2 + 2K_2O$, 1,000 K $2K_2O \rightarrow K_2O_2 + 2K$, 1,100 K	27

(continued)

Table 5.10 (continued)

Cycle: Elements	T_{max} (K)	Reactions	η(%)
KIER-1: Cl, Cu	1,100	$2H_2O + 2Cl_2 \rightarrow 0.5O_2 + 4HCl + 0.5O_2$, 873 K $Cu_2O + 4HCl \rightarrow H_2 + 2CuCl_2 + H_2O$, 500 K $2CuCl_2 + 0.5O_2 \rightarrow 2Cl_2 + Cu_2O$, 1,100 K	N/A
KIER-2: Cl, Cu	1,100	$H_2O + CuCl_2 \rightarrow 2HCl + CuO$, 1,100 K $2HCl + 4CuCl_2 \rightarrow H_2 + 4CuCl + 3Cl_2$, 1,100 K $4CuCl + 3Cl_2 + CuO \rightarrow 0.5O_2 + 5CuCl_2$, 873 K	N/A
KIER-A: Cl, Cu	1,100	$H_2O + 2CuCl_2 \rightarrow 0.5O_2 + 2CuCl + 2HCl$, 900 K $4CuCl_2 + 2HCl \rightarrow H_2 + 4CuCl + 3Cl_2$, 1,100 K $6CuCl + 3Cl_2 \rightarrow 6CuCl_2$, 873 K	N/A
KIER-3: Cu, S	1,100	$H_2O + CuSO_4 \rightarrow CuO + H_2SO_4$, 1,100 K $CuO + SO_2 + H_2O \rightarrow H_2 + CuSO_4$, 473 K $H_2SO_4 \rightarrow 0.5O_2 + SO_2 + H_2O$, 1,100 K	N/A
UNLV-159: C, S	1,023	$H_2O + SO_2 + C_2H_4 \rightarrow C_2H_6 + H_2SO_4$, 623 K $C_2H_6 \rightarrow H_2 + C_2H_4$, 1,073 K $H_2SO_4 \rightarrow 0.5O_2 + SO_2 + H_2O$, 1,123 K	58
Theme-S3 I, S	1,073	$H_2O + I_2 + SO_2 \rightarrow 2HI + SO_3$, 373 K $2HI \rightarrow H_2 + I_2$, 573 K $SO_3 \rightarrow 0.5O_2 + SO_2$, 1,073 K	36
Euratom: Cl, Fe	973	$H_2O + Cl_2 \rightarrow 0.5O_2 + 2HCl$, 973 K $2HCl + 2FeCl_2 \rightarrow H_2 + 2FeCl_3$, 873 K $2FeCl_3 \rightarrow 2FeCl_2 + Cl_2$, 723 K	30
Hitachi: C, N, Na, I	973	$H_2O + 2NaI + 2NH_3 + CO_2 \rightarrow Na_2CO_3 + 2NH_4I$, 298 K $2NH_4I \rightarrow H_2 + I_2 + 2NH_3$, 773 K $Na_2CO_3 + I_2 \rightarrow 0.5O_2 + CO + 2NaI$, 973 K	29
Yokohama Mark-3: Fe, I, S	973	$2H_2O + I_2 + 2FeSO_4 \rightarrow 2HI + 2Fe(OH)SO_4$, 973 K $2HI \rightarrow H_2 + I_2$, 573 K $2Fe(OH)SO_4 \rightarrow 0.5O_2 + 2FeSO_4 + H_2O$, 523 K	42
GIRIO: Cu, I, N	973	$H_2O + 0.5O_2 + 2CuI + 2NH3 \rightarrow 2CuO + 2NH_4I$, 298 K $2NH_4I \rightarrow H_2 + I_2 + 2NH_3$, 773 K $2CuO + I_2 \rightarrow 0.5O_2 + 2CuI + 0.5O_2$, 973 K	46
KIER-B: Cl, Cu	873	$H_2O + 2CuCl_2 \rightarrow 0.5O_2 + 2CuCl + 2HCl$, 873 K $2HCl + 6CuCl + 2Cl_2 \rightarrow H_2 + 6CuCl_2$, 500 K $4CuCl_2 \rightarrow 4CuCl + 2Cl_2$, 773 K	N/A
Shell: Cu, S	773	$H_2O + 2Cu \rightarrow H_2 + Cu_2O$, 773 K $1/3\,Cu_2O + 4/3\,CuSO_4 \rightarrow 0.5O_2 + Cu + 4/3\,SO_2 + O_2$, 773 K $2/3\,Cu_2O + 4/3\,SO_2 + O_2 \rightarrow 4/3\,CuSO_4$, 573 K	26

GA General Atomics, *GDF* Gas de France, *GIRIO* Government Industrial Research Institute at Osaka, *GM* General Motors, *IGT* Institute of Gas technology, *Jülich* Jülich Research Centre of the Rheinisch-Westfaelische Technische Hochschule Aachen, *KIER* Korean Institute of Energy Research, *LASL* Los Alamos Scientific Laboratory, *ONRL* Oak Ridge National Laboratory, *UNLV* University of Nevada Las Vegas, *N/A* not available

Sources: Abanades et al. (2006), McQuillan et al. (2010), Williams (1980), Lewis and Masin (2009)

Another four-step cycle comprises hydrolysis with oxygen evolution followed by a hydrogen evolution reaction and two chemical steps for reagent recycling. Funk and Reinstrom (1964) developed the first four-step thermochemical cycle based on this type of reaction sequence, known as a vanadium–chlorine cycle. In a similar manner, there are cycles which consist of hydrolysis and hydrogen evolution plus a separate oxygen generation reaction followed by two chemical steps for reagent recycling. Some significant four-step cycles are listed in Table 5.11.

Further thermochemical cycles are given in Table 5.12—for five-step processes, and Table 5.13—for processes with six reaction steps. Note that the cycles compiled in Tables 5.10, 5.11, 5.12, and 5.13 were selected from several databases, namely:

- IGT database compiled by the former Institute of Gas Technology (Carty et al. 1981) comprising more than 200 thermochemical cycles.
- Compilation by Bamburger (1978) which includes 125 pure thermochemical cycles.
- Book of Williams entitled "Hydrogen Power" which comprises a compilation of significant processes developed in the 1970s (Williams 1980).
- Recent database compiled in France by Abanades et al. (2006) including 280 processes.
- The UNLV database developed for the US Department of Energy and compiled by a consortium of institutions within the USA (Sandia National Laboratory, ANL, Savannah River National Laboratory, National Renewable Energy Laboratory, University of Colorado, University of Nevada at Las Vegas, General Atomics) lists 375 cycles as indicated in McQuillan et al. (2010) and Perret (2011).

It is interesting to examine the frequency of particular chemical elements in thermochemical cycles. For the cycles listed in Tables 5.2, 5.4, and 5.10, 5.11, 5.12, 5.13—a list which includes in total 122 significant cycles—the bar chart from Fig. 5.22 is obtained, which correlates the occurrence of chemical elements in thermochemical cycles with the atomic number (Z) of the elements. It shows that iron is the element with the highest occurrence of 29 %, followed by sulfur and chlorine with 26 %, iodine 19 %, carbon with 13 %, magnesium with 10 %, copper and nitrogen with 9 %, sodium with 8 %, bromine and calcium with 7 %, and zinc with 6 %. All other elements have an occurrence of less than 5 %.

The cycle most developed to date is the sulfur–iodine cycle (Mark 16) with prototypes (pilot scale) in several countries as discussed in detail in Section 5.10. Another promising cycle for nuclear hydrogen production is the UT-3 cycle developed at the University of Tokyo, which includes four reaction steps, requires a heat source of 1,023 K, and has a theoretical efficiency of 45 %.

Among the "pure" thermochemical cycles with the lowest temperature of heat source, the prominent examples are the 5-step cycle of General Electric with Cl, Cu, and Ng compounds (773 K), the 3-step cycle of Shell operating with Cu and S compounds (773 K), the UNLV-82 cycle of 4 steps which requires 710 K, and the heavy element halide cycle of the ANL with Br, Eu, and U which requires 573 K.

The last mentioned cycle (Argonne—heavy element halide) is currently under development, but not yet well documented. This cycle is attractive due to its low temperature requirement. However, other aspects must be considered when selecting a thermochemical cycle, including efficiency, element availability, cost, etc.

Table 5.11 Significant four-step thermochemical cycles

Cycle: Elements	T_{max} (K)	Reactions	η (%)
Aerojet General: Cs	1,473	$2H_2O + 2Cs \rightarrow H_2 + 2CsOH$, 373 K $Cs_2O \rightarrow 0.5O_2 + 2Cs$, 1,473 K $2CsOH + 1.5O_2 \rightarrow 2CsO_2 + H_2O$, 773 K $2CsO_2 \rightarrow Cs_2O + 1.5O_2$, 973 K	48
UNLV-92: Mg, S, W	1,400	$H_2O + WO_2 \rightarrow H_2 + WO_3$, 400 K $MgSO_4 \rightarrow 0.5O_2 + MgO + SO_2$, 1,400 K $MgO + SO_2 \rightarrow MgSO_3$, 610 K $WO_3 + MgSO_3 \rightarrow WO_2 + MgSO_4$, 800 K	37
UNLV-96: Mg, Na, S	1,400	$H_2O + Na_2SO_3 \rightarrow H_2 + Na_2SO_4$, 500 K $MgSO_4 \rightarrow 0.5O_2 + MgO + SO_2$, 1,400 K $MgO + SO_2 \rightarrow MgSO_3$, 610 K $MgSO_3 + Na_2SO_4 \rightarrow MgSO_4 + Na_2SO_3$, 298 K	38
UNLV-115: I, Mg, S	1,400	$H_2O + MgI_2 \rightarrow 2HI + MgO$, 673 K $2HI \rightarrow H_2 + I_2$, 573 K $MgSO_4 \rightarrow 0.5O_2 + MgO + SO_2$, 1,400 K $2MgO + SO_2 + I_2 \rightarrow MgSO_4 + MgI_2$, 373 K	28
UNLV-99: Ag, Cr, K	1,300	$2/3\,(H_2O + Ag_2O + K_2CrO_4) \rightarrow 2/3\,(Ag_2CrO_4 + 2KOH)$, 298 K $1/3(H_2O + Cr_2O_3 + 4KOH) \rightarrow H_2 + 2/3K_2CrO_4$, 1,300 K $8/3\,Ag_2CrO_4 \rightarrow 0.5O_2 + 16/3\,Ag + 1/3\,Cr_2O_3 + 1/3\,O_2$, 980 K $4/3Ag + 1/3\,O_2 \rightarrow 2Ag_2O$, 410 K	45
UNVL-140: Cl, Fe	1,300	$H_2O + 3FeO \rightarrow H_2 + Fe_3O_4$, 673 K $H_2O + Cl_2 \rightarrow 0.5O_2 + 2HCl$, 1,300 K $Fe_3O_4 + 8HCl \rightarrow 3FeCl_2 + Cl_2 + 4H_2O$, 653 K $3FeCl_2 + 3H_2O \rightarrow 3FeO + 6HCl$, 963 K	42
UNVL-144: Ba, Fe, S	1,293	$H_2O + 0.25BaS \rightarrow H_2 + 0.25BaSO_4$, 1,143 K $FeSO_4 \rightarrow 0.5O_2 + 0.5Fe_2O_3 + 2SO_2$, 1,200 K $Fe_2O_3 + 2.5SO_2 \rightarrow 2Fe_{SO_4} + 0.5S$, 700 K $0.25BaSO_4 + 0.5S \rightarrow 0.25BaS + 0.5SO_2$, 1,293 K	48
NCLI: Ca, I	1,273	$H_2O + CaI_2 \rightarrow 2HI + CaO$, 1,273 K $2HI \rightarrow H_2 + I_2$, 573 K $0.2Ca(IO_3)_2 \rightarrow 0.5O_2 + 0.2I_2 + 0.2CaO$, 973 K $1.2CaO + 1.2I_2 \rightarrow CaI_2 + 0.2Ca(IO_3)_2$, 300 K	31
UNLV-150: Co, Ge	1,273	$H_2O + Ge \rightarrow H_2 + GeO$, 1,173 K $Co_3O_4 \rightarrow 0.5O_2 + 3CoO$, 1,273 K $GeO + SO_2 \rightarrow Ge + SO_3$, 1,073 K $3CoO + SO_3 \rightarrow Co_3O_4 + SO_2$, 673 K	41
UNVL-30: Fe, S	1,173	$H_2O + 0.5Fe_3O_4 + 1.5SO_2 \rightarrow H_2 + 1.5FeSO_4$, 298 K $SO_3 \rightarrow 0.5O_2 + SO_2$, 1,073 K $0.75Fe_2O_3 + 0.25SO_2 \rightarrow 0.5Fe_3O_4 + 0.25SO_3$, 298 K $1.5FeSO_4 \rightarrow 0.75Fe_2O_3 + 0.75SO_2 + 0.75SO_3$, 1,173 K	49
UNLV-81: Cu, I Mg	1,173	$2H_2O + MgI_2 \rightarrow 2HI + Mg(OH)_2$, 500 K $2HI \rightarrow H_2 + I_2$, 573 K $2CuO \rightarrow 0.5O_2 + Cu_2O$, 1,173 K $Cu_2O + I_2 + Mg(OH)_2 \rightarrow 2CuO + MgI_2 + H_2O$, 298 K	27
UNLV-88: Co, I, Sc	1,160	$H_2O + 2/3\,ScI_3 \rightarrow 2HI + 1/3\,Sc_2O_3$, 500 K $2HI \rightarrow H_2 + I_2$, 573 K $Co_3O_4 \rightarrow 0.5O_2 + 3CoO$, 1,160 K $3CoO + I_2 + 1/3\,Sc_2O_3 \rightarrow Co_3O_4 + 2/3\,ScI_3$, 300 K	28

(continued)

Table 5.11 (continued)

Cycle: Elements	T_{max} (K)	Reactions	η(%)
UNVL-127: Cl, V	1,123	$H_2O + Cl_2 \rightarrow 0.5O_2 + 2HCl$, 1,123 K $2HCl + 2VOCl \rightarrow H_2 + 2VOCl_2$, 443 K $4VOCl_2 \rightarrow 2VOCl + 2VOCl_3$, 873 K $2VOCl_3 \rightarrow Cl_2 + 2VOCl_2$, 473 K	48
UNVL-142: C, S	1,123	$H_2O + CH_4 \rightarrow 3H_2 + CO$, 1,100 K $H_2SO_4 \rightarrow 0.5O_2 + SO_2 + H_2O$, 1,123 K $CO + 2H_2 \rightarrow CH_2OH$, 500 K $CH_3OH + SO_2 + H_2O \rightarrow CH_4 + H_2SO_4$, 473 K	52
GA-23: S	1,123	$H_2O + 1.5SO_2 \rightarrow H_2SO_4 + 0.5S$, 298 K $H_2S \rightarrow H_2 + S$, 1,073 K $H_2SO_4 \rightarrow 0.5O_2 + SO_2 + H_2O$, 1,123 K $1.5S + H_2O \rightarrow H_2S + 0.5SO_2$, 973 K	42
UNLV-85: As, I, Mg	1,100	$2H_2O + MgI_2 \rightarrow 2HI + Mg(OH)_2$, 500 K $2HI \rightarrow H_2 + I_2$, 573 K $0.5As_2O_5 \rightarrow 0.5O_2 + 0.25As_4O_6$, 1,100 K $0.25As_4O_6 + I_2 + Mg(OH)_2 \rightarrow 0.5As_2O_5 + MgI_2 + H_2O$, 300 K	28
UNLV-95: Ce, S, W	1,100	$H_2O + WO_2 \rightarrow H_2 + WO_3$, 400 K $1.5Ce(SO_4)_2 \rightarrow 0.5O_2 + 1.5CeO_2 + 3SO_2 + O_2$, 1,100 K $2CeO_2 + 3SO_2 + O_2 \rightarrow Ce_2(SO_4)_3$, 825 K $Ce_2(SO_4)_3 + WO_3 \rightarrow 0.5CeO_2 + 1.5Ce(SO_4)_2 + WO_2$, 680 K	35
UT-3: Br, Ca, Fe	1,023	$H_2O + CaBr_2 \rightarrow CaO + 2HBr$, 1,023 K $3FeBr_2 + 4H_2O \rightarrow H_2 + Fe_3O_4 + 6HBr$, 873 K $CaO + Br_2 \rightarrow 0.5O_2 + CaBr_2$, 873 K $Fe_3O_4 + 8HBr \rightarrow 3FeBr_2 + Br_2 + 4H_2O$, 573 K	45
UNLV-3: Br, Ca, Hg	1,023	$H_2O + CaBr_2 \rightarrow CaO + 2HBr$, 1,023 K $2HBr + Hg \rightarrow H_2 + HgBr_2$, 473 K $2HgO \rightarrow 0.5O_2 + 2Hg$, 773 K $CaO + HgBr_2 \rightarrow CaBr_2 + HgO$, 298 K	26
UNVL-164: C, S, Zn	1,000	$H_2O + CH_4 \rightarrow 3H_2 + CO$, 973 K $ZnSO_4 \rightarrow 0.5O_2 + SO_2 + ZnO$, 1,000 K $CO + 2H_2 \rightarrow CH_3OH$, 500 K $CH_3OH + SO_2 + ZnO \rightarrow CH_4 + ZnSO_4$, 600 K	42
ANL: Ag, Br, C, N, Na	1,000	$2(H_2O + NH_3 + CO_2 + NaBr) \rightarrow 2(NaHCO_3 + NH_4Br)$, 298 K $2NH_4Br + 2Ag \rightarrow H_2 + 2NH_3 + 2AgBr$, 750 K $2AgBr + Na_2CO_3 \rightarrow 0.5O_2 + 2Ag + 2NaBr + CO_2$, 1,000 K $2NaHCO_3 \rightarrow Na_2CO_3 + H_2O$, 400 K	25
NCLI: I, Mg	973	$H_2O + MgI_2 \rightarrow 2HI + MgO$, 673 K $2HI \rightarrow H_2 + I_2$, 573 K $0.2Mg(IO_3)_2 \rightarrow 0.5O_2 + 0.2I_2 + 0.2MgO$, 973 K $1.2I_2 + 1.2MgO \rightarrow MgI_2 + 0.2Mg(IO_3)_2$, 423 K	37
GM (Funk): Cl, V	973	$H_2O + Cl_2 \rightarrow 0.5O_2 + 2HCl$, 883 K $2HCl + 2VCl_2 \rightarrow H_2 + 2VCl_3$, 298 K $2VCl_3 \rightarrow VCl_4 + VCl_2$, 973 K $2VCl_4 \rightarrow Cl_2 + 2VCl_3$, 298 K	30

(continued)

Table 5.11 (continued)

Cycle: Elements	T_{max} (K)	Reactions	η(%)
UNLV-43: Ag, Br, N, Na	900	$2H_2O + 2CO_2 + 2NH_3 + 2NaBr \rightarrow 2NaHCO_3 + 2NH_4Br$, 298 K $2Ag + 2NH_4Br \rightarrow H_2 + 2AgBr + 2NH_3$, 700 K $2AgBr + Na_2CO_3 \rightarrow 0.5O_2 + 2Ag + CO_2 + 2NaBr$, 900 K $2NaHCO_3 \rightarrow CO_2 + Na_2CO_3 + H_2O$, 400 K	25
LASL: Ce, Cl	883	$2H_2O + 2CeClO \rightarrow H_2 + 2CeO_2 + 2HCl$, 723 K $H_2O + Cl_2 \rightarrow 0.5O_2 + 2HCl$, 883 K $2CeO_2 + 8HCl \rightarrow 2CeCl_3 + Cl_2 + 4H_2O$ 298 K $2CeCl_3 + 2H_2O \rightarrow 2CeClO + 4HCl$, 700 K	21
LLL-UoC: Cs, Hg	873	$2H_2O + Cs_2Hg \rightarrow H_2 + Hg + 2CsOH$, 873 K $HgO \rightarrow 0.5O_2 + Hg$, 773 K $2CsOH \rightarrow Cs_2O + H_2O$, 683 K $Cs_2O + 2Hg \rightarrow Cs_2Hg + HgO$, 573 K	34
UNLV-162: C, U	873	$H_2O + CO \rightarrow H_2 + CO_2$, 573 K $3UO_3 \rightarrow 0.5O_2 + U_3O_8$, 873 K $4CO_2 + U_3O_8 \rightarrow CO + 3UO_2CO_3$, 423 K $3UO_2CO_3 \rightarrow 3CO_2 + 3UO_3$, 873 K	39
UNLV-82: I, Mg, Mn	710	$2H_2O + MgI_2 \rightarrow 2HI + Mg(OH)_2$, 500 K $2HI \rightarrow H_2 + I_2$, 573 K $2MnO_2 \rightarrow 0.5O_2 + Mn_2O_3$, 710 K $Mg(OH)_2 + I_2 + Mn_2O_3 \rightarrow MgI_2 + 2MnO_2 + H_2O$, 300 K	38 (10)
ANL: Br, Eu, U	573	$2H_2O + Br_2 + UO_3 \cdot H_2O \rightarrow 0.5O_2 + UO_2 \cdot 3H_2O$, 298 K $2EuBr_2 + 2HBr \rightarrow H_2 + 2EuBr_3$, 298 K $2EuBr_3 \rightarrow 2EuBr_2 + Br_2$, 573 K $UO_2 \cdot 3H_2O \rightarrow 2UO_3 \cdot H_2O + 2HBr + H_2O$, 573 K	N/A

ANL Argonne National Laboratory, *GA* General Atomics, *GM* General Motors, *ASL* Los Alamos Scientific Laboratory, *LLL-UoC* Lawrence Livermore Laboratory, University of California, *NCLI* National Chemical Laboratory for Industry (Tokyo), *UoK* University of Kentucky, *UNLV* University of Nevada at Las Vegas, *UT* University of Tokyo, *N/A* not available

Sources: Abanades et al. (2006), Carty et al. (1981), McQuillan et al. (2010), Williams (1980)

The heavy element halide cycle has not yet shown a reported efficiency and therefore, although promising, it is not mentioned widely in the literature among recommended cycles for nuclear hydrogen production.

5.6 Hybrid Thermochemical Cycles

Hybrid thermochemical cycles can be generally defined as closed loop processes that split water using thermal and chemical energy in conjunction with another form of energy. Three types of hybrid processes have been proposed as follows:

- Thermo-electrochemical cycles, which include at least one electrochemical reaction.
- Thermo-photo-electrochemical cycles, which include, besides thermochemical reactions, at least one photochemical and one electrochemical reaction.
- Thermo-radiochemical cycles, which comprise a radiochemical process besides other thermochemical reactions.

Table 5.12 Significant five-step thermochemical cycles

Cycle: Elements	T_{max} (K)	Reactions	η (%)
NCLI: I, Mg, S	1,400	$H_2O + MgI_2 \rightarrow 2HI + MgO$, 673 K $2HI \rightarrow H_2 + I_2$, 573 K $MgSO_4 \rightarrow 0.5O_2 + MgO + SO_2$, 1,400 K $I_2 + SO_2 + 2H_2O \rightarrow 2HI + H_2SO_4$, 373 K $2HI + H_2SO_4 + 2MgO \rightarrow MgI_2 + MgSO_4 + 2H_2O$, 350 K	26
LLL-UoC: Mg, Se	1,373	$0.5H_2O + 0.25MgSe \rightarrow 0.25H_2Se + 0.25Mg(OH)_2$, 373K $0.5H_2O + 1/8\,MgSe \rightarrow 0.5H_2 + 1/8\,MgSeO_4$, 673 K $0.5H_2Se \rightarrow 0.5H_2 + 0.5Se$, 473 K $0.25MgSeO_4 \rightarrow 0.5O_2 + 0.25MgSe$, 1,373 K $0.25Mg(OH)_2 + 0.5Se \rightarrow 0.25H_2Se + 1/8\,MgSe +$ $1/8\,MgSeO_4$, 773 K	31
IGT, A-2: Cl, Fe	1,273	$4H_2O + 3Fe \rightarrow H_2 + Fe_3O_4 + 3H_2$, 773 K $Fe_3O_4 + 4.5Cl_2 \rightarrow 0.5O_2 + 3FeCl_3 + 1.5O_2$, 1,273 K $3FeCl_3 \rightarrow 3FeCl_2 + 1.5Cl_2$, 623 K $3FeCl_2 + 3H_2 \rightarrow 3Fe + 6HCl$, 1,273 K $6HCl + 1.5O_2 \rightarrow 3Cl_2 + 3H_2O$, 773 K	29
LLL-UoC: Cl, S, Se, Zn	1,200	$H_2O + ZnCl_2 \rightarrow 2HCl + ZnO$, 883 K $H_2Se \rightarrow H_2 + Se$, 473 K $ZnSO_4 \rightarrow 0.5O_2 + SO_2 + ZnO$, 1,200 K $2HCl + ZnSe \rightarrow H_2Se + ZnCl_2$, 350 K $Se + SO_2 + 2ZnO \rightarrow ZnSe + ZnSO_4$, 883 K	28
JAERI: I, Ni, S	1,123	$2H_2O + I_2 + SO_2 \rightarrow 2HI + H_2SO_4$, 298 K $2Ni + 2HI + H_2SO_4 \rightarrow 2H_2 + NiI_2 + NiSO_4$, 600 K $NiSO_4 \rightarrow 0.5O_2 + NiO + SO_2$, 1,123 K $NiO + H_2 \rightarrow Ni + H_2O$, 673 K $NiI_2 \rightarrow Ni + I_2$, 873 K	26
MHI: Cl, Cu, Fe	1,073	$4H_2O + 3FeCl_2 \rightarrow H_2 + Fe_3O_4 + 6HCl$, 873 K $H_2O + Cl_2 \rightarrow 0.5O_2 + HCl$, 1,073 K $Fe_3O_4 + 8HCl \rightarrow FeCl_2 + 2FeCl_3 + 4H_2O$, 373 K $2FeCl_3 + 2CuCl \rightarrow 2FeCl_2 + 2CuCl_2$, 373 K $2CuCl_2 \rightarrow 2CuCl + Cl_2$, 773 K	27
Hitachi: C, I, N, Na, Ni	1,073	$2H_2O + 2NaI + 2NH_3 + 2CO_2 \rightarrow 2NaHCO_3$ $+ 2NH_4I$, 733 K $2NH_4I + Ni \rightarrow H_2 + NiI_2 + 2NH_3$, 773 K $Na_2CO_3 + I_2 \rightarrow 0.5O_2 + 2NaI + CO_2$, 973 K $2NaHCO_3 \rightarrow Na_2CO_3 + CO_2 + H_2O$, 573 K $NiI_2 \rightarrow Ni + I_2$, 1,073 K	38
GE-Agnes process: Cl, Fe, Mg	1,023	$4H_2O + 3FeCl_2 \rightarrow H_2 + Fe_3O_4 + 6HCl$, 1,023 K $Mg(OH)_2 + Cl_2 \rightarrow 0.5O_2 + MgCl_2 + H_2O$, 363 K $MgCl_2 + 2H_2O \rightarrow Mg(OH)_2 + 2HCl$, 623 K $Fe_3O_4 + 8HCl \rightarrow FeCl_2 + 2FeCl_3 + 4H_2O$, 383 K $2FeCl_3 \rightarrow 2FeCl_2 + Cl_2$, 573 K	25
ANL: I, K, N	973	$H_2O + NO_2 + 0.5O_2 \rightarrow 2HNO_3$, 298 K $2NH_4I \rightarrow H_2 + 2NH_3 + I_2$, 773 K $I_2 + 2KNO_3 \rightarrow 0.5O_2 + 2KI + 2NO_2 + 0.5O_2$, 973 K $2KI + 2NH_4NO_3 \rightarrow 2KNO_3 + 2NH_4I$, 298 K $2HNO_3 + 2NH_3 \rightarrow 2NH_4NO_3$, 298 K	22

(continued)

Table 5.12 (continued)

Cycle: Elements	T_{max} (K)	Reactions	η (%)
LLL-UoC: K, Se, V	973	$2H_2O + K_2Se \rightarrow 2KOH + H_2Se$, 373 K $H_2Se \rightarrow H_2 + Se$, 473 K $V_2O_5 \rightarrow 0.5O_2 + V_2O_4$, 773 K $V_2O_4 + 0.5SeO_2 \rightarrow Ve_2O_5 + 0.5Se$, 500 K $1.5Se + KOH \rightarrow K_2Se + 0.5SeO_2 + H_2O$, 973 K	26
LLL-UoC: As, C	973	$H_2O + CH_4 \rightarrow 3H_2 + CO$, 973 K $0.5As_2O_5 \rightarrow 0.5O_2 + 0.5As_2O_3$, 973 K $CO + 2H_2 \rightarrow CH_3OH$, 500 K $CH_3OH + As_2O_4 \rightarrow CH_4 + AS_2O_5$, 500 K $0.5As_2O_3 + 0.5As_2O_5 \rightarrow As_2O_4$, 298 K	31
GE: Cl, Cu, Mg	773	$2H_2O + MgCl_2 \rightarrow 2HCl + Mg(OH)_2$, 723 K $2HCl + 2Cu \rightarrow H_2 + 2CuCl$, 373 K $Mg(OH)_2 + Cl_2 \rightarrow 0.5O_2 + MgCl_2 + H_2O$, 298 K $4CuCl \rightarrow 2Cu + 2CuCl_2$, 373 K $2CuCl_2 \rightarrow 2CuCl + Cl_2$, 773 K	13

ANL Argonne National Laboratory, *GE* General Electric, *IGT* Institute of Gas technology, *JAERI* Japanese Atomic Energy research Institute, *LLL-UoC* Lawrence Livermore Laboratory, University of California, *MHI* Mitshibishi Heavy Industries, *NCLI* National Chemical Laboratory for Industry (Tokyo), *UT* University of Tokyo

Sources: Abanades et al. (2006), Carty et al. (1981), McQuillan et al. (2010), Williams (1980)

Table 5.13 Significant thermochemical cycles with six steps

Cycle: Elements	T_{max} (K)	Reactions	η (%)
Hitachi: Cl, Cu, N, S	1,223	$H_2O + 2NH_3 + 2CuCl \rightarrow Cu_2O + 2NH_4Cl$, 353 K $2NH_4Cl + 2Cu \rightarrow H_2 + 2CuCl + 2NH_3$, 873 K $2CuO \rightarrow 0.5O_2 + Cu_2O$, 1,223 K $2Cu_2O + 2H_2SO_4 \rightarrow 2Cu + 2CuSO_4 + 2H_2O$, 373 K $2CuSO_4 \rightarrow 2CuO + 2SO_3$, 1,123 K $2SO_3 + 2H_2O \rightarrow 2H_2SO_4$, 573 K	41
UNLV-116: C, I, S	1,123	$H_2O + CH_4 \rightarrow H_2 + CO + 2H_2$, 973 K $H_2SO_4 \rightarrow 0.5O_2 + SO_2 + H_2O$, 1,123 K $CH_3I + HI \rightarrow CH_4 + I_2$, 298 K $CH_3OH + HI \rightarrow CH_3I + H_2O$, 298 K $CO + 2H_2 \rightarrow CH_3OH$, 500 K $I_2 + SO_2 + 2H_2O \rightarrow 2HI + H_2SO_4$, 373 K	45
GE, Catherine process: I, K, Li, N	973	$2H_2O + 2LiI \rightarrow 2HI + 2LiOH$, 873 K $2HI + Ni \rightarrow H_2 + NiI_2$, 423 K $1/3\,KIO_3 \rightarrow 0.5O_2 + 1/3\,KI$, 923 K $I_2 + 2LiOH \rightarrow 5/3\,LiI + 1/3\,LiIO_3 + H_2O$, 463 K $1/3\ LiO_3 + 1/3\,KI \rightarrow 1/3KIO_3 + 1/3\,LiI$, 298 K $NiI_2 \rightarrow Ni + I_2$, 973 K	22

GE General Electric, *UNLV* University of Nevada at Las Vegas

Sources: Williams (1980), McQuillan et al. (2010)

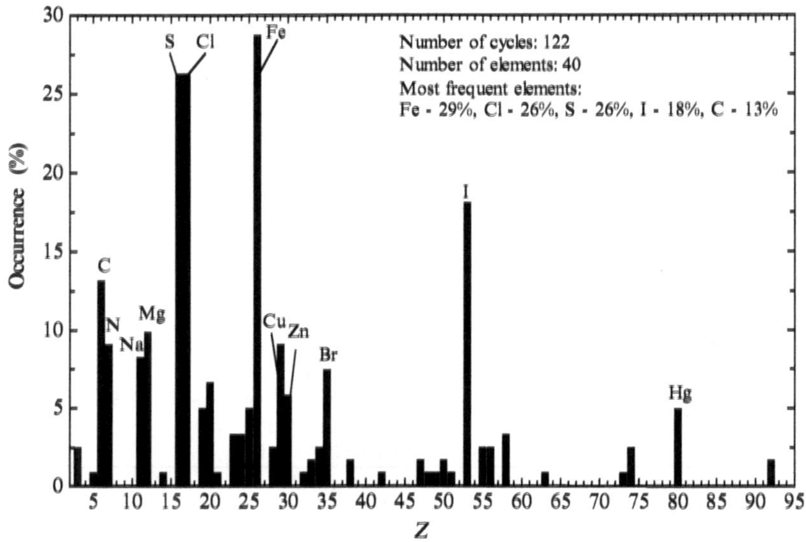

Fig. 5.22 Occurrence of chemical elements in "pure" thermochemical cycles as a function of their atomic number (*Z*)

Among all classes of hybrid cycles, the thermo-electrochemical cycles are the most developed. They are generally called "hybrid cycles." There are very few number of cycles that use photonic energy or nuclear radiation, in addition to thermal energy. In this section, the hybrid thermochemical cycles are reviewed and discussed.

5.6.1 Hybrid Electro-ThermoChemical Cycles

With reference to Fig. 5.18, it can be observed that some thermochemical reactions have a positive Gibbs free energy even at high temperatures. For example, thermal decomposition of halides in the gas phase shows a positive slope of Gibbs free energy vs. temperature (see curves 3, 4, and 5 from Fig. 5.18). A comparative analysis of thermochemical decomposition of hydrogen halides, with the assumption of no side reactions, is presented in Fig. 5.23. In the figure, the variation of reaction yield with temperature is illustrated. The reaction temperature must be selected such that the yield is satisfactory.

It can be observed that hydrochloric acid requires temperatures higher than 1,000 K to achieve a significant yield. Chao (1974) analyzed the thermochemical decomposition of hydrogen chloride at 1,073 K although the equilibrium yield occurs at the limit of practicality (see Fig. 5.23). Other cycles comprise thermochemical reactions of decomposition of hydrogen bromide or hydrogen iodide. Hydrogen bromide has also a modest equilibrium yield, which as predicted in Fig. 5.23 is ~1 % at 1,073 K.

On the opposite extreme, thermochemical decomposition of HI gives an excellent equilibrium yield even at lower temperatures, e.g., ~38 % at 500 K and ~67 %

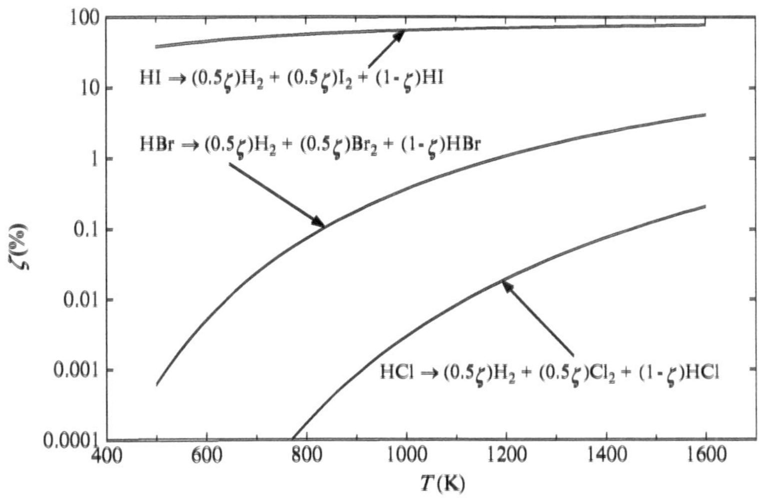

Fig. 5.23 Comparative analysis of equilibrium conversion (ζ) for thermal decomposition of three hydrogen halides (HCl, HBr, HI) in the gas phase

at 1,073 K. More accurate results, based on experiments and accounting for various side reactions, report the following decomposition yields of hydrogen iodide: 19 % at 300 K, 22 % at 723 K, and 33 % at 1,273 K (see Sato 1979). In many thermochemical cycles, the thermochemical decomposition of hydrogen iodide is conducted at 773 K. However, decomposition of HCl and HBr is preferably performed electrochemically in solution, at a temperature close to the ambient condition. This approach makes the process practically realizable, although some electrical energy is required—with an inherent penalty for conversion.

In an electrochemical process, the Gibbs free energy is provided to the reaction in the form of electricity. Electrochemical reactions have three main advantages that make them preferable (in some conditions) from a technical and economical point of view: (1) the temperature of the reaction may be lowered with respect to a pure thermochemical process, (2) reactions which have infeasible thermochemical conversion can be made possible when performed electrochemically, and (3) the reaction products are intrinsically separated.

Table 5.14 shows some thermochemical cycles that have hybrid corresponding versions. For example, the two-step hydrochloric acid cycle can be performed either thermochemically or electro-thermochemically (McQuillan et al. 2010). In the thermochemical version, proposed by Chao (1974), this cycle comprises the reverse Deacon reaction conducted at 873 K, followed by subsequent thermolysis of hydrochloric acid at 1,073 K (see Table 5.2, cycle #24). This process requires a complicated separation, namely, with chlorine and hydrogen in the gaseous phase. In principle, this separation can be performed at high temperature on palladium membranes which are selective to hydrogen only, but with a marginal hydrogen yield. When the thermolysis is replaced with electrochemical hydrochloric acid decomposition in solution, the reaction temperature is reduced from 1,073 to 363 K.

Table 5.14 Thermochemical and hybrid closure reactions for selected thermochemical cycles

#	First thermochemical step(s)	Cycle closure reactions: Purely thermochemical	Cycle closure reactions: Hybrid thermo-electrochemical[a]
1	$H_2O + Cl_2 \rightarrow 0.5O_2 + 2HCl$, 873 K	*Cycle*: Chao (1974), Table 5.2, cycle #24 $2HCl \rightarrow H_2 + Cl_2$, 1,073 K *Cycle*: GM (Funk), Table 5.10 $2HCl + 2TaCl_2 \rightarrow H_2 + 2TaCl_3$, 298 K $2TaCl_3 \rightarrow 2TaCl_2 + Cl_2$, 1,366 K	*Cycle*: APC/UNLV-53 E: $2HCl \rightarrow H_2 + Cl_2$, 363 K, 0.99 V
2	$H_2O + 0.5Sb_2O_3 + I_2 \rightarrow$ $\rightarrow 2HI + 0.5Sb_2O_5$, 298 K $0.5Sb_2O_5 \rightarrow 0.5O_2 + 0.5Sb_2O_3$, 1,273 K	*Cycle*: Miura, Table 5.10 $2HI \rightarrow H_2 + I_2$, 823 K	*Cycle*: KU E: $2HI \rightarrow H_2 + I_2$, 373 K, 0.5 V
3	$H_2SO_4 \rightarrow 0.5O_2 + SO_2 + H_2O$, 1,173 K	*Cycle*: Mark 16/GA, Table 5.4 $2H_2O + SO_2 + I_2 \rightarrow H_2SO_4 + 2HI$, 373 K $2HI \rightarrow H_2 + I_2$, 773 K	*Cycle*: Mark 11/Westinghouse E: $2H_2O + SO_2 \rightarrow H_2 + H_2SO_4$, 310 K, 22 bar, 0.48 V *Cycle*: NCLI E: $2H_2O + SO_2 + I_2 \rightarrow H_2SO_4 + 2HI$, 353 K T: $2HI \rightarrow H_2 + I_2$, 773 K *Cycle*: Mark 13/LASL E: $2HBr \rightarrow H_2 + Br_2$, 298 K, 0.58 V T: $2H_2O + Br_2 + SO_2 \rightarrow H_2SO_4 + 2HBr$, 373 K
4	$H_2O + CaBr_2 \rightarrow CaO + 2HBr$, 1,023 K $CaO + Br_2 \rightarrow 0.5O_2 + CaBr_2$, 873 K	*Cycle*: UT-3, Table 5.11 $3FeBr_2 + 4H_2O \rightarrow H_2 + Fe_3O_4 + 6HBr$, 873 K $Fe_3O_4 + 8HBr \rightarrow 3FeBr_2 + Br_2 + 4H_2O$, 573 K	*Cycle*: ANL E: $2HBr \rightarrow H_2 + Br_2$, 323 K, 0.58 V
5	$CuCl_2 \rightarrow 2CuCl + Cl_2$, 773 K $H_2O + Cl_2 \rightarrow 0.5O_2 + HCl$, 1,073 K	*Cycle*: MHI, Table 5.12 $4H_2O + 3FeCl_2 \rightarrow H_2 + Fe_3O_4 + 6HCl$, 873 K $Fe_3O_4 + 8HCl \rightarrow FeCl_2 + 2FeCl_3 + 4H_2O$, 373 K $2FeCl_3 + 2CuCl \rightarrow 2FeCl_2 + 2CuCl_2$, 373 K	*Cycle*: UNLV-56 E: $2CuCl + 2HCl \rightarrow H_2 + 2CuCl_2$, 473 K

T thermochemical, *E* electrochemical, *ANL* Argonne National Laboratory, *GA* General Atomics, *KU* Kyushu University, *LASL* Los Alamos Scientific Laboratory, *MHI* Mitsubishi Heavy Industry, *NCLI* National Chemical Laboratory for Industry (Tokyo), *UNLV* University of Nevada at Las Vegas

[a]See hybrid cycle description in Table 5.15

The electrochemical version of the hydrochloric acid cycle was proposed by Air Products and Chemicals and it is known as the Hallett cycle (see cycle #53 in McQuillan et al. 2010; see also Chao 1974, Williams 1980, and Lewis and Masin 2009). Thus, the hybrid hydrochloric acid cycle requires a temperature at a heat source of 873 K, reduced by 200 K as compared with the thermochemical version of the cycle. Table 5.14 shows a second thermochemical closure for the hydrochloric acid cycle, namely, Funk's cycle with tantalum. This cycle closure requires a heat source of 1,366 K for thermolysis of tantalum tri-chloride.

The second example in Table 5.14 shows two corresponding water-splitting cycles using antimony (Sb)-based compounds; one with thermochemical closure and the other with hybrid thermo-electrochemical closure. The cycle by Miura uses thermochemical decomposition of hydrogen iodide as a closure reaction at 823 K. The cycle developed at Kyushu University (KU) performs the closure reaction electrochemically at 373 K at which the reversible potential is 0.5 V. The third example in the table is based on the reaction of thermochemical decomposition of sulfuric acid which evolves oxygen at 1,173 K. One of the well-known closure processes for this reaction is thermochemical as with the Mark 16/General Atomics S–I cycle. It requires a heat source of 773 K and necessitates two reaction steps: (1) hydrolysis of sulfur dioxide in the presence of iodine and (2) hydrogen iodide thermolysis. Three examples are given as electrochemical closure for this process. The first is a Mark 11/Westinghouse cycle with a single-step hydrogen process and sulfuric acid generated from H_2O and SO_2. The second example is the process by the National Chemical Laboratory for Industry (NCLI) in Japan which includes similar chemical steps as in the Mark 16 cycle with a difference that the sulfuric acid production step is performed electrochemically (see Table 5.14). The third example is the Mark 13/LASL process where bromine is used as an intermediate chemical. Hydrogen bromide is formed as an intermediate which is decomposed electrochemically at 298 K.

Two corresponding cycles having in common the reaction of calcium bromide hydrolysis at 1,023 K as a first step and bromination of calcium oxide as an oxygen-evolving reaction are also shown in Table 5.14. The purely thermochemical cycle is the well-known UT-3 cycle listed in Table 5.11, which uses iron-bromide and iron oxide intermediates to recycle the reagents and to evolve hydrogen in two reaction steps. The hybrid version of this cycle requires a single electrochemical reaction of hydrobromic acid dissociation which evolves hydrogen at 323 K and regenerates bromine.

The last example in Table 5.14 compares two copper–chlorine-based cycles, which have in common a reverse Deacon reaction and thermal decomposition of cupric chloride. The pure thermochemical cycle, proposed by Mitsubishi Heavy Industry (MHI), uses iron-chloride and iron oxide intermediates to recycle the reagents and to generate hydrogen in a process with three reaction steps. The hybrid version is listed as the UNVL-56 cycle and requires only one electrochemical step of CuCl disproportionation which evolves hydrogen and recycles the cupric chloride. It is interesting to observe that the UNVL-56 cycle has itself a purely thermochemical version mentioned by Williams (1980), which includes exactly the same reaction, but all thermochemical; this cycle is also listed in Table 5.10. When chlorination of CuCl, through $2CuCl(s) + 2HCl(g) \rightarrow H_2(g) + 2CuCl_2(s)$,

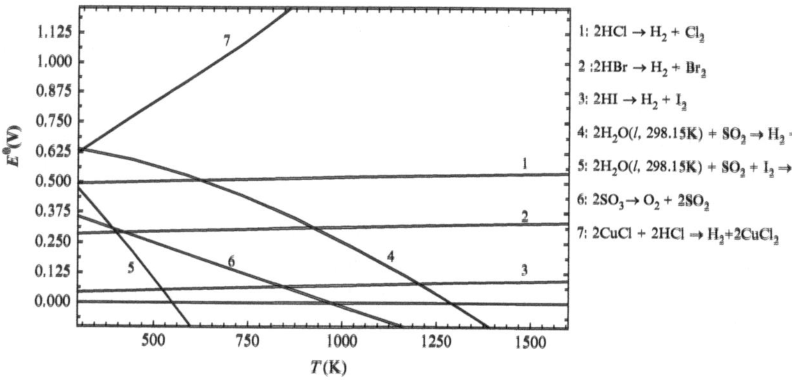

Fig. 5.24 Reversible cell potential of some significant electrochemical processes as a function of temperature, assuming reactions in gas phase

is performed thermochemically, purging of solid phases is necessary in order to shift the reaction equilibrium forward. It is however rather complicated to separate hydrogen from hydrochloric acid.

Figure 5.24 presents the reversible potential of some significant electrochemical processes as a function of temperature, assuming that reactions occur in the gas phase. It can be observed that some reactions, such as 1–3 and 7, are favored at low temperature where their reversible potential is reduced. Some other reactions (4–6) are favored at higher temperatures as they have thermo-neutral points.

The reaction #1 is the hydrochloric acid decomposition step which is a known industrial process, referenced as Bayer–Hoechst–Uhde (patented in 1964). It uses a feed of 22 % aqueous HCl by weight and generates wet chlorine gas with 1–2 % moisture (see McQuillan et al. 2010). The moisture does not need to be removed from chlorine. The process uses a reverse Deacon reaction in conjunction with electrolytic decomposition of hydrochloric acid. The electrochemical process requires inexpensive electrodes in graphite. Current technological developments have shown improvement of the process with a 25 % reduction of reversible potential as compared to the original process, 1.5 V vs. 2 V, respectively (see Lewis and Masin 2009). The electrolyzer can generate hydrogen at 5 bars with an efficiency of around 88 % and a requirement of electricity input of 289.5 MJ per kg of hydrogen (McQuillan et al. 2010). An older approach, investigated by Air Products and Chemicals, performed the electrochemical decomposition of hydrochloric acid at 21 bar and 473 K (Sato 1979).

The reaction #2 is the electrochemical decomposition of hydrobromic acid; typically, in the electrolyte, the chemical activity of HBr is unity, while bromine evolves under vacuum at 0.28 bar absolute. At such operating conditions, the reversible cell potential becomes 1.07 V. A remarkable property of bromine is that it is highly soluble, as it forms the group Br_3^- in the electrolyte according to $Br_2 + Br^- \rightarrow Br_3^-$. Therefore, gaseous bromine evolves hardly. The concentration of HBr in solution is maintained at ~50 %, while the current density with graphite electrodes approaches 1 kA/m^2. When platinum is used, the current density can surpass 12 kA/m^2 (Sato 1979). Hydrogen iodide electrolysis is similar to that of HBr. When HI is in aqueous solution, the reversible cell potential is 0.535 V. This electrochemical reaction must be conducted over the melting point of iodide (321 K) to avoid deposition of solid iodine on the electrode due to the reaction: $I_2(aq) + I^- \rightarrow I_3^-(s)$.

Electrolysis of sulfurous acid (H_2SO_3) is a crucial process in many hybrid cycles, e.g., Mark 11/Westinghouse cycle and the NCLI I–S cycle (NCLI = National Chemical Laboratory for Industry at Tokyo). When sulfur dioxide dissolves in water, sulfurous acid is formed according to

$$SO_2 + H_2O \rightarrow H_2SO_3(aq). \tag{5.25}$$

The sulfurous acid exists in solution only; it cannot be insulated as it evolves sulfur dioxide. Therefore, an aqueous solution of sulfur dioxide can be electrolyzed in an acid electrolyte according to the following half-reactions:

$$\begin{aligned} H_2O + H_2SO_3(aq) \rightarrow H_2SO_4(aq) + 2H^+ + 2e^-, \quad \text{at anode, } E^0 = -0.17\,\text{V} \\ 2H^+ + 2e^- \rightarrow H_2(g), \quad \text{at cathode, } E^0 = 0.0\,\text{V}. \end{aligned} \tag{5.26}$$

The overall process described in (5.25) and (5.26) therefore becomes

$$2H_2O + SO_2(aq) \rightarrow H_2 + H_2SO_4(aq) \tag{5.27}$$

which reduces the electricity requirement for water splitting to one-third with respect to direct water electrolysis (Dokyia et al. 1977). Precautions are made to impede the migration of sulfurous acid from the anode (where it is generated) toward the cathode. The possible presence of sulfurous acid at the cathode favors side reactions which eventually lead to sulfur deposition on the electrode or emanation of hydrogen sulfide. The main side reactions at the cathode are described by (see Sato 1979)

$$\begin{cases} H_2SO_3 + 4H^+ + 4e^- \rightarrow 3H_2O + S, \quad E^0 = +0.45\,\text{V} \\ S + 2H^+ + 2e^- \rightarrow H_2S, \quad E^0 = +0.141\,\text{V}. \end{cases} \tag{5.28}$$

The practical experience with the electrolytic process described in equation (5.27) shows that the current density of 2 kA/m^2 is readily obtained with a cell voltage less than 0.8 V. It uses diaphragms around the cathode where the electrolyte is slightly over-pressurized to reduce the probability of H_2SO_3 migration and the occurrence of side reactions.

A variant of the sulfurous acid electrolysis process has been proposed by the NCLI in Tokyo, according to Dokyia et al. (1977). Aqueous sulfurous acid is electrolyzed in the presence of iodine according to the overall reaction:

$$2H_2O + SO_2 + I_2 \rightarrow H_2SO_4 + 2HI \quad \text{at 353 K}, \tag{5.29}$$

which is performed based on the following half-reactions:

$$\begin{cases} H_2O + H_2SO_3(aq) \rightarrow H_2SO_4 + 2H^+ + 2e^- \quad \text{at anode} \\ I_2 + 2H^+ + 2e^- \rightarrow 2HI \quad \text{at cathode}. \end{cases} \tag{5.30}$$

This electrolysis process uses activated carbon as the anode and a platinum electrode at the cathode. It does not require electrical energy input; actually this cell behaves like a separator of sulfuric and hydroiodic acid. In order to accelerate the reaction to obtain a better production rate, some potential difference must be applied at the electrodes. Dokyia et al. (1977) show that the electricity consumption of the process described in (5.30) is less than the case of the Westinghouse/Mark 11 cycle at about 45 %.

Cuprous chloride electrolysis is also an important process. It occurs in the Cu–Cl hybrid cycle family. One variant of this reaction is mentioned in Table 5.14, namely:

$$2CuCl(aq) + 2HCl(aq) \rightarrow H_2(g) + 2CuCl_2(aq). \tag{5.31}$$

This process has been actively researched recently at Atomic Energy of Canada Limited (AECL) and ANL. The copper (I) oxidation reaction is presently conducted in a 0.5 mol/L CuCl/11 mol/L HCl electrolyte—anode reaction. In 11 mol/L HCl, copper metal deposition at the cathode appears to be absent at the 0.1 A $\cdot$ cm^{-2} current density used in the experiments. In 6 mol/L HCl, the $CuCl^+$ that is formed can be transported across the membrane by diffusion and migration. In 11 mol/L HCl, $CuCl_2$ is transported across the membrane by diffusion only. Thus, the flux of copper species through the membrane is dependent upon the HCl concentration. Experimental data shows that for similar experimental conditions, the catholyte copper species concentration decreases with increasing HCl concentration. This is consistent with the dependence of the flux on HCl concentration.

Results from AECL's experiments—see Naterer et al. (2011a)—show very promising stable performance at a cell voltage of about 0.57 V over a period of about 50 h. The catholyte flow rate was 1.5 L/min and the anolyte flow rate was 1.25 L/min. The operating temperature and current density were 45 °C and 0.1 A cm^{-2}, respectively. No anode catalyst was used and there was a Pt cathode catalyst. Stable cell performance at voltages down to 0.477 V has been observed.

ANL developed an electrolytic disproportionation system for CuCl(l) which generates hydrogen at 24 bar. Lewis and Masin (2009) reported that the ANL electrolysis reactor operates at 0.5 V with 0.5 A/cm^2. This reaction is attractive because it generates compressed hydrogen. Consequently, the work required to compress hydrogen from standard pressure—to 350 bars for example—reduces approximately by 95 %.

Another version of CuCl(aq) electrolysis occurs based on the following reaction:

$$2CuCl(aq) \rightarrow 2CuCl_2(aq) + Cu(s) \tag{5.32}$$

Experimental investigations were performed on reaction (5.32) at University of Ontario Institute of Technology (UOIT) using 1 mM solution of CuCl in 1.5 M HCl and platinum and/or glassy carbon electrodes. Both electrodes show similar potentials although the platinum electrode showed a three times higher limiting current density, namely, 130 μA cm^{-2} for platinum and 40 μA cm^{-2} for the glassy carbon electrode. However, the surface area of the carbon electrode is much higher than a typical carbon-supported Pt catalysts; therefore, glassy carbon appears to be the recommended choice.

Table 5.15 lists some significant hybrid thermo-electrochemical cycles. This compilation lists in total 25hybrid cycles, of which four have two steps, 13 have three steps, seven have four steps, and one has five steps. It is interesting to compare the occurrence of chemical elements in hybrid thermo-electrochemical cycles with the same for pure thermochemical cycles. Based on the analysis in Fig. 5.25, it shows that the occurrence of chlorine is the highest among hybrid cycles, namely, 40 %, followed by copper with 32 %, then sulfur with 28 %, and then Br and I with 16 and 12 %, respectively.

The above analysis shows that iron, which is the most encountered element in pure thermochemical cycles, is often not preferred in hybrid cycles; hence it has only a 4 % occurrence. Chlorine and sulfur however show a high occurrence in both thermochemical and hybrid cycles.

The process efficiencies reported in Table 5.15 are extracted from literature data and they are based on the following equation, similar with (5.18):

$$\eta = \frac{\mathrm{HHV}}{Q_{\mathrm{in}} + W_{\mathrm{el}}/\eta_{\mathrm{el}}} \tag{5.33}$$

where Q_{in} is the heat input to the process for one mole of hydrogen produced, HHV is the molar higher heating value of hydrogen, W_{el} is the electrical (or mechanical) heat input, and η_{el} is a conventionally adopted conversion efficiency of power generation from high-temperature heat. For the efficiency data in Table 5.15, the power conversion efficiency is taken 50 %. The electrical heat input is given by the equation

$$W_{\mathrm{el}} = W_{\mathrm{elchem}} + W_{\mathrm{sep}} + W_{\mathrm{aux}}, \tag{5.34}$$

where W_{elchem} is the electrical energy needed in electrochemical reactions of hybrid cycles per mole of hydrogen generated. Also, $W_{\mathrm{sep}} = \Delta G_{\mathrm{sep}}$ is the energy for separation of chemicals—see (5.18)—and W_{aux} is any other auxiliary energy required for mechanical/electrical processes such as for material conveying, pumping, compressing, chemical mixing, etc.

The first hybrid cycle listed in Table 5.15 is the well-known sulfuric acid process invented at ISPRA (Mark 11) and under development at Westinghouse. This cycle is discussed in detail in Sect. 5.9. The second process (APC/UNLV-53) was discussed above as the Hallett hybrid cycle developed at Air Products and Chemicals. This cycle was recently assessed by Lewis and Masin (2009) and McQuillan et al. (2010) and shows good potential as a promising technology for further development.

The third cycle mentioned in the table, proposed by Dokyia and Kotera (1976) at NCLI, is a two-step version of the copper–chlorine cycle; this cycle is denoted as CuCl-C by Lewis and Masin (2009). It uses an electrochemical step similar to that

Table 5.15 Significant hybrid thermochemical water-splitting cycles

N_s	Cycle ID	Elements	T_{max} (K)	Reactions	η
2	Mark 11, Westinghouse	S	1,173	E: $2H_2O + SO_2 \rightarrow H_2 + H_2SO_4$, 310 K, 22 bar, 0.48 V T: $H_2SO_4 \rightarrow 0.5O_2 + SO_2 + H_2O$, 1,173 K	51
	APC, UNLV-53	Cl	873	E: $2HCl \rightarrow H_2 + Cl_2$, 363 K T: $H_2O + Cl_2 \rightarrow 0.5O_2 + 2HCl$, 873 K	49
	NCLI, CuCl-C	Cl, Cu	823	E: $2CuCl + 2HCl \rightarrow H_2 + 2CuCl_2$, 473 K T: $H_2O + 2CuCl_2 \rightarrow 0.5O_2 + 2HCl + 2CuCl$, 823 K	46
2	ANL	Bi, K	850	E: $KOH + 1/3\,Bi \rightarrow 0.5H_2 + 0.5O_2 + 1/3\,K_3Bi$, 850 K T: $H_2O + 1/3\,K_3Bi \rightarrow 0.5H_2 + KOH + 1/3Bi$, 850 K	46
3	IGT, UNLV-5	Cd	1,473	E: $2H_2O + Cd \rightarrow H_2 + Cd(OH)_2$, 298 K, 0.02 V T: $CdO \rightarrow 0.5O_2 + Cd$, 1,473 K T: $Cd(OH)_2 \rightarrow CdO + H_2O$, 650 K	53
	UNLV-184	Br, Sb	1,273	E: $2HBr \rightarrow H_2 + Br_2$:, 373 K, 0.58 V T: $H_2O + 0.5Sb_2O_3 + Br_2 \rightarrow 2HBr + 0.5Sb_2O_5$, 353 K T: $0.5Sb_2O_5 \rightarrow 0.5O_2 + 0.5Sb_2O_3$, 1,273 K	36
	KU	I, Sb	1,273	E: $2HI \rightarrow H_2 + I_2$, 373 K, 0.5 V T: $H_2O + 0.5Sb_2O_3 + I_2 \rightarrow 2HI + 0.5Sb_2O_5$, 298 K T: $0.5Sb_2O_5 \rightarrow 0.5O_2 + 0.5Sb_2O_3$, 1,273 K	36
	NCLI	I, S	1,173	E: $2H_2O + I_2 + SO_2 \rightarrow 2HI + H_2SO_4$, 353 K $2HI \rightarrow H_2 + I_2$, 773 K T: $H_2SO_4 \rightarrow 0.5O_2 + SO_2 + H_2O$, 1,173 K	
	Mark 13, LASL	Br, S	1,123	E: $2HBr \rightarrow H_2 + Br_2$, 298 K, 0.58 V T: $2H_2O + Br_2 + SO_2 \rightarrow H_2SO_4 + 2HBr$, 373 K T: $H_2SO_4 \rightarrow 0.5O_2 + SO_2 + H_2O$, 1,123 K	39
	NCLI, CuCl-D, UNLV-56	Cl, Cu	1,073	E: $2CuCl + 2HCl \rightarrow H_2 + 2CuCl_2$, 473 K T: $H_2O + Cl_2 \rightarrow 0.5O_2 + 2HCl$, 1,073 K T: $2CuCl_2 \rightarrow 2CuCl + Cl_2$, 773 K	31
	ANL	Br, Ca	1,043	E: $2HBr \rightarrow H_2 + Br_2$, 323 K, 0.58 V (or plasma) T: $H_2O + CaBr_2 \rightarrow CaO + 2HBr$, 1,043 K T: $CaO + Br_2 \rightarrow 0.5O_2 + CaBr_2$, 853 K	30

	UNLV-4	Cl, Fe	923	E: $1.5Cl_2 + Fe_3O_4 + 6HCl \rightarrow 0.5O_2 + 3FeCl_3 + 3H_2O$, 423 K T: $3FeCl_2 + 4H_2O \rightarrow H_2 + Fe_3O_4 + 6HCl$, 823 K T: $3FeCl_3 \rightarrow 1.5Cl_2 + 3FeCl_2$, 693 K	44
	CuCl, UNLV-191	Cl, Cu	823	E: $4CuCl \rightarrow 2Cu + 2CuCl_2$, 353 K T: $H_2O + 2CuCl_2 \rightarrow 0.5O_2 + 2CuCl + 2HCl$, 823 K T: $2Cu + 2HCl \rightarrow H_2 + 2CuCl$, 700 K	49
	ANL, CuCl-B	Cl, Cu	800	E: $2CuCl + 2HCl \rightarrow H_2 + 2CuCl_2$, 373 K, 24 bar T: $H_2O + 2CuCl_2 \rightarrow Cu_2OCl_2 + 2HCl$, 700 K T: $Cu_2OCl_2 \rightarrow 0.5O_2 + 2CuCl$, 800 K	45
	JAEA-HHLT	S	773	E: $2H_2O + SO_2 \rightarrow H_2 + H_2SO_4$, 350 K E: $SO_3 \rightarrow 0.5O_2 + SO_2$, 773 K T: $H_2SO_4 \rightarrow SO_3 + H_2O$, 673 K	44
	ANL, INL	Cl, Mg	773	E: $2HCl \rightarrow H_2 + Cl_2$, 363 K T: $H_2O + MgCl_2 \rightarrow 2HCl + MgO$, 723 K T: $MgO + Cl_2 \rightarrow 0.5O_2 + MgCl_2$, 773 K	50
	ANL, UNLV-24	Li, N	750	E: $H_2O + I_2 + LiNO_2 \rightarrow LiNO_3 + 2HI$, 300 K, 1.2 V T: $2HI \rightarrow H_2 + I_2$, 573 K T: $LiNO_3 \rightarrow 0.5O_2 + LiNO_2$, 700 K	38
4	LASL	Bi, S	1,173	E: $2H_2O + SO_2 \rightarrow H_2 + H_2SO_4$, 350 K, 0.17 V T: $SO_3 \rightarrow 0.5O_2 + SO_2$, 1,073 K T: $H_2SO_4 + Bi_2O_3 \rightarrow Bi_2O_3SO_3 + H_2O$, 350 K T: $Bi_2O_3SO_3 \rightarrow Bi_2O + SO_3$, 1,173 K	24
4	UNLV-185	Br, Co	1,123	E: $2HBr \rightarrow H_2 + Br_2$, 298 K, 0.58 V T: $H_2O + CoBr_2 \rightarrow CoO + 2HBr$, 1,023 K T: $Co_3O_4 \rightarrow 0.5O_2 + 3CoO$, 1,023 K T: $Br_2 + 4CoO \rightarrow CoBr_2 + Co_3O_4$, 773 K	26
	IGT	Cu, S	1,100	E: $2H_2O + SO_2 \rightarrow H_2 + H_2SO_4$, 298 K, 0.17 V T: $CuSO_4 \rightarrow 0.5O_2 + CuO + SO_2$, 1,100 K T: $H_2SO_4 + CuO + 3H_2O \rightarrow CuSO_4 \cdot 4H_2O$, 298 K T: $CuSO_4 \cdot 4H_2O \rightarrow CuSO_4 + 4H_2O$, 800 K	55

(continued)

Table 5.15 (continued)

N_s	Cycle ID	Elements	T_{max} (K)	Reactions	η
	ISPRA	S	1,073	E: $H_2O + H_2SO_3 \rightarrow H_2 + H_2SO_4$, 300 K T: $SO_3 \rightarrow 0.5O_2 + SO_2$, 1,073 K T: $H_2SO_4 \rightarrow SO_3 + H_2O$, 1,073 K T: $SO_2 + H_2O \rightarrow H_2SO_3$, 298 K	
	IUU, UNLV-114	I, K, N	973	E: $2HNO_3 + 2KI \rightarrow H_2 + I_2 + 2KNO_3$, 298 K T: $H_2O + 3NO_2 \rightarrow 2HNO_3 + NO$, 373 K T: $I_2 + 2KNO_3 \rightarrow 0.5O_2 + 2KI + 2NO + 1.5O_2$, 973 K T: $3NO + 1.5O_2 \rightarrow 3NO_2$, 373 K	33
	IGT, CuCl-A	Cl, Cu	800	E: $4CuCl \rightarrow 2Cu + 2CuCl_2$, 353 K T: $H_2O + 2CuCl_2 \rightarrow Cu_2OCl_2 + 2HCl$, 700 K T: $2Cu + 2HCl \rightarrow H_2 + 2CuCl$, 700 K T: $Cu_2OCl_2 \rightarrow 0.5O_2 + 2CuCl$, 800 K	50
	UOIT, CuCl	Cl, Cu	800	E: $2CuCl + 2HCl \rightarrow H_2 + 2CuCl_2(aq)$, 353 K, 1 bar T: $H_2O + 2CuCl_2(s) \rightarrow Cu_2OCl_2 + 2HCl$, 700 K T: $Cu_2OCl_2 \rightarrow 0.5O_2 + 2CuCl$, 800 K T: $CuCl_2(aq) \rightarrow CuCl_2(s)$, 400 K	
5	UOIT, CuCl-5	Cl, Cu	773	E: $4CuCl(aq) \rightarrow 2Cu(s) + 2CuCl_2(aq)$, 353 K T: $H_2O + 2CuCl_2(s) \rightarrow CuO \cdot CuCl_2 + 2HCl$, 648 K T: $2Cu + 2HCl \rightarrow H_2 + 2CuCl$, 723 K T: $CuO \cdot CuCl_2 \rightarrow 0.5O_2 + 2CuCl(l)$, 773 K T: $2CuCl_2(aq) \rightarrow 2CuCl_2(s)$, 373 K	63

N_s number of steps, *T* thermochemical, *E* electrochemical, *APC* Air Products and Chemicals (Hallett Cycle), *INL* Idaho National Laboratory *IUU* Iwate University of Ueda (Japan), *JAEA* Japan Atomic Energy Agency, *HHLT* hybrid hydrogen process at low temperature, *KU* Kyushu University (Japan), *NCLI* National Chemical Laboratory for Industry (Tokyo), *UIC* University of Illinois at Chicago
Sources: Chao (1974), McQuillan et al. (2010)

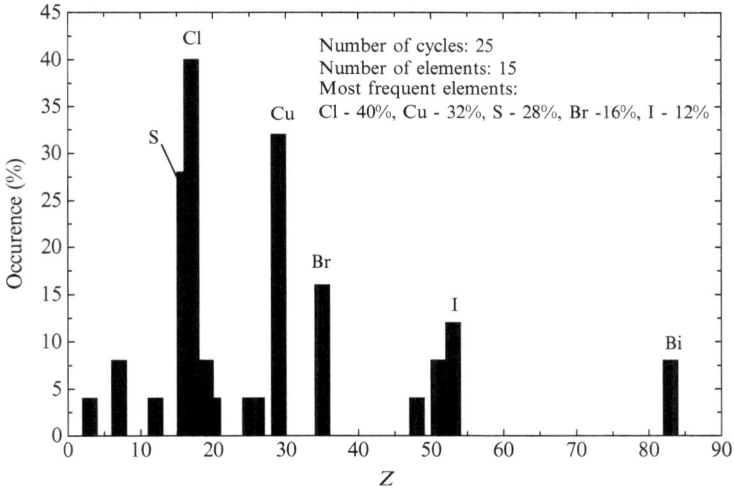

Fig. 5.25 Occurrence of chemical elements in hybrid thermo-electrochemical cycles as a function of their atomic number (*Z*)

under development by the international consortium led by the UOIT; see Naterer et al. (2011a). The electrochemical reaction is coupled with a thermochemical reaction at higher temperature, which is the hydrolysis of $CuCl_2$ at 823 K. Other versions of the Cu–Cl cycle are listed in the table. Dokyia and Kotera (1977) also proposed a three-step version of the Cu–Cl cycle which was later denoted as CuCl-D by Lewis and Masin (2009). It is called the process UNLV-56 in McQuillan et al. (2010). The assessment by McQuillan et al. (2010) reports an overall efficiency of the process of 31 %. The reverse Deacon reaction is used to generate hydrogen at 1,073 K. In the third step, cupric chloride is decomposed thermally at 773 K to regenerate chlorine and cuprous chloride. Another three-step version from the Cu–Cl family uses the electrochemical process described in (5.32) to generate solid copper. This process is the UNLV-191 cycle assessed in McQuillan et al. (2010), which reports a higher energy efficiency of 49 % (assuming that the electricity generation is made with 50 % efficiency). The maximum temperature requirement is 823 K.

ANL studied a three-step version of the Cu–Cl cycle based on an electrochemical reaction described in (5.31) and denoted in Lewis and Masin (2009) as the CuCl-B cycle. In their implementation, ANL performs the electrochemical process at 24 bar absolute pressure and 373 K to generate pressurized hydrogen. This cycle is also currently under study at the University of Illinois at Chicago (UIC). The Institute of Gas Technology (IGT) proposed a four-step version of the Cu–Cl cycle based on the electrolytic process (5.32) followed by four thermochemical reactions and requiring a source of heat at 800 K. According to Lewis and Masin (2009), this cycle is denoted as CuCl-A. Its efficiency—assuming full conversion of chemicals—is estimated to be 50 % (see Lewis and Masin 2009).

At UOIT, there is currently a scale-up project of a four-step Cu–Cl cycle which is based on electrochemical reaction (5.31) and includes a hydrolysis reaction of $CuCl_2$, thermolysis of copper oxychloride (Cu_2OCl_2), and a drying step of aqueous cupric chloride. Another cycle under study at UOIT is the five-step Cu–Cl cycle based on electrochemical reaction (5.32), a drying step, and three other chemical reaction steps (see Table 5.15).

The two-step cycle with Bi and K developed at ANL is mentioned by Lewis and Masin (2009). This cycle is under investigation; a preliminary analysis shows that its theoretical efficiency is around 46 %. At ANL, there are other hybrid cycles under development such as the three-step Br–Ca cycle at 1,043 K maximum temperature, three-step Cl–Mg cycle (studied in conjunction with Idaho National Laboratory), and the three-step $LiNO_3$ cycle with a 750 K temperature requirement (see UNVL-24 in the table). IGT—in the 1970s—proposed a cadmium-based cycle with 53 % energy efficiency, which is considered as promising in recent studies by McQuillan et al. (2010). Other cycles inventoried and assessed at the University of Nevada at Las Vegas are the UNLV-184 with barium and antimony, the UNLV-4 which is a hybrid version of the Mark 9 cycle, and the UNLV-185 of four steps working with bromine and cobalt compounds.

In the mid 1970s, cycles with iodine and antimony were studied in Japan (at Kyushu University), and iodine, potassium, and nitrogen (at Iwate University of Ueda, IUU). The Los Alamos Scientific Laboratory proposed a four-step bismuth–sulfur cycle. ISPRA proposed a four-step hybrid version of the sulfur–iodine cycle which operates at 1,073 K. A recent development was reported at the Japan Atomic Energy Agency, which developed a lower temperature version of the Westinghouse process which includes two electrochemical steps. In this cycle, H_2SO_4 is decomposed thermally at 673 K in steam and SO_3; thereafter, SO_3 is decomposed electrochemically in a solid oxide membrane electrolysis cell operating at 773 K. The cell generates oxygen gas and SO_2. The second electrochemical process is the same as the Westinghouse cycle: it performs the electrolysis of sulfurous acid. The process efficiency is estimated by Chikazawa et al. (2006) at 44 %.

5.6.2 Thermo-Photo-ElectroChemical Water-Splitting Cycles

A large number of photochemical reactions exist for hydrogen production. A photochemical reaction uses light as an energy source to promote chemical reactions. Hybrid water-splitting systems comprise photochemical, thermochemical, and electrochemical reactions with recycling of all intermediate reagents. The block diagram of a thermo-photo-electrochemical system for water splitting is shown in Fig. 5.26. A hypothetical cycle is shown with three reaction steps, one purely thermochemical (hydrolysis), one photochemical (hydrogen generation), and one electrochemical (oxygen evolution). Hybrid processes of this type may be advantageous with respect to purely thermochemical processes in what concerns the reduction of the maximum process temperature, improvement of the yield and

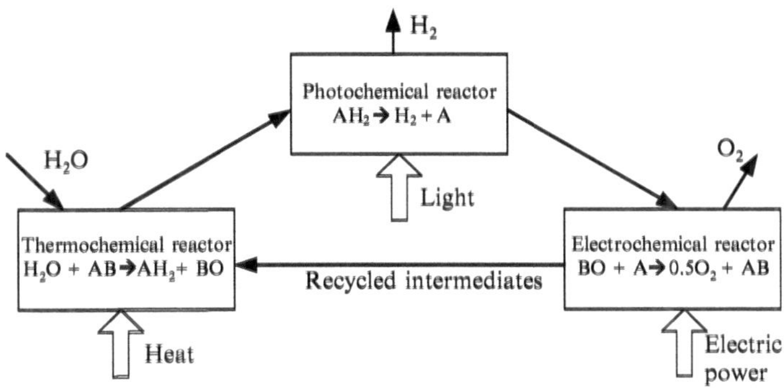

Fig. 5.26 Block diagram of a hypothetical hybrid thermo-photo-electrochemical water-splitting cycle

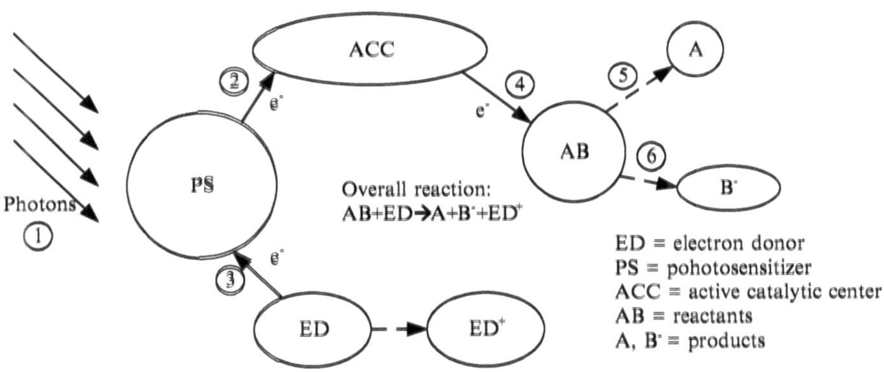

Fig. 5.27 Main elements and functioning of a molecular system for photocatalysis

the overall reaction kinetics, as well as improvement of chemical separation. Light-induced photochemical reactions relevant to hydrogen evolution from water were recently reviewed by Zamfirescu et al. (2012). The process of light–matter interactions may generate molecular excited states which are very reactive and as a consequence, chemical reactions can be induced.

A molecular system for photocatalysis in general must comprise several functional elements such as photosensitizer (the molecule that can absorb photonic radiation), an electron donor or acceptor, and an active catalytic center to promote redox reactions. Figure 5.27 indicates the functioning of a general system for photocatalysis. The photon absorption by the photosensitizer (PS)—process (1)—initiates a sequence of events, first being the molecular excitation:

$$PS + h\nu \rightarrow PS^*,$$

which is followed by a fast charge transfer process (2) from PS to the active center:

$$PS^* + ACC \rightarrow PS^+ + ACC^-.$$

Then the photosensitizer recovers the electron from a neighboring electron donor according to a photo-catalytic charge transfer process (3):

$$PS^+ + ED \rightarrow PS + ED^+$$

The activated catalytic center catalyzes the redox reaction—process (4)—according to

$$ACC^- + AB \rightarrow A + B^- + ACC.$$

Therefore, the overall photoreaction becomes

$$AB + ED \xrightarrow{h\nu} A + B^- + ED^+$$

In a subsequent reaction—not necessary photo-induced—one can have a charge transfer as follows: $B^- + ED^+ \rightarrow B + ED$; therefore, the electron donor is recycled and the A and B are obtained as end products according to $AB \rightarrow A + B$, in which no sacrificial reagents (e.g., electron donors) are used. In some classes of photoreactions, no photosensitizers or electrons are used, because the reactants themselves have the capacity to absorb light. Once the photons are absorbed, the photoreaction occurs and converts irreversibly the reactants into products which are insensitive to light.

Some hybrid thermo-photochemical cycles are discussed next. One of the first hybrid cycles of this type was proposed in the first half of the 1970s at Yokohama National University in Japan and it is called Yokohama Mark 2A. This cycle was further developed and three other versions were proposed, namely, Yokohama Mark 2A, Yokohama Mark 6, and Yokohama Mark 7 (Ohta et al. 1976). The cycle Yokohama Mark 2A operates according to the following reactions:

$$\begin{cases} 2H_2O + I_2 + 2FeSO_4 \xrightarrow{h\nu} 2HI + 2Fe(OH)SO_4 \\ 2HI \rightarrow H_2 + I_2 \\ Fe(OH)SO_4 \rightarrow 0.5O_2 + 2FeSO_4 + H_2O. \end{cases} \tag{5.35}$$

The iodine in aqueous solution dissociates according to $I_2(aq) + I^- \rightarrow I_3^-(s)$, while the ferrous sulfate dissociates according to $FeSO_4 \rightarrow Fe^{2+} + (SO_4)^{2-}$. The photochemical reaction in solution (first equation from the set (5.35)) is essentially equivalent to

$$2Fe^{2+} + I_3^- \xrightarrow{h\nu} 2Fe^{3+} + 3I^-. \tag{5.36}$$

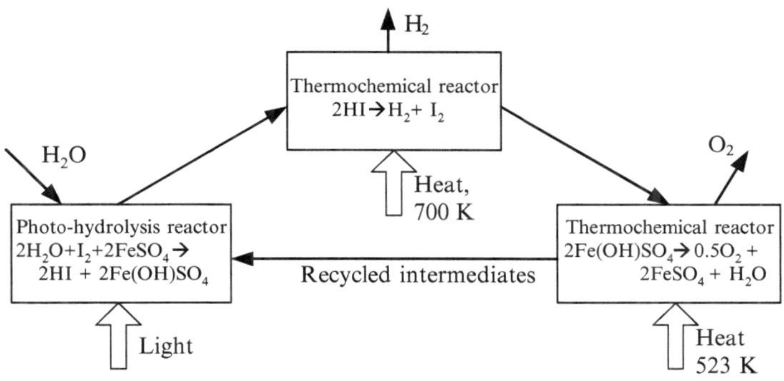

Fig. 5.28 Schematic representation of Yokohama Mark 2A thermo-photochemical cycle

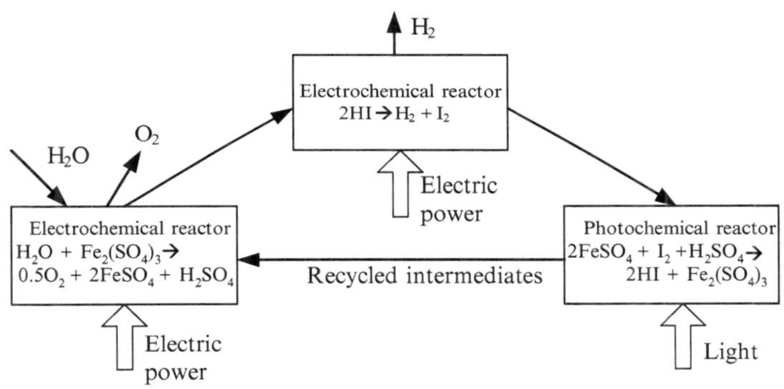

Fig. 5.29 Schematic representation of Yokohama Mark 5 electro-photochemical cycle

The second reaction in (5.35) is accomplished thermochemically with a 700 K heat source. The third reaction is also thermochemical at 523 K. The schematic of the Yokohama Mark 2A cycle is shown in Fig. 5.28. The experiments done with the Yokohama Mark 2A cycle showed that the oxygen evolution process is relatively inefficient. Therefore the cycle was modified into a Yokohama Mark 5 version which uses an electrochemical hydrogen evolution process.

The intermediate reagent was changed in Yokohama Mark 5, from $Fe(OH)SO_4$ as in Yokohama Mark 2A to $Fe_2(SO_4)_3$. The photochemical process is essentially the same for both cycles, as described in (5.34). For practical implementation, Yokohama Mark 5 uses a thermoelectric converter with concentrated solar radiation, which generates the electricity required to drive the electrochemical process. Visible light from the solar spectrum is used to drive the photochemical process. The processes in the Yokohama Mark 5 cycle, which are slightly different than previous versions, are presented in Fig. 5.29.

The hydrogen-evolving reaction in the Yokohama Mark 5 cycle is also performed electrochemically. In order to supply electricity to this cycle, an external generator must be used, in addition to the thermoelectric generator integrated into the process. The efficiency of the cycle is reported as around 20 % (see Sato, 1979). The chemical processes in the Yokohama Mark 5 cycle are given in (5.37):

$$\begin{cases} H_2SO_4 + I_2 + 2FeSO_4 \xrightarrow{h\nu} 2HI + 2Fe_2(SO_4)_3, \text{ photochemical, } \lambda \le 350\,\text{nm} \\ 2HI \rightarrow H_2 + I_2, \text{ electrochemical, } \Delta G = 109.7\,\text{kJ/mol} \\ Fe(OH)SO_4 \rightarrow 0.5O_2 + 2FeSO_4 + H_2O, \text{ electrochemical, } \Delta G = 82.5\,\text{kJ/mol.} \end{cases} \tag{5.37}$$

The Yokohama Mark 5 cycle represents an intermediate stage of development which was abandoned because of the efficiency losses with electrical energy generation. In the Yokohama Mark 6 implementation, the hydrogen-evolving reaction is performed thermochemically at 700 K. This cycle is described by the following set of reactions:

$$\begin{cases} H_2SO_4 + I_2 + 2FeSO_4 \xrightarrow{h\nu} 2HI + 2Fe_2(SO_4)_3, \text{ photochemical, } \lambda \le 350\,\text{nm} \\ 2HI \rightarrow H_2 + I_2, \text{ thermochemical, } \Delta H = 158.3\,\text{kJ/mol} \\ Fe(OH)SO_4 \rightarrow 0.5O_2 + 2FeSO_4 + H_2O, \text{ electrochemical, } \Delta G = 82.5\,\text{kJ/mol.} \end{cases} \tag{5.38}$$

The thermo-photo-electrochemical water-splitting cycle, Yokohama Mark 6, is described in Fig. 5.30. According to the published assessments, the efficiency of this cycle is 30 % (see Sato, 1979). Another hybrid reaction scheme has been also developed, namely, Yokohama Mark 7—see Kamita et al. (1984). The new process includes a photochemical reaction which is equivalent to that described in (5.36), although it uses a phosphate intermediate rather than a sulfate. The chemical reactions of the process are given according to (5.39).

$$\begin{cases} H_2O + \frac{2}{3}FePO_4 \rightarrow 0.5O_2 + \frac{2}{3}H_3PO_4 + \frac{2}{3}Fe_3(PO_4)_2, \text{ electrochemical} \\ 2HI \rightarrow H_2 + I_2, \text{ thermochemical, } \Delta H = 158.3\,\text{kJ/mol} \\ \frac{2}{3}H_3PO_4 + I_2 + \frac{2}{3}Fe_3(PO_4)_2 \xrightarrow{h\nu} 2HI + 2FePO_4, \text{ photochemical.} \end{cases} \tag{5.39}$$

A four-step thermo-photochemical process called the "photolytic sulfur ammonia cycle" is discussed in Perret (2011). This process includes a photolysis reaction (a photochemical reaction of hydrolysis) which evolves hydrogen at 353 K, and a thermal decomposition reaction with oxygen evolution at 1,143 K. There are two thermochemical reactions that recycle the intermediate reagents, one at a low temperature of 393 K and the other at 773 K. The simplified cycle diagram is presented schematically in Fig. 5.31. The chemical reactions of the cycle are described in (5.40):

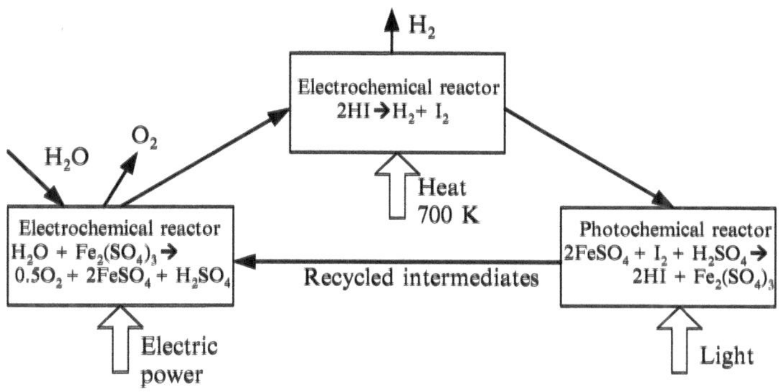

Fig. 5.30 Representation of the Yokohama Mark 6 thermo-photo-electrochemical cycle

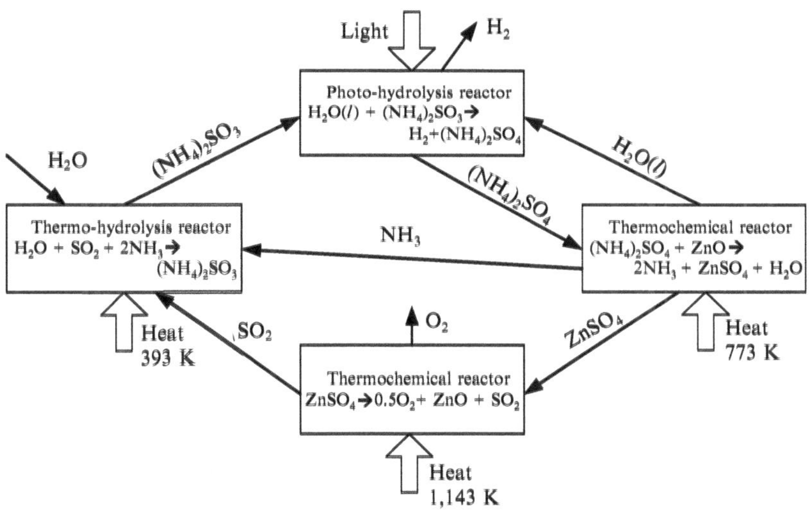

Fig. 5.31 Photolytic sulfur–ammonia cycle

$$\begin{cases} H_2O + SO_2 + 2NH_3 \rightarrow (NH_4)_2SO_3, \ 393\,K \\ H_2O + (NH_4)_2SO_3 \rightarrow H_2 + (NH_4)_2SO_4, \ \text{photochemical} \\ ZnSO_4 \Rightarrow 0.5O_2 + ZnO + SO_2, \ 1{,}143\,K \\ (NH_4)_2SO_4 + ZnO \rightarrow 2NH_3 + ZnSO_4 + H_2O, \ 773\,K. \end{cases} \tag{5.40}$$

All independent steps of this cycle were successfully demonstrated and all reactions went to completion. The photo-hydrolysis reaction requires an expensive photo-catalyst made in noble metals based on cadmium sulfide and a platinum–palladium–ruthenium complex. The photo-catalyst requires light at a wavelength shorter than 520 nm. The photo-reactor may require a source of green/blue light which can be derived from solar radiation with adequate spectral splitters. As mentioned in Perret (2011), the cost of a solar spectral splitter may alleviate

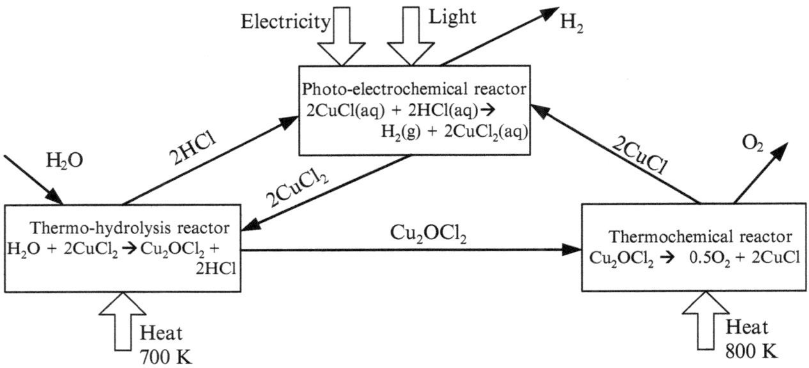

Fig. 5.32 Simplified diagram of a Cu–Cl cycle with a photo-electrochemical reaction step

the advantage of using photo-catalysis. Another practical problem arises from the necessity of solid material transport between reactors which must have a coordinated operation, although this type of problem is common to many thermochemical cycles which include solid–solid or solid–gas reactions. The energy efficiency of this cycle was derived based on flow-sheet modeling and it is reported as 48 % (see McQuillan et al. 2010).

A conceptual method for photochemical disproportionation of CuCl was recently proposed by Zamfirescu et al. (2012) as a potential method to reduce the electricity consumption and enhance the efficiency of hybrid copper–chlorine cycles. In this approach, the following reaction—representing one of the steps of the cycle—is driven photo-electrochemically:

$$2\mathrm{CuCl(aq)} + \mathrm{HCl(aq)} \rightarrow 2\mathrm{CuCl_2(aq)} + \mathrm{H_2(g)}. \tag{5.41}$$

The block diagram of a copper–chlorine cycle in a thermo-photo-electrochemical hybridization version is shown in Fig. 5.32. The photo-electrochemical reaction is conducted in a cell comprising two reactors separated by a proton-conducting membrane. One of the reactors is a half electrochemical cell where CuCl liquid is disproportionated with hydrochloric acid by releasing protons, according to

$$2\mathrm{CuCl(aq)} + \mathrm{HCl(aq)} \rightarrow 2\mathrm{CuCl_2(aq)} + 2\mathrm{H^+} + 2\mathrm{e^-}. \tag{5.42}$$

The second half cell is a hybrid photo-catalytic/electrochemical reactor which comprises an electrode and dissolved supramolecular photo-catalysts where the following concurrent reactions occur in parallel:

$$\begin{cases} 2\mathrm{H^+} + 2\mathrm{e^-} \rightarrow \mathrm{H_2(g)}, \text{ electrochemical, at electrode surface} \\ 2\mathrm{H_2O(l)} + 2\mathrm{e^-} \xrightarrow{h\nu} \mathrm{H_2(g)} + 2\mathrm{OH^-(aq)}, \text{ photocatlytic} \\ \\ 2\mathrm{H^+} + 2\mathrm{OH^-} \rightarrow 2\mathrm{H_2O(l)}. \end{cases} \Bigg\} \text{ in solution.} \tag{5.43}$$

The photocatalytic process uses supramolecular photo-catalysts based on ruthenium photosensitizers coupled to a single rhodium active center. According to Zamfirescu et al. (2012) the hydrogen production efficiency of the photo-electrochemical reactor, calculated based on the higher heating value of hydrogen and absorbed photonic radiation, is about6% assuming that the electrical energy portion that drives the cell is derived also from solar energy using a dye-sensitized solar cell with ~10 % efficiency. If the electrical component of the energy supply is derived from a next-generation nuclear reactor (50 % power generation efficiency) for a bias voltage of 0.3–0.4 V, then the cell efficiency can potentially reach 20 %. Additional results on this system are presented in Chap. 6, which focuses on the copper–chlorine cycle only.

5.6.3 Thermo-RadioChemical Water Splitting

The IGT (USA) developed in 1974 a thermochemical water-splitting cycle comprising one radiochemical reaction step of carbon dioxide decomposition. Spent fuel elements from nuclear reactors are used as a radiation source for the radiochemical reaction step of the cycle. A detailed flow-sheet diagram of the process is presented in Von Federsdorff (1974). This process was listed also in the thermochemical cycle database compiled by Bamberger (1978). The energy for carbon dioxide decomposition is derived from fission fragments with high kinetic energy in a special reactor that includes neutron shielding at 500 K according to the reaction

$$CO_2(g) \xrightarrow{\text{radiation}} 0.5O_2 + CO(g) \text{ at } 500\,\text{K}. \tag{5.44}$$

Further, carbon monoxide is used to reduce iron oxides at around 1,000 K; the reduced iron is then combined with steam to release hydrogen at about 700 K. The overall thermo-radiochemical cycle is expressed by the reactions

$$\begin{cases} H_2O(g) + 3FeO(s) \rightarrow H_2(g) + Fe_3O_4(s),\ 673\,\text{K} \\ CO_2(g) \xrightarrow{\text{radiation}} 0.5O_2(g) + CO(g),\ 500\,\text{K} \\ Fe_3O_4(s) + CO(g) \rightarrow 3FeO(s) + CO_2(g),\ 1{,}000\,\text{K}. \end{cases} \tag{5.45}$$

It appears that the IGT cycle described in (5.45) has a major disadvantage of a high temperature requirement due to utilizing iron oxides. In principle, it is possible to use another closure reaction for the carbon dioxide decomposition, namely, the water-gas-shift reaction. In this case, a possible low-temperature two-step process can be formulated as

$$\begin{cases} H_2O(g) + CO(g) \rightarrow H_2(g) + CO_2(g),\ 573\,\text{K} \\ CO_2(g) \xrightarrow{\text{radiation}} 0.5O_2 + CO(g),\ 500\,\text{K}. \end{cases} \tag{5.46}$$

Although the process (5.42) appears attractive for nuclear hydrogen production, it poses difficult technological problems. The first major problem was signaled by Von Federsdorff (1974) and refers to possible side reactions of radiochemical processes (5.44). The following elementary reactions may occur in a reactor performing radiolysis of carbon dioxide:

$$\left.\begin{array}{l} CO_2 \rightarrow C + 2O \\ CO_2 \rightarrow CO + O \end{array}\right\}\text{radiolytic}$$

$$\left.\begin{array}{l} CO + C \rightarrow C_2O \\ C_2O + CO \rightarrow C_3O_2 \end{array}\right\}\text{thermochemical,}$$

which lead to two main side products, namely, dicarbon monoxide (C_2O) and tricarbon dioxide, (C_3O_2) which are stable carbonaceous deposits that form scale buildup on the reactor walls.

A second major problem refers to separation of carbon monoxide and oxygen in the gas phase. The method originally proposed by Von Federsdorff (1974) uses mercury as an intermediate to chemically separate oxygen from a carbon monoxide + oxygen gas mixture; further, the formed HgO is decomposed thermally and mercury is condensed to separate oxygen. This method was deemed unacceptable for industrial scale applications. Other methods to consider may be based on physical processes such as those using absorption columns; these are extensively used in chromatography.

Although thermo-radiochemical water-splitting processes are less developed, they may present some potential for further consideration due to their unique possibility to directly use nuclear radiation to promote chemical reactions—either at a nuclear reactor site or at nuclear fuel processing plants or spent fuel processing facilities.

5.7 Down Selection of Thermochemical Cycles for Nuclear Hydrogen Production

In the above Sects. 5.1–5.6, thermochemical water-splitting cycles were introduced and most issues related to the status of their design and development were analyzed in detail. A comprehensive list of significant pure and hybrid cycles was provided based on screening of past literature. In this section, the down selection of thermochemical cycles for water splitting and hydrogen generation is analyzed. The down selections aim to recommend promising thermochemical cycles for specific nuclear reactors.

The general down selection process of thermochemical cycles was discussed in several papers published in the open literature, such as Perret (2011), McQuillan et al. (2010), Lewis et al. (2009), Dincer and Balta (2011), Abanades et al. (2006), and Brown et al. (2002). Some of these papers analyze thermochemical cycles for other applications except nuclear, namely, solar based and geothermal-based hydrogen production. However, the methodology of top-down selection is rather general, and applicable to nuclear reactors, provided that the correct scale of the application and the temperature level is considered.

A flowchart of a down selection of thermochemical hydrogen generation by water splitting is presented in Fig. 5.33. This process includes several necessary phases. In the first phase of the methodology, a database of significant thermochemical cycles must be compiled for the assessment. Primarily, the development of such a database can be performed based on literature screening. Some major compilations published in the literature are mentioned in Sect. 5.5.

The database of thermochemical cycles may be completed with novel potential cycles. Systematic identification (discovery) of novel processes has generated active interest from early times of thermochemical cycle research. Computer-assisted searches for thermochemical cycles were pursued since the early 1970s (see Beghi). For example, the OPTIMO code developed at ISPRA is able to assist synthesis of new thermochemical cycles. A revision of past efforts of development of computer algorithms for generation of potential chemical reactions for thermochemical cycles is discussed in Fishtik and Datta (2008).

The computer-assisted approach is based on the fact that unique chemical reactions may be identified starting from stoichiometric equations. In this respect, a set of chemical reactions is sought under some constraints such as an imposed selection of chemical elements. A flowchart which details the computer-assisted synthesis step of novel thermochemical cycles is shown in Fig. 5.33. The set of chemical elements can be selected based on past experience. For example, in Figs. 5.21 and 5.24, the chemical elements that dominate thermochemical cycles are mainly Fe, Cl, S, I, C, Cu, and Br. In addition, the number of chemical species may be restricted; for example, one can identify nine compounds relevant to thermochemical cycles with copper–chlorine: Cu(s), Cu_2(s), CuO(s), CuCl(s,l), $CuCl_2$(s), HCl(g), Cl_2(g), Cu_2OCl_2(s), $CuCl_2 \cdot 3CuO$. In total, 44 reactions were identified with these compounds and their Gibbs energy calculated to determine the reaction "spontaneity" at various temperatures (Fishtik and Datta 2008, Andress and Martin 2010). From these, only 16 reactions were selected by Fishtik and Datta (2008) based on chemical thermodynamics analysis as viable for thermochemical or hybrid cycles. The computer program determined some viable reactions that were not considered before, for example, $Cu_2O + 4HCl \rightarrow H_2 + 2CuCl_2 + H_2O$ conducted at 373 K. The first step in a thermodynamic analysis of a chemical reaction is to compute the reaction enthalpy and Gibbs free energy in standard conditions. There are two main cases to observe as follows.

- The Gibbs energy increases with temperature, a case in which the reaction may not be favorable in the forward direction; if the yield is found to be marginal, electrolysis can be a solution. A practical rule adopter in Lewis et al. (2009) is if

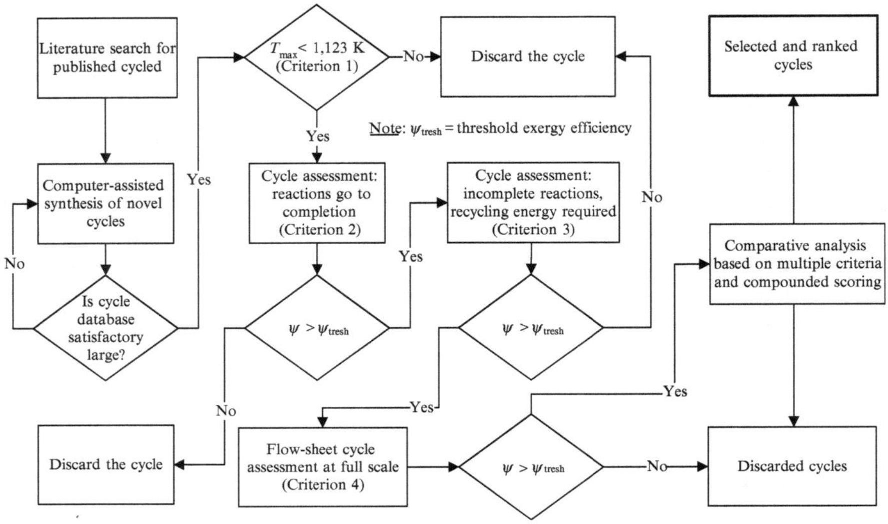

Fig. 5.33 Top-down selection process of thermochemical cycles for nuclear hydrogen generation by water splitting

the Gibbs energy of the reactions is higher than 100 kJ/mol and if the reaction cannot be driven electrochemically, it must be discarded as unfavorable; see the flowchart from Fig. 5.33.

- The Gibbs energy decreases with temperature, a case in which the operational temperature must be established on a technical-economic basis such that the Gibbs energy of the reaction is low enough to assure a reasonable yield.

Because the thermochemical processes operate cyclically, only the reactions with good yield should be selected; this is illustrated in the algorithm from Fig. 5.34. If the yield is marginal, more energy is required for recycling of chemicals which makes the overall process competitively less efficient. The yield of each reaction can be determined based on the Gibbs energy and reaction stoichiometry. Examples given in previous sections in conjunction with Figs. 5.16, 5.17, 5.19, 5.20, and 5.22 show how the yield can be estimated as a function of reaction temperature. Reactions that do not have good yield are discarded.

Once a set of feasible reactions is established, an algorithm must be implemented to assist for the identification of a possible closed cycle. Some methods specific to computational chemistry and mathematics (reaction route method, graph theory, etc.) can be applied to generate cycles which are feasible from a stoichiometric point of view. Next, a cycle feasibility assessment based on a thermodynamic analysis similar to that presented in conjunction with Fig. 5.10 must be performed. If the cycle is a two-step cycle, then it has to comply with

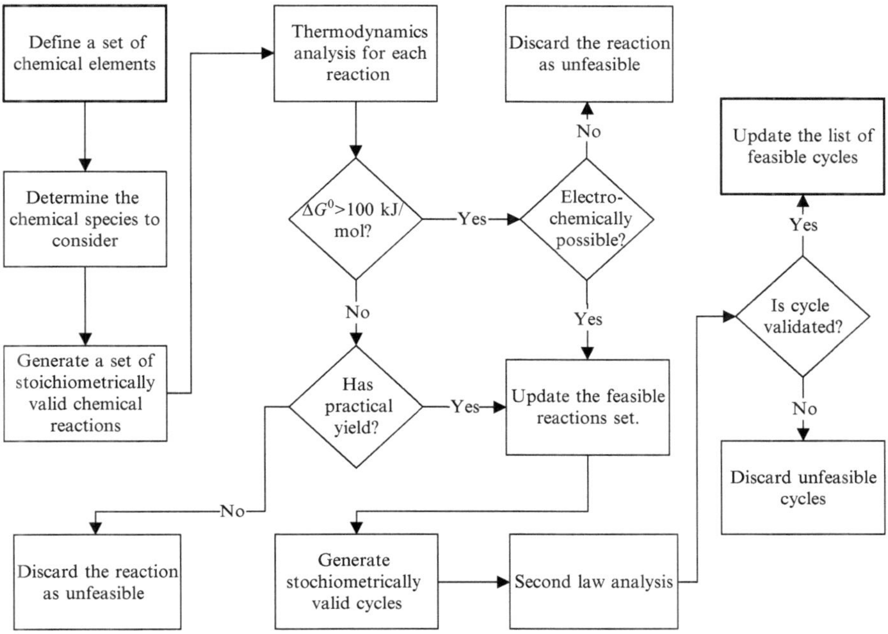

Fig. 5.34 Methodology of computer-assisted synthesis of novel thermochemical cycles

conditions expressed in (5.16). Similar conditions may be determined for multistep cycles. A second law analysis is performed with the thermodynamic system presented in Fig. 5.35.

The entropy balance on a thermochemical cycle with n steps is written as follows:

$$\frac{Q_1}{T_1} + \frac{Q_2}{T_2} + \cdots + \frac{Q_n}{T_n} + S_{\text{gen}} = S_{\text{H}_2} + S_{\text{O}_2} \text{ and } S_{\text{gen}} > 0 \tag{5.47}$$

If (5.45) is not satisfied, this is a key indication that the thermochemical cycle is not thermodynamically feasible. Although the experience accumulated after ~45 years of worldwide research on thermochemical cycles led to the establishment of a relatively limited number of thermochemically feasible cycles (around 300), recent computer-assisted efforts have discovered some new processes. For example, the following new Cu–Cl cycle, having a theoretically estimated energy efficiency of 24.5 %, has been formulated after a computer-assisted search by Andress and Martin (2010):

$$\begin{cases} \text{H}_2\text{O} + 2\text{CuCl} \rightarrow 2\text{H}_2 + \text{Cu}_2\text{OCl}_2,\ 298\,\text{K},\ \text{electrochemical} \\ 0.5\,\text{CuCl}_2 \cdot 3\text{CuO} + 0.5\text{Cl}_2 \rightarrow 0.5\text{O}_2 + \text{CuCl} + 0.5\,\text{Cu}_2\text{OCl}_2,\ 800\,\text{K} \\ 1.5\text{Cu}_2\text{OCl}_2 \rightarrow \text{CuCl} + 0.5\text{Cl}_2 + 0.5\text{CuCl}_2 \cdot 3\text{CuO},\ 800\,\text{K}. \end{cases} \tag{5.48}$$

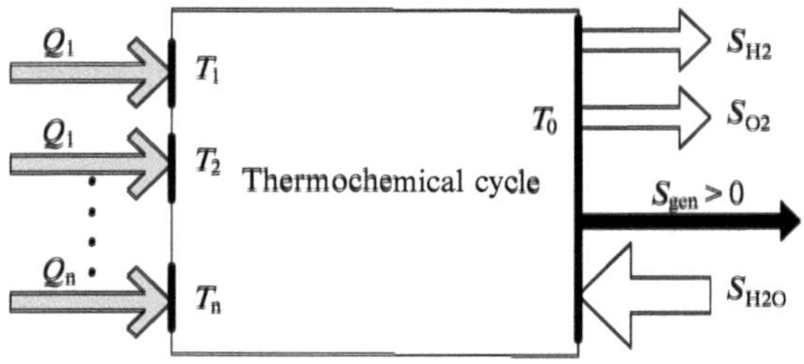

Fig. 5.35 Overall entropy balance analysis of a thermochemical cycle

Once the database of thermochemical cycles is considered large enough, the down selection continues by selecting only cycles with a temperature level requirement compatible with next-generation nuclear reactors. According to Lewis et al. (2009), the threshold level for this selection should be 1,123 K. Call this Criterion 1 for the selection (see flowchart from Fig. 5.33). The second selection criterion—applied successively—must be based on an efficiency assessment. Both energy and exergy efficiency of the thermochemical cycles may be included in the analysis. For Criterion 2, one considers the idealized cases when all reactions of the cycle go to completion. There is no chemical lost, and there is no energy spent for chemical recycling. However, the thermodynamic analysis considers the work needed for chemical separation. Full recovery and internal reuse of the rejected heat by exothermic reactions are considered. The heat input into the cycle is determined based on a pinch analysis. Due to the assumption of completed heat recovery and reuse, the heat amount needed to drive the cycle is given as the difference between the heat required by all endothermic reactions and the heat rejected by all exothermic reactions. The resulting efficiency of the thermochemical cycle must be compared to that of a water electrolysis plant which is driven by electric power produced at the nuclear reactor. The energy efficiency (expressed on hydrogen's HHV basis) of advanced large-scale electrolysis plants is ~70 %. The projected energy efficiency of future nuclear reactors is ~50 %. Therefore, generation of hydrogen via electrolysis at next-generation reactors will have a probable efficiency of 35%.

There are debatable ways to compare technologies such as thermochemical water splitting and nuclear-based electrolysis using efficiency. Historically, assessments of thermochemical cycles considered energy efficiency as expressed in (5.33). This definition is adopted by many researchers because it relates to the existence of foreseen power generation systems, which are included in the equation via power generation efficiency η_{el}. In other words, this definition is applicable to particular cases, when the power generation efficiency can be specified. Another particular case of definition (5.33) occurs when the work-heat equivalency factor is taken as unity. This is equivalent to an extremely high-temperature source and power generation efficiency that becomes 1 ($\eta_{el} = 1$); in general this definition is taken as the "energy efficiency" as it is not related to the power generation efficiency:

$$\eta = \frac{\text{HHV}}{Q_{in} + W_{el}}. \tag{5.49}$$

A generally applicable efficiency formulation is that based on exergy. Equation (5.18) defines an exergy efficiency ψ under the assumption that the cycle is supplied with high-temperature heat at a fixed temperature, T_H. In the general case when heat is supplied at several temperatures, the exergy efficiency becomes

$$\psi = \frac{\mathrm{ex}^{\mathrm{ch}}}{\sum \mathrm{Ex}^{\mathrm{Q}} + \sum \mathrm{Ex}_{\mathrm{in}} + W_{\mathrm{el}}}, \tag{5.50}$$

where $\mathrm{ex}^{\mathrm{ch}}$ is the molar specific chemical exergy of hydrogen; and for one mole of hydrogen generated, one has $\sum \mathrm{Ex}^{\mathrm{Q}}$ as the exergy associated to all heat inputs, W_{el} is the total electric power input, and $\sum \mathrm{Ex}_{\mathrm{in}}$ is exergy input in other forms.

For the down selection algorithm, one can set a threshold of exergy efficiency of the thermochemical process under which the respective cycle is discarded. Based on the ~35 % energy efficiency of nuclear-based water electrolysis (see above), the corresponding exergy efficiency—assuming heat input at 800–1,200 K (for next-generation nuclear reactors) with a Carnot factor of 0.63–0.75—becomes ~38–46 % (chemical exergy of hydrogen $\mathrm{ex}^{\mathrm{ch}} = 236.12$ is kJ/mol and its HHV is 283.6 kJ/mol). Therefore, in a flowchart from Fig. 5.33, one can take $\psi_{\mathrm{tresh}} = 38 - 46\%$ for Criterion 2 for the down selection (this figure is consistent with past work such as Rosen and Scott 1992). Alternatively, one can consider energy efficiency as a criterion with the conventional adopted energy efficiency of power generation. This approach was taken in recent studies by Lewis et al. (2009), McQuillan et al. (2010), and Brown et al. (2002). The exergy efficiency approach is included in studies such as Dincer and Balta (2011), Rosen (2008), and Naterer et al. (2011a).

The down selection process continues—in a sieve-like fashion—with deeper assessment of chemical reactions by considering the fact that they do not go to completion. Moreover, there is formation of competing chemical species. In Lewis et al. (2009), this type of assessment is referred to as a "Level 2" efficiency calculation. Recirculation is allowed and its energy penalty is estimated based on a thermodynamic analysis. Actual kinetic data for chemical reactions are included in the efficiency assessment. The assessment is more laborious but it is applied to a smaller number of cycles than the initial selection. This justifies the adoption of a down selection approach: only the cycles that are selected as competitive based on simpler analysis are further assessed more elaborately. Due to the auxiliary energy needed for recycling, the efficiency of most cycles is lower than for estimation at "Criterion 2"—reactions at completion. In Lewis et al. (2009), this assessment step is called a "chemical viability" assessment. At this step, the cycles which show many side products and low recyclability of reagents are filtered. In a flowchart from Fig. 5.33, this step is denoted as selection Criterion 3, which is expressed by $\psi > 44\%$.

The next level of the analysis is the so-called flow-sheet assessment of the cycle (see the discussion in Levis et al. 2009). This analysis can be performed with specialized software—such as ASPEN PLUS—or customized software and it includes a heat exchanger network analysis, thorough reactor analysis, optimization of process parameters, accounting of mechanical power needs for mixing, chemical conveying, pumps, blowers, compressors, etc. Integrated systems may be considered in the design; for example, an integrated heat pump with the copper–chlorine cycle is proposed in Zamfirescu et al. (2010). A convention regarding the hydrogen and oxygen delivery and as well for water input must be adopted; for example, in Levis et al. (2009) it is proposed that hydrogen is delivered at 21 bar and oxygen at 1 bar, while water is an input in a liquid form at 298.15 K. The flow-sheet analysis must account for the experimental data obtained for different processes and reactions, which are available either as a result of test bench experimentation with the analyzed cycle or relevant work developed in other contexts. The design must be optimized for a conventional industrial production scale. In general, it is assumed that the scale of the plant corresponds to the linkage with a dedicated 600 MW nuclear reactor. The exergy efficiency obtained from a flow-sheet-based assessment is then compared with the threshold value again; this selection step is denoted as Criterion 4 in the flowchart presented in Fig. 5.33.

Further assessment of the cycles can be performed based on various practical criteria and comparative analyses. Some of the criteria considered in Brown et al. (2002) are given here:

- Abundance of chemicals involved
- Corrosiveness of involved chemicals
- Degree of solid flow
- Number of difficult separations (e.g., solid–solid, liquid–liquid, gas–gas, aqueous/nonaqueous)
- Use (or not) of radiant heat transfer to/from solids
- Number of major parts of the process flow sheet
- Toxicity of chemicals involved
- Chemical reactivity of air and water
- Environmental considerations of plant
- Hydrogen end-cost

Other parameters for comparative assessment of thermochemical cycles were proposed and studied in Dincer and Balta (2011), such as the sustainability index and the improvement potential. When judging a system based on multiple criteria, it is important to develop metrics (or scoring system) which can define a compounded single criterion. Brown et al. (2002) conceived a scoring system for criteria to assign for each cycle a single score which ultimately led to a selection of 25 cycles recommended for nuclear hydrogen production. In a second stage of down selection, they recommended three top cycles, namely, the SI cycle (Mark 16/GA, Table 5.4), UT-3 cycle (Table 5.11), and Westinghouse/Mark 11 cycle (Table 5.15). Note that the analysis by Brown et al. (2002) did not consider an intermediate-temperature nuclear reactor and hybrid thermochemical cycles. In the down selection flowchart in Fig. 5.33, the last phase of the process has a generic box of "comparative analysis based on multiple criteria and compounded scoring." It appears that a comparative analysis of selected cycles is an essential tool for ranking them, vis-à-vis of a specific application (e.g., high-temperature or intermediate-temperature nuclear reactors).

It appears that efficiency is the key step of cycle down selection, comparative assessment, and ranking. Efficiency determination (either energy or exergy based) is made primarily based on thermodynamic analysis of chemical reactions. In the thermodynamic analysis, one can consider 1 kmol of hydrogen produced to report all quantities such as heat input/output or work input in terms of a per kmol basis. For the analysis, the following can be assumed (these assumptions can be relaxed for further analysis):

- The values for the reference environment (deadstate) temperature (T_0) and pressure (P_0) are 298.15 K and 1 bar, respectively.
- In all chemical reactions, reactants and products are at the reaction temperature and a pressure of 1 bar.
- All processes are considered to be of steady state and steady flow with negligible potential and kinetic energy effects.
- Some heat loss fraction is assumed (e.g., 30 % of all heat flows at temperatures higher than the ambient).

For a general steady-state, steady-flow process, three balance equations, namely, mass, energy, and exergy balances, are written. For example, in a chemical reaction

$$\sum m_R = \sum m_P,$$

where m is mass and subscripts R and P refer to reactants and products, respectively. The energy balance for a steady-state reaction reduces to

$$Q = \sum n_P \left(H_f^0 + H - H^0\right)_P - \sum n_R \left(H_f^0 + H - H^0\right)_R,$$

where H is molar enthalpy, subscript f refers to formation enthalpy, superscript 0 indicates the reference state, and Q is the heat transferred to the reaction (positive or negative). In the above equation, the case of a purely thermochemical reaction is considered; similar energy balance equations can be written for electrochemical reactions, a case in which one must account for the Gibbs free energy. The exergy balance for the overall process can be written as follows:

$$\sum \mathrm{Ex_{in}} = \mathrm{Ex_{H_2}} + \mathrm{Ex_{loss}} + \mathrm{Ex_d},$$

where the subscript "loss" indicates lost energy (e.g., in the form of oxygen out) and subscript d indicates destroyed exergy due to irreversibilities. The exergy retrieved in hydrogen has two components (valid for any exergy stream), namely, thermomechanical and chemical (other forms of exergy may be involved in special cases, e.g., magnetic, photonic, etc.); on a mass basis, $\mathrm{Ex} = m \times \mathrm{ex}$, where ex is the mass specific exergy, given by

$$\mathrm{ex} = H - H^0 + T_0 \times (S - S_0) + \mathrm{ex^{ch}},$$

where s is the mass specific entropy and $\mathrm{ex}^{\mathrm{ch}}$ is the chemical exergy of the chemical compound which can be calculated based on standard chemical exergy of its chemical elements based on the formula (see Dincer and Zamfirescu 2011)

$$\mathrm{ex}^{\mathrm{ch}} = -\Delta G_{\mathrm{f}}^{0} + \sum_{\mathrm{el}} n \times \mathrm{ex}_{\mathrm{el}}^{\mathrm{ch}},$$

where n represents the number of moles of each chemical element forming the chemical compound and index "el" identifies an involved chemical element.

A way to account for the incompleteness of chemical reactions is proposed in Dincer and Balta (2011) via a recycling ratio, r. For a "perfect reaction" with 100 % yield, the recycling ratio is zero; on the contrary, a reaction with no yield has a recycling ratio of unity. A preliminary analysis may assume an equal recycling ratio for all involved reactions; in this case, the exergy efficiency, previously defined in (5.50), becomes

$$\psi = \frac{(1-r)\mathrm{ex}^{\mathrm{ch}}}{\sum \mathrm{Ex}^{\mathrm{Q}} + \sum \mathrm{Ex}_{\mathrm{in}} + W_{\mathrm{el}}} \tag{5.51}$$

The heat input used in the efficiency equation and its exergy should be derived from a pinch analysis. The pinch analysis proceeds by calculating the heat input and outputs for each process involved. Then temperature vs. total steam enthalpy profiles are determined by joining all streams that require heat input. The same approach is used for streams that require heat output. In a preliminary analysis, it is assumed that all heat rejected by the plant can be recovered and reused. Based on this consideration, the temperature profile is determined at the source. A typical coolant must be assumed depending on the nuclear reactor. Certain temperature differences must be assumed at the pinch point. A pinch point analysis for the five-step copper–chlorine cycle is exemplified in Fig. 5.36. Once the temperature profile at the heat source is determined, one can find the required heat input into the process, Q_{in}. Based on the coolant temperature and pressure, one can determine the enthalpy ($\Delta H = Q_{\mathrm{in}}$) and entropy ($\Delta S$) variation during heat transfer from the heat source. Then the exergy associated with Q_{in} is given by the equation $\mathrm{Ex}^{\mathrm{Q}} = Q_{\mathrm{in}} - T_0 \times \Delta S = Q_{\mathrm{in}} \times (1 - T_0/(\Delta S/Q_{\mathrm{in}}))$. Denoting $T_{\mathrm{eq}} = \Delta S/Q_{\mathrm{in}}$ as the equivalent heat source temperature, one obtains the equivalent Carnot factor of the heat source.

Besides energy and exergy efficiency, other quantitative parameters may be defined for comparative thermochemical cycle assessment and ranking purposes. As it was mentioned above, Dincer and Balta (2011) introduced the sustainability index (SI) and improvement potential (IP) in the assessment of thermochemical cycles. Such parameters are expected to give a comparative perspective about the process to quantify how they will contribute to a cleaner environment and sustainable development. The sustainability increases with the exergy efficiency of a process as indicated by its definition relationship, namely,

$$\mathrm{SI} = 1/(1-\psi). \tag{5.52}$$

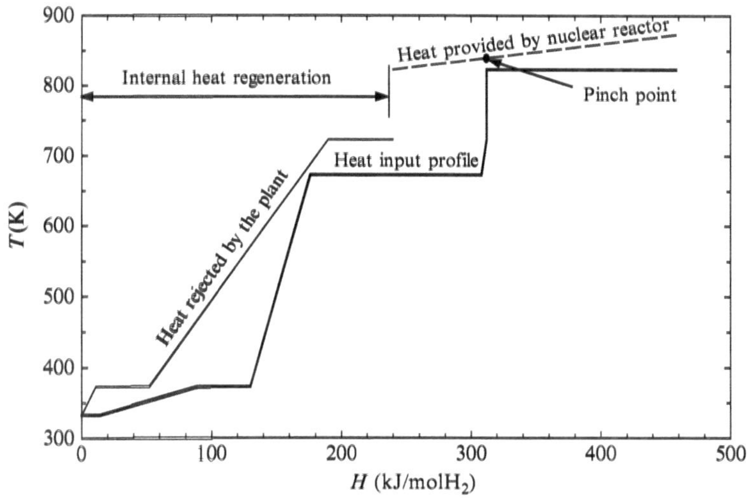

Fig. 5.36 Pinch point analysis for the five-step copper–chlorine cycle

Moreover, the maximum improvement in the exergy efficiency for a system is achieved when the exergy loss or irreversibility is minimized, $\min\{Ex_{in} - Ex_{out}\}$. Based on this observation, the "exergetic" improvement potential is defined according to

$$IP = (1 - \psi) \times (Ex_{in} - Ex_{out}). \tag{5.52}$$

5.8 Comparative Assessment of Thermochemical Cycles

A comparative assessment of thermochemical cycles is a general method that can be applied for several purposes such as a cycle ranking in view of a selection of the best technology for a specific application (e.g., hydrogen production at intermediate- vs. high-temperature nuclear reactors), identification of the most suited applications for each thermochemical cycle, understanding of the general features, differences and similarities among cycles, etc. This section presents some general guidelines of comparative assessments of thermochemical cycles for nuclear hydrogen production, exemplified with case studies.

A comparative assessment of some intermediate-temperature cycles is illustrated in Fig. 5.37. A base of comparison is the energy efficiency defined through (5.33). This example is adapted from Lewis and Masin (2009). The following cycles with a maximum temperature requirement smaller than 1,123 K were considered for comparative analysis:

- Ce-Cl, cerium-chloride (IGT < Ce, Cl>, Table 5.10)
- Ca-Br, hybrid calcium-bromide (ANL < Br, Ca>, Table 5.15)
- Cu-Cl-A, copper-chloride (CuCl-A, CuCl-B, CuCl-C, CuCl-D, Table 5.15)

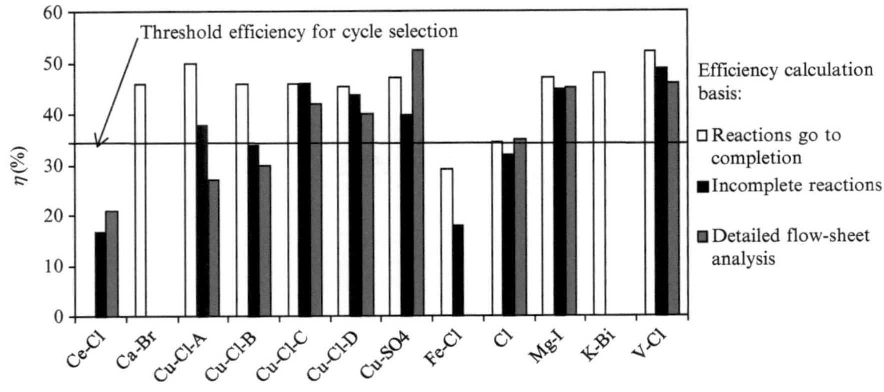

Fig. 5.37 Comparative analysis of energy efficiency of selected thermochemical cycles for three assessment criteria; efficiency definition according to (5.33) [data from Lewis and Masin (2009)]

- Cu-SO_4, copper-sulfate (IGT, Table 5.15)
- Fe-Cl, iron-chlorine (Mark 15, Table 5.4)
- Cl, hybrid chloride (APC, Table 5.15)
- Mg-I, magnesium-iodide (NCLI < I, Mg>, Table 5.11)
- K-Bi, potassium-bismuth cycle (ANL < Bi, K>, Table 5.15)
- V-Cl, vanadium-chlorine (GM < Funk>, Table 5.11)

The efficiency formula adopted in this analysis is based on (5.33). The plot compares the cycle's efficiency for three cases. In the first case (white bars), the cycle efficiency is calculated assuming that all chemical reactions go to full completion; there is no energy spent for chemical recycling, although species separation energy is considered. This type of calculation corresponds to Criterion 2 of the down selection process in the previous section (see flowchart from Fig. 5.33). In the second case of comparison, the efficiencies are calculated under the assumption that none of the chemical reactions go to completion. Equilibrium and kinetics calculations are performed; experimentally determined parameters relevant to the chemical analysis are used to improve modeling accuracy; and the energy required for chemical recycling is considered. This type of calculation exemplifies a down selection based on Criterion 3. In the third case, a detailed flow sheet, modeling, and cycle optimization are performed to determine the efficiency with maximum possible accuracy. This case corresponds to Criterion 4 above. Efficiencies for some cases are not indicated in the figure because they are not available in the literature. It is observed that Ce-Cl and Fe-Cl cycles show the lowest efficiency while the Cu-SO_4 and V-Cl have the highest. In the analysis by Lewis and Masin (2009), it was assumed that the electrical power supplied to the thermochemical plant is sourced from a next-generation nuclear reactor, characterized by an energy efficiency of 50 %. Therefore, one can consider for comparison the base water electrolysis with 35 % efficiency.

Based on the above comparative analysis, one can conclude that two cycles have an efficiency too low to justify further development: Ce-Cl and Fe-Cl. Six cycles show satisfactory efficiency when the analysis accounts for incomplete reactions: Cu-Cl-A, Cu-Cl-C, Cu-CL-D, Cu-SO_4, Mg-I, and V-Cl. A detailed flow-sheet analysis shows that Cu-Cl-B is not satisfactory to be competitive. The possible cycle ranking based on a flow-sheet analysis yields (in decreasing order) Cu-SO_4, V-Cl, Mg-I, Cu-Cl-C, and Cu-Cl-D.

The next example illustrates a comparative assessment of four thermochemical cycles based on both energy (5.49) and exergy (5.50) efficiency. The example is based on the analysis by Dincer and Balta (2011). The cycles considered are $LiNO_3$ (UNVL-24 in Table 5.15), Cu-Cl-5 (Table 5.15), Mg-Cl (ANL-INL in Table 5.15), and Mark 11 (or Westinghouse in Table 5.15). The efficiencies of these cycles are first compared in Fig. 5.38. The study assumes that all chemical reactions go to completion (Criterion 2-based assessment). In addition, in the pinch analysis, it is assumed that heat losses represent 30 % of the heat carried by high-temperature streams (see the note in Fig. 5.38).

Further comparative assessment of the same cycles is performed by assuming incomplete reactions. Figure 5.39 exemplifies a comparative parametric study which assesses the impact of the recycling ratio on exergy efficiency of the cycles. The next comparative assessment for the same cycle is shown in Fig. 5.40 and refers to the sustainability index. It is seen in this figure that the sustainability index of the analyzed cycles decreases exponentially with the recycling ratios. In this regard, at lower recycling ratios, the corresponding sustainability of cycles increases. The highest sustainability index, such as 2.85, is obtained by the Cu–Cl cycle, which is based on complete reactions. The sustainability index for the analyzed cycles varies from 1.0 to 2.85 with the recycling ratios varying between 0 and 0.9.

The improvement potential of the assessed cycles—as defined in (5.50)—is compared next in the bar chart in Fig. 5.41. It is found that the Mark11/Westinghouse cycle has the highest IP value with about 78 kJ/mol, followed by $LiNO_3$, Mg-Cl, and Cu-Cl-5 with about 61, 60, and 42 kJ/mol, respectively.

The next illustrative example of comparative assessment refers to a one-to-one comparison of sulfur–iodine (S–I, Mark 16, Table 5.4) and copper–chlorine cycles, based on the work of Wang et al. (2010). Both of these cycles are promising methods of nuclear hydrogen production, yet both are at an advanced stage of research and development, although the S–I cycle has a longer development history. These two cycles can be compared from the perspectives of heat quantity, heat grade, thermal efficiency, related engineering challenges, and hydrogen production cost. The five-step version of the Cu–Cl cycle is considered in this comparative assessment (see cycle CuCl-5 in Table 5.15).

The thermodynamic analysis is performed under the assumption that all reactions go to completion. This allows for a simplified calculation of reaction enthalpies assuming stoichiometric reactions. The heat input to be supplied to both cycles is calculated first. Then the energy efficiency is estimated based on (5.49). In the original paper of Wang et al. (2010), the energy efficiency is calculated based on the numerator of (5.49), the formation enthalpy of water instead of HHV of

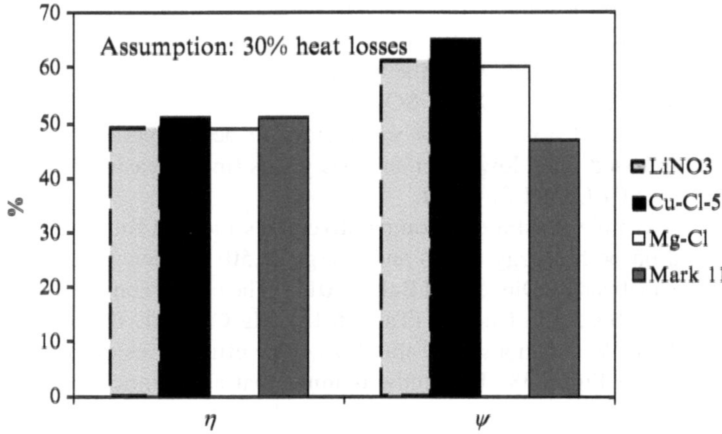

Fig. 5.38 Comparative assessment of some cycles, based on energy and exergy efficiencies. Equations (5.49), (5.50) are used [data from Dincer and Balta (2011)]

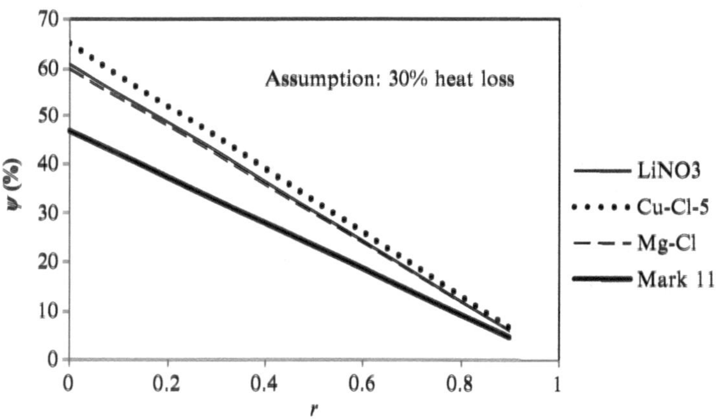

Fig. 5.39 Comparative assessment of recycling ratio impact for four thermochemical cycles [data from Dincer and Balta (2011)]

hydrogen. The efficiency results reported in Wang et al. (2010) are converted to remain consistent with (5.49) by multiplying them by 0.99 which is the ratio between the water formation enthalpy and the conventional higher heating value of hydrogen. For the S–I cycle, it is assumed that there is no need of electricity generation because there is no electrochemical reaction. In other words, the work for chemical separations is neglected. This is an acceptable approach for preliminary assessment.

The Cu–Cl-5 cycle requires power to drive the electrochemical reactions. This is estimated based on the Gibbs free energy calculated for the "stoichiometric" electrochemical reaction of the cycle (see Table 5.15; chemical equations for each reactions). Every thermochemical cycle rejects heat. There is an engineering

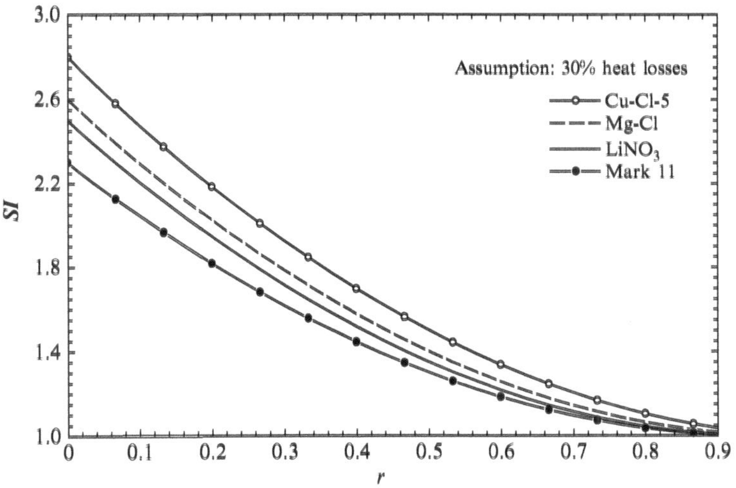

Fig. 5.40 Comparative assessment of recycling ratio impact on sustainability index of four thermochemical cycles [data from Dincer and Balta (2011)]

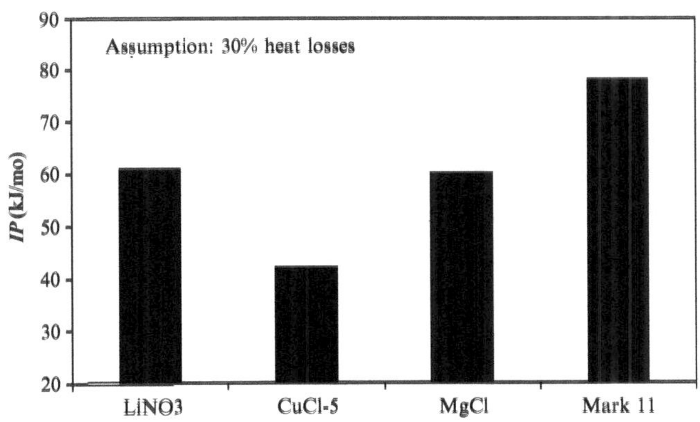

Fig. 5.41 Comparative assessment of thermochemical cycles based on the improvement potential defined in (5.52) [data from Dincer and Balta (2011)]

challenge to recover and internally reuse the rejected heat within a cycle. In particular, it is more challenging to reuse the low-grade heat. The amount of heat input, heat output, and percentage of low-grade heat recovery for S–I and Cu–Cl cycles is compared in the bar chart of Fig. 5.42. The low-grade heat is conventionally assumed to be at temperatures below 343 K. It can be seen from the figure that heat input for the Cu–Cl cycle is slightly higher than the S–I system. However, the Cu–Cl cycle has the ability to recover and recycle ~30 % of low–grade heat; this is used for the drying step of $CuCl_2(aq)$.

The next comparison—shown in Fig. 5.43—refers to cycle efficiency assuming the ideal situation of no heat losses and 100 % heat recycled within the cycle.

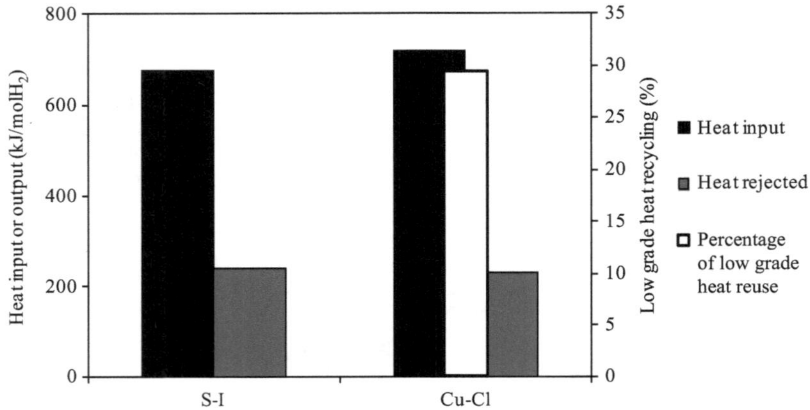

Fig. 5.42 Comparative analysis of heat requirements and heat recycling potential for S–I and Cu–Cl cycles [data from Wang et al. (2010)]

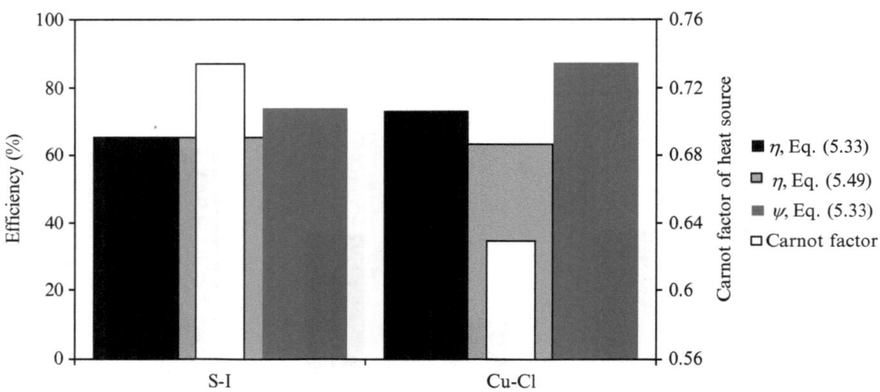

Fig. 5.43 Comparative assessment of S–I and Cu–Cl-5 cycles based on various efficiencies

The analysis of Wang et al. (2010) is expanded to include an energy efficiency analysis based on (5.33) and exergy efficiency based on (5.50); the power generation efficiency η_{el} used in (5.33) is taken as 50 %. The exergy efficiency is calculated based on the assumption that the heat input is provided at a maximum temperature within the cycle which is 1,123 K for the S–I cycle and 803 K for the Cu–Cl-5 cycle. It can be concluded that the energy efficiency of both cycles is very similar; however, the exergy efficiency within the Cu–Cl cycle is more than 10 % higher with respect to the S–I cycle. This is due to the ability of the Cu–Cl cycle to use lower grade heat, despite the fact that some electrical power is required for the electrochemical reaction. The same plot indicates the efficiency range of advanced water electrolysis (gray band). It follows a conclusion that the Cu–Cl cycle becomes competitive at over 70 % internal heat recycled, whereas the S–I cycle requires to achieve more than 90 % internal recycling.

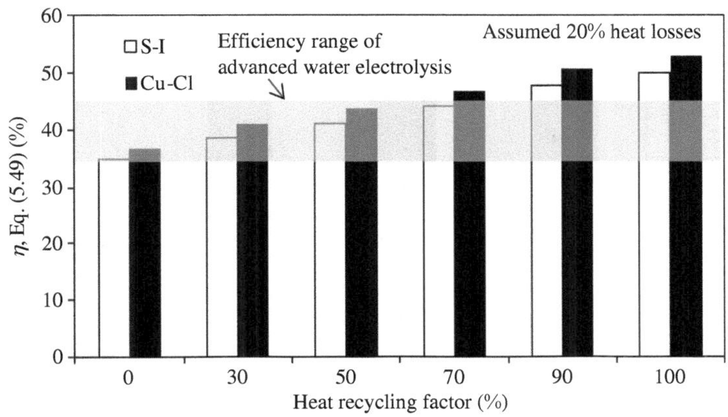

Fig. 5.44 Influence of heat recycling factor on energy efficiency of S–I and Cu–Cl cycles [data from Wang et al. (2010)]

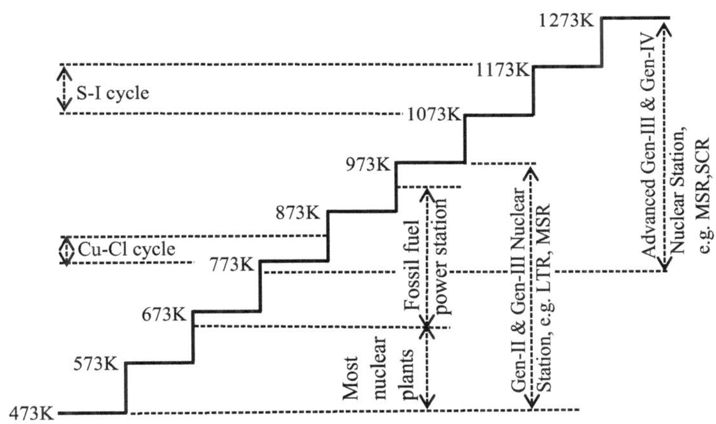

Fig. 5.45 Compatibility of S–I and Cu–Cl cycles with some relevant power stations [adapted from Wang et al. (2010)]

The third comparative plot for the cycles is presented in Fig. 5.44 and studies the influence of heat recycled within the cycle on the energy efficiency—defined in (5.49). In addition, it is assumed that 20 % heat losses occur for both cases. In general, the energy efficiency of the Cu–Cl cycle appears to be a few percentages higher than the S–I cycle.

Figure 5.45 shows the heat grade linkage of S–I and Cu–Cl cycles with the output heat of various nuclear power generation stations. The temperature range of fossil fuel power plants is also illustrated, since these will continue to have an important role. The steam temperature of this type of power station lies in the range of 643–853 K, depending on the fuels used such as natural gas or pulverized coal. The typical temperature of steam entering the turbine is 813 K. In some advanced

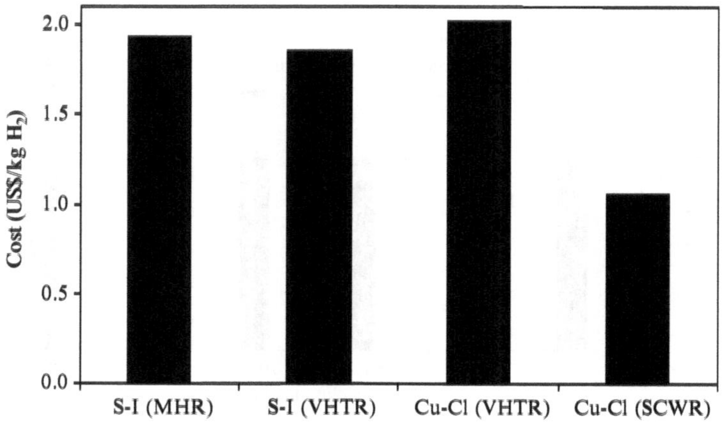

Fig. 5.46 Cost comparison of nuclear hydrogen production with S–I and Cu–Cl cycle technologies (reference year 2003) [data from Wang et al. (2010)]

fossil fuel power stations, high pressure is applied and the temperature of steam or supercritical water can reach 923 K. Therefore, the temperature of many fossil fuel stations may match the maximum required heat grade of the Cu–Cl cycle, more readily than the S–I cycle. The reaction heat of step 2 at 1,123 K in the S–I cycle is 371.0 kJ/mol, which is approximately three times the reaction heat (129.2 kJ/mol) of step five in the Cu–Cl cycle (see Table 5.15). This indicates another advantage of the Cu–Cl cycle: a much smaller quantity of high-grade heat requirement.

It can be found that the usage of low-grade heat (<343 K) in the Cu–Cl cycle can take about 30 % of the total heat input (not net input), which implies a promising possibility to use waste heat from power generation stations or recover more heat within the thermochemical cycle. By comparison, the S–I cycle cannot use this type of low-grade heat. Also, the maximum grade heat is the majority of heat required by the sulfur–iodine cycle, while the copper–chlorine cycle has a more attractive feature of less high-grade heat that is required.

The cost comparison of two technologies is presented in Fig. 5.46 for a production capacity of 200 t hydrogen per day. The factors that influence hydrogen production cost are (1) local heat and electricity price, (2) production scale, (3) energy and exergy efficiency of the processes, (4) energy requirement for the thermochemical cycle (heat only or heat and power), (5) type of nuclear reactor to which the thermochemical plant is linked, (6) storage and distribution methods for hydrogen, (7) local/regional application of a carbon tax or not, (8) assumed years for capital recovery, (9) operating cost, (10) costs related to toxicity of chemicals and safety issues, and (11) sale market for oxygen by-product. The complexity of a cost analysis also lies in the mutual influence of those factors on each other, because factors cannot be completely isolated from each other. For example, there are mutual interdependencies: the costs of hydrogen storage and distribution depend on production scale; toxicity prevention cost depends on the type of thermochemical cycle; production scale and energy efficiency of a cycle are restricted by the energy availability of a nuclear power station, such as off-peak length of time and output temperature.

Three types of nuclear reactors are relevant for this cost analysis. The modular helium reactor (MHR) which can be applied to the S–I cycle only, the very-high-temperature reactor (VHTR) in which both the S–I and Cu–Cl cycles can be linked, and the supercritical water reactor (SCWR) which is compatible only with the Cu–Cl cycle.

Regarding the price of electricity, it will be corrected to the same pricing year, region, country, and currency. The data of cost analysis is based on the prices of past several years. Industrial rather than residential and commercial retail prices are used in the analysis. The reason to use the retail price, instead of wholesale price, is that a conservative evaluation can be assumed and the retail price is often easier to obtain from information sources. The energy price can be determined by using the ratio of electricity and heat prices, since electricity and heat prices often have similar influencing factors. The ratio could vary from 2 to 5 depending on the peak and off-peak time use, nuclear reactor type, and so on.

According to the results presented in Fig. 5.46, future Cu–Cl and S–I cycles may produce hydrogen at costs as low as 1.60–1.93 US\$/kgH_2. By comparison, HTE yields 2.25–4.80 US\$/kgH_2 (see Wang et al. 2010). When comparing the Cu–Cl cycle with the S–I cycle, the Cu–Cl cycle yields 2.02 US\$/kgH_2, thus higher than 1.85 US\$/kgH_2 for the S–I cycle, if both cycles use VHTR as the heat and electricity source. This can be explained by the use of more electricity in the Cu–Cl cycle, since it is a hybrid thermochemical cycle. However, the Cu–Cl cycle yields 1.60 US\$/kgH_2 using SCWR while the S–I cycle yields 1.90 US\$/kgH_2 when coupled with MHR. This is due to the higher costs of MHR than SCWR. For example, the operating temperature of MHR is 1,123 K (using helium) while SCWR is 923 K with water. Higher temperatures and helium correspond to higher investment and operating costs than lower temperatures with water.

5.9 Flow-Sheet Assessment of Thermochemical Plants: Mark 10 Case Study

A comprehensive flow-sheet cycle simulation of a thermochemical water-splitting plant is necessary as the ultimate assessment prior to full-scale implementation. Flow-sheet simulations are performed with advanced process modeling software such as ASPEN Plus and Cycle Tempo or with customized software. These types of analyses extend beyond the thermodynamic modeling of the cycle itself. The objective is a comprehensive simulation of the overall plant including its main units and all auxiliary units such as oxygen and hydrogen reactors, other equipment such as pumps, conveyors, mixers, etc.

Most of the published flow-sheet simulations are based on energy and mass balance analyses. Relatively few comprehensive exergy analyses of thermochemical plants have been published. This section discusses some general aspects related to flow-sheet simulation of thermochemical water-splitting plants based on a case study, namely, the six-step thermochemical cycle Mark 10 (see Table 5.4). The hydrogen production plant is linked to a helium-cooled high-temperature gas reactor. The analyses are adapted from Rosen (2008). The thermochemical cycle ISPRA Mark 10 includes the following six reactions:

- Hydrolysis:

$$2H_2O(l) + I_2(s) + SO_2(g) + 4NH_3 \rightarrow 2NH_4I(s) + (NH_4)_2SO_4(s), \text{ at } 323\,K \tag{5.53}$$

- Hydrogen-evolving reaction:

$$2NH_4I(s) \rightarrow H_2(g) + 2NH_3(g) + I_2(g), \text{ at } 900\,K \tag{5.54}$$

- Oxygen-evolving reaction:

$$SO_3(g) \rightarrow 0.5O_2(g) + SO_2(g), \text{ at } 1{,}123\,K \tag{5.55}$$

- Ammonia regeneration reaction:

$$(NH_4)_2SO_4(s) + Na_2SO_4(s) \rightarrow 2NaHSO_4(s) + 2NH_3(g), \text{ at } 673\,K \tag{5.56}$$

- Steam regeneration reaction:

$$2NaHSO_4(s) \rightarrow Na_2S_2O_7(s) + H_2O(g), \text{ at } 673\,K \tag{5.57}$$

- Sulfur trioxide regeneration reaction:

$$Na_2S_2O_7(s) \rightarrow Na_2SO_4(s) + SO_3(g), \text{ at } 825\,K \tag{5.58}$$

This cycle is purely thermochemical; thus under idealized operation, it does not utilize electrical energy. The approximate energy balance on the idealized cycle is deduced from the analysis reported in Rosen (2008), i.e., the cycle requires 886 kJ of heat input to generate 1 mol of hydrogen with an energy efficiency of 32 %. Assuming that heat is provided at 1,123 K (the highest temperature within the cycle), the exergy efficiency of the idealized system becomes 36 %.

A potential process flow-sheet diagram of the cycle is presented in Fig. 5.47. The streams are indicated by numbers on the diagram. Table 5.16 specifies the composition of each stream. The equipment is indicated with letters in the figure. Table 5.17 explains the meaning of each letter. According to the design in Fig. 5.47, the plant has 23 components connected by 34 material streams. The flow sheet illustrates the complexity of the thermochemical plant, although some details are not shown, namely, a heat exchanger network for internal heat recycling. There are

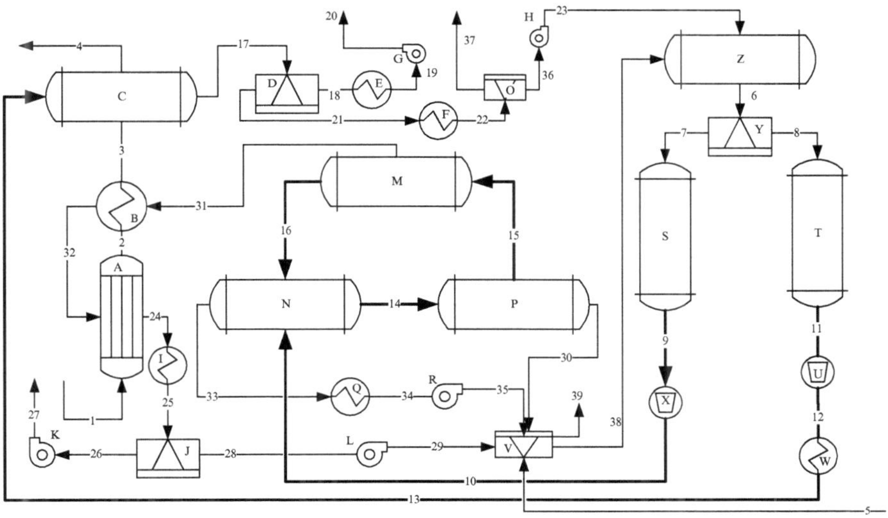

Fig. 5.47 Flow-sheet diagram for a Mark 10 thermochemical cycle

Table 5.16 Stream definitions for Mark 10 plant shown in Fig. 5.47

Str.	Composition	Str.	Composition	Str.	Composition
1	Hot helium input	14	$Na_2S_2O_7(s)$ preheated	27	Oxygen, blown-out
2	Helium	15	Heat transfer flux	28	$SO_2 + O_2$ cold
3	Helium	16	$2NaHSO_4(s)$ cooled	29	$SO_2 + O_2$ blown
4	Colder helium return	17	$H_2 + 2NH_3 + I_2$ hot	30	Steam, cooled
5	Liquid water input	18	Hydrogen purified, hot	31	$SO_3(g)$ colder
6	$2NH_4I(s) + (NH_4)_2SO_4(s)$	19	Cooled hydrogen	32	$SO_3(g)$ heated
7	$(NH_4)_2SO_4(s)$ cold	20	Pressurized hydrogen	33	$2NH_3(g)$ hot
8	$2NH_4I(s)$ cold	21	$2NH_3 + I_2 + H_2$ hot	34	$2NH_3(g)$ cooled
9	$(NH_4)_2SO_4(s)$ preheated	22	$2NH_3 + I_2 + H_2$ cooled	35	$2NH_3(g)$ blown
10	$(NH_4)_2SO_4(s)$ conveyed	23	$2NH_3 + I_2$ blown	36	$2NH_3 + I_2$ recovered
11	$2NH_4I(s)$ preheated	24	$0.5O_2(g) + SO_2(g)$ hot	37	Unrecoverable H_2
12	$2NH_4I(s)$ conveyed	25	$0.5O_2 + SO_2$ cooled	38	NH_3, H_2O, SO_2 recycled
13	$2NH_4I(s)$ hot	26	Oxygen, cold	39	Unrecoverable O_2

Str. stream index

a significant number of blowers and solid material conveyers; these consume power which must be included.

Within the thermochemical plant, water is supplied at state #5 and hydrogen is delivered at #20. Water must be supplied in excess, because neither all water can be converted nor all resulting hydrogen and oxygen can be recovered. The thermochemical plant also rejects heat at various points as it is not possible to recover and reuse all heat internally. The process simulation of this plant was performed with ASPEN Plus and reported by Rosen (2008). Figure 5.48 summarizes the overall

Table 5.17 Component list for flow-sheet diagram from Fig. 5.47

S.	Component	S.	Component	S.	Component
A	Oxygen-evolving reactor	J	Absorption separator for SO_2	S	$(NH_4)_2SO_4(s)$ preheater[a]
B	$SO_3(g)$ preheater	K	Oxygen blower	T	$2NH_4I(s)$ preheater[a]
C	Hydrogen-evolving reactor	L	SO_2 blower	U	$2NH_4I(s)$ conveyer
D	Hydrogen separation reactor	M	Sulfur trioxide reactor	V	NH_3, H_2O, SO_2 recovery reactor
E	Hydrogen cooler[b]	N	Ammonia regeneration reactor	W	$2NH_4I(s)$ heater[a]
F	Cooler for $2NH_3(g) + I_2(g)$[b]	O	NH_3, I_2 recovery reactor	X	$(NH_4)_2SO_4(s)$ conveyer
G	Hydrogen compressor	P	Steam regeneration reactor	Z	Hydrolysis reactor
H	Blower $2NH_3(g) + I_2(g)$	Q	Ammonia cooler[b]		
I	Cooler for $0.5O_2(g) + SO_2(g)$[b]	R	Ammonia blower		

[a]Internal heat reuse, [b]heat recovery applied

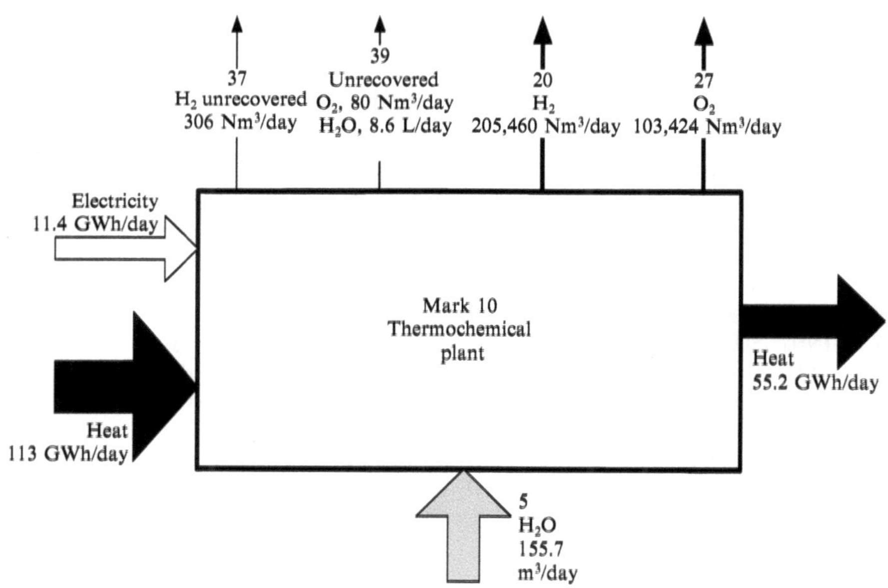

Fig. 5.48 Mass and energy balances for a Mark-10 plant of 205,000 Nm^3 H_2/day

mass and energy balances for the thermochemical planton the assumption that the plant is supplied with high-temperature heat in the form of hot helium derived from HTTR and in addition, electric power, necessary to drive all auxiliary units.

Based on the overall mass and energy balance, the energy and exergy flows can be calculated for the thermochemical plant itself. Figure 5.49 illustrates these balances with numerical results from the flow sheet by Rosen (2008). Once the

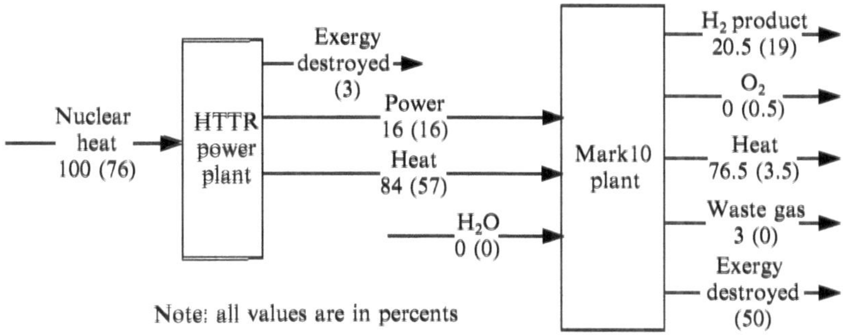

Fig. 5.49 Energy and exergy flows for a Mark 10 plant integrated with an HTTR [data from Rosen (2008)]

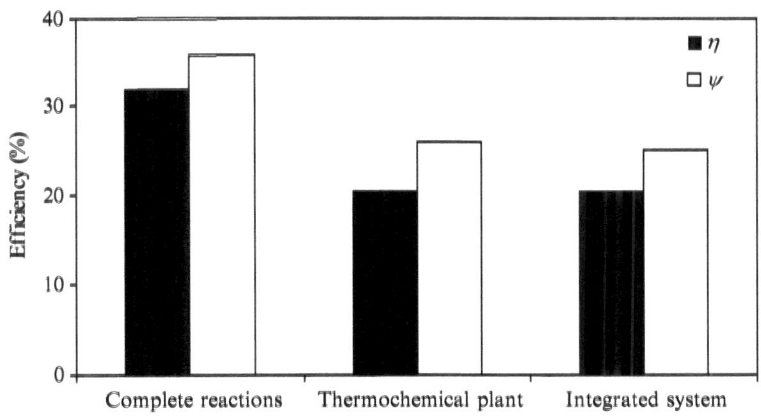

Fig. 5.50 Comparison of hydrogen production efficiencies as a function of the level of the process analysis

energy and exergy of each input/output flow is determined, the thermochemical plant is integrated with the nuclear plant which supplies both electrical and thermal energy to the water-splitting process. Figure 5.49 illustrates also the energy/exergy stream flows for the integrated system. Based on those figures, the energy efficiency is 20.5 % and the exergy efficiencies are 26 and 25 % for the thermochemical plant and integrated plant, respectively.

A comparison of various efficiencies as a function of the level of the process analysis is presented in Fig. 5.50. For the first level of analysis—only the thermochemical cycle is studied, assuming complete reactions—the predicted efficiencies are highest. No electricity is found necessary to drive the processes since the system is idealized. A flow-sheet analysis of the thermochemical plant identifies the needs of auxiliary power. Consequently, the efficiencies drop sharply. Further integration with the power generation system does not reduce the energy efficiency. In the integrated system, all of the heat rejected by the power plant is recovered and reused. However, there is a 1 % penalty on exergy efficiency because the integrated system adds more equipment and therefore more irreversibilities.

5.10 Development Aspects of Thermochemical Plants

After a comprehensive screening analysis, only very few thermochemical cycles have been pursued for further development at a flow-sheet plant simulation level for nuclear hydrogen plants (see Sect. 5.7).Brown et al. (2002) shortlisted 25 cycles from which they selected only seven for further flow-sheet development of a full-scale plant. Very few cycles were developed up to lab-scale demonstrations; even fewer have bench-scale or pilot plant proof-of-concept facilities. There has been no full-scale thermochemical plant constructed until the present day, although the technology shows much future promise.

Three major cycles are considered to be the most significant for thermochemical plant development. These are (1) sulfur–iodine, (2) hybrid-sulfur (H_2SO_4, or Westinghouse process), and (3) Cu–Cl cycle. In this section, the main technical issues and the status of development of thermochemical plants for nuclear hydrogen production based on S–I, H_2SO_4, and Cu–Cl cycles and the development issues are discussed.

5.10.1 Sulfur–Iodine Thermochemical Plant

5.10.1.1 Main Cycle Steps

The nuclear-based sulfur–iodine cycle is a leading example that has been scaled up from proof-of-principle tests to a larger engineering scale by the Japan Atomic Energy Agency. General Atomics was also investigating the S–I cycle and a capacity of 2 kg/day was developed. Commissariat à l'Energie Atomique (CEA, France) and the Sandia National Laboratory (SNL) have also actively developed the S–I cycle. The flow-sheet simulation of this cycle showed that 47 % can be obtained (Brown et al. 2002). The S–I cycle is very attractive because all chemicals are either in the gas or liquid phase; therefore the auxiliary power to transport the chemicals is minor with respect to other processes that need to transport solids.

Although there are several types of S–I cycles, the following three-step cycle for thermal decomposition of water is the most common (see also Fig. 5.51).

- Step 1: Hydrolysis step (exothermic):

$$2H_2O(l) + SO_2(g) + I_2(s) \rightarrow H_2SO_4(aq) + 2HI(aq), \text{ at } 393\,K \tag{5.59}$$

- Step 2: Oxygen production step (endothermic):

$$H_2SO_4(g) \rightarrow 0.5O_2(g) + SO_2(g) + H_2O(g), \text{ endothermic, at } 1{,}123\,K \tag{5.60}$$

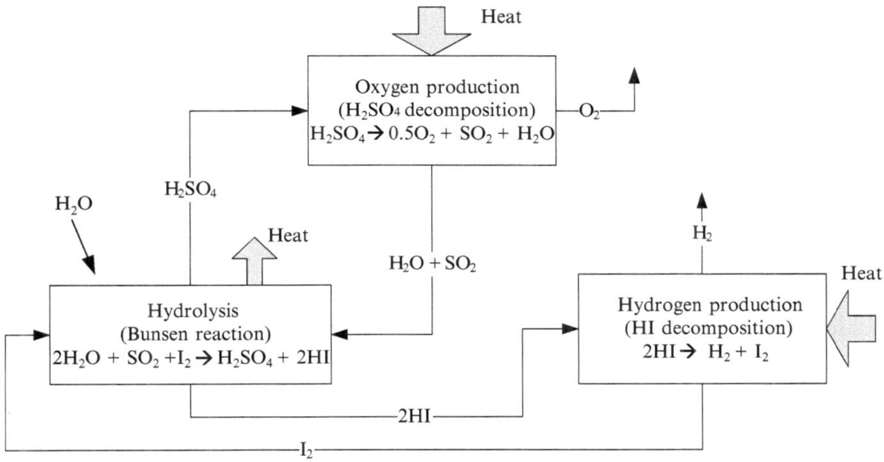

Fig. 5.51 Simplified representation of the S–I cycle

- Step 3: Hydrogen production step (endothermic):

$$2HI(g) \rightarrow H_2(g) + I_2(g), \text{ at } 723\,K \tag{5.70}$$

Figure 5.51 presents the process diagram in the most simplified manner. This diagram illustrates clearly the three main processes, namely: Bunsen process, H_2SO_4 decomposition, and HI decomposition. The separation of chemicals is not emphasized in Fig. 5.51. Each of the three reaction steps poses special problems regarding the chemical separation, avoiding side products, and in addition there are important issues regarding heat recycling within the cycle. All main processes—chemical and physical—which are involved in the operation of a S–I plant are discussed in following subsections.

5.10.1.2 Bunsen Reaction

Reaction (5.59) is known as the Bunsen reaction, which leads to mixed acid formation, namely, hydroiodic and sulfuric acids in the aqueous phase which are difficult to separate. According to the chemical equation (5.59), gaseous sulfur dioxide is absorbed chemically by liquid water and aqueous iodine. The Bunsen reaction is attractive because it is spontaneous and decomposes water at low temperature in the liquid phase. However, there are some difficulties to conduct the process in the desired way because (1) side reactions may occur, and (2) there may be difficult liquid–liquid separation of mixed acids.

The Bunsen process has many potential side reactions, depending on the reactor type and operation parameters. In particular, the iodine, water, and sulfate dioxide concentrations are important. The main side reactions are

$$6HI + H_2SO_4 \rightarrow S + 3I_2 + 4H_2O$$
$$H_2SO_4 + 8HI \rightarrow H_2S + 4I_2 + 4H_2O$$
$$SO_2 + 6HI \rightarrow H_2S + 3I_2 + 2H_2O$$
$$SO_2 + 4HI \rightarrow S + 2I_2 + 2H_2O.$$

Formation of sulfur or hydrogen sulfide in the Bunsen reactor is very undesirable within a thermochemical cycle, because any side products must be recycled in order to form a closed chemical process. Researchers at General Atomics determined that the desired products, namely, sulfuric acid and hydroiodic acid, are obtained with reasonable yield if iodine is added in excess to the reaction. More exactly, for a feasible process, the molar fraction of iodine must be larger than 10 % and the sulfur dioxide molar fraction must be higher than 5–7 % (Brown et al. 2002), while the molar fraction of water is ~90 %. Moreover, in such conditions, the products are spontaneously separated as two liquid phases with different densities. A light phase containing sulfuric acid accumulates at the top of the reactor while the heavier HI accumulates at the bottom.

The Bunsen reaction with liquid–liquid phase separation (as proposed by General Atomics) can be represented, according to Ewan and Allen (2005), by the following two-reaction set evolving at 398 K:

$$\begin{cases} SO_2(g) + 9I_2(l) + 16H_2O(l) \rightarrow H_2SO_4 \cdot 4H_2O(l) + 2HI \cdot 10H_2O \cdot 8I_2(l) \\ H_2SO_4 \cdot 4H_2O(l) + 2HI \cdot 10H_2O \cdot 8I_2(l) \rightarrow H_2SO_4 \cdot 4H_2O(l) + 8I_2 \cdot 10H_2O + 2HI(g). \end{cases} \quad (5.71a)$$

An alternative way to perform the Bunsen reaction is in a short-circuit electrolytic cell. Electrolytic cells that operate in a short circuit are devices similar to fuel cells that allow for ion exchange and electron transfer in a process that eventually conducts a spontaneous reaction with intrinsic separation of chemicals. Figure 5.52 shows the short-circuit electrolytic cell which was proposed by Dokyia et al. (1979). The half-reaction at the cathode is

$$I_2 + 2H^+ + 2e^- \rightarrow 2HI \quad (5.71b)$$

while the half-reaction at the anode is

$$2H_2O + SO_2 \rightarrow H_2SO_4 + 2H^+ + 2e^-. \quad (5.72)$$

Thus the overall reaction is the Bunsen reaction, (5.59). The anode described by Dokyia et al. (1979) was a composite made as a platinized electrode and activated charcoal with very low over potentials. The cathode is an electro-catalytic electrode with a platinum substrate doped with various metal oxides (e.g., TiO_2, Fe_3O_4).

Diluted hydroiodic acid must be supplied at the anode while diluted sulfuric acid must be supplied at the cathode for optimized cell operation. Some typical values

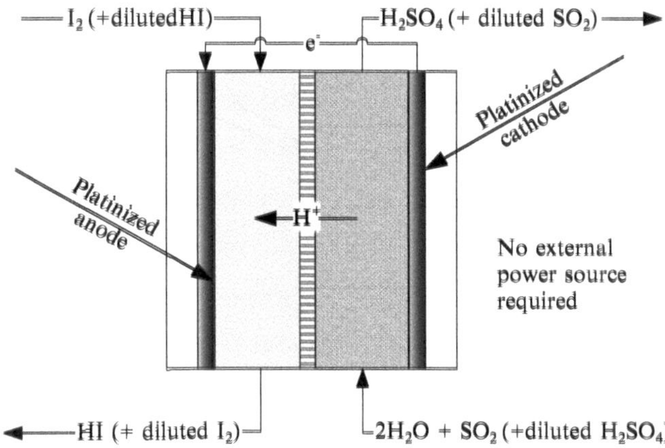

Fig. 5.52 Short-circuit electrochemical Bunsen reaction cell [adapted from Dokyia et al. (1979)]

from Dokyia et al. (1979) indicate that the cell must be supplied with 20 % sulfuric acid by weight and it will concentrate the output stream to 60 %. At the anode, the feed must be about 115/19.2/10.6 at the inlet and it produces a stream that will achieve a composition of 115/21.2/9.6, both given in molecular ratios of the mixture $H_2O/HI/I_2$. Using these actual feed streams, the chemical representation of the short-circuit electrochemical cell at 298 K is given from Dokyia et al. (1979) as

$$\begin{aligned} &SO_2 + 0.25(H_2SO_4 \cdot 26H_2O) + 19.2(HI \cdot 6H_2O) + 10.6I_2 \\ &\quad \rightarrow 1.25(H_2SO_4 \cdot 3.6H_2O) + 21.2(2HI \cdot 5.43H_2O) + 9.6I_2. \end{aligned} \tag{5.73}$$

Nafion membranes were used in construction of the electrochemical cell. This allows for sulfur dioxide crossover which leads to formation of elemental sulfur; this is reported as a potential drawback of the electrochemical cell (Nomura et al. 2004).

5.10.1.3 Decomposition of Hydrogen Iodide

The gas-phase hydrogen iodide decomposition reaction—(5.70)—was analyzed in Fig. 5.23. At standard pressure, the is satisfactory at ~600 K. Some improvement can be obtained by performing the reaction under pressurized conditions. Dokyia et al. (1979) conducted experiments of catalytic decomposition of hydrogen iodide at 523 K under increased pressure. In Fig. 5.52, the yield may be increased from 19 % up to 45 % when the pressure is varied from 3 to 10 MPa; simultaneously, the production rate of hydrogen increased 2.3 times. In Fig. 5.53, $\dot{n}_{ref}$ is the production rate of hydrogen corresponding to the reaction conducted at 3 MPa. Platinum deposited on activated carbon supports showed good catalytic activity. With membrane reactors developed in Japan at NCLI, it was found that the yield may reach >65 % molar for one-pass conversion.

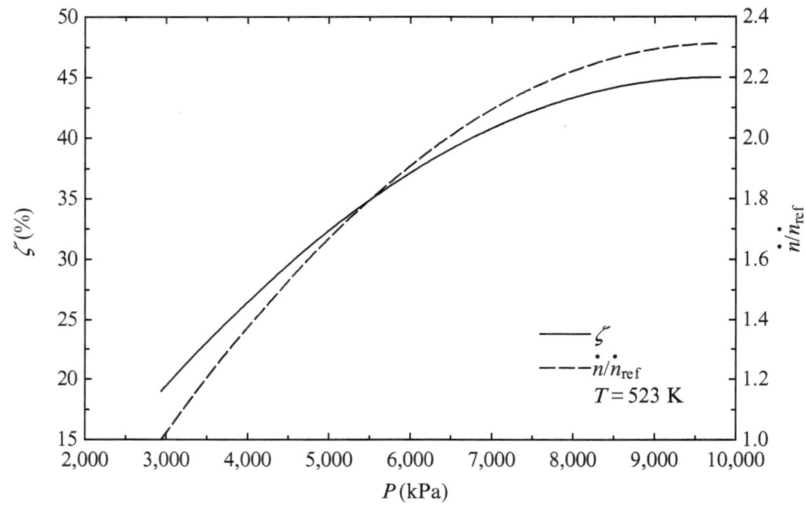

Fig. 5.53 Improvement of reaction yield (ζ) and hydrogen production rate of catalytic hydrogen iodide decomposition at elevated pressures [data from Dokyia et al. (1979)]

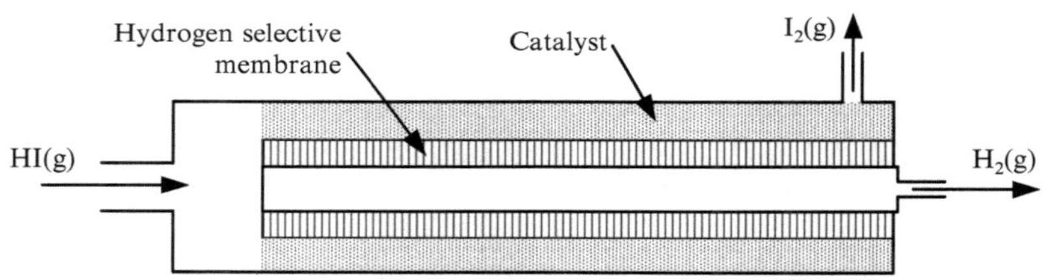

Fig. 5.54 Membrane reactor for hydrogen iodide decomposition in the gas phase

Figure 5.54 depicts a reactor configuration for hydrogen iodide in the gas phase. Membrane technology is proposed using hydrogen-selective membranes to shift the reaction equilibrium forward (Dokyia et al. 1979). Recent research by Kasahara et al. (2007) reported that silica membranes can be used for selective hydrogen extraction and enhanced HI decomposition conversion up to a demonstrated conversion of 61.3 % at 723 K.

A concurrent technology was developed at General Atomics to decompose hydrogen iodide in the liquid phase by using noble metal catalysts at 423 K. The liquid phase acts as an iodine absorber for the system $HI/H_2/I_2$. Thus it is possible to devise a reactive distillation column which extracts hydrogen from an HI_x solution which performs three functions, namely, extraction of HI from HI_x, decomposition of HI, and separation of H_2 from I_2 (see Brown et al. 2002).

The concentration of HI prior to its decomposition presents major challenges. The distillation columns normally have low efficiency. Extraction of HI using phosphoric acid and membrane-based pre-concentrations were explained by

Kasahara et al. (2007). Another technology development recently was also reported by Kasahara et al. (2007), namely, the concentration of HI from HI_x by electrodialysis technology. For an electrodialysis system, a detailed flow sheet of the S–I process predicts 40 % energy efficiency.

5.10.1.4 Sulfuric Acid Decomposition

The reaction of sulfuric acid decomposition, (5.60), is performed catalytically at 1,123 K using platinum catalysts on titanium dioxide supports, although cheaper iron-based catalysts may be used with lower activity. Aqueous sulfuric acid is first concentrated up to 90 % and heated up to 675 K to release water and SO_3 according to (5.74) which occurs spontaneously; thereafter, the final step of decomposition occurs according to (5.74) at 1,123 K:

$$H_2SO_4(g) \rightarrow SO_3(g) + H_2O(g), \text{ at } 675\,K \tag{5.74}$$

$$SO_3(g) \rightarrow 0.5O_2(g) + SO_2(g), \text{ at } 1{,}123\,K \tag{5.75}$$

The actual process in a flow-sheet diagram is more complicated as it requires excess reactants for better kinetics. According to Dokyia et al. (1979), it includes the dehydration of aqueous sulfuric acid after the Bunsen reaction, which can be expressed according to

$$1.25\,(H_2SO_4 \cdot 3.6H_2O) \rightarrow 1.25H_2SO_4 + 4.5H_2O, \text{ at } 298\,K \tag{5.76}$$

which is followed by dehydration of sulfur trioxide (at General Atomics, this reaction is conducted at 700 kPa, Brown et al. (2002)):

$$1.25H_2SO_4 + 4.5H_2O \rightarrow 5.25SO_3 + 5.75H_2O, \text{ at } 298\,K \tag{5.77}$$

and a heating step of water/sulfur trioxide mixture up to 1,123 K where the following reaction is performed:

$$5.25SO_3 + 5.75H_2O \rightarrow 0.5O_2 + SO_2 + 0.25SO_3 + 5.75H_2O, \text{ at } 1{,}123\,K \tag{5.78}$$

A sulfuric acid decomposition system—with a minimal number of functional units—is shown in Fig. 5.55 (pumps and auxiliary heat exchangers and units are not shown). After extraction from the Bunsen reaction system, sulfuric acid is directed to an evaporator which concentrates the solution by heat addition. A vapor phase, mostly containing steam, is removed and directed back to the Bunsen reactor, for recycling. A concentrated solution of 90 % H_2SO_4 is obtained and fed into an acid boiler that heats, boils, and partially decomposes acid by releasing sulfur trioxide.

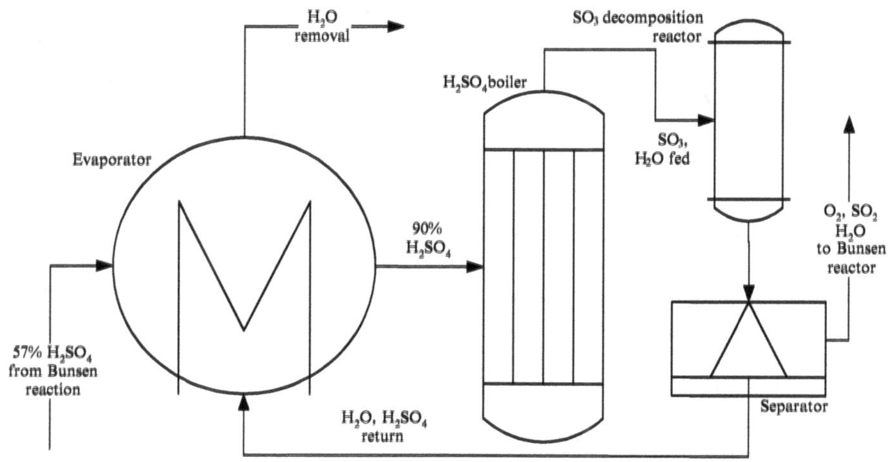

Fig. 5.55 Sulfuric acid decomposition system [adapted from Brown et al. (2002)]

Special heat exchanger designs were adopted for this boiler at General Atomics, comprising printed channels coated with catalysts for reaction rate enhancement.

After partial decomposition at 675 K—according to (5.76)—the product gases comprising sulfur trioxide and steam are fed in a catalytic reactor for complete decomposition, where sulfur trioxide is converted to sulfur dioxide and steam. A separation step is necessary to recycle the unconverted sulfur trioxide. After cooling (not detailed in Fig. 5.55), it is converted back to sulfuric acid and recycled to the evaporator. The final products of the sulfuric acid decomposition step are extracted in the form of hot oxygen, sulfur dioxide, and steam in the gas phase; this mixture, after cooling, is fed back into the Bunsen reactor.

5.10.1.5 Flow-Sheet Development and Pinch Point Analysis

Some simplified flow sheets of the S–I cycle were developed with the aim to perform a pinch point analysis and determine the plant efficiency. Kasahara et al. (2007) present a simplified flow sheet of the General Atomics (GA) process. The fully detailed GA flow sheet is outlined by Brown et al. (2002).

A simplified flow sheet of the GA process is presented in Fig. 5.56. At least nine processing units are needed to perform the three reactions of the cycle. Water is fed into a Bunsen reactor where it reacts with SO_2 and I_2 to form mixed acids. At the bottom of the Bunsen reactor, these acids are collected. There is a need for active mixing in a Bunsen reactor, which means that mechanical power must be provided to this reactor. It follows a liquid–liquid separation column which separates the HI fraction at the bottom. Prior to decomposing HI, this should be distilled. The distillation of HI is necessary. In an aqueous solution, HI exists as HI_x (in other words, in the liquid mixture, there are heavier phases of the hydroiodic acid that are present).

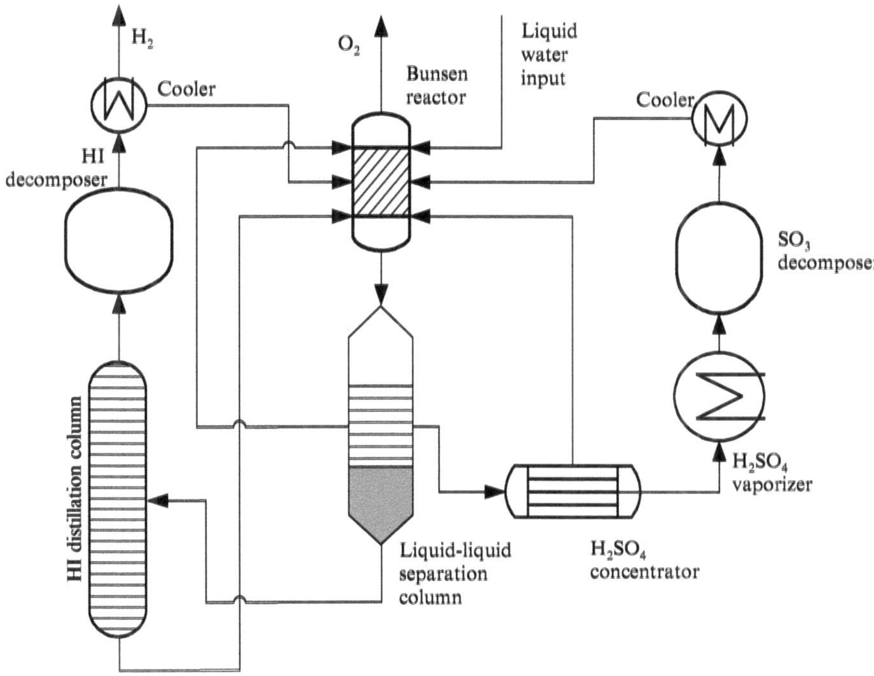

Fig. 5.56 Simplified flow-sheet diagram of the S–I cycle processes [adapted from Kasahara et al. (2007)]

Distillation allows for the extraction of the lighter phase, which is the phase with a maximum molar content of hydrogen. It is the phase that is to be decomposed to generate hydrogen. Distillation requires heat addition. The product iodine is returned to a Bunsen reactor after cooling it. On the other side, the sulfuric acid is concentrated by removing the water which is returned to the Bunsen reactor. Thereafter, sulfuric acid is vaporized, during which it decomposes thermally: $H_2SO_4 \rightarrow SO_3 + H_2O$. By heat addition, SO_3 is further decomposed to generate oxygen. After cooling, the mixture of oxygen and SO_3 is returned in the Bunsen reactor where SO_3 is consumed in the reaction while oxygen evolves as a gas and becomes withdrawn from above the reactor. The flow sheet from Fig. 5.52 does not show the heat recovery and its internal reuse in the system, but is understood that such a system is applied to the maximum possible extent to improve the plant efficiency.

A pinch analysis of the S–I cycle in Fig. 5.56 is presented in Fig. 5.57 and Table 5.18. The heat input required by the process is 565 kJ/mol H_2; the heat recovered is 130 kJ/mol, while the recovered heat for 20 % heat losses is 104 kJ/mol H_2. It is assumed that the temperature of helium from a high-temperature gas-cooled reactor is made available at 1,123 K. During the heat transfer process the helium temperature decreases to 900 K.

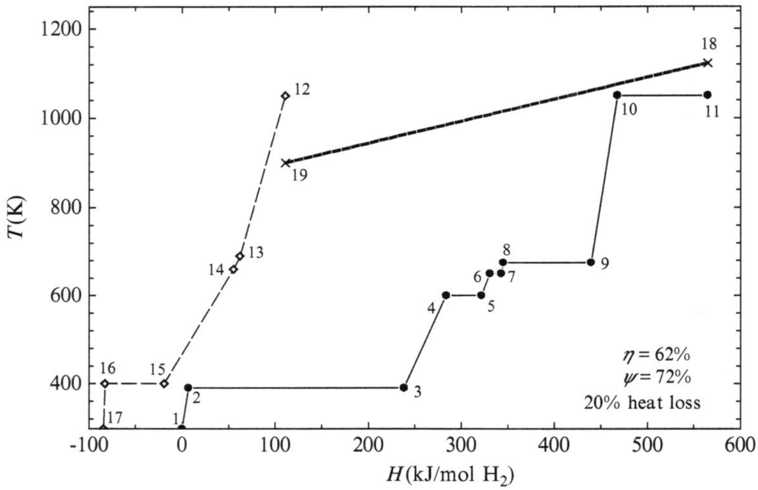

Fig. 5.57 Pinch diagram for S–I cycle based on simplified flow diagram from Fig. 5.56 [enthalpy data is approximated from Ewan and Allen (2005)]

Table 5.18 Process enthalpies for the pinch point analysis in Fig. 5.57

Process states	Process description	ΔH (kj/mol H_2)
$1 \rightarrow 2$	Heating pressurized water (3 bar) from 298 K to 390 K	6.9
$2 \rightarrow 3$	Liquid–liquid separation of mixed acids (5.71b)	232
$3 \rightarrow 4$	Heating Bunsen reaction products from 390 K to 600 K	45
$4 \rightarrow 5$	Fractional distillation of HI at 600 K	38
$5 \rightarrow 6$	Heating of all reaction products from 600 K to 650 K	9
$6 \rightarrow 7$	Hydrogen iodide decomposition at 650 K	12.1
$7 \rightarrow 8$	Heating of H2SO4/H2O mixture from 650 K to 675 K	2
$8 \rightarrow 9$	Sulfuric acid decomposition at 675 K, (5.74)	94.6
$9 \rightarrow 10$	Heating of sulfur trioxide/steam mixture from 675 K to 1,050 K	28
$10 \rightarrow 11$	Sulfur trioxide decomposition at 1,050 K, (5.75)	97.3
$12 \rightarrow 13$	Cooling of $O_2/SO_2/H_2O$ mixture from 1,050 K to 690 K	−49
$13 \rightarrow 14$	Cooling ofH_2SO_4/H_2O and $O_2/SO_2/H_2O$ from 690 K to 660 K	−7
$14 \rightarrow 15$	Cooling of H_2SO_4/H_2O, $O_2/SO_2/H_2O$, and H_2 streams from 660 K to 400 K	−74
$15 \rightarrow 16$	Heat recovery from Bunsen reaction at 400 K, (5.71a)	−64
$16 \rightarrow 17$	Cooling H_2 and O_2 streams from 400 K to 298 K	−1.4
$18 \rightarrow 19$	Heat transfer from pressurized helium at 4 bar, 1,123 K to 900 K	428

Note: For enthalpies, difference (+) sign means heat addition, (−) is heat rejection
Data approximated from Ewan and Allen (2005)

Based on $T_{eq} = \Delta S / Q_{in}$, where Q_{in} is the heat input, the equivalent temperature of the heat source is 1,007 K and its Carnot factor is 0.7; therefore the exergy of the heat source is 319.6 kJ/mol H_2. In these conditions, the energy efficiency of the process becomes 62 % while the exergy efficiency is 72 %. A detailed flow-sheet

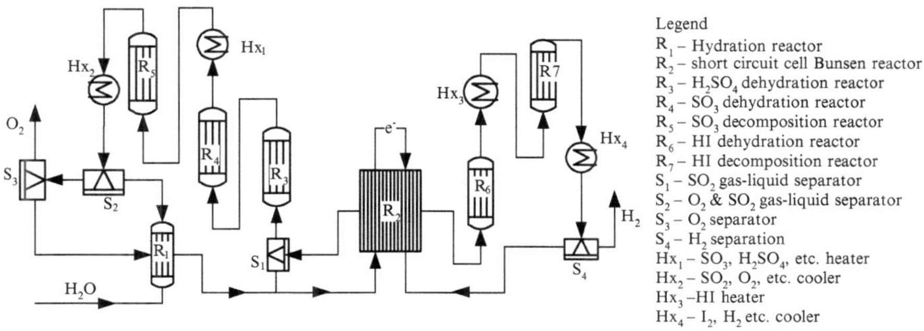

Fig. 5.58 Flow sheet of S–I plant based on short-circuit electrochemical Bunsen reactor [adapted from Dokyia et al. (1979)]

analysis of the same system, which accounts more thoroughly on heat transfer analysis of the heat exchangers, separation, pumping, gas compression, and incomplete chemical conversion, indicates 47 % energy and 55 % exergy efficiencies.

An alternative flow sheet, based on the short-circuit electrolytic cell for the Bunsen reaction, is illustrated in Fig. 5.58. The diagram is based on the study by Dokyia et al. (1979) and comprises 15 functional units: seven reactors, four separators, and four heat exchangers. The heat recovery system is not shown, but for the pinch point analysis, it is assumed that heat losses are 20 % (so 80 % of rejected heat can be recovered).

Liquid water is fed into a hydrating reactor R_1 where it hydrates the unreacted sulfur trioxide, and the unreacted sulfuric acid in the presence of dissolved sulfur dioxide. All resulting streams are fed at the cathode of the short-circuit electrochemical cell where the Bunsen reaction is performed. The Bunsen electrochemical reactor (R_2) is fed in addition with iodine at the anode. The reactor generates two separate streams as output, one is aqueous sulfuric acid with unreacted sulfur dioxide, and the other is aqueous hydroiodic acid. Sulfur dioxide dissociates as a gas from the liquid mixture at the output of the Bunsen reactor. This dissociation process is conducted in the separation unit (S_1) which returns the unreacted SO_2 (gaseous) to the Bunsen reactor (blower not shown in the figure) and the separated liquid phase, concentrated in sulfuric acid, which goes toward the sulfuric acid decomposition process.

The concentrated sulfuric acid is passed to a dehydration reactor (R_3) where heat is added such that the reaction described in (5.76) occurs. By further heat addition in an isothermal process, sulfur trioxide is dehydrated next in R_4 as per (5.77). In heat exchanger H_1, the sulfur trioxide and water present in molar proportions $1.25SO_3$ and $5.75H_2O$ are heated up to a decomposition temperature of 1,100 K. The sulfur trioxide decomposes according to the reaction described in (5.78), which occurs in reactor R_5. The reaction products are cooled to a low temperature (assumed 298 K for the idealized analysis) with heat recovery; the process occurs in heat exchanger H_2. The flow is then directed to the separator S_2 where the gas phase comprising oxygen and sulfur trioxide is separated from the liquid phase of aqueous sulfuric acid.

Some sulfur trioxide remains residual in the liquid phase but it is further recycled in a separator by hydration in reactor R_1 (see description above). The remaining sulfur dioxide in the gas phase is separated from oxygen by liquefaction. At 25 °C, the saturation pressure of sulfur dioxide is 392.2 kPa. This process is illustrated schematically by the separator block S_3.

At the electrochemical cell anode, hydroiodic acid is produced in aqueous solution. There is no need of any fractional distillation as the case of the GA process, because there is little or no HI_x, $x > 1$, produced by the electrochemical cell. The products are first dehydrated by isothermal heat addition in reactor R_6 according to

$$3HI \cdot H_2O \rightarrow 3HI + 16.2H_2O. \tag{5.79}$$

The products from R_6 are heated up to the HI decomposition temperature which is about 900 K; heating occurs in the heat exchanger denoted by H_3 in the figure. After heating, hydrogen iodide is decomposed in R_7 according to

$$3HI + 1.5I_2 + 16.2H_2O \rightarrow H_2 + HI + 2.5I_2 + 16.2H_2O. \tag{5.80}$$

The product stream resulting from R_7 is cooled to an ambient temperature in the heat exchanger H_4 with heat recovery. A liquid–gas separator follows where the gas phase separates as (assumed) pure hydrogen. In the liquid phase, it remains hydroiodic acid, iodine, and water. This separator is illustrated in the figure as S_4. The liquid is fed thereafter at the anodic side of the electrochemical cell.

Note also that in R_1, the unreacted sulfur trioxide is rehydrated by isothermal heat absorption according to the following chemical equation:

$$0.25SO_3 + 0.25H_2O \rightarrow 0.25H_2SO_4. \tag{5.81}$$

In the same reactor (R_1), the unreacted sulfuric acid is rehydrated as described by

$$0.25H_2SO_4 + 26H_2O \rightarrow 0.25\ (H_2SO_4 \cdot 26H_2O). \tag{5.82}$$

In addition, once the recycled hydrogen iodide is fed into the electrochemical cell (R_2), a spontaneous hydration process occurs isothermally with heat rejection, according to

$$HI + 16.2H_2O \rightarrow HI \cdot 16.2H_2O. \tag{5.83}$$

The temperature and heat inputs/outputs for each process in a flow-sheet diagram from Fig. 5.58 are listed in Table 5.19 in view of a pinch point analysis. The temperature–enthalpy diagram for this S–I process is presented in Fig. 5.59. The energy and exergy efficiencies based on the pinch analysis of this system are high, namely, 55 % energy and 65 % exergy efficiency. There is an assumed 20 % heat loss. As opposed to the previous flow sheet, in this case, there is major heat

Table 5.19 Process enthalpies for the pinch point analysis presented in Fig. 5.59

Process states	Process description	ΔH (kj/molH_2)
$1 \rightarrow 2$	Dehydration of H_2SO_4 at 298 K, (5.76)	65.3
$2 \rightarrow 3$	Dehydration of SO_3 at 298 K, (5.77)	165.8
$3 \rightarrow 4$	Dehydration of HI at 298 K, (5.79)	1,563
$4 \rightarrow 5$	Heating [$1.25SO_3 + 5.75H_2O$ and $21.2HI + 115H_2O + 9.6I_2$] from 298 K to 900 K	9,099
$5 \rightarrow 6$	Hydrogen iodide decomposition at 900 K, (5.80)	13.8
$6 \rightarrow 7$	Heating $1.25SO_3 + 5.75H_2O$ from 900 K to 1,100 K	122.6
$7 \rightarrow 8$	Sulfur trioxide decomposition at 1,100 K, (5.78)	93.4
$9 \rightarrow 10$	Heat rejection $0.25SO_3 + SO_2 + 0.5O_2 + 5.75H_2O$ from 1,100 K to 900 K	−121.9
$10 \rightarrow 11$	Heat rejection [$0.25SO_3 + SO_2 + 0.5O_2 + 5.75H_2O$ and $16.2HI + 115H_2O + 9.6I_2$] from 900 K to 298 K	−9,163
$11 \rightarrow 12$	Heat rejection from Bunsen reaction at 298 K, (5.73)	−64.5
$12 \rightarrow 13$	Heat rejection at hydration of SO_3, (5.81)	−33.1
$13 \rightarrow 14$	Heat rejection at hydration of H_2SO_4, (5.82)	−18
$14 \rightarrow 15$	Heat rejection at hydration of HI, (5.83) Bunsen reaction at 298 K, (5.72)	−1,439

Note: For enthalpies, difference (+) sign means heat addition, (−) is heat rejection
Data from Dokyia et al. (1979)

transfer process for preheating the reactant. There is a large quantity of water that is recycled in the system. For example, according to (5.83) one mole of HI is hydrated with 16.2 mol of water. Hydration of 1 mol H_2SO_4 is made with more than 4 mol of water. Therefore, there is a large mass of recirculated chemicals. However, because all chemicals are in the fluid phase, high-performance heat exchangers can be devised to obtain a very high level of heat recovery and reduced exergy destruction (temperature profiles can be more easily matched when fluids transfer heat than in the case where solid phases are present). This feature makes the "Japanese" process proposed by Dokyia et al. (1979) very attractive. More recent investigations by Nomura et al. (2004) found that the S–I process with a short-circuit electrochemical cell can reach 42 % efficiency with an optimum HI concentration on the catholyte of ~14.5 mol/kg H_2O.

5.10.2 Hybrid Sulfur Thermochemical Plant

5.10.2.1 Main Cycle Steps

Westinghouse developed a thermochemical water-splitting plant based on a hybrid sulfur cycle described in Table 5.15 (see Mark 11) by two equations of which the electrochemical process is

$$2H_2O + SO_2 \rightarrow H_2 + H_2SO_4. \tag{5.84}$$

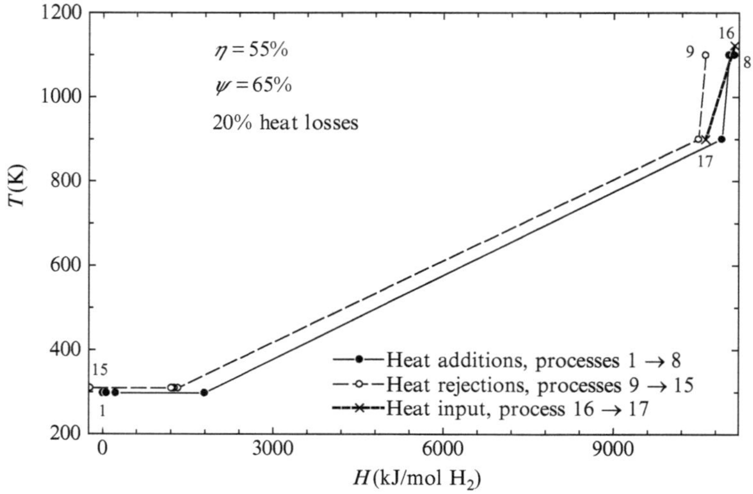

Fig. 5.59 Pinch diagram for S–I cycle based on simplified flow diagram from Fig. 5.58 [data from Dokyia et al. (1979)]

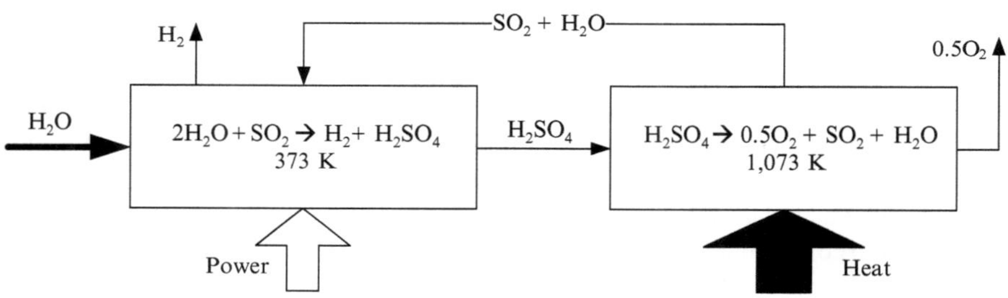

Fig. 5.60 Simplified schematic of hybrid sulfur cycle

The high-temperature thermochemical reaction is endothermic decomposition of sulfuric acid, namely,

$$H_2SO_4 \rightarrow 0.5O_2 + SO_2 + H_2O. \quad (5.85)$$

The electrolysis of sulfur dioxide in an aqueous solution—through formation of sulfurous acid $H_2SO_3(aq)$—is described in Sect. 5.6.1 in conjunction with (5.25) and (5.26). The simplified process schematic is shown in Fig. 5.60 (no separation and auxiliary devices are detailed). The main processes in a hybrid sulfur cycle are analyzed next.

5.10.2.2 Electrolytic Cell

A depolarized electrolytic process is conducted in a parallel plate electrolyzer operating under pressure, invented by Westinghouse and using diaphragms for

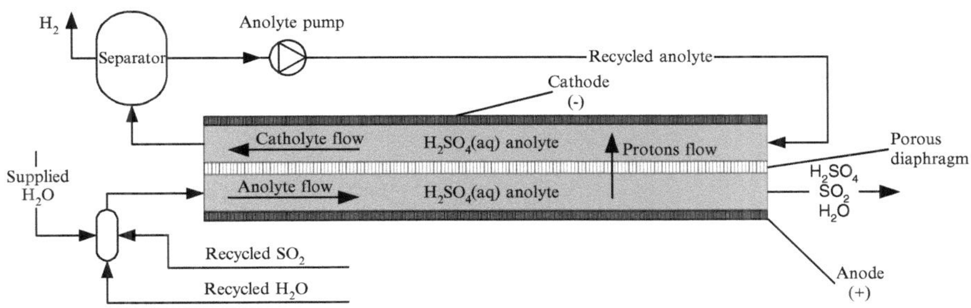

Fig. 5.61 Conceptual configuration of the Westinghouse electrolytic system [adapted from Lu et al. (1981)]

gas product separation as described by Brecher et al. (1977). The term "depolarized" refers to addition of sulfur dioxide in the anodic compartment so that the reversible potential decreases (anode is depolarized) with respect to basic electrolysis of water. Proton exchange membranes (PEM) are proposed in the most recent research concerning the electrochemical process in a hybrid sulfur cycle. The electrolysis cell and its immediate auxiliaries are illustrated in Fig. 5.61 as in the early Westinghouse design. Both the anolyte and catholyte are recirculated (separately) and formed from a sulfuric acid aqueous solution (about 70 % by weight concentration). The process operates at ~25 bar with a slightly pressurized catholyte in order to avoid sulfur dioxide penetration in the catholyte. The two-half reactions are described in (5.26). Recycled sulfur dioxide and water are mixed with supply water to form the anolyte. Under the influence of an electric field and electro-osmotic force, protons transfer to the anolyte and form dissolved molecular hydrogen. At the exit of the anolyte, a gas–liquid separator is placed where hydrogen is separated and released while the remaining anolyte is recirculated by a pump.

The detailed flow sheet of the early Westinghouse electrolysis cell and its auxiliaries comprises 15 operational units as follows (see Carty and Conger 1980): electrolyzers, cathode surge tank, anode mixing drum, four electrolyzer feed pumps, anode effluent cooler, sulfur dioxide flash drum for the anode effluent, sulfur dioxide condenser, acid pump and surge tank, sulfur dioxide removal tower, sulfur dioxide/oxygen compressor, sulfur dioxide condenser, and knockout drum separator of liquid SO_2 from gaseous SO_2/O_2 mixture.

According to Westinghouse researchers, the electrolytic cell shows effective measured potentials of 0.6–1.05 V for current densities of 2–4 kA/m^2. The purity of hydrogen produced from the cell is 98.7 %. The optimum concentration of sulfur dioxide in the feed stream is 30 %. Platinum-based electrodes were used with various substrates. Figure 5.62 reports the potential vs. current density curve of the Westinghouse cell, compared with more recent results by Sivasubramanian et al. (2007). The recent design consists of a PEM-based cell with a configuration as depicted in Fig. 5.63. As opposed to the Westinghouse cell where water is fed at the cathode, and the electrolyte—consisting of PEM—is traversed by protons from the anode to cathode, there is an effective flow of water from the cathode to the anode.

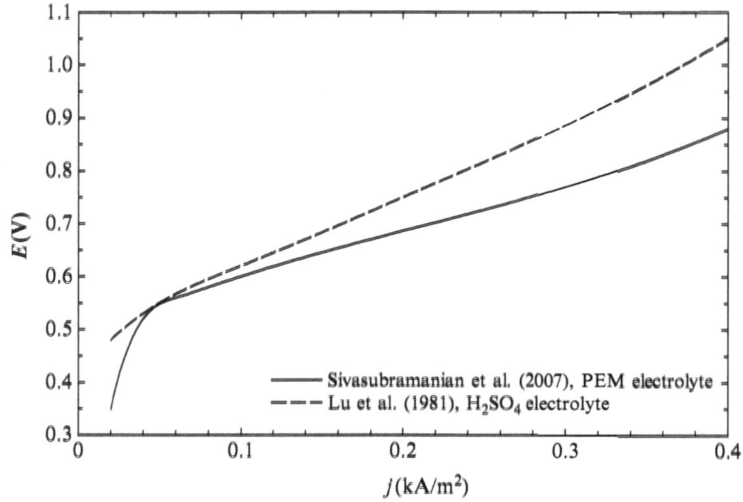

Fig. 5.62 Cell potential vs. current density for early Westinghouse design compared with newer PEM cell design [data from Sivasubramanian et al. (2007)]

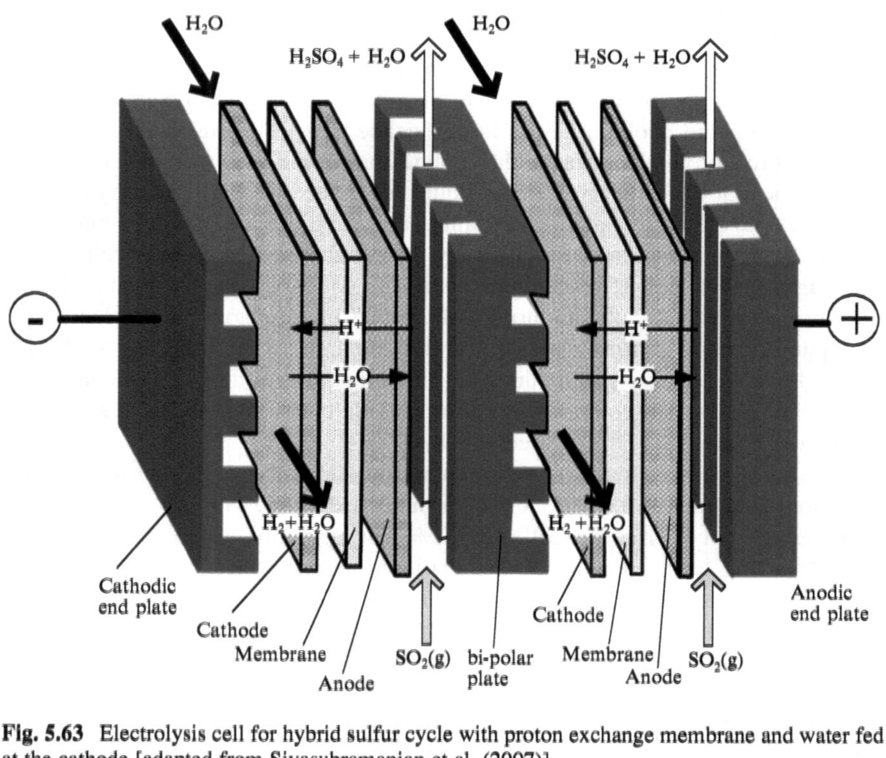

Fig. 5.63 Electrolysis cell for hybrid sulfur cycle with proton exchange membrane and water fed at the cathode [adapted from Sivasubramanian et al. (2007)]

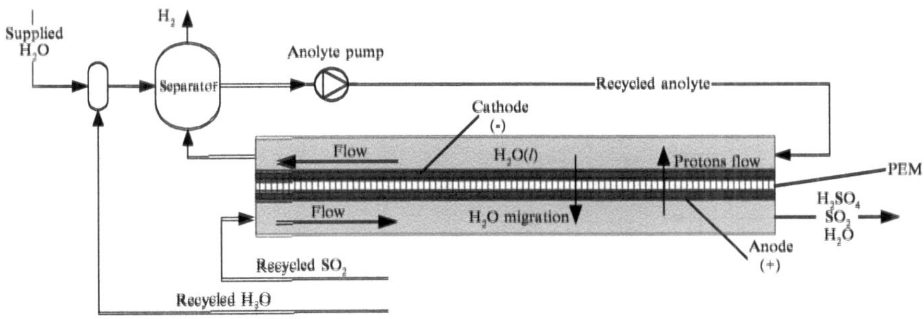

Fig. 5.64 Electrolytic process for hybrid sulfur cycle with a PEM electrolysis cell

It can be observed from the potential vs. current density diagram that a novel design has better efficiency because voltages are lower for the same current density.

The constructive design of depolarized PEM electrolytic cell for sulfuric acid production—as proposed in Sivasubramanian et al. (2007)—is similar to a PEM fuel cell for water electrolysis, as represented in Fig. 4.13. The only difference involves the chemical flows in the two cases. In PEM electrolysis, water is the feed and the products are oxygen and hydrogen. In the present case, gaseous sulfur dioxide is the only feed stream at the bottom of the anodic side of the bipolar plate or the anodic end plate. Liquid water migrates through the PEM and at the anode encounters activated electro-catalysts under adequate electrode polarization which produces water decomposition according to the half-reaction

$$2H_2O(l) + SO_2(g) \rightarrow H_2SO_4(aq) + 2H^+ + 2e^-, \qquad (5.86)$$

which is a depolarized electrochemical reaction (with respect to water dissociation) due to the presence of sulfur dioxide. The protons migrate toward the cathode across the PEM and combine with electron reduction in a half-reaction described in (5.26). The complete description of the electrochemical PEM reactor and its main auxiliary is presented in Fig. 5.64.

The electrolytic process with a PEM, depicted in Fig. 5.64, supposes that the supplied water and the recycled water are fed at the cathode. The anolyte consists of demineralized water only in which the hydrogen product is dissolved. Hydrogen is separated with a gas–liquid separator. The recycled sulfur dioxide is fed at the anode. While flowing through the anode's channels, the dissolved sulfur dioxide enters in a reaction with protons and forms sulfuric acid. The anode's product consists of unreacted sulfur dioxide, water, and sulfuric acid.

A third type of electrochemical cell and associated process was recently proposed by Gorensek and Summers (2009) as shown in Fig. 5.65. The cell is also a PEM electrolysis cell; however, in this case, water is fed at the anode. Consequently, there is no net migration of water toward the cathode, as required by electrochemical reactions wherein only protons need to transfer to cathodic side. However, due to the functioning issues of the PEM, protons migrate as H_3O^+ and water transferred to the cathodic side must be transferred back to the anode.

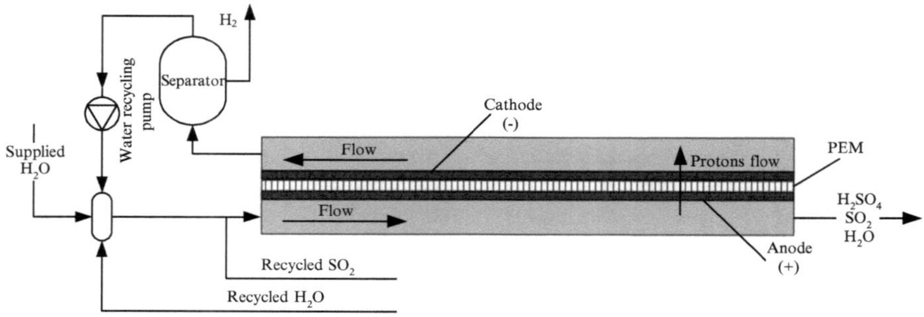

Fig. 5.65 PEM-based electrolytic cell and its system when water is fed at the anode [adapted from Gorensek and Summers (2009)]

5.10.2.3 Thermochemical Decomposition of Sulfuric Acid and Pinch Point Analysis

Decomposition of sulfuric acid is similar as in the S–I cycle. It requires heating of the sulfuric acid which first dehydrates and eventually sulfur dioxide is formed and decomposes thermally. The initial process proposed by Westinghouse comprises—according to Brecher et al. (1977)—ten functional units as follows: surge tank (ST), acid separator (AS), stream condenser (SC), compressor (C), primary acid vaporizer (PAV), secondary acid vaporizer (SAV), sulfur trioxide reduction reactor (RR), primary and secondary SO_2 condenser (PCND, SCND), oxygen separator (OS), and a turbo-expander (TE). This decomposition system is illustrated in Fig. 5.66.

A detailed flow sheet for sulfuric acid decomposition with bayonet-type reactors was developed recently in ASPEN Plus software by Gorensek and Summers (2009). A silicon carbide bayonet decomposition reactor technology was developed at SNL and it aims to decompose sulfuric acid at 9,000 kPa and 1,073 K. Silicon carbide was found as the best (ceramic) material choice which confers good heat transfer properties, satisfactory mechanical stability, and corrosion resistance. A bayonet reactor has a tube-in-tube configuration and it is immersed in a hot fluid. The temperature in the annulus increases from ~400 K at the fed inlet to 950 K when the catalytic decomposition process starts. The flow reaches 1,143 K after the catalytic bed and then returns through the core. It exits at the cold side with a temperature of 528 K, which corresponds to the pinch point.

The schematic of the bayonet reactor-based system is illustrated in Fig. 5.67. The pinch point diagram of the system is shown. The bayonet reactor facilitates the heat transfer between the feed gas and the return gas at lower temperatures, as well as between the heat transfer fluid of the nuclear reactor (e.g., helium) and the feed stream at higher temperature. According to data from the figure, more than 80 % of heat rejected is recovered and used for preheating of a cold stream delivered by the electrolytic subsystem.

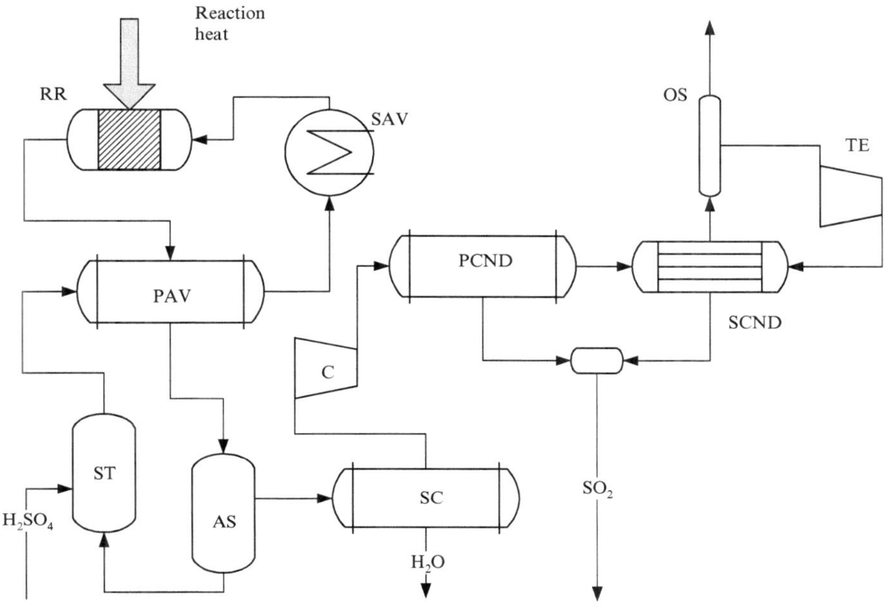

Fig. 5.66 Westinghouse flow sheet for sulfuric acid decomposition (simplified) [adapted from Brecher et al. (1977)]

Based on a detailed flow-sheet analysis, under the assumption that electricity is generated with 45 % efficiency, the net energy efficiency of the hybrid sulfur cycle is ~42 %, thus superior to water electrolysis (Gorensek and Summers 2009).

5.10.3 Hybrid Copper–Chlorine Plant

5.10.3.1 Main Cycle Steps

There are several configurations of the Cu–Cl cycle as indicated in Table 5.15. There are two-, three-, four-, and five-step configurations. This cycle is a hybrid cycle, in the sense that it needs high-temperature heat and electricity to operate. The most relevant configurations, currently under study, are included in Table 5.20. Two main parameters are important when selecting a hybrid cycle version for development, namely, the maximum temperature and the electricity requirement. From Table 5.20, the electricity required for the Cu–Cl cycle can be as low as 53 % of that for water electrolysis with a three-step cycle and an electrolytic process evolving hydrogen at 24 bars. Another implementation that generates atmospheric pressure hydrogen and requires 57 % electricity with respect to water electrolysis is being developed at UOIT (indicated with bold lines in Table 5.20). Approximately a 43 % net efficiency

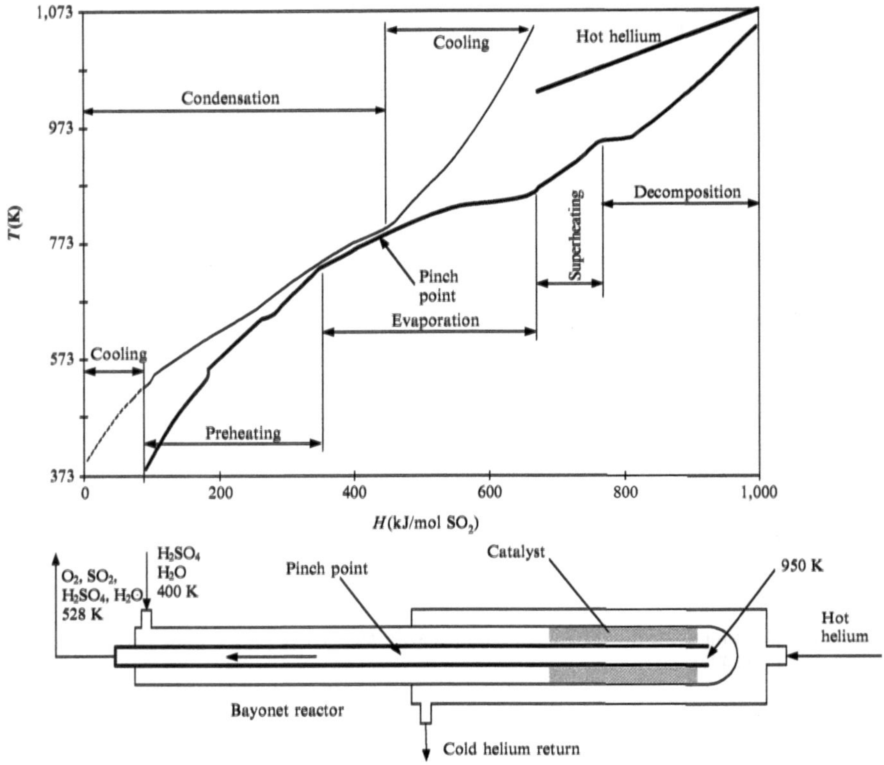

Fig. 5.67 Pinch analysis for the bayonet-type sulfuric acid decomposition reactor [data from Gorensek and Summers (2009)]

is envisioned with this Cu–Cl cycle, which is a significant margin of improvement over water electrolysis. Option 2 from the three-step cycle is studied at the ANL.

5.10.3.2 Copper–Chlorine Cycle International Consortium

Over the past decade, the copper-chlorine cycle has been extensively developed by AECL, the ANL, and a UOIT-led team of Canadian universities and other institutions, as a potential candidate for nuclear hydrogen production by water splitting. This cycle is suitable to be linked with the generation IV CANDU-SCWR (supercritical water-cooled reactor) as it operates at 800 K (maximum temperature).

There is a Canadian-led international consortium in connection with the Generation IV International Forum since 2006 on development and scale-up of this technology. The consortium consists of UOIT, Memorial University of Newfoundland, AECL, ANL, University Network of Excellence in Nuclear Engineering

Table 5.20 Variations of Cu–Cl cycle and their electricity requirements

Number of steps		Processes	Electricity[a]
Three	Option 1	$4CuCl(aq) \rightarrow 2Cu(s) + 2CuCl_2(aq)$ at 80 °C, electrolytic $2CuCl_2(aq) + H_2O(g) \rightarrow 2CuCl(l) + 2HCl(g) + 1/2O_2(g) + H_2O(g)$ at 400–600 °C $2Cu(s) + 2HCl(g) \rightarrow 2CuCl(l) + H_2(g)$ at 430–475 °C	89 %
	Option 2	$2CuCl(aq) + 2HCl(aq) \rightarrow 2CuCl_2(aq) + H_2(g)$ at 100 °C, electrolytic at 24 bar $2CuCl_2(s) + H_2O(g) \rightarrow Cu_2OCl_2(s) + 2HCl(g)$ $Cu_2OCl_2(s) \rightarrow 1/2O_2(g) + 2CuCl(s)$	53 %
Four	Option 1	$4CuCl(aq) \rightarrow 2Cu(s) + 2CuCl_2(aq)$ at 80 °C, electrolytic $2CuCl_2(aq) + 2H_2O(g) \rightarrow Cu_2OCl_2(s) + 2HCl(g) + H_2O(g)$ $Cu_2OCl_2(s) \rightarrow 1/2O_2(g) + 2CuCl(l)$ $2Cu(s) + 2HCl(g) \rightarrow 2CuCl(l) + H_2(g)$	89 %
	Option 2	$2CuCl(aq) + 2HCl(aq) \rightarrow 2CuCl_2(aq) + H_2(g)$ at 100 °C, electrolytic at 1 bar $CuCl_2(aq) \rightarrow CuCl_2(s)$ $2CuCl_2(s) + H_2O(g) \rightarrow Cu_2OCl_2(s) + 2HCl(g)$ $Cu_2OCl_2(s) \rightarrow 2CuCl(l) + 1/2O_2(g)$	57 %
Five		$4CuCl(aq) \rightarrow 2Cu(s) + 2CuCl_2(aq)$ at 80 °C, electrolytic $CuCl_2(aq) \rightarrow CuCl_2(s)$, drying, no chemical reaction $2Cu(s) + 2HCl(g) \rightarrow 2CuCl(l) + H_2(g)$ $2CuCl_2(s) + H_2O(g) \rightarrow CuO{*}CuCl_2(s) + 2HCl(g)$ $CuO{*}CuCl_2(s) \rightarrow 2CuCl(l) + 1/2O_2(g)$	89 %

[a]Given in a percentage of electrical input of water electrolysis (calculated based on Gibbs free energy of the reaction)

(UNENE, Canada), University of Guelph, University of Western Ontario, University of Waterloo, Pennsylvania State University, University of Maribor, Technical University of Liberec, Czech Academy of Science, National Hydrogen and Fuel Cell Centre of Romania, and the Atomic Energy National Commission of Argentina—see Table 5.21.

A comprehensive review of research and development on the Cu–Cl cycle and its progress was presented by Naterer et al. 2011a, b. In this section, the main progress of the international consortium is described, along with achievements of each member institution and scale-up activities of a Cu–Cl pilot plant at UOIT. The technical/scientific details of this cycle and its version are presented in detail in Chap. 6 of this book, dedicated only to this technology.

The first test tube-scale experiments of individual steps of the Cu–Cl cycle were studied at ANL in around 2004 for a production of hydrogen of about 5 g per day. An integrated lab engineering-scale demonstration project of 5 kg/day is currently under development at UOIT. It is envisioned that a future larger pilot-scale facility will be built with a capacity of 500 kg/day by around 2017–2020, followed by a small pre-commercial plant of about 50 ton hydrogen per day and ultimately ending

Table 5.21 Participants in the international consortium of Cu–Cl cycle development

Canadian institutions and partners	International partnerships
University of Ontario Institute of Technology (UOIT)	Generation IV International Forum
Memorial University of Newfoundland	Argonne National Laboratory (ANL)
University of Toronto	Pennsylvania State University
University of Guelph	Gas Technology Institute (GTI)
Western University	Generation IV International Forum
University of Waterloo	University of Maribor
McMaster University	The Czech Academy of Science
University Network of Excellence in Nuclear Engineering (UNENE)	Atomic Energy National Commission of Argentina
Atomic Energy of Canada Limited (AECL)	National Hydrogen and Fuel Cell Centre (Romania)
Ontario Power Generation (OPG)	
Regional Municipality of Durham	
Marnoch Thermal Power Inc.	
Gerdau	
St. Mary's Cement	

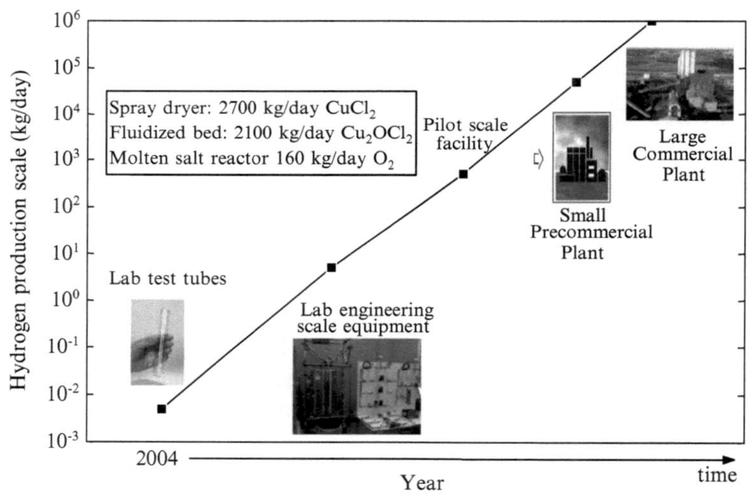

Fig. 5.68 Vision of Cu–Cl technology development

with a large commercial hydrogen plant of 1,000 ton per day. Figure 5.68 illustrates the UOIT-led team's future vision of Cu–Cl technology development.

Some of the key current research activities include the development and scale-up of each unit operation reactor, electrochemical cell, system integration studies, exergo-economic modeling and life cycle analysis, heat exchanger analysis for heat recovery, safety issues, thermochemical data development, and advanced materials and applications. A comprehensive list of the research topics of the UOIT-led team on the Cu–Cl cycle and related issues is presented in Table 5.22.

Table 5.22 Research and development areas of the Cu–Cl technology

Research topic	Description
Electrochemical cell	Required development of cost-effective electrodes and separation technology
	Selection/development of a cost-effective membrane without significant crossover issues
	Membrane characterization: permeability and conductivity determination
Drying aqueous $CuCl_2$	Spray drying system development
	Characterization of $CuCl_2$ particle morphology
	Development of crystallization process for drying
Hydrolysis reaction	Experiments to characterize the reaction, $H2O + 2CuCl_2 \rightarrow Cu_2OCl_2 + 2HCl$
	Fluidized bed reactor design and analysis
Copper oxychloride decomposition	Thermochemical characterization of Cu_2OCl_2
	Performing decomposition experiments and improving the reactor
	Heat recovery system design from molten CuCl
System integration	All independent experimental units for each reaction step must be integrated
Aspen-Plus simulation	Simulation aiming to accurately predict the system performance and efficiency
	Design of heat exchanger network for internal heat recovery
Exergo-economic modeling	Application of design tools such as exergy–cost–energy–mass
Life cycle assessment	Determination of environmental impact and life cycle cost of the technology
Waste heat recovery	Investigating ways to use waste heat and to upgrade temperature level
Linkage to nuclear plant	Investigation of possible ways to couple Cu–Cl cycle and nuclear reactors
Safety and reliability	Development of dynamic flow graph methodology to model system behavior
Thermochemical data	Chemical potentials and solubility data
	Formation of Cu(I) and Cu(II) complexes
	Synthesis of copper oxychloride
	Properties of HCl and copper–chlorine compounds
	Determination of reaction rate constants
Advanced materials	Anode electrodes of electrochemical cell
	Development of thermal spray coatings
	Characterization and development of corrosion resistance coatings
Economics and applications	Economics of nuclear hydrogen production
	Applications to oil sand upgrading
	Hydrogen for the transportation sector

5.10.3.3 Research and Development Status

This section describes the progress on the Cu–Cl cycle development—and its specific equipment/reactors/processes—at various institutional members of the UOIT-led consortium. The pilot plant development at UOIT is briefly addressed also.

The electrochemical cell is under investigation and development at AECL and ANL. AECL has successfully demonstrated the production of hydrogen over a period of several days with a CuCl/HCl electrolyzer. Metallic copper at the cathode acts as a poison, greatly reducing the activity of the Pt catalyst and causing a major decrease in cell performance, so active research is under way to eliminate any possible copper crossover. The copper (I) oxidation reaction is presently conducted in a 0.5 mol/L CuCl/11 mol/L HCl electrolyte—anode reaction. In 11 mol/L HCl, copper metal deposition at the cathode appears to be absent at the 0.1 A cm^{-2} current density used in the experiments. In 6 mol/L HCl, the $CuCl^+$ that is formed can be transported across the membrane by diffusion and migration. In 11 mol/L HCl, $CuCl_2$ is transported across the membrane by diffusion only. Thus, the flux of copper species through the membrane is dependent upon the HCl concentration. Experimental data shows that for similar experimental conditions, the catholyte copper species concentration decreases with increasing HCl concentration. This is consistent with the dependence of the flux on HCl concentration.

Results from AECL's experiments show very promising stable performance at a cell voltage of about 0.57 V over a period of about 50 h. The catholyte flow rate was 1.5 L/min and the anolyte flow rate was 1.25 L/min. The operating temperature and current density were 45 °C and 0.1 A cm^{-2}, respectively. No anode catalyst was used and there was a Pt cathode catalyst. Stable cell performance at voltages down to 0.477 V has been observed.

ANL developed an electrolytic disproportionation system for CuCl(l) which generates hydrogen at 24 bar. Lewis and Masin (2009) reported that the ANL electrolysis reactor operates at 0.5 V with 0.5 A/cm^2. This reaction is attractive because it generates compressed hydrogen. Consequently, the work required to compress hydrogen from standard pressure—to 350 bars for example—reduces approximately by 95 %.

In collaborations between ANL, Pennsylvania State University (PSU), and Gas Technology Institute (GTI), studies have been conducted to identify membranes that have lower copper diffusion rates than Nafion, but with similar proton conductivity. Three types of membranes were investigated. At ANL, physical modifications of Nafion were conducted to impede the diffusion of copper (II). A cross-linked flexible polymer matrix was developed with inter-winding Nafion chains and sulfonic acid groups to maintain adequate conductivity. Polysulfone membranes, based on Radel NT-5500, a commercially available polymer from Solvay Advanced Polymers, were developed at PSU. These are typically characterized by low rates of ion diffusion and tunable properties that should prevent Cu crossover. These membranes were first tested in AECL's electrolyzer and showed higher copper diffusion than Nafion. However, permeability measurements showed copper diffusion about ten times lower than Nafion. Based on the permeability test results, further work was conducted and the polysulfone membranes were modified by cross-linking to improve thermal stability. The cross-linking allowed for tuning of the water uptake and conductivity by the addition of varying equivalents of cross-linker.

Collaborations between ANL and GTI have also identified commercially available membranes which are characterized by their manufacturers as having very low metal ion diffusion, high proton conductivity, and high thermal stability.

They are also chemically stable under highly oxidative and high reductive conditions. The membranes were treated by GTI to improve wettability and remove skin effects by an etching process. The pretreatment process is still under development. They are very inexpensive at about $1 per square meter. Membrane permeability is determined in the range of about $1 - 2 \times 10^{-8}$ cm^2/s with a target of 0.15 cm^2/s which is a tenth of that of NAFION 117 membrane. At PSU, the membrane conductivity is measured in a planar cell in 2 mol/l HCl solution obtaining conductivities smaller than 0.061 S/cm (where 0.061 S/cm is the conductivity of NAFION 117).

ANL and UOIT performed Aspen Plus simulations of the Cu–Cl cycle to predict the behavior of process reactions and steps using standard engineering relationships, mass and energy balances, as well as phase and chemical equilibrium data. An "Electrolyte Non Random Two Liquid" (ELECNTRL) activity coefficient model was used for modeling of the electrolyte step in the stoichiometry reactor. The Soave–Redlich–Kwong (SRK) cubic equation of state was used to evaluate component properties and phase equilibria in the steps that involve vapor–liquid phase change. The liquid-phase equilibrium in the unit operation models manages physical property and phase equilibrium calculations. Aspen Plus has the capability of estimating physical property parameters for components that are not present in the Aspen Plus database, based on a combination of experimental data and molecular formulae.

The linkage of a nuclear reactor and the Cu–Cl cycle has also been studied at AECL in collaboration with UOIT. SCWR is a generation IV nuclear reactor under development by AECL that will operate at temperatures up to 625 °C, which would facilitate the thermochemical production of hydrogen with the Cu–Cl cycle. Integrating an SCWR into a process heating application, through an interface with a hydrogen production plant based on the Cu–Cl cycle, enables hydrogen production during off-peak electrical demand. A fraction of the high-temperature SCW coolant may be diverted to an HX (heat exchanger), where it exchanges heat with an intermediate fluid (such as steam or helium). It is then returned to a location on the feed water heating line or reactor inlet. To link the Cu–Cl cycle with an SCWR, the HX would be in a location where the inlet coolant would be at temperatures exceeding the maximum temperature requirement of the Cu–Cl cycle, approximately 530 °C.

The University of Western Ontariois collaborating with UOIT on safety and reliability studies concerning the Cu–Cl technology. To facilitate the control system design, a linear model was developed by linearizing the system model at the desired operating point. The linearized model was verified by numerical simulations. A feedback compensator was designed and successfully implemented.

Experimental studies conducted at the University of Guelph and UOIT have allowed the identification of copper complex species in HCl solutions that are responsible for copper crossover in the electrolysis cell. Experimental measurements on the $CuCl_2/LiCl/H_2O$ system for ionic strengths up to 13 mol/kg and temperatures between 100 and 200 °C were conducted at the University of Guelph in a high-temperature, high-pressure titanium cell with sapphire windows (1.72 cm optical path length; 0.34 cm^3 volume).

At the University of Maribor, there have been studies conducted on theoretical predictions of various copper/chlorine compounds in the Cu–Cl cycle, based on a formulation of nonequilibrium statistical thermodynamics. This includes HCl/steam mixtures, Cu(I) and Cu(II) chloride, and cupric oxide.

X-ray absorption near edge structure (XANES) studies were performed at the Advanced Photon Source (APS) at ANL to determine reaction rate constants. Relevant kinetic and mechanistic data for the decomposition of $CuCl_2$ and Cu_2OCl_2 and the hydrolysis of $CuCl_2$ was obtained.

Collaboration between UOIT and the University of Toronto has identified coatings that provide corrosion resistance to high-temperature molten CuCl. The research has aimed to characterize the coatings, as well as test their surface integrity in molten CuCl. Two different substrates were selected for corrosion-resistant spray coatings: (1) Inconel 625, which is a nickel-based superalloy used both for its high strength and excellent aqueous corrosion resistance, and (2) an austenitic stainless steel AL-6XN with high nickel (24 %) and molybdenum (6.3 %) content that has outstanding resistance to oxidizing chloride solutions.

Electrochemical characterization has been performed at the University of Toronto in collaboration with UOIT for Inconel 625 and AL6XN Stainless Steel samples, samples coated with Diamalloy 4006 metallic layer, and Diamalloy 4006 specimens top coated with either zirconia or alumina. Characterization of the corrosion behavior of surfaces was performed at room temperatures in neutral NaCl and acidified NaCl (up to 5 mol/L HCl). The examination of the as-received specimens will establish a baseline characterization of the passive layer and corrosion characteristics.

AECL in collaboration with UOIT studied the economics of nuclear hydrogen production with the Cu–Cl cycle. For hydrogen production capacity between 10 and 20 t/day it is found that water electrolysis using off peak electricity from nuclear power plants is economically advantageous. However, at higher capacities, the thermochemical plant becomes increasingly advantageous over electrolysis. Electrolysis can take advantage of reduced demand and pricing of electricity during off-peak hours, such as evenings and weekends. A constant base-load production of hydrogen can be maintained with SCWR, but a bypass heat exchanger could redirect steam to the Cu–Cl plant during periods of low electricity demand and price.

The UOIT-led team is developing separate experimental units—one for each process step—that will be integrated to obtain a small pilot-scale demonstration plant at a capacity of about 5 kg hydrogen per day. There are four main units, namely, the hydrolysis reactor (fluidized bed), molten salt reactor, electrolyzer (which generates hydrogen), and spray dryer. Through a spray drying process, water is removed from the aqueous $CuCl_2$ that exits the electrolysis cell to produce solid $CuCl_2(s)$, which is then supplied to the hydrolysis reactor (step 3) to produce copper oxychloride (Cu_2OCl_2) and HCl gas. Spray drying experiments have been performed with a Yamato D-41 spray dryer to produce solid Cu(II) chloride particles from an aqueous $CuCl_2$ solution. The dryer was modified to avoid corrosion caused by the corrosive cupric chloride solution. A titanium two-fluid nozzle was used and stainless steel parts downstream of the nozzle were coated with a thin layer of Viton. The spray-drying chamber has a diameter of 0.45 m, and a 1.0 m cylindrical height. Also, a scrubber was used to clean the flue gases of cupric chloride particles, which were not captured inside the cyclone. The drying air entered the dryer at the top and exited through the cyclone separator and finally the scrubber. A peristaltic pump was used to transport the atomization liquid to the two-fluid nozzle. Flow meters were installed to measure the atomized airflow rate. The mean cupric chloride particle size was about 31 μm. A design challenge has been the low solid concentration in the feed and high energy requirement for powder production.

Crystallization was also investigated for drying $CuCl_2$. Dimensions of dentritic crystals are typically in the range of 0.1–6 cm. About 20 % of the $CuCl_2$ can be crystallized in 4 h from 80 to 20 °C. Hydrolysis experiments were performed in a loop successfully tested for stable operation and production of copper oxychloride from solid feed of Cu(II) chloride. Decomposition experiments of copper oxychloride were also conducted at a capacity of about 1,000 times larger than the initial tests at ANL. The Gibbs free energy of the reaction from ambient temperature up to 550 °C was determined with results in the range of −10 to 70 kJ/mol. Heat recovery from hot streams of substances specific to the Cu–Cl cycle was thoroughly investigated with ASPEN Plus; a network of heat exchangers was developed. Additional research performed up to the present time includes system unit integration, exergo-economic modeling, life cycle assessment, utilizing waste heat, upgrading temperature levels using heat pumps, chemical potentials and solubility data, synthesis of copper oxychloride, anode electrode structures, applications of the Cu–Cl cycle for upgrading oil sand bitumen, hydrogen for the transportation sector, and eventual transition to a hydrogen economy.

5.11 Conclusions

In this chapter, thermochemical water-splitting systems are discussed as one of the most promising methods of nuclear hydrogen production at a large scale. A historical perspective is presented on thermochemical cycle development, as well as the original ideas and the main concepts. The fundamental concepts regarding identification, design, and analysis of these cycles are introduced. First, thermochemical water splitting in a single reaction step (thermolysis) is discussed and the necessary chemical thermodynamics and reaction kinetic concepts are explained. Thermolysis is difficult to apply in practice because it requires high temperatures and complicated separation processes, although the chemical reaction occurs in a single step. Two-step thermochemical cycles were introduced as a means to split water at significantly lower temperatures than thermolysis and with intrinsic hydrogen and oxygen separation. Pure thermochemical cycles of two steps still have a high process temperature that cannot be coupled to nuclear reactors to generate hydrogen.

Multistep thermochemical cycles are treated next as an effective means to generate nuclear hydrogen at a large scale. It was found that the most frequent chemical elements encountered in pure thermochemical cycles are iron, chlorine, and sulfur. Not the same chemical elements are dominant in hybrid cycles. Chlorine and copper appear to be the most encountered chemical elements in hybrid cycles. Some special hybrid cycles which use photonic or nuclear radiation as inputs are also presented. A comprehensive treatment is included for chemical processes, including aspects of identification of chemical reactions with good yield and elimination of those with unpractical yield. The main thermochemical and electrochemical reactions are analyzed individually or as classes. The important issues are explained and discussed. The Deacon reaction appears to be one of the most used hydrolysis processes, coupled with oxygen evolution. Other important classes of reactions are acid decomposition (e.g., H_2SO_4 or HI), chlorination, halogenation, and metal oxidation with hydrogen evolution.

Separation of chemicals and issues regarding chemical recycling are discussed and modeled in several parts of the chapter. Accounting for these processes is significantly important for determining realistic values of cycle efficiency. In a down selection process—as explained in detail in Sect. 5.7—thermochemical cycles are assessed gradually by starting with simpler analyses which assume complete reactions, full recovery and reuse of the rejected heat, as well as full recycling of chemicals. These efficiency estimates offer a first cycle ranking criterion for down selection. More elaborate analyses such as the pinch point or flow sheets are applied to a selected number of cycles that are considered to be the most promising cycles. Other methods for cycle selection are introduced such as "comparative assessment" based on various criteria, flow-sheet analysis and cycle assessment via energy and exergy efficiencies, or assessment by cost analyses.

Development aspects of three main thermochemical plants—based on the sulfur–iodine, hybrid sulfur, and copper–chlorine cycles—are presented also. For the copper–chlorine cycle, which is among the emerging and most promising cycles, an upcoming full chapter will describe the cycle in more detail.

Nomenclature

A	Pre-exponential factor
c	Molar concentration, kmol/m^3
ex	Specific molar exergy, kJ/mol
Ex	Exergy, kJ
G	Molar Gibbs free energy, kJ/mol
H	Molar enthalpy, kJ/mol
HHV	Molar based higher heating value, kJ/mol
IP	Improvement potential, kJ
k	Rate constant, s^{-1}
K_{eq}	Equilibrium constant
m	Mass, kg
n	Number of moles
$\dot{n}$	Molar flow rate, mol/s
P	Pressure, Pa
Q	Heat flux, kJ
$\dot{Q}$	Heat flux, kW
r	Recycling ratio
R	Universal gas constant, J/mol K

S	Molar entropy, kJ/mol K
SI	Sustainability index
T	Temperature, K
v	Molar volume, m^3/kmol
W	Work, kJ
y	Molar fraction

Greek Letters

η	Energy efficiency
μ	Chemical potential, kJ/mol
ψ	Exergy efficiency
ξ	Extent of reaction

Subscripts

0	Reference state
act	Activation
aux	Auxiliary
b	Backward
d	Destruction
el	Electric
elchem	Electrochemical
eq	Equivalent
f	Forward, formation
gen	Generation
in	Inlet, input
loss	Losses
out	Output
P	Products
R	Reactants
sep	Separation
tresh	Threshold

Superscripts

0	Reference state
circ	Circulation
ch	Chemical
Q	Heat

References

Abanades S, Charvin P, Flamant G, Neveu P (2006) Screening of water-splitting thermo-chemical cycles potentially attractive for hydrogen production by concentrated solar energy. Energy 31: 2469–2486

Andress RJ, Martin LL (2010) On the synthesis of hydrogen production alternative thermochemical cycles with electrochemical steps. Int J Hydrogen Energy 35:958–965

Balta MT, Dincer I, Hepbasli A (2010) Potential methods for geothermal-based hydrogen production. Int J Hydrogen Energy 35:4949–4961

Bamberger CE (1978) Hydrogen production from water by thermochemical cycles; a 1977 update. Cryogenics 18:170–183

Beghi GE (1986) A decade of research on thermochemical hydrogen at the joint research centre, Ispra. Int J Hydrogen Energy 11:761–771

Bilgen E, Bilgen C (1982) Solar hydrogen production using two-step thermochemical cycles. Int J Hydrogen Energy 7:637–644

Brecher LE, Spewock S, Warde CJ (1977) The Westinghouse sulfur cycle for the thermochemical decomposition of water. Int J Hydrogen Energy 2:7–15

Brown LC, Besenbruch GE, Schultz KR, Showalter SK, Marshall AC, Pickard PS, Funk JF (2002) High efficiency generation of hydrogen fuels using thermochemical cycles and nuclear power. General Atomics Report GA-A24326

Carty RH, Cogner WL (1980) A heat penalty and economic analyses of the hybrid sulfuric acid process. Int J Hydrogen Energ 5:7–20

Carty RH, Mazumder MM, Schreider JD, Pangborn JB (1981) Thermochemical hydrogen production, vols. 1–4. Gas Research Institute, Chicago, IL, Report GRI-80/0023

Chao RE (1974) Thermochemical water decomposition process. Industrial Eng Chem, Process Res Develop 13:94–101

Chikazawa Y, Nakagiri T, Konomura M (2006) A system design study of fast breeder reactor hydrogen production plant using thermochemical and electrolytic hybrid process. Nuclear Technol 155:340–349

Dincer I, Balta MT (2011) Potential thermochemical and hybrid cycles for nuclear-based hydrogen production. Int J Energy Res 35:123–137

Dincer I, Zamfirescu C (2011) Sustainable energy systems and applications. Springer, New York

Dokyia M, Kotera Y (1976) Hybrid cycle with electrolysis using Cu-Cl system. Int J Hydrogen Energy 1:117–121

Dokyia M, Fukuda K, Kameyama T, Kotera Y, Asakura S (1977) The study of thermochemical hydrogen preparation. (II) Electrochemical hybrid cycle using sulphur-iodine system. Denki Kagaku (Electrochemistry, Jpn) 45:139–143

Dokyia M, Kameyama T, Fukuda K (1979) Thermochemical hydrogen preparation—Part V. A feasibility study of the sulphur iodine cycle. Int J Hydrogen Energy 4:267–277

Engineering Vilage (2012) http://www.engineeringvillage2.org. Accessed in March 2012

Ewan BCR, Allen RWK (2005) Assessing the Efficiency Limits for Hydrogen Production by Thermochemical Cycles. AIChE annual meeting, Cincinnati 30 October–4 November 2005, Paper 210c

Fishtik I, Datta R (2008) Systematic generation of thermochemical cycles for water splitting. Comput Chem Eng 32:1625–1634

Funk JE (2001) Thermochemical hydrogen production: past and present. Int J Hydrogen Energy 26:158–190

Funk JE, Reinstrom RM (1964) Final report energy depot electrolysis systems study. TID 20441 (EDR 3714), Vol. 2, Suppl. A.

Funk JE, Reinstrom RM (1966) Energy requirement in the production of hydrogen from water. Ind Eng Chem Process Des Develop 5:336–342

Gorensek MG, Summers WA (2009) Hybrid sulfur flowsheets using PEM electrolysis and a bayonet decomposition reactor. Int J Hydrogen Energy 34:4097–4114

Grimes PG (1966) Energy depot fuel production and utilization. SAE Transactions 74:65001

Kamita N, Ohta T, Asano N (1984) Hybridized hydrogen production system with Fe-I photochemical reaction in concentrated phosphoric acid. Int J Hydrogen Energy 9:563–570

Kasahara S, Kubo S, Hino R, Onuki K, Nomura M, Nakao S (2007) Flow-sheet study of the thermochemical water splitting iodine-sulfur process for effective hydrogen production. Int J Hydrogen Energy 32:489–496

Klein SA (2011) Engineering equation solver. http://www.fchart.com/assets/downloads/ees_manual.pdf. Accessed in March 2012

Knoche KF, Cremer H, Breywisch D, Hegels S, Steinborn G, Wüster G (1978) Experimental and theoretical investigation of thermochemical hydrogen production. Int J Hydrogen Energy 3:209–216

Lede J, Lapicque F, Villermaux J, Gales B, Ounalli A, Baumard JF, Anthony AM (1982) Production of hydrogen by direct thermal decomposition of water: preliminary investigations. Int J Hydrogen Energy 7:939–950

Lewis MA, Masin JG, O'HAre PA (2009) Evaluation of alternative themochemical cycles, Part I: the methodology. Int J Hydrogen Energ 34:4115–4124

Lewis MA, Masin JG (2009) The evaluation of alternative thermochemical cycles – Part II: The down-selection process. Int J Hydrogen Energy 34:4125–4135

Lu PWT, Garcia ER, Ammon RL (1981) Recent developments in the technology of sulfur dioxide depolarized electrolysis. J Appl Electrochem 11:347–355

Marin GD, Wang Z, Naterer GF, Gabriel K (2011) Byproducts and reaction pathways for integration of the CueCl cycle of hydrogen production. Int J Hydrogen Energy 36:13414–13424

McQuillan BW, Brown LC, Besenbruch GE, Tolman R, Cramer T, Russ BE, Vermillion BA, Earl B, Hsieh H-T, Chen Y, Kwan K, Diver R, Siegal N, Weimer A, Perkins C, Lewandowski A (2010) High efficiency generation of hydrogen fuels using solar thermal-chemical splitting of water. General Atomics Project 3022

Nakamura T (1977) Hydrogen production from water utilising solar heat at high temperatures. Solar Energy 19:467–475

Naterer GF, Suppiah S, Stolberg L, Lewis M, Ferrandon M, Wang Z, Dincer I, Gabriel K, Rosen MA, Secnik E, Easton EB, Trevani L, Pioro I, Tremaine P, Lvov S, Jiang J, Rizvi G, Ikeda BM, Luf L, Kaye M, Smith WR, Mostaghimi J, Spekkens P, Fowler M, Avsec J (2011a) Clean hydrogen production with the Cu-Cl cycle – Progress of international consortium, I: Experimental unit operations. Int J Hydrogen Energy 36:15472–15485

Naterer GF, Suppiah S, Stolberg L, Lewis M, Ferrandon M, Wang Z, Dincer I, Gabriel K, Rosen MA, Secnik E, Easton EB, Trevani L, Pioro I, Tremaine P, Lvov S, Jiang J, Rizvi G, Ikeda BM, Luf L, Kaye M, Smith WR, Mostaghimi J, Spekkens P, Fowler M, Avsec J (2011b) Clean hydrogen production with the Cu-Cl cycle – Progress of international consortium, II: Simulations, thermochemical data and materials. Int J Hydrogen Energy 36:15486–15501

Nomura M, Nakao S, Okuda H, Fujiwara S, Kasahara S, Ikenoya K, Kubo S, Onuki K (2004) Development of an electrochemical cell for efficient hydrogen production through the IS process. AIChE J 50:1991–1998

Ohta T, Asakura S, Yamaguchi M, Kamiya N, Gotoh H, Otagawa T (1976) Photochemical and thermoelectric utilisation of solar energy in a hybrid water splitting system. Int J Hydrogen Energy 1:113–116

Perret R (2011) Solar thermochemical hydrogen production. Thermochemical cycle selection and investment priority. Sandia National Laboratory, Report 3622

Rosen MA (2008) Exergy analysis of hydrogen production by thermochemical water decomposition using the Ispra Mark-10 Cycle. Int J Hydrogen Energy 33:6921–6933

Rosen MA (2010) Advances in hydrogen production by thermochemical water decomposition: A review. Energy 35:1068–1076

Rosen MA, Scott DS (1992) Exergy analysis of hydrogen production from heat and water by electrolysis. Int J Hydrogen Energy 17:199–204

Sakurai M, Bilgen E, Tsutsumi A, Yoshida K (1996) Adiabatic UT-3 thermochemical process for hydrogen production. Int J Hydrogen Energy 21:865–870

Sato S (1979) Thermochemical hydrogen production. In: Otha T (ed) Solar-hydrogen energy systems. Pergamon, New York

Savage RL, Blank L, Cady T, Cox K, Murray R, Dee Williams R (1973) A hydrogen energy carrier. Systems Design Institute, NASA Grant NGT 44-005-114

Sivasubramanian P, Ramasamy RP, Freire FJ, Holland CE, Weidner JW (2007) Electrochemical hydrogen production from thermochemical cycles using a proton exchange membrane electrolyzer. Int J Hydrogen Energ 32:463–468

Verfonden K (2007) Nuclear energy for hydrogen production. Writings of Research Center Jülich, Energy Technology, Volume 58, Research Center Jülich GmbH: Jülich(Germany). http://juwel.fz-juelich.de:8080/dspace/bitstream/2128/2518/1/Energietechnik_58.pdf. Accessed in March 2012

Von Federsdorff CG (1974) Non-fossil fuel process for production of hydrogen and oxygen. US Patent 3,802,993

Williams LO (1980) Hydrogen power. An Introduction to hydrogen energy and its applications. Pergamon, New York

Wang ZL, Naterer GF, Gabriel KS, Gravelsins R, Daggupati V (2010) Comparison of sulfur-iodine and copper-chlorine thermochemical hydrogen production cycles. Int J Hydrogen Energy 35:4820–4830

Yan XL, Hino R (2011) Nuclear hydrogen production. CRC Press, Boca Raton

Zamfirescu C, Naterer GF, Dincer I (2010) Novel CuCl vapor compression heat pump integrated with a thermochemical water splitting cycle. Thermochimica Acta 512:40–48

Zamfirescu C, Naterer GF, Dincer I (2012) Solar light-based hydrogen production systems. In: Anwar S (eds) Encyclopedia of energy engineering and technology. Taylor Francis http://www.tandfonline.com/doi/abs/10.1081/E-EEE-120047413. Accessed in February 2013

Chapter 6
Hybrid Copper–Chlorine Cycle

Abstract In this chapter, the hybrid copper–chlorine (Cu–Cl) cycle for hydrogen production is examined in detail. The historical perspective of this cycle development is presented in Sect. 6.1. A precursor of the cycle was proposed in 1974, which uses a non-electrochemical, non-thermochemical disproportionation of cuprous chloride; this process is based on complexation and chelating schemes that generate the desired products. Electrochemical hydrogen generation from hydrochloric acid and cuprous chloride electrolysis is one of the latest cycle developments for engineering scale-up. This process simplifies the separation steps and it has been proven by test-bench experiments. Two reactors were mainly studied for the hydrolysis reaction, which is a crucial cycle step: fluidized bed and spray reactor. Both are interesting schemes proposed for scaling up the cycle. At the University of Ontario Institute of Technology, a scaled up laboratory facility has been developed for each cycle step.

In total, seven cycle variants are examined in this chapter. The variants, including a copper electrowinning step, were studied mostly since 2003; much progress has been made in the development of the processes. Because electrowinning implies difficult separation of chemicals, it appears less feasible for large-scale implementation.

This chapter presents a detailed treatment of all relevant processes, such as electrochemical disproportionation, electrochemical chlorination, complexation, dehydration, drying, crystallization, hydrolysis, fluidized bed hydrodynamics and heat transfer, spray drying hydrodynamics and heat transfer, multiphase processes in reactors, thermochemical chlorination, thermolysis, heat recovery from molten salt, and special heat exchangers, as well as system integration of the unit operations.

Integration of the chemical plant with heat pumps and heat engines may have promising potential to substantially increase the overall hydrogen production efficiency. An interesting option is to use integrated heat pumps based on thermochemical processes such as a steam–methane reaction, or vapor compression heat pumps with CuCl as a working fluid.

G.F. Naterer et al., *Hydrogen Production from Nuclear Energy*,
DOI 10.1007/978-1-4471-4938-5_6,

Material research is also important for cycle development. Several corrosion-resistant coatings were developed—as explained in the chapter—and assessed experimentally by various methods. Various auxiliary systems are required for a full-scale plant. One item of major interest is separation of hydrochloric acid from an HCl/steam mixture exiting the hydrolysis reactor for purposes of recycling; this aspect is also detailed in a section of this chapter. Various concepts for a full-size plant and its equipment are presented, as well as simulation results with ASPEN Plus software. Other aspects are also discussed such as reliability, control, safety, environmental impact, and life cycle assessment. The overview in this chapter concludes that the Cu–Cl cycle is a promising candidate for large-scale hydrogen production with nuclear reactors.

6.1 Introduction

The analysis presented in the previous chapter shows that the most commonly encountered chemical compounds in hybrid thermochemical cycles are those of chlorine and copper. According to Fig. 5.24, chlorine compounds are present in 40 % of significant hybrid cycles while copper in 32 % of them. In general, the raw materials from which copper and chlorine can be produced are relatively cheap and widely available. Electrolytic processes involving copper or chlorine are relatively well understood and used in various industries. These are some of the reasons in favor of selecting a hybrid copper–chlorine thermochemical water-splitting cycle to generate hydrogen with nuclear energy. The low temperature requirement of these cycles—around 800 K—represents their strongest advantage. A lower temperature level allows for linking the thermochemical plant with a wider range of nuclear reactors including those of intermediate temperature—such as supercritical water-cooled reactor (SCWR)—and those of high temperature as well—for example, the helium-cooled high-temperature reactor (HTGR), modular temperature reactor (MHR), very-high-temperature reactor (VHTR), etc.

The copper–chlorine thermochemical cycle—with several variants—is under development in several institutions around the world. The US Department of Energy envisages the use of this cycle in conjunction with concentrated solar power systems to generate hydrogen. In Canada, Atomic Energy of Canada Limited (AECL) is interested in this cycle development for their planned SCWR which generates heat at 800–900 K. In France, CEA (Commisariat à Energy Atomique et aux Energies Alternatives) has conducted research on this cycle both for concentrated solar and nuclear energy purposes.

A precursor of the hybrid copper–chlorine cycle was proposed by Wentorf from General Electric and a collaborator Hanneman (1974). They developed a process that includes a cuprous chlorine disproportionation step which is described by the same chemical reaction as the electrolytic step in the standard 5-step copper–chlorine cycle (the Cu–Cl-A cycle listed in Chap. 5, Table 5.16). However, in the approach proposed by Wentorf and Hanneman (1974), this reaction is conducted by physicochemical means rather than electrochemically as per Cu–Cl-A. Namely, the disproportionation of CuCl is performed in two steps: (1) complexation of CuCl according to $4CuCl + \chi \rightarrow 2Cu + 2CuCl_2 \cdot \chi$, followed by (2) thermal decomposition of the $CuCl_2$ complex $2CuCl_2 \cdot \chi \rightarrow 2CuCl_2 + \chi$, where χ is a chelating agent. As proposed, the complexation is performed with propylenediamine which acts as a chelating agent that selectively dissolves $CuCl_2$. Another proposed agent suggested by Wentorf and Hanneman (1974) for complexation of CuCl is aqueous hydrochloric acid with a moderate concentration so that hydrolysis is prevented and the solubility of CuCl is reduced. The third suggested process in Wentorf and Hanneman (1974) is to use an ion-exchange resin capable of absorbing copper ions and to regenerate $CuCl_2$ by a hydrochloric acid solution. Such physicochemical processes for the disproportionation step may have the advantage of not requiring electricity, although some mechanical work consumption is still necessary for separation and extraction processes. The possibility of using an electrochemical process for the disproportionation step was not envisioned in the early work by Wentorf and Hanneman (1974).

A copper–chlorine cycle comprising an electrochemical process was first proposed by Dokyia and Kotera (1976) as a two-step cycle comprising an electrolytic process of cathodic reduction of hydrochloric acid to generate hydrogen, and oxidation of metallic copper at the anode from Cu^{+} to Cu^{2+}. The closure reaction is thermo-hydrolysis of cupric chloride which evolves oxygen at 823 K. This hydrolysis was demonstrated experimentally by Dokyia and Kotera (1976) in a small glass reactor with an attached chromatograph that showed the oxygen evolution. Dokyia and Kotera (1976) suggested also that the hydrolysis reaction can be performed as a two-step process, namely, by thermolysis of cupric chloride followed by a reverse Deacon reaction. If these reaction steps are implemented in two separate reactors, then the cycle becomes a three-step hybrid cycle with one electrochemical and two thermochemical steps.

Further analysis of the hydrolysis reaction of cupric chloride was performed at the Institute of Gas Technology (IGT) by the end of the 1970s. As reported by Carty et al. (1981), they studied hydrolysis of cupric chloride with the aim of reducing the reaction temperature of ferrous chloride hydrolysis within their iron–copper–chloride cycle B-16. In this cycle, ferrous chloride is hydrolyzed in the presence of solid cupric chloride to form a mixed copper oxide–cupric chloride which is a coordination complex, namely, $CuO{\cdot}CuCl_2$. In their first set of experiments with this reaction, ferric chloride was not used and only the hydrolysis of cupric chloride was performed. It was found that the conversion of cupric chloride to mixed copper oxide–cupric chloride (also denoted as copper oxychloride) was essentially complete at 675 K. Thereafter, IGT proposed an oxygen evolution step by thermolysis of copper oxychloride at 800 K. However, according to Carty et al. (1981), oxygen evolution was not confirmed experimentally at IGT at that time, although the thermodynamic analysis showed that the reaction appears feasible. This process was confirmed by experimental work performed much later at the Argonne National Laboratory (ANL) by Lewis et al. (2003).

Furthermore, an electrochemical process was proposed by IGT for cuprous chlorine disproportionation which generates solid copper and cupric chlorine. The process performs electrochemically the same reaction as suggested by Wentorf and Hanneman (1974) through complexation or ion-absorption techniques. This reaction was not tested experimentally by IGT, which only stated that the reaction might be workable with electrochemistry, although the reaction configuration is unknown (Carty et al. 1981). With this electrochemical reaction, the hydrolysis of cupric chloride, thermolysis of mixed copper oxychloride, and a hydrogen evolution reaction, IGT formulated the hybrid cycle denoted as H-6, which is essentially the same as the five-step copper–chlorine cycle adopted later for development at ANL and UOIT. The hydrogen evolution reaction in cycle H-6 is a chlorination of metallic copper with hydrochloric acid which also produces cuprous chloride.

IGT screened 200 over thermochemical cycles and found that only five thermochemical and six hybrid cycles were potential viable candidates for linkage to nuclear energy. The copper–chlorine cycle, H-6, was selected as a promising cycle. Carty et al. (1981) estimated that the maximum attainable cycle efficiency of a Cu–Cl plant coupled with high-temperature nuclear reactors is 38 %, whereas the overall hydrogen production process efficiency is 26 % when the thermochemical plant is coupled to a helium-cooled high-temperature reactor (helium is supplied at 875 K and returned at 365 K). If molten lithium is used as a coolant for the nuclear reactor with a 975 K supply temperature and 625 K return flow temperature, then the overall hydrogen production process efficiency estimated by Carty et al. (1981) becomes 28 %.

In his book, Williams (1980) proposed a purely thermochemical copper–chlorine cycle (also called US-Chlorine) which involves three chemical reactions: a reverse Deacon reaction as hydrolysis which evolves oxygen, cuprous chloride chlorination as the hydrogen evolving reaction, and thermolysis of cupric chlorine as a reaction for recycling of intermediate reagents. The reactions of this cycle are listed in Table 5.10. In the 1990s, the Korean Institute of Energy Research (KIER) identified pure thermochemical cycles with copper and chlorine compounds. Three of the cycles proposed at KIER by Sim et al. (1993) have reactions similar to those in hybrid versions (e.g., hydrolysis of $CuCl_2$ and chlorination of CuCl) but all steps are thermochemical (see cycles KIER-2, KIER-A, KIER-B in Table 5.10). However, none of the purely thermochemical copper–chlorine cycles are suitable for linkage with nuclear reactors due to their higher temperature requirement (above 1,100 K).

Beginning in the 2000s, research on the hybrid copper–chlorine cycle was reinvigorated at ANL where laboratory investigations were performed for each of the reaction steps. In a seminal paper, Lewis et al. (2003) conducted a comprehensive experimental study to establish the thermodynamic feasibility of the cycle, determine the preliminary kinetic data, and assess the importance of side reactions.

The first interest of ANL was toward the IGT H-5 version of the cycle. Vilim (2004) at ANL studied an integrated system consisting of a sodium-cooled fast reactor of 840 MWth (design PRISM/Mod B) and the copper–chlorine thermochemical cycle. The nuclear heat is used in a proportion of 25 % to drive the thermochemical reactions and 75 % to generate power via a steam-based Rankine plant. According to the results of Vilim (2004), the nuclear reactor vessel is 1,400 m^3 while the chlorination reactor represents about half and the oxygen-evolving reactor is about one-quarter in size, whereas the associated costs with the reactor vessels and copper inventory represent less than 10 % of the power plant cost.

A comparative study of four hybrid cycles by ANL concluded that the copper–chlorine cycle is a main candidate for further development because the efficiency is reasonable high, the chemistry experiments have proven the reactions as feasible, and the temperature level for heat input is compatible with next-generation reactors (see Lewis et al. 2005a, b) collaborated with IGT to develop a proof-of-principle demonstration of the electrolytic step which produces metallic copper at the cathode. The main challenges that were identified for electrochemical copper generation are related to the membrane development (to impede undesired crossover of ionic species), finding of suitable electrode materials, and determination of operating conditions to generate a well-concentrated cupric chloride solution at the anode and small dendritic copper particles at the cathode.

The solubility of CuCl in water can be increased by the addition of HCl in order to concentrate the $CuCl_2$ in the output stream. A flow-sheet analysis developed in ASPEN Plus, which included model data validated by chemical lab experiments, was presented in Lewis et al. (2005b). For a production scale of 40,000 N m^3 hydrogen per day, the flow-sheet analysis revealed an efficiency of 63 % based on the higher heating value (HHV) of hydrogen. This figure corresponds to the assumption that electricity is generated with 50 % efficiency and all heat rejected by the cycle is recovered and reused internally. A more realistic flow-sheet efficiency was also presented by Lewis et al. (2005b) based on a pinch analysis which reveals that the cycle efficiency is around 50 % (based on HHV).

Miller (2005) reported AECL's interest in the development of the copper–chlorine cycle in view of integration with future CANDU reactors in collaboration with ANL. The focus at AECL was primarily on the development of the electrolytic step for direct production of hydrogen by CuCl/HCl electrolysis. A first set of results on the electrolytic step developed at AECL was reported by Suppiah et al. (2005). They obtained a current density of ~800 A/m^2 at ~0.7 V with a copper cathode and platinum anode when a solution of 0.3 mol/L CuCl in 4 mol/L HCl was electrolyzed.

The researchers in University of Nevada at Las Vegas (UNLV) performed simulation work and process analysis (Khan and Chen 2005). According to the APEN Plus simulations performed at UNLV, the flow sheet-based efficiency of hydrogen production was 33 %, assuming that the electricity is produced with 50 % energy efficiency.

The University of Ontario Institute of Technology (UOIT) began a long-term program on the copper–chlorine thermochemical cycle development in 2006. The UOIT team published numerous papers since that time, covering general analyses of the hybrid cycle, experimental unit operations, and a scaled-up hydrogen production plant. An integrated engineering-scale demonstration unit of 50 Nm^3

hydrogen per day is presently under development at UOIT. A UOIT-led international team on this cycle development comprises eight institutions, five countries, and six industrial partners (see Table 5.21). This consortium and some selected main research results are summarized in Chap. 5, Sect. 5.10.3.3.

In this chapter, the hybrid copper–chlorine cycle is analyzed in detail. Seven versions of the cycle are identified. The physicochemical properties and their prediction are discussed. Each major cycle step is examined extensively. Heat recovery and reuse are reported. Also, scaled-up design and plant simulations are included.

6.2 Copper–Chlorine Cycle Versions

In past literature, a total of seven versions of the copper–chlorine cycle were reported: 1 cycle with two-step, 2 versions of three-step, 2 cycles with four-step, and 1 five-step cycle. In this section, each of these cycle versions is introduced and diagrams are presented. The knowledge of the processes is essential for cycle analysis and design. There are seven main classes of processes that are encountered in the copper–chlorine cycles. Their nature is thermochemical, electrochemical, physicochemical, or thermophysical. Table 6.1 lists the main classes of processes encountered in copper–chlorine cycles and shows definitions. These classes of processes are complexation, chlorination, oxychlorination, hydrolysis, thermolysis, and drying.

Past literature results show that for the copper–chlorine cycles, the following 12 processes occur: (1) electrochemical disproportionation of CuCl, (2) four distinct hydrolysis reactions, (3) two chlorination reactions of which one is thermochemical and one is electrochemical, (4) one oxychlorination reaction, (5) two thermolysis reactions, and (6) two dehydration processes of hydrated cupric chloride, of which one occurs via spray drying and one is based on crystallization. Table 6.2 presents the 12 distinct processes based on which copper–chlorine cycles can be formed.

Process #1 in Table 6.2 is the electrochemical disproportionation of cuprous chloride which is conducted in an aqueous solution in the presence of hydrochloric acid. In this process, a complexation of cupric chloride occurs, namely, hydrated cupric chloride is produced in the form of the complex $CuCl_2{\cdot}nH_2O$ which is present in the aqueous phase (dissolved). At the same time, copper precipitates as solid particles of micron size.

Processes #2 to #5 are hydrolysis processes. The hydrolysis that evolves at the lowest temperature is process #2 which requires 650 K and combines $2CuCl_2{\cdot}nH_2O(s)$, a hydrated cupric chloride in solid form, with steam to form copper oxychloride and hydrochloric acid. During the process, the moisture from $CuCl_2$ is released as steam. Process #3 is similar to #2; however it uses dried cupric chloride as a reactant and thus no steam is released. The highest temperature hydrolysis process of $CuCl_2$ is process #3 which is conducted at 823 K. This process uses hydrated cupric chloride as reactant and generates oxygen, hydrochloric acid, and molten cuprous chloride.

Table 6.1 Main processes encountered in copper–chlorine thermochemical cycles

Process	Nature	Definition and remarks
Complexation	Physicochemical	Reaction that forms a chemical complex by loose association of two or more molecular entities with bonds weaker than a covalent bond. In the Cu–Cl cycle, complexation of CuCl and $CuCl_2$ occurs.[a]
Chlorination	Thermochemical Electrochemical	Reaction of a chemical compound with chlorine. In the Cu–Cl cycle, chlorine is sourced from HCl. Electrolysis of CuCl/HCl is an electrochemical chlorination of CuCl. Chlorination of Cu is conducted thermochemically.
Oxychlorination	Thermochemical	Chlorination in the presence of oxygen. In the Cu–Cl cycle, oxygen is derived from water during oxychlorination of the complex $CuCl_2{\cdot}H_2O$.
Disproportionation	Electrochemical[b]	Reaction during which a molecule is transformed in two dissimilar chemical compounds. In the Cu–Cl cycle, CuCl is disproportionated electrochemically to form metallic copper and cupric chloride.
Hydrolysis	Thermochemical	Reaction of a chemical compound with water. Depending on the version, Cu–Cl cycle hydrolysis can be conducted at lower temperature when $CuO{\cdot}CuCl_2$ is formed, or at higher temperatures when oxygen is released.
Thermolysis	Thermochemical	Chemical reaction of thermal decomposition of a molecule. In Cu–Cl cycles, thermolysis of $CuO{\cdot}CuCl_2$ or $CuCl_2$ is encountered.
Drying	Thermophysical	Removing moisture from a material. In the Cu–Cl cycle, water is removed from the coordination complexes of cupric chloride with water ($CuCl_2{\cdot}nH_2O$) either by spray drying or crystallization methods.

[a]Thermochemically conducted disproportionation exists, but is not in the scope of the hybrid Cu–Cl cycle

[b]In the Cu–Cl cycle, a coordination complex $CuO{\cdot}CuCl_2$ is also formed during hydrolysis or oxychlorination; also the hydrated complex of $CuCl_2$, namely, $CuCl_2{\cdot}nH_2O$, is generated in the anolyte of the electrochemical cell

In addition, steam from the moisture content of $CuCl_2$ is released. There is also a hydrolysis process of chlorine—encountered in the three-step cycle of Dokyia and Kotera (1976)—which is a reverse Deacon reaction at 1,073 K, noted in Table 6.2 as process #5. A thermochemical chlorination of copper is used in three of the versions of the copper–chlorine cycle as will be described below. Chlorination of copper is performed with hydrochloric acid according to process #6.

Process #7 is an electrochemical process of chlorination in which cuprous chloride is reacted with hydrochloric acid in an aqueous electrolyte. The electrochemical reaction generates hydrogen at the cathode. A complexation process occurs in the electrolyte and forms a hydrated cupric chloride complex in the aqueous phase at the anode, namely, $CuCl_2 \cdot nH_2O(aq), n > 0$. Electrochemical cells for this reaction are in development at UOIT, ANL, AECL, and Gas Technology Institute (GTI, USA). Process #8 is oxychlorination, having as the sole reactant

Table 6.2 Main processes and chemical reactions in copper–chlorine thermochemical cycles

ID	Process	Chemical reactions	*T* (K)
1	CuCl disproportionation	$E: 4CuCl(s) + nH_2O(l) \rightarrow 2Cu(s) + 2CuCl_2 \cdot nH_2O(aq)$, in presence of HCl	350
2	$CuCl_2$ hydrolysis, low *T*	$T: H_2O(g) + 2CuCl_2 \cdot nH_2O(s) \rightarrow CuO \cdot CuCl_2(s) + 2HCl(g) + nH_2O(g)$	650
3	$CuCl_2$ hydrolysis, dry	$T: H_2O(g) + 2CuCl_2(s) \rightarrow CuO \cdot CuCl_2(s) + 2HCl(g)$	673
4	$CuCl_2$ hydrolysis, high *T*	$T: H_2O(g) + 2CuCl_2 \cdot nH_2O(s) \rightarrow 0.5O_2(g) + 2HCl(g) + 2CuCl(l) + nH_2O(g)$	823
5	Chlorine hydrolysis	$T: H_2O(g) + Cl_2(g) \rightarrow 0.5O_2(g) + 2HCl(g)$	1,073
6	Copper chlorination	$T: 2Cu(s) + 2HCl(g) \rightarrow H_2(g) + 2CuCl(l)$	723
7	CuCl chlorination[a]	$E: nH_2O(l) + 2CuCl(s) + 2HCl(aq) \rightarrow H_2(g) + 2CuCl_2 \cdot nH_2O(aq)$	<473[b]
8	Oxychlorination	$T: 2CuCl_2 \cdot nH_2O(s) \rightarrow CuO \cdot CuCl_2(s) + 2HCl(g) + (n-1)H_2O(g)$	800
9	CuO-$CuCl_2$ thermolysis	$T: CuO \cdot CuCl_2(s) \rightarrow 0.5O_2(g) + 2CuCl(l)$	800
10	$CuCl_2$ thermolysis	$T: 2CuCl_2 \cdot nH_2O(s) \rightarrow 2CuCl(l) + Cl_2(g) + nH_2O(g)$	773
11	Drying (spray)	$T: 2CuCl_2 \cdot nH_2O(s) \rightarrow 2CuCl_2 \cdot mH_2O(s) + (n-m)H_2O(g), \quad n > m$	<530
12	Drying (crystallization)	$T: 2CuCl_2 \cdot nH_2O(aq) \rightarrow 2CuCl_2(s) + nH_2O(l)$	<373

ID process identification number, *E* electrochemical, *T* thermochemical

[a]This is an electrochemical chlorination process, also denoted as CuCl/HCl electrolysis

[b]473 K in process by Dokyia and Kotera (1976); ~350 K in other implementations(see Naterer et al. 2011a)

the complex $CuCl_2 \cdot nH_2O(s)$. In this reaction, copper oxychloride is formed together with gaseous hydrochloric acid. The thermolysis of copper oxychloride is described by process #9 in Table 6.2. This is a thermal decomposition that releases oxygen and forms molten cuprous chloride. Thermolysis of hydrated cupric chloride is also encountered as described by process #10.

Dehydration of wet cupric chloride can be achieved by spray drying, in which the moisture content is reduced partially or totally as indicated in Table 6.2 by process #11. This process requires heat at temperature below 573 K. A crystallization process is also possible at lower temperatures, namely, below 373 K. In this case, anhydrous cupric chloride is obtained according to process #12, listed in Table 6.2.

There are some simplified forms to describe processes #1 to #12 listed in Table 6.2 which can be found in various sources. For example, the electrolytic process #1—disproportionation—is described by the following simplified equation:

$$4CuCl(aq) \rightarrow 2Cu(s) + 2CuCl_2(aq). \tag{6.1}$$

The reaction is conducted in an aqueous solution of HCl. Equation (6.1) does not indicate how many molecules of water are made in the complexation of cupric chloride. This detail is not necessary for a screening of the cycle; however it is important for accurate analysis and design. In the same manner, processes #2, #3, #4, and #8 can be simplified by the following equation, which corresponds exactly to process #3:

$$H_2O(g) + 2CuCl_2(s) \rightarrow CuO \cdot CuCl_2(s) + 2HCl(g). \tag{6.2}$$

Process #7 can be written as

$$2CuCl(s) + 2HCl(aq) \rightarrow H_2(g) + 2CuCl_2. \tag{6.3}$$

Moreover, the dehydration processes #10 and #11 can be represented by the following simplified equation, namely:

$$2CuCl_2(aq) \rightarrow 2CuCl_2(s). \tag{6.4}$$

The five-step hybrid cycle can be described in a non-detailed manner using (6.1)–(6.4) and (6.5) as follows:

$$CuO \cdot CuCl_2(s) \rightarrow 0.5O_2(g) + 2CuCl(l). \tag{6.5}$$

Therefore, the five-step hybrid copper–chlorine cycle is characterized by the following set of simplified equations:

Table 6.3 Definitions of copper–chlorine thermochemical cycles and their main processes

	Process ID (see Table 6.2)											
Cycle ID	1	2	3	4	5	6	7	8	9	10	11	12
Cu–Cl-2	No	No	No	Yes	No	No	Yes	No	No	No	No	No
Cu–Cl-3A	Yes	No	No	Yes	No	Yes	No	No	No	No	No	No
Cu–Cl-3B	No	No	No	No	No	No	Yes	Yes	Yes	No	No	No
Cu–Cl-3C	No	No	No	No	Yes	No	Yes	No	No	Yes	No	No
Cu–Cl-4A	Yes	No	No	No	No	Yes	No	Yes	Yes	No	No	No
Cu–Cl-4B	No	No	Yes	No	No	No	Yes	No	Yes	No	No	Yes
Cu–Cl-5	Yes	Yes	No	No	No	Yes	No	No	Yes	No	Yes	No
Enc. (%)	43	14	14	28	14	43	57	28	57	14	14	14

Enc. processed encountered with respect to the total number of cycles (7); e.g., process 1 encountered in three cycles which is 43 % of the total number of cycles

$$\begin{cases} H_2O(g) + 2CuCl_2(s) \rightarrow CuO \cdot CuCl_2(s) + 2HCl(g) \\ 2CuCl(s) + 2HCl(aq) \rightarrow H_2(g) + 2CuCl_2 \\ CuO \cdot CuCl_2(s) \rightarrow 0.5O_2(g) + 2CuCl(l) \\ 4CuCl(aq) \rightarrow 2Cu(s) + 2CuCl_2(aq) \\ 2CuCl_2(aq) \rightarrow 2CuCl_2(s). \end{cases} \tag{6.6}$$

In this five-step thermochemical cycle, water is fed in the gaseous phase (as steam) to the hydrolysis reaction. In the process, all the intermediate reagents are recycled, but steam, which is consumed, and hydrogen and oxygen are delivered as products. All versions of the copper–chlorine cycle proposed in past literature are described compactly in Table 6.3 where the processes are indicated for each cycle. An identification coding for the cycles is shown in Table 6.3 which consists of "Cu–Cl" followed by the number of steps and a letter "A," "B," or "C" to indicate the cycle version. For example, as indicated in the table, cycle Cu–Cl-2 is a two-step version consisting of processes #4 and #7; the processes are those listed in Table 6.2. Cycle Cu–Cl-3A is a three-step cycle with the steps being processes #1, #4, and #6. The Cu–Cl-5 cycle, also described by simplified equation (6.6), comprises processes #1, #2, #6, #9, and #11.

It can be observed from Table 6.3 that the processes which are encountered mostly in copper–chlorine cycle versions are #7, namely, the electrochemical chlorination of CuCl with hydrochloric acid, and process #9, namely, the thermolysis of copper oxychloride. These processes occur in four of seven cycle versions. The electrochemical process of CuCl/HCl electrolysis (namely, #7) is used in four cycles while the other electrochemical process (CuCl disproportionation, process #1) is encountered in only three cycles.

Table 6.4 lists for each of the cycles the proponent, main references, and alternative cycle identifications as proposed in past literature. As indicated in the table, the cycle Cu–Cl-2 was proposed by Dokyia and Kotera (1976) and it has an alternative identification in Lewis and Masin (2009) as Cu–Cl-C. The three-step cycle Cu–Cl-3A is also called ANL-287 in Abanades et al. (2006), as UNLV-191 in

Table 6.4 Cycle proponents and alternative identifications of Cu–Cl cycle versions

ID	T_{max} (K)	Proponent	Alternative identifications
Cu–Cl-2	823	NCLI, Dokyia and Kotera (1976)	Cu–Cl-C, in Lewis and Masin (2009)
Cu–Cl-3A	800	ANL, Perret et al. (2005)	ANL-287 in Abanades et al. (2006) UNLV-191 in Perret et al. (2005) Three-step cycle (UOIT) in Wang et al. (2009)
Cu–Cl-3B	800	AECL and ANL, Suppiah et al. (2005)	Cu–Cl-B in Lewis and Masin (2009) Three-step cycle (UOIT) in Naterer et al. (2010)
Cu–Cl-3C	1,073	NCLI, Dokyia and Kotera (1976)	UNLV-56, in McQuillan et al. (2010) Cu–Cl-D in Lewis and Masin (2009)
Cu–Cl-4A	800	UOIT, Wang et al. (2009)	Four-step cycle (UOIT) in Wang et al. (2009)
Cu–Cl-4B	800	UOIT, Naterer et al. (2010)	Four-step cycle (UOIT) in Naterer et al. (2010)
Cu–Cl-5	800	IGT, Carty et al. (1981)	H-5 in Carty et al. (1981) ALTC-1 in Lewis et al. (2003) Cu–Cl-A in Lewis and Masin (2009) Five-step cycle (UOIT) in Wang et al. (2009)

NCLI - National Chemical Laboratory for Industry (Tokyo), ANL - Argonne National Laboratory, AECL - Atomic Energy Canada Limited, UOIT - University of Ontario Institute of Technology, IGT - Institute of Gas Technology

Perret et al. (2005). Other alternative identifications of the copper–chlorine cycles are Cu–Cl-A, Cu–Cl-B, Cu–Cl-C, UNLV-56, and H-5.

The full description of all cycles is presented in Table 6.5 which indicates the cycle identification (CID), the reactions for each step, their IDs, and their associated temperatures. In the table, the first step for each cycle represents the chemical reaction to which water is supplied. It can be observed that for cycles Cu–Cl-2, Cu–Cl-3A, Cu–Cl-3C, Cu–Cl-4B, and Cu–Cl-5, water is supplied as steam to the hydrolysis process. This implies that water is preheated, boiled, and superheated separately or in an integrated system prior to its supply to the reaction in the gaseous phase. The other two cycles, namely, Cu–Cl-3B and Cu–Cl-4A, do not require water boiling because water is supplied as a liquid to the electrochemical process where complexation occurs with cupric chloride and then it is heated up to the required temperature for oxychlorination.

The operating principle of each cycle is presented schematically in the diagrams shown in Fig. 6.1. These diagrams indicate the processes and main flows of chemicals among the processes. The diagrams in Fig. 6.1 present in a visual manner the same information as presented through the chemical equations in Table 6.5. The useful value of the diagram is to visually show the flow of chemicals between processes.

The conceptual flowcharts for the cycles, which are introduced next, describe the cycle in a more detailed manner than the schematic representations presented above. The conceptual flowcharts include chemical reactors, electrochemical

Table 6.5 Hybrid copper–chlorine thermochemical cycles for water splitting

CID	Reactions	ID	Chemical equation	T (K)
Cu–Cl-2	Hydrolysis	4	$T: H_2O(g) + 2CuCl_2 \cdot nH_2O(s) \rightarrow 0.5O_2(g) + 2HCl(g) + 2CuCl(l) + nH_2O(g)$	823
	CuCl chlorination	7	$E: 2CuCl(s) + 2HCl(aq) + nH_2O(l) \rightarrow H_2(g) + 2CuCl_2 \cdot nH_2O(aq)$	473
Cu–Cl-3A	Hydrolysis	4	$T: H_2O(g) + 2CuCl_2 \cdot nH_2O(s) \rightarrow 0.5O_2(g) + 2HCl(g) + 2CuCl(l) + nH_2O(g)$	823
	Cu chlorination	6	$T: 2Cu(s) + 2HCl(g) \rightarrow H_2(g) + 2CuCl(l)$	723
	Disproportionation	1	$E: nH_2O(l) + 4CuCl(s) \rightarrow 2Cu(s) + 2CuCl_2 \cdot nH_2O(aq)$	350
Cu–Cl-3B	CuCl chlorination	7	$E: nH_2O(l) + 2CuCl(s) + 2HCl(aq) \rightarrow H_2(g) + 2CuCl_2 \cdot nH_2O(aq)$	350
	Thermolysis	9	$T: CuO \cdot CuCl_2(s) \rightarrow 0.5O_2(g) + 2CuCl(l)$	800
	Oxychlorination	8	$T: 2CuCl_2 \cdot nH_2O(s) \rightarrow CuO \cdot CuCl_2(s) + 2HCl(g) + (n-1)H_2O(g)$	673
Cu–Cl-3C	Cl_2 hydrolysis	5	$T: H_2O(g) + Cl_2(g) \rightarrow 0.5O_2(g) + 2HCl(g)$	1,073
	CuCl chlorination	7	$E: 2CuCl(s) + 2HCl(aq) + nH_2O(l) \rightarrow H_2(g) + 2CuCl_2 \cdot nH_2O(aq)$	473
	$CuCl_2$ thermolysis	10	$T: 2CuCl_2 \cdot nH_2O(s) \rightarrow 2CuCl(l) + Cl_2(g) + nH_2O(g)$	773
Cu–Cl-4A	Disproportionation	1	$E: nH_2O(l) + 4CuCl(s) \rightarrow 2Cu(s) + 2CuCl_2 \cdot nH_2O(aq)$	350
	Cu chlorination	6	$T: 2Cu(s) + 2HCl(g) \rightarrow H_2(g) + 2CuCl(l)$	723
	Thermolysis	9	$T: CuO \cdot CuCl_2(s) \rightarrow 0.5O_2(g) + 2CuCl(l)$	800
	Oxychlorination	8	$T: 2CuCl_2 \cdot nH_2O(s) \rightarrow CuO \cdot CuCl_2(s) + 2HCl(g) + (n-1)H_2O(g)$	650
Cu–Cl-4B	Hydrolysis, dry	3	$T: H_2O(g) + 2CuCl_2(s) \rightarrow CuO \cdot CuCl_2(s) + 2HCl(g)$	673
	CuCl chlorination	7	$E: 2CuCl(s) + 2HCl(aq) + nH_2O(l) \rightarrow H_2(g) + 2CuCl_2 \cdot nH_2O(aq)$	350
	Thermolysis	9	$T: CuO \cdot CuCl_2(s) \rightarrow 0.5O_2(g) + 2CuCl(l)$	800
	Crystallization	12	$T: 2CuCl_2 \cdot nH_2O(aq) \rightarrow 2CuCl_2(s) + nH_2O(l)$	<373
Cu–Cl-5	Hydrolysis	2	$T: H_2O(g) + 2CuCl_2 \cdot mH_2O(s) \rightarrow CuO \cdot CuCl_2(s) + 2HCl(g) + mH_2O(g)$	650
	Cu chlorination	6	$T: 2Cu(s) + 2HCl(g) \rightarrow H_2(g) + 2CuCl(l)$	723
	Thermolysis	9	$T: CuO \cdot CuCl_2(s) \rightarrow 0.5O_2(g) + 2CuCl(l)$	800
	Disproportionation	1	$E: 4CuCl(s) + nH_2O(l) \rightarrow 2Cu(s) + 2CuCl_2 \cdot nH_2O(aq)$	350
	Drying (spray)	11	$T: 2CuCl_2 \cdot nH_2O(aq) \rightarrow 2CuCl_2 \cdot mH_2O(s) + (n-m)H_2O(g), \quad n > m$	<530

CID cycle identifier format is Cu–Cl-Ns(-V), where Ns is the number of process steps and V is subversion identified with letters A, B, or C; *g* gas, *l* liquid (or molten), s solid, · (bullet) indicates a chemical complex, *E* electrochemical reaction, *T* thermochemical reaction or physical process driven by heat

Fig. 6.1 Schematic representation of hybrid copper–chlorine cycles

reactors, dehydration reactors, heat exchangers, and separators. Various devices such as conveyors, pumps, compressors, and other auxiliary components are omitted. Also, the heat recovery and recycling systems are not shown. Full details regarding the cycles and their processes are given in subsequent sections of this chapter.

The conceptual flowchart for the two-step water splitting cycle Cu–Cl-2 is presented in Fig. 6.2. In this system, liquid water is supplied in #1, which is converted to superheated steam in #2 at a temperature that corresponds to the hydrolysis reaction. In the reactor R, the hydrolysis of $CuCl_2$ occurs at a high temperature according to process #4 (see Table 6.5). Hot gases, namely, oxygen, hydrochloric acid, and steam, are extracted at location #3. This mixture is cooled to a temperature close to the ambient in Hx2 and delivered as stream #4 to the liquid

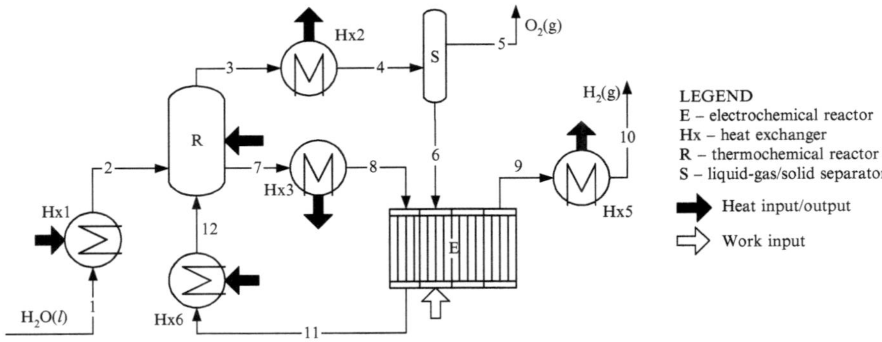

Fig. 6.2 Conceptual flowchart for Cu–Cl-2 cycle (two steps)

gas separator S. Then water is present as liquid and hydrochloric acid in the aqueous phase, while oxygen separates as a gas and it is extracted at #5. The liquid of the aqueous hydrochloric solution and water are transferred to the electrochemical cell, E. In the hydrolysis reactor R, molten cuprous chloride is generated at the outlet #7, which is thereafter cooled and solidified in the heat exchanger system Hx3, and then delivered as solid particles to the electrolytic cell E as stream #8. In the electrolytic cell, the solid CuCl particles dissolve and form an aqueous phase. The electrolytic cell E generates gaseous hydrogen in #9 at around 473 K, which is cooled in Hx5 and delivered as a product at a temperature close to the ambient in #10. The other product of the electrolytic step is aqueous cupric chloride which results at #11; this is heated in Hx6 to the reaction temperature of the hydrolysis process and supplied at point #12 of the hydrolysis reactor R.

The conceptual flowchart for the three-step cycle Cu–Cl-3A is presented in Fig. 6.3. This cycle comprises three reactors: R1 (hydrolysis reactor, process #4), R2 (chlorination reactor, process #6), and E (electrolytic disproportionation cell, process 1). Liquid water is fed at #1 of liquid–gas separator (LGS) S1 where it helps scrubbing of oxygen gas, which is delivered at #2 and mixes with aqueous hydrochloric acid and recycled water, which are then supplied to #2 of the separator. A liquid solution of water and aqueous HCl leaves the separator from the bottom location #4, and mixes in mixer M with a slurry formed of solid copper particles in liquid water, shown on the flowchart as stream #5. The resulting slurry in #5 comprises copper particles, water, and hydrochloric acid. This mixture is heated in heat exchanger Hx1 to a suitable temperature for chlorination (process #6 listed in Table 6.2).

The stream #7 consists of steam, gaseous hydrochloric acid, and copper particles. This is fed into chlorination reactor R1. At the top of the reactor, a gaseous product is generated, consisting of hydrogen and steam. This is extracted from topmost #8. The gaseous mixture is cooled in the heat exchanger system Hx2 where water condenses fully while hydrogen remains gaseous. This two-phase mixture is fed at #9 into the gas–liquid separator S2 where hydrogen separates at the top and it is extracted at port #10. Liquid water is extracted from the bottom #11

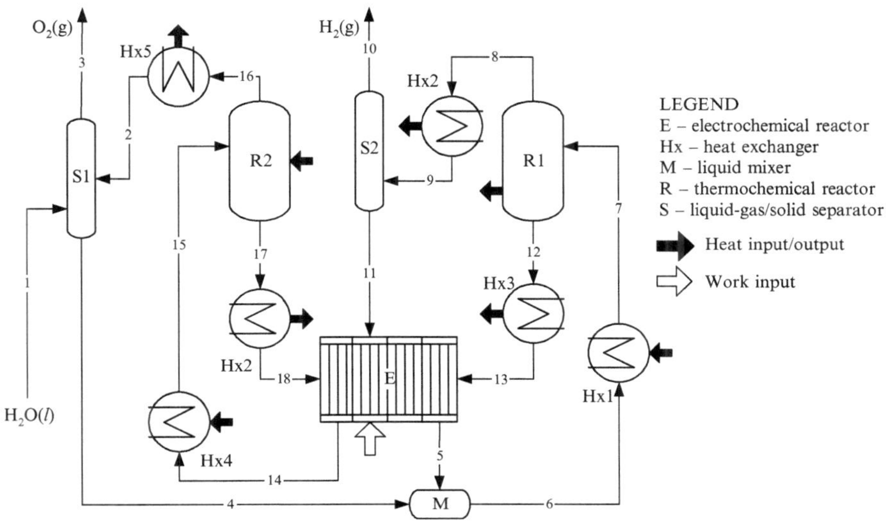

Fig. 6.3 Conceptual flowchart for Cu–Cl-3A cycle (three steps)

and fed to the electrolytic cell E. At the bottom of chlorination reactor R1, molten cuprous chloride is formed, then collected, and extracted at #12.

The molten CuCl is cooled in the heat exchanger system Hx3 such that cuprous chloride solidifies and moves as solid particles—stream #13—to the electrochemical cell E. The electrochemical cell generates solid copper in an aqueous slurry in #5. In addition, an aqueous mixture of wet cupric chloride, $CuCl_2{\cdot}nH_2O(aq)$, is formed, which is extracted at #14. This aqueous solution is heated in a heat exchanger system Hx4 such that all water boils and forms a mixture of solid $CuCl_2{\cdot}nH_2O(s)$ and steam, represented on the flowchart as stream #15. This stream is fed to the hydrolysis reactor R2 where process #4 occurs (see Table 6.2). In this process, oxygen is released together with other gaseous products which are steam and hydrochloric acid. This gas mixture is indicated in the diagram as stream #16, which after cooling in heat exchanger system Hx5 becomes a two-phase mixture in #2 which comprises liquid water, aqueous hydrochloric acid, and gaseous oxygen.

The three-step cycle Cu–Cl-3B is represented by the flowchart diagram indicated in Fig. 6.4. Liquid water is supplied as stream #1 directly to the electrochemical cell E where process 7 occurs. During this process, hydrogen is generated and delivered as product, stream #2. In addition, wet cupric chloride is generated in aqueous solution, namely, $CuCl_2{\cdot}nH_2O(aq)$, which is extracted as stream #3. The wet cupric chloride is heated in heat exchanger system Hx1 such that all free water boils and superheats while cupric chloride remains in hydrated form as solid, $CuCl_2{\cdot}nH_2O(s)$. This stream, namely, stream #4, is directed toward the oxychlorination reactor R1. In this reactor, process #8 occurs (Table 6.2), which forms solid copper oxychloride, gaseous hydrochloric acid, and steam. The gaseous phase is extracted from the top of the reactor as stream #5 and further directed

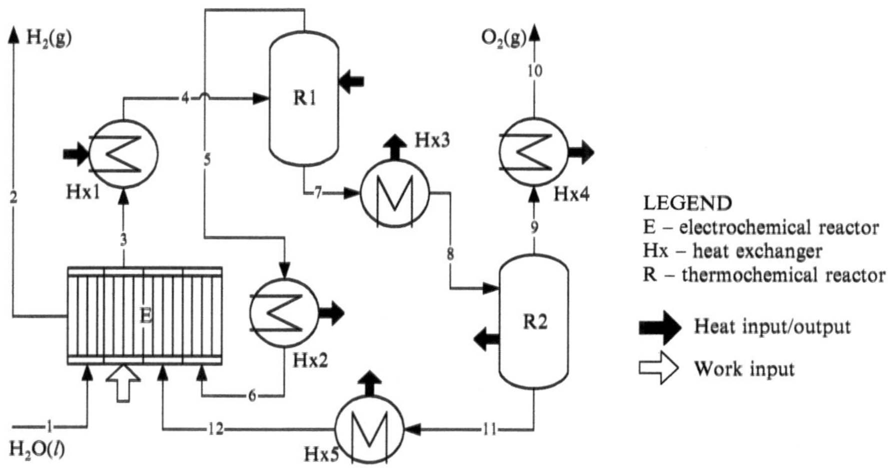

Fig. 6.4 Conceptual flowchart for Cu–Cl-3B cycle (three steps)

toward the heat exchanger system Hx2 where water condenses and hydrochloric acid completely dissolves in the aqueous solution. The resulting liquid solution is fed to the electrochemical cell E as stream #6. The solid particles of copper oxychloride, resulting from the oxychlorination reactor, are extracted as the bottom stream #7 and heated in heat exchanger system Hx3 to the temperature level compatible with the thermolysis reaction in reactor R2. The copper oxychloride is supplied to the thermolysis reactor R2 as stream #8. In the thermolysis reactor, process #9 occurs and releases gaseous oxygen and forms liquid cuprous chloride. The two phases—gas and liquid—separate inside the reactor. Oxygen is extracted as gas at the top of the reactor—stream #9—and further cooled to the ambient temperature in heat exchanger system Hx4 which delivers the oxygen product as stream #10. The molten cuprous chloride, collected from the bottom of the reactor as stream #11, is cooled and solidified in the heat exchanger system Hx5, and then supplied to the electrolytic cell E as stream #12.

The conceptual flowchart of Cu–Cl-3C cycle is presented in Fig. 6.5. In this cycle, water fed as liquid in #1 is heated and boiled in Hx1 to around 1,000 K to state #2 prior to entering into the hydrolysis reactor R1 that performs the reverse Deacon reaction, process #5. From this reactor, a stream #3 of gaseous products comprising oxygen and hydrochloric acid is generated. The gas mixture is cooled down in Hx2 to the ambient temperature and fed into a LGS S1 where liquid water is also supplied. The hydrochloric acid separates as an aqueous solution while gaseous oxygen is collected from the top of the separator as product stream #5. The aqueous hydrochloric acid solution—stream #6—is heated to around 473 K in Hx3 prior to feeding the electrochemical cell E at #7. The electrochemical cell generates pure hydrogen gas in #8 which is cooled to ambient temperature in Hx4 and delivered at #9. The other product of the electrochemical process is the hydrated cupric chlorine—stream #10—which is heated in Hx5 prior to entering

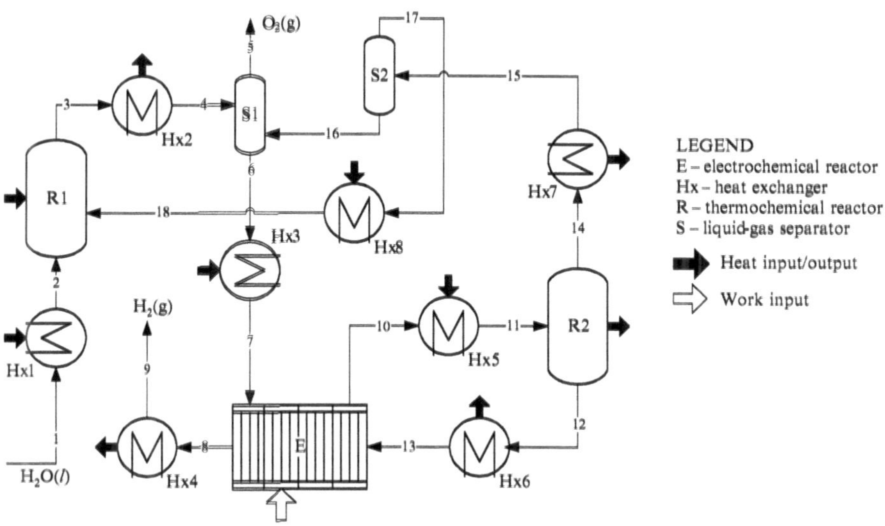

Fig. 6.5 Conceptual flowchart for Cu–Cl-3C cycle (three steps)

the thermolysis reactor R2 at #11. The reaction in the thermolysis reactor occurs according to process #10, Table 6.2. It generates molten cuprous chloride at #12 which is cooled and solidified in heat exchanger system Hx6 prior to flowing into the electrochemical cell as stream #13. The gaseous products of thermolysis consist of steam and chlorine collected from the top of the reactor as stream #14. This stream is cooled in Hx7 such that water condenses completely while chlorine remains as gas. The separation of chlorine from liquid water can be performed by distillation-based methods performed as successive boiling/condensation and gas–liquid separations, or other methods. In the flowchart, the separation process is shown in a simplified manner, such as stream #15 of water + gaseous chlorine fed to the LGS S2 where water separates at the bottom (stream #16) and chlorine at the top (stream #17). Water is fed to separator S1 where it dissolves the hydrochloric acid, while chlorine is heated to around 1,000 K and supplied to the reverse Deacon reaction reactor at #18.

The cycle Cu–Cl-4A with four steps has a conceptual flowchart presented in Fig. 6.6. Supply water in the liquid phase is fed to the electrochemical cell E (process #1, Table 6.2) as stream #1. The electrochemical cell produces hydrated cupric chloride in the aqueous phase represented by stream #2 in the flowchart. This is heated in the heat exchanger system Hx1 and fed into oxychlorination reactor R1 (process #8). In this reactor, a stream of gaseous products consisting of hydrochloric acid and steam is released as stream #4. This is cooled at lower temperature in heat exchanger system Hx2 and provided as stream #5 to a separator where hydrochloric acid is separated from water. This separator presents significant engineering challenges. Dehydrated hydrochloric acid results in #6 which is heated in Hx3 to a temperature level for the chlorination reactor R3 where it enters as

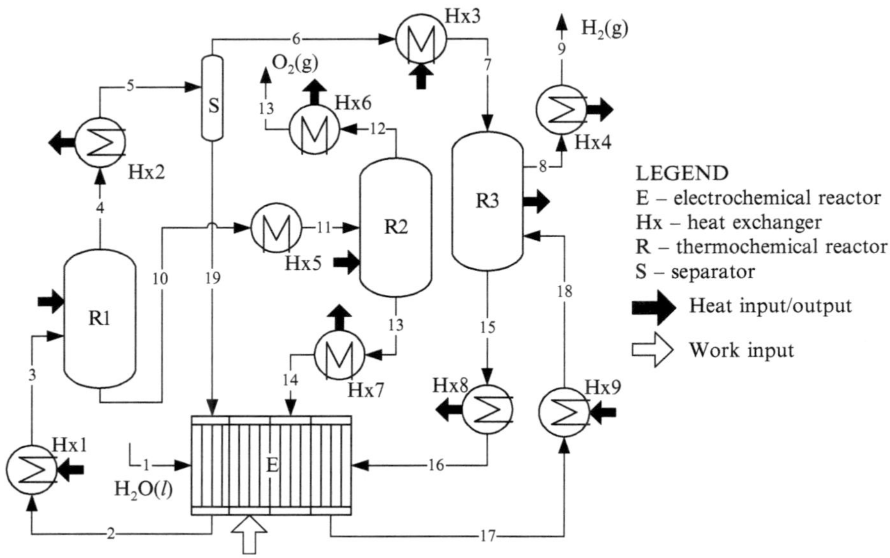

Fig. 6.6 Conceptual flowchart for Cu–Cl-4A cycle (four steps)

stream #7. In this reactor, process #6 occurs (Table 6.2) and gaseous hydrogen is formed, which is extracted as stream #8. This stream is cooled in Hx4 and delivered as product in #9. The other product of oxychlorination is copper oxychloride in a solid phase which separates at the bottom of the reactor as stream #10. This is heated in Hx5 and provided to the thermolysis reactor R2 at port #11. In the thermolysis reactor, the process #9 (Table 6.2) is conducted which generates hot oxygen in #12. Furthermore, the oxygen stream is cooled by Hx6 and delivered at #13. The other product of thermolysis is molten cuprous chloride which is collected at the bottom of the reactor as stream #13. This stream is cooled by Hx7 and supplied as solid to the electrochemical cell in #14. In the chlorination reactor R3, molten CuCl is also formed at the bottom and collected as stream #15 which is cooled and solidified in Hx8 before providing it to the electrochemical cell as stream #16. The electrochemical process generates particulate copper in #17 which is heated in Hx9 to the temperature level corresponding to the chlorination reaction and further supplied to reactor R3. Water separated from hydrochloric acid—stream #19—is returned to the electrochemical cell.

The conceptual flowchart for Cu–Cl-4B cycle is presented in Fig. 6.7. Water is supplied in #1 as liquid and heated in Hx1 to a temperature level corresponding to the hydrolysis reaction which occurs in reactor R1 according to process #3 (Table 6.2). In this process, solid copper oxychloride is generated—stream #3—and further heated in Hx2, and then supplied as stream #4 to the thermolysis reactor R1 where it decomposes thermally according to the chemical equation described in Table 6.2 as process #9. During decomposition, oxygen gas is released in #5, which is cooled in Hx3 and delivered at #6. The other decomposition process is molten

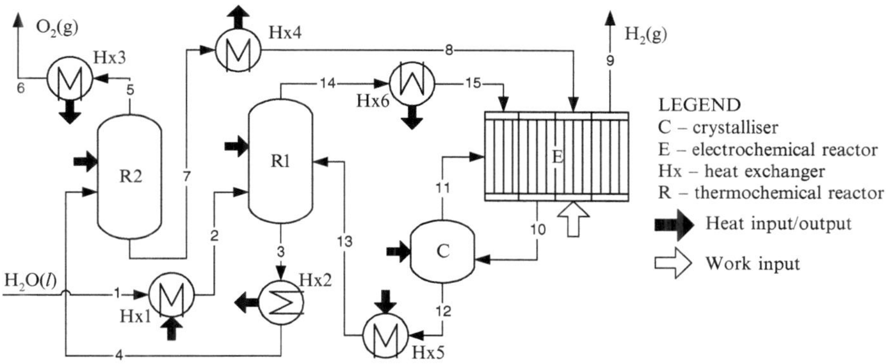

Fig. 6.7 Conceptual flowchart for Cu–Cl-4B cycle (four steps)

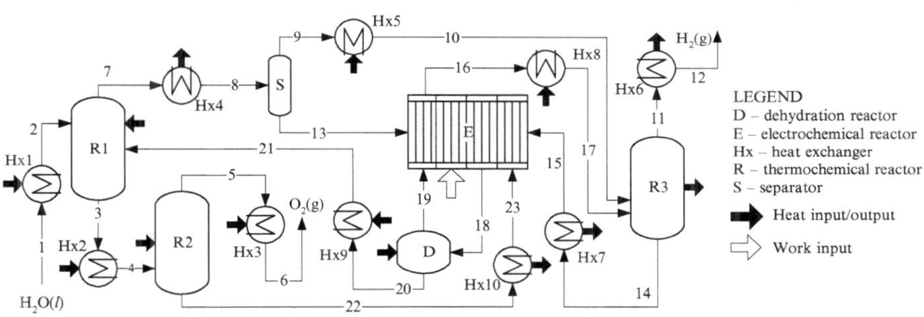

Fig. 6.8 Conceptual flowchart for Cu–Cl-5 cycle (five steps)

cuprous chloride which is collected at the bottom of R1—stream #7. Furthermore, the cuprous chloride is cooled in Hx4. It solidifies—stream #8—and it is fed to the electrochemical cell E. The electrochemical process (#7 in Table 6.2) generates a hydrogen product stream #9. In addition, the electrochemical process produces a hydrated cupric chloride output stream, namely, #10, which is directed toward a crystallization system C. The moisture is removed and dehydrated cupric chloride is obtained according to process #12 (Table 6.2). Water is returned to the electrochemical cell as stream #11. The dehydrated cupric chloride in stream #12 is heated in Hx5 and supplied as stream #13 to the hydrolysis reactor R1.

Figure 6.8 illustrates the conceptual flowchart of the copper–chlorine cycle with five steps, namely, Cu–Cl-5. This cycle comprises a hydrolysis reactor R1, a thermolysis reactor R2, a chlorination reactor R3, a dehydration reactor D, a separator S, and nine heat exchangers. Water is supplied as liquid in #1; then heated, boiled, and superheated in Hx1; and provided as gas—stream #2—to hydrolysis reactor R1. In the hydrolysis reactor, solid copper oxychloride is formed and collected at the reactor bottom as stream #3. Additional heating is supplied to the copper oxychloride stream in Hx2 prior to the supply of the thermolysis reactor

R3 as stream #4. Oxygen evolves from the thermolysis process—stream #5—which is cooled in heat exchanger system Hx3 and delivered in #6. The hydrolysis reaction also produces a gaseous mixture of hydrochloric acid and steam in #7 which is cooled in Hx4 in view of separation. It is assumed that stream #8 comprises liquid water and aqueous hydrochloric acid which are separated by a selected method (e.g., cascaded distillation) in separator S. Hydrochloric acid results as stream #9, further heated in Hx5 and supplied to the chlorination reactor R3 in a gas phase as stream #10. The chlorination process releases pure hydrogen which is extracted at #11, cooled in Hx6, and delivered as product in #12. Water from separator S—stream #13—is supplied to the electrochemical cell E. In the chlorination process, molten cuprous chloride is generated and collected at the bottom of the reactor in #14. It is further cooled and solidified in Hx7 and supplied as solid particles to the electrochemical cell in #15. The electrochemical cell disproportionates CuCl and generates particulate copper—stream #16—which is heated to the chlorination reaction temperature in Hx and supplied as stream #17 to R3. The electrochemical process also produces hydrated cupric chloride in #18 which is dehydrated in the spray drier D from which water is returned as stream #19 to the electrochemical cell E. The dehydrated cupric chloride—stream #20—is heated in Hx9 to the temperature level of the hydrolysis reaction and supplied as hot stream #21 of metallic particles to R1. The thermolysis reactor R2 produces molten cuprous chloride in #22 which is cooled and solidified in Hx10. A stream of solid CuCl particles is provided to the electrochemical cell.

Although the cycles are simplified in the figures, the actual flowcharts are more complicated as they involve additional heat exchange systems and separation systems. Heat recovery and recycling systems must be carefully designed. Also, the type and number of auxiliary devices for each chemical reactor depend on the reaction type, conversion factor, and the need for separation and chemical recycling. Physicochemical properties of chemicals—copper and chlorine compounds—must be well understood in order to effectively design the reactors, heat exchange systems, separators, and other devices. Moreover, all chemical reactions must be thoroughly analyzed based on thermodynamics and kinetics.

6.3 Physicochemical Properties of Copper and Chlorine Compounds in Cu–Cl Cycles

Four chemical elements are involved in copper–chlorine thermochemical cycles, namely, hydrogen, oxygen, chlorine, and copper. These elements form a number of chemical compounds which occur in different steps of the Cu–Cl cycle. The relevant compounds are water (H_2O), hydrochloric acid (HCl), copper oxychloride ($CuO{\cdot}CuCl_2$), cupric chloride ($CuCl_2$), and cuprous chloride (CuCl). In addition, cupric oxide (CuO) may be formed in reactions involving copper oxychloride. In this section, physicochemical properties of the main elements and compounds in the hybrid copper–chlorine cycles are presented. Physicochemical properties relate to physical and chemical types of properties of substances. Thermophysical and thermochemical properties are included in the larger category of physicochemical properties. For example, valence and oxidation states of a chemical element are chemical properties; density is a physical property; formation enthalpy, entropy, or Gibbs free energy are physicochemical properties (also denoted as thermochemical); and specific heat, thermal conductivity, and expansion coefficient are thermophysical properties.

The main properties of chemical elements in the copper–chlorine cycle are summarized in Table 6.6. Hydrogen and its properties have been discussed extensively in Chap. 2. Oxygen and its compounds represent around 49 % by weight from the earth's crust. It also represents nine-tenths part of water. Oxygen is very reactive; in the copper–chlorine cycle, it reacts with copper–chlorine compounds during the hydrolysis, chlorination, and oxychlorination reactions. The cost of oxygen is roughly \$1.7/N m^3 (Haynes 2012). Copper is reddish in color, ductile, and malleable and it is mined from minerals such as azurite, bornite, chalcopyrite, cuprite, and malachite, and it also occurs naturally. One of the main methods of copper extraction from minerals is electrochemical. Commercial copper can be found in high purity, namely, higher than 99.999 % at a price of around \$1.5/kg (Haynes 2012). Copper has two oxidation states, namely, +1 and +2. Chlorine is largely available in nature as salt (NaCl), carnallite, or sylvite minerals. Chlorine is toxic in large quantities; it becomes fatal at 1,000 ppm but it can be detected by the human nose at as little as 3.5 ppm. Chlorine is highly soluble in water where it is used as a disinfectant. One volume of liquid water dissolves 3.1 volumes of chlorine gas at 10 °C (Haynes 2012). Chlorine is very reactive and it has four oxidation states.

Physicochemical properties of hydrogen and oxygen are well understood and predicted by advanced equations of state which are incorporated in process simulation software such as ASPEN Plus or general engineering software such as Engineering Equation Solver (EES). Some relevant properties of copper such as linear thermal expansion coefficient, density, specific heat, thermal conductivity, molar enthalpy, Gibbs free energy, and molar entropy are presented graphically as functions of temperature in Figs. 6.9 and 6.10, which are obtained by EES software. The Shomate equation is used for estimating C_p, H, S, and G as

$$\begin{cases} C_p = a_1 + a_2\theta + a_3\theta^2 + a_4\theta^3 + a_5/\theta^2 \\ H - H_0 = a_1\theta + 0.5a_2\theta^2 + 1/3a_3\theta^3 + 0.25a_4\theta^4 - a_5/\theta + a_6 - a_8 \\ S = a_1\ln(\theta) + a_2\theta + 0.5a_3\theta^2 + 1/3a_4\theta^3 - 0.5a_5/\theta^2 + a_7 \\ G = H - T \times S \\ \theta = T/1{,}000. \end{cases} \tag{6.7}$$

In (6.7), a reduced temperature is used—denoted θ—which is defined based on the process temperature T in Kelvin. The molar specific quantities are expressed with (6.7) and the coefficients listed in Table 6.7. The table shows available

Table 6.6 Thermochemical properties of chemical elements in the Cu–Cl cycle

Chemical element	Z	Valence	Oxidation states	M	MP	NBP	T_c	ρ^0	c_p^0	S^0	ex^{ch}
Chlorine, Cl_2	17	1, 3, 5, 7	+1, +5, +7, −1	35.4	171.7	239.1	417	3.214	33.9	2,23.1	124.0
Copper, Cu	29	1, 2	+1, +2	63.5	1,358	2,835	X	8,934	24.4	33.15	134.2
Hydrogen, H_2	1	1	+1, −1	1	14.05	20.39	32.97	0.00898	28.8	130.7	236.1
Oxygen, O_2	8	2	−2	16	54.36[a]	90.2	154.6	1.31	29.4	205.1	3.97

Z atomic number, *M* molecular mass (kg/kmol), *MP* melting point, *NBP* normal boiling point (K), T_c critical temperature (K), ρ^0 standard density (kg/m^3), c_p^0 standard specific heat (J/mol K), S^0 standard molar entropy (J/mol K), ex^{ch} molar specific chemical exergy (J/mol)

Source: Haynes (2012)

[a]Triple point

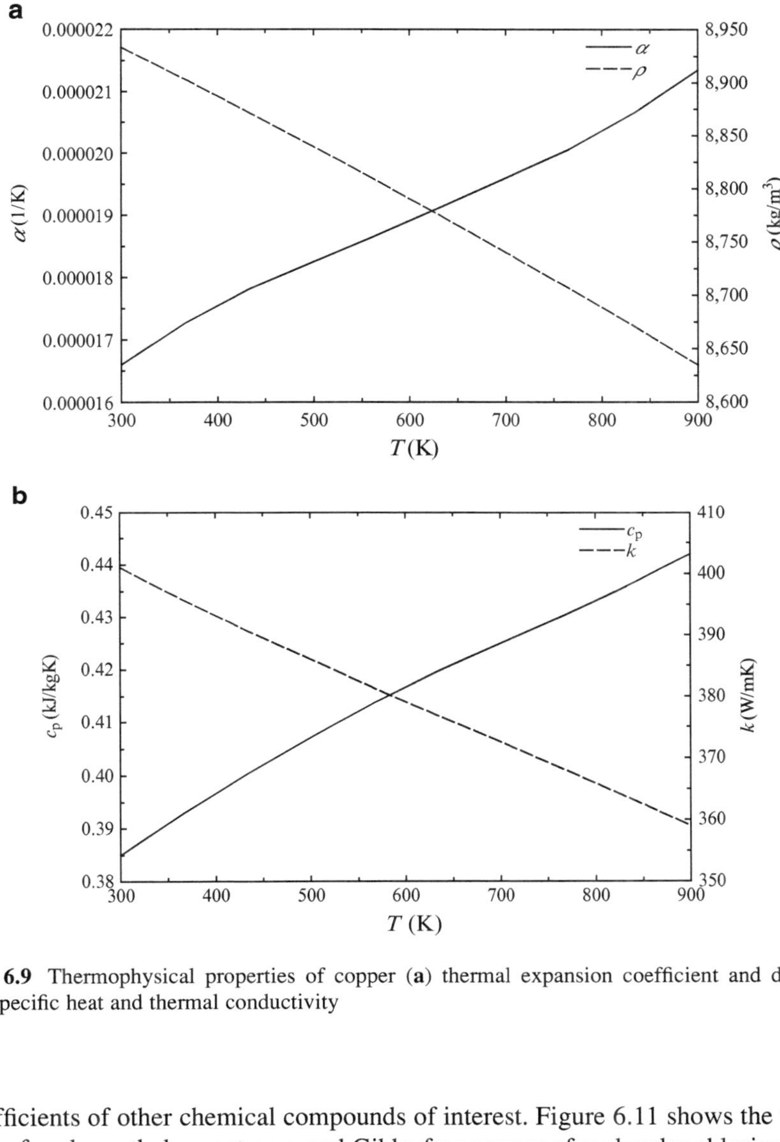

Fig. 6.9 Thermophysical properties of copper (**a**) thermal expansion coefficient and density, (**b**) specific heat and thermal conductivity

coefficients of other chemical compounds of interest. Figure 6.11 shows the variation of molar enthalpy, entropy, and Gibbs free energy of molecular chlorine (Cl_2) with temperature. Additional physicochemical properties of relevant compounds are summarized in Table 6.8.

The Shomate equation can be used for the determination of reaction enthalpies, entropies, and Gibbs free energy. However, in other cases, such as heat balance analyses of heat exchangers or other equipment when accurate estimation of specific heat is required, more rigorous models must be used. The importance of thermodynamic model selection for physicochemical properties of analyzed

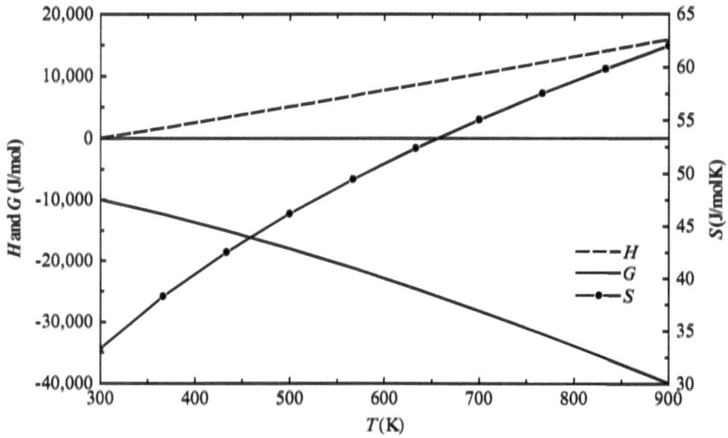

Fig. 6.10 Molar enthalpy, Gibbs energy, and entropy of metallic copper

compounds is discussed in Avsec et al. (2009) for the case of hydrochloric acid. The results predicted by the Shomate equation for the specific heat are compared with those generated by statistical thermodynamic simulations—which are considered more accurate. The results are shown in Fig. 6.12. A significant discrepancy has been observed among property data for copper oxychloride, copper oxide, and cupric and cuprous chloride from various sources. In Zamfirescu et al. (2009a), a careful selection of literature sources for thermochemical data of these compounds in modeling and design of the cycle is presented.

Two reaction processes of the Cu–Cl cycles have as a result the formation of cuprous chloride (CuCl, CAS 7758-89-6) in the form of a crystal (copper chlorination reactor), and in molten salt form (copper oxychloride thermolysis reactor). Cuprous chloride crystals occur naturally in the form of a mineral: nantokite (Stunz Class 3/A.01-10). Cuprous chloride in the solid phase has two crystalline forms: a γ cubic structure up to temperatures close to the melting point, and a β-hexagonal form up to the melting point. The properties of this chemical compound are summarized in Zamfirescu et al. (2009a).

Cuprous chloride is a salt, and molten salts are commonly assumed to be ionic liquids. However, there is little structural evidence that molten salts actually must consist of ions. As reported in Zamfirescu et al. (2009a), cuprous chloride is a molecular liquid above its melting point. Empirical equations for viscosity and thermal conductivity of low-temperature molecular liquids can be used in the absence of experimental data, whereby the low temperature of liquids is understood for temperatures between the melting point and about $T_r = T/T_c = 0.7$. For copper–chlorine cycle operation, CuCl occurs at reduced temperatures T_r of 0.1–0.4. The following equation is recommended for the dynamic viscosity (Zamfirescu et al. 2009a):

$$\mu = 0.365 \times \exp(-6.95 + 1,418/T), \tag{6.8}$$

Table 6.7 Shomate coefficients for chemical compounds in the Cu–Cl cycle

	T (K)	a_1	a_2	a_3	a_4	a_5	a_6	a_7	a_8
Cl_2(g)	298–1,000	33.0506	12.2294	−12.0651	4.38533	−0.159494	−10.8348	259.029	0.0
Cu(s)	298–1,358	17.72891	28.0987	−31.25289	13.97243	0.068611	−6.056591	47.89592	0.0
H_2(g)	298–1,000	33.066178	−11.363417	11.432816	−2.772874	−0.158558	−9.980797	172.70797	0.0
O_2(g)	100–700	31.32234	−20.23531	57.86644	−36.50624	−0.007374	−8.903471	246.7945	0.0
O_2(g)	700–2,000	31.32234	−20.23531	57.86644	−36.50624	−0.007374	−8.903471	246.7945	0.0
H_2O(l)	298–500	−203.606	1,523.29	−3,196.413	2,474.455	3.855326	−256.5478	−488.7163	−285.8304
H_2O(g)	500–1,700	30.092	6.832514	6.793435	−2.53448	0.082139	−250.881	223.3967	−241.8264
HCl(g)	298–1,200	32.12392	−13.45805	19.86852	−6.853936	−0.049672	−101.6206	228.6866	−92.31201
CuCl(s)	298–703	75.271	−26.83212	25.69156	−7.357982	−1.847747	−165.7299	174.6644	−138.072
CuCl(l)	703—3,000	66.944	-3.69×10^{-10}	2.16×10^{-10}	-3.9×10^{-11}	-9.8×10^{-12}	−151.1374	174.7653	−131.178
$CuCl_2$(s)	298–1,500	70.21882	23.36132	−14.86876	4.053899	−0.366203	−228.9405	184.6378	−205.8532
CuO(s)	298–2,000	48.56494	7.498607	−0.05598	0.013851	−0.760082	−173.4272	94.85128	−156.0632

Data from NIST (2012)

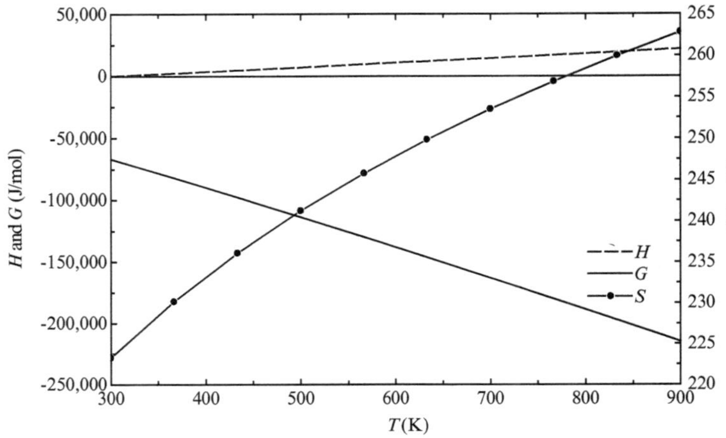

Fig. 6.11 Molar enthalpy, Gibbs energy, and entropy of molecular chlorine in the gas phase

Table 6.8 Properties of chemical compounds relevant to hybrid copper–chlorine cycles

Compound	M	MP	NBP	T_c (K)	P_c	ρ^0	ΔH^0	S^0	ΔG^0	ex^{ch}
HCl(g)	36.46	158.98	188.15	324.61	8.31	1.49	−92.3[a]	186.9	−95.3	84.73
H_2O(l)	18.015	0.0	99.97	647.10	22.06	997	−285.83	69.95	−237.1	0.77
CuCl(s)	98.99	703	1,485	2,435	2.65	4,139	−136.8	87.4	−119.4	75.0
$CuCl_2$(s)	134.45	871	1,266	2,010	N/A	3,400	−218.0	108.1	−173.8	82.4
CuO(s)	79.54	1,299[b]	N/A	N/A	N/A	6,450	−156.06	42.59	−128.3	6.27
CuO·$CuCl_2$	214	625[c]	N/A	N/A	N/A	4,080	−384.65	154.3	−369.7	21.08

ex^{ch} molar specific chemical exergy (kJ/mol), *M* molecular mass (kg/kmol), *MP* melting point (K), *NBP* normal boiling point (K), P_c critical pressure (MPa), T_c critical temperature (K), ΔH^0 standard formation enthalpy (kJ/mol), S^0 standard entropy (J/mol K), ΔG^0 standard Gibbs free energy (kJ/mol), ρ^0 standard density (kg/m^3), *N/A* not available

Sources: Haynes (2012), Zamfirescu et al. (2010a), Parry (2008)

[a]Enthalpy in solution electrolyte of HCl(aq) is −74.84 kJ/mol

[b]Decomposition starts at slow rate at 1,000 K; it starts melting with simultaneous decomposition at 1,299 K

[c]There is no melting point for copper oxychloride; it starts a thermal decomposition at this temperature

where T is temperature in K and for the range of temperatures of interest (709–900 K), the dynamic viscosity varies from 2.6 to 1.7 mPa s. For the thermal conductivity of molten CuCl, the empirical equation presented by Zamfirescu et al. (2009a) is

$$k = 0.19 \times (1 - T_r)^{0.38} / T_r^{1/6}, \tag{6.9}$$

where T_r is the reduced temperature. At the melting point, the thermal conductivity of molten CuCl is estimated to be 0.21 W/mK while at 900 K, it is 0.19 W/mK.

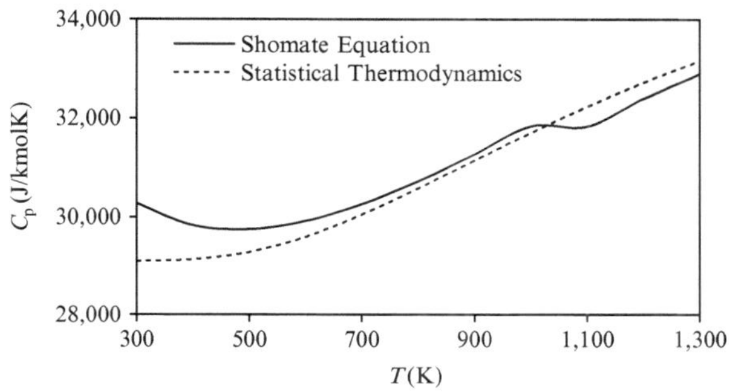

Fig. 6.12 Comparison of specific heat estimation for HCl with Shomate equation and statistical thermodynamic-based models [data from Avsec et al. (2009)]

In Zamfirescu et al. (2009a), the vapor pressure data of CuCl are fitted by the following Riedel-like equation:

$$P_{sat} = \exp\left(273.8096 - 39,240.52/T + 0.27747 \times 10^{-5} \times T^2 - 33.261 \times \ln(T)\right), \tag{6.10}$$

where P_{sat} is in Pa and T is in K.

For a reduced temperature of 0.7 (i.e., 1,704 K), the vapor pressure calculated with (6.10) is 0.845 bar. Using this value and an estimation of the critical pressure, the acentric factor can be found. This factor is an important parameter that allows for establishing an equation of state for the fluid phase. A preliminary estimation of critical pressure, $P_c = 2.65$ MPa, can be obtained by setting $T = T_c$ in (6.10), where T_c is the critical temperature of CuCl listed in Table 6.8. The acentric factor can be calculated from its definition formula as follows:

$$\begin{cases} \omega = -\log_{10}(P_{r,0.7}) - 1 \\ P_{r,0.7} = P_{sat}(T_{r,0.7})/P_c \\ T_{r,0.7} = T/T_c = 0.7. \end{cases} \tag{6.11}$$

With the estimated $P_{r,07} = 0.0323$, the calculated acentric factor is $\omega = 0.49$. There is a lack of data regarding the properties of the liquid phase, such as density and expansion coefficient. There is also a discrepancy in the melting point temperature, which was reported in various studies as 696, 703, and 709 K; the recommended value is 703 K (see Zamfirescu et al. 2009). High discrepancy also exists with past reported values for the boiling point: 1,763, 1,485, and 1,482 K; the recommended value is 1,485 K (Zamfirescu et al. 2009).

Development of a T–S diagram for cuprous chloride is important because this substance can have a role of a working fluid within the copper–chlorine thermochemical cycle. Cuprous chlorine has a relevant vapor pressure. Processes such as throttling, expansion with work generation, or compression of CuCl vapor may be considered within a Cu–Cl plant. The calculation of specific entropy and enthalpy at

Table 6.9 Thermodynamic functions for enthalpy and entropy as functions of pressure

Formula	Remarks
$\left(\frac{\partial H}{\partial P}\right)_T = v(1 - \alpha T)$	H—molar enthalpy, P—pressure, T—temperature, v—molar specific volume, $\alpha = \left(\frac{\partial v}{\partial T}\right)_P$ is the volumetric thermal expansion coefficient
$\left(\frac{\partial C_P}{\partial P}\right)_T = -T\left(\frac{\partial^2 v}{\partial T^2}\right)_P$	C_P—molar specific heat, defined by $C_P = \left(\frac{\partial H}{\partial T}\right)_P$
$\left(\frac{\partial S}{\partial P}\right)_T = -v\alpha$	S—molar entropy, defined by $\frac{C_P}{T} = \left(\frac{\partial S}{\partial T}\right)_P$

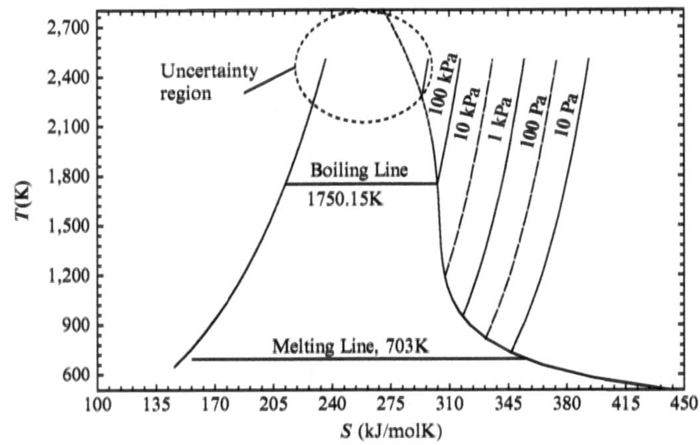

Fig. 6.13 The T–S diagram of cuprous chloride (CuCl)

any given temperature and pressure can be made based on thermodynamic functions. Relevant thermodynamic functions are presented in Table 6.9.

The vapor and gas phases over the range of temperatures of interest exist at low pressures, mostly at vacuum, and high temperatures, which means that the gas and vapor can be treated as ideal gases. For an ideal gas ($Pv = RT$), one has $\alpha = 1/T$ and thus $(\partial H/\partial P)_T = 0$ and $(\partial S/\partial P)_T = -P/R$. Therefore, entropy in the vapor phase is $S(T,P) = S(T,P_0) + R\ \ln(P_0/P)$, and the enthalpy is $H(T,P) = H(T,P_0)$ for all phases, where P_0 is the reference pressure from thermodynamic tables. The estimated T–S diagram of CuCl is shown in Fig. 6.13. The Riedel equation (6.10) has been extrapolated outside of the range of pressures. However, at higher pressures, the uncertainty of the vapor pressure and the vapor line cannot be accurately quantified.

The recommended correlations for density of CuCl, CuO, and $CuO{\cdot}CuCl_2$ are listed in Table 6.10. The density of solid $CuCl_2$ is listed in Table 6.8 for standard temperature.

The cupric chloride is produced in the electrolysis step of the Cu–Cl cycle, where copper is extracted electrochemically from solid cuprous chloride (CuCl). Cupric chloride is produced in its hydrated form ($CuCl_2{\cdot}nH_2O$), which may be dried or crystallized (depending on the cycle version). Cupric chloride may result in one

Table 6.10 Recommended density correlations for CuCl, CuO, and CuO-$CuCl_2$

Compound	Density correlation	Validity range (K)
CuCl(s) (γ-crystal)	$\rho = \rho_0 \times \exp(0.000415281 - 0.0000311T + 1.082325 \times 10^7 T^2), \quad \rho_0 = 4,136\,\mathrm{kg/m^3}$	273–687
CuCl(l)	$\rho = \rho_\mathrm{m} + 0.000769 \times (T - 728), \quad \rho_\mathrm{m} = 3,692\,\mathrm{kg/m^3}$	728–883
CuO(s)	$\rho = \rho_0 \times (a + bT + cT^2); \quad \rho_0 = 6,450\,\mathrm{kg/m^3}$	298–675
	$a = 1.01143, \quad b = 9.3396 \times 10^{-5}, \quad c = 1.91537 \times 10^{-7}$	
CuO-$CuCl_2$(s)	$\rho = \rho_0 \times \exp(a\Delta T + b\Delta T^2); \quad \Delta T = T - 298.15; \quad \rho_0 = 4,080\,\mathrm{kg/m^3}$	298–675
	$a = -2.660832 \times 10^{-24}, \quad b = -1.221274397 \times 10^{-13}$	

Note: Density values are in kg/m^3
Source: Zamfirescu et al. (2009a)

of the possible decomposition products (undesired) of copper oxychloride in the thermolysis reactor.

In hydrolysis or oxychlorination reactors, cupric chloride is used as a reactant. Cupric chloride has the Chemical Abstract Service (CAS) number CAS-7447-38-4 and occurs naturally in mineral form—known under the name tobachite—and called Strunz Class 3/A.07-30. Under atmospheric conditions, cupric chloride releases chlorine at 766 K, while at 1,266 K, it starts boiling and begins decomposition to form Cu_2Cl_2.

Cupric chloride has a solid–solid transition at 675 K and a solid–liquid transition at 871 K. There is a γ-crystal which exists at 398–685 K. At 685 K, there is a phase transition between γ and β crystals accompanied by a latent heat release of 6 kJ/mol (see Avsec et al. 2009). The melting occurs at 696 K with a latent heat of melting of 7.08 kJ/mol (Avsec et al. 2009).

Cupric oxide (CuO) may be one of the decomposition by-products (undesired) of copper oxychloride. The knowledge of its properties is important in conjunction with the modeling of copper oxychloride decomposition and cupric chloride hydrolysis or oxychlorination processes. Cupric oxide is a recorded chemical (CAS 1317-38-0) and mineral (tenorite, Strunz Class IV/A.05-10). Solid cupric oxide is a gray-black monoclinic crystal having a density of 6,450 $kg/m^3$and Vickers hardness of 190–300 kg/mm^2. At high temperatures (over 1,000 K), cupric oxide decomposes into cuprous oxide (Cu_2O) and releases oxygen. This aspect is important for the study of thermal decomposition of copper oxychloride. Recent experiments showed that the heat capacity is highly influenced by the uncommon magnetic behavior of cupric oxide at high temperatures (see the literature review compiled by Zamfirescu et al. (2010a), Table 3). Due to magnetic components influencing the internal energy of a crystalline structure, the specific heat is almost constant between 600 and 700 K and it slightly decreases after 750 K.

One of the key substances in the hydrolysis reactor is copperoxychloride ($CuO{\cdot}CuCl_2$ or also denoted with Cu_2OCl_2). This is a classified mineral, melanothallite (Strunz Class III/D.01-60), which is part of the category of oxyhalogenides. It has an orthorhombic, dipyramidal crystalline structure with interlaced atoms of copper, chlorine, and oxygen. Limited experimental determinations of copper oxychloride have been published in the archival literature. There is no critical point of copper oxychloride because it decomposes thermally and releases oxygen. According to Parry (2008), the thermal decomposition of copper oxychloride starts at ~625 K. Copper oxychloride is a coordination complex with a Fddd *ortho*-rhombing cell. The determination of the standard enthalpy of formation of oxychloride has been performed in a reaction calorimeter, as reported by Parry (2008). Kinetic analyses of copper oxychloride synthesis and decomposition are presented in Lewis et al. (2003).

As it results from the analysis presented in Zamfirescu et al. (2009a), the specific heat of copper oxychloride at 298.15 K is 116.77 kJ/mol which agrees well with theoretical estimations and modeling. In Zamfirescu et al. (2009a), the estimation of specific heat at 298.15 K is made with the "element contribution method" by

$$C_p = \sum_{i=1}^{N} N_i \Delta E_i, \tag{6.12}$$

where N_i represents the number of atoms of species i, N is the total number of atomic species in the molecule, and ΔE_i is the contribution of each atomic species. According to the details given in Zamfirescu et al. (2009a), each copper atom has a contribution of $\Delta E_{\mathrm{Cu}} = 26.92\,\mathrm{kJ/kmol\,K}$. An oxygen atom contributes with $\Delta E_{\mathrm{O}} = 13.42\,\mathrm{kJ/kmol\,K}$ and the atom of chlorine with $\Delta E_{\mathrm{Cl}} = 24.69\,\mathrm{kJ/kmol\,K}$. The prediction accuracy of (6.12) is about $\pm 10\ \%$.

Specific heat measurements at high temperatures are available only from Parry (2008). The value at 673 K is 114 kJ/kmol K. Using data from measurements, the following Debye–Einstein correlation for specific heat has been developed by Parry (2008) for a temperature range of 14–700 K:

$$C_p(T) = 3R\left[\frac{13.962}{\theta_{\mathrm{D}}^3}\int_0^{\theta_{\mathrm{D}}} \frac{xe^x}{(e^x - 1)^2}\mathrm{d}x + 0.945\theta_{\mathrm{E}}^2 \times \frac{e^{\theta_{\mathrm{E}}}}{(e^{\theta_{\mathrm{E}}} - 1)^2}\right], \tag{6.13}$$

where R is the ideal gas constant, $\theta_{\mathrm{D}} = T_{\mathrm{D}}/T$ is the dimensionless temperature calculated with $T_{\mathrm{D}} = 373.41\,\mathrm{K}$ (Debye temperature), $\theta_{\mathrm{E}} = T_{\mathrm{E}}/T$ is calculated with the Einstein temperature, $T_{\mathrm{E}} = 1{,}193\,\mathrm{K}$, and T is temperature in K.

Since copper oxychloride is not commercially available, methods of synthesis are needed. The method adopted by Ikeda and Kaye (2008) involved the use of stoichiometric amounts of CuO and $CuCl_2$. The process was carried out in a glove box under argon to avoid the presence of moisture. The reaction was performed in a sealed quartz ampoule with precursors (CuO and $CuCl_2$), dried overnight under 5×10^{-8} mbar vacuum at ambient temperature, and then inserted into a tubular oven. Crystal growth was achieved by holding the ampoule at 718 K for 2 days. The sample appeared as small black crystalline lumps that broke up easily when shaken with no visible CuCl spots. Characterization and synthesis of copper oxychloride are also reported in Nixon et al. (2011).

Trevani (2011) prepared copper oxychloride in a conventional horizontal tube furnace with a quartz tube and stainless-steel connections for gas injection. The copper precursor, CuCl or $CuCl_2$, was loaded into an alumina crucible and exposed to a flow of high-purity dry air at 643 K for 48 h. Analysis with the XRD method confirmed that highly pure copper oxychloride samples were obtained. The method is easily scalable and it can be used to produce the amounts required for large-scale oxygen production experiments. Figure 6.14 illustrates the copper oxy- and hydroxyl-chloride samples obtained at UOIT and reported by Naterer et al. (2011b).

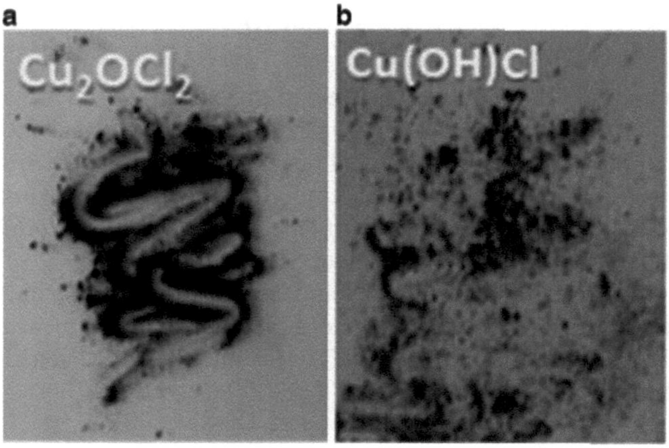

Fig. 6.14 Copper oxychloride powder (**a**) and side product Cu(OH)Cl obtained at UOIT (courtesy of Prof. L. Trevani)

6.4 Electrochemical Step in the Hybrid Copper–Chlorine Cycles

6.4.1 Classification of Electrochemical Cells

The electrochemical process within the Cu–Cl cycle has several variants depending on the cycle version and the type of membrane used within the cell. The cell can produce at the cathode either metallic copper or molecular hydrogen; also one can use a cation exchange membrane (CEM) that conducts protons or an anion exchange membrane (AEM) conductive to chlorine ions. Regardless of the cell variant, the anodic half-cell reaction is the same for all cases. Hydrochloric acid must be present in the anolyte and catholyte where it has multiple roles.

- It impedes precipitation of CuCl which otherwise is sparingly soluble in water.
- It facilitates complexation of copper ions with water molecules with an eventual formation of a hydrated cupric chloride within the catholyte.
- It contributes to an ionic balance in electrolyte, being either a source of protons or Cl^-.

Table 6.11 categorizes the electrochemical cell variants based on the half-cell reaction at the cathode and the type of membrane. A coded identification of the cell type is also proposed for presenting them in this chapter. The cell variants are illustrated schematically in Fig. 6.15. The cell variants are given as follows.

- Cu-Cl^-, see Fig. 6.15a which produces copper at the cathode and it has an AEM permeable to chlorine ions.
- Cu-H^+, see Fig. 6.15b which produces copper at the cathode and it has a CEM permeable to protons.

Table 6.11 Electrochemical cell classifications based on cathodic reaction and membrane type

ID	Electrolyte/membrane processes	Anodic reaction	Cathodic reaction	Overall[a]
$Cu\text{-}Cl^-$	AEM (Cl^-)	$2CuCl + 2Cl^- \rightarrow 2CuCl_2 + 2e^-$	$2CuCl + 2e^- \rightarrow 2Cu + 2Cl^-$	Process #1
$Cu\text{-}H^+$	Anolyte: $2HCl \rightarrow 2H^+ + 2Cl^-$ CEM (H^+) Catholyte: $2H^+ + 2Cl^- \rightarrow 2HCl$			
$H_2\text{-}Cl^-$	AEM (Cl^-) Catholyte: $2HCl \rightarrow 2H^+ + 2Cl^-$		$2H^+ + 2e^- \rightarrow H_2$	Process #7
$H_2\text{-}H^+$	Anolyte: $2HCl \rightarrow 2H^+ + 2Cl^-$ CEM (H^+)			

ID cell identification, the cell is identified by the product obtained at the cathode, namely, Cu or H_2, and by ionic species transferred between the anode and cathode, namely, Cl^- or H^+

AEM anion exchange membrane, *CEM* cation exchange membrane

[a]Overall process as described in Table 6.2

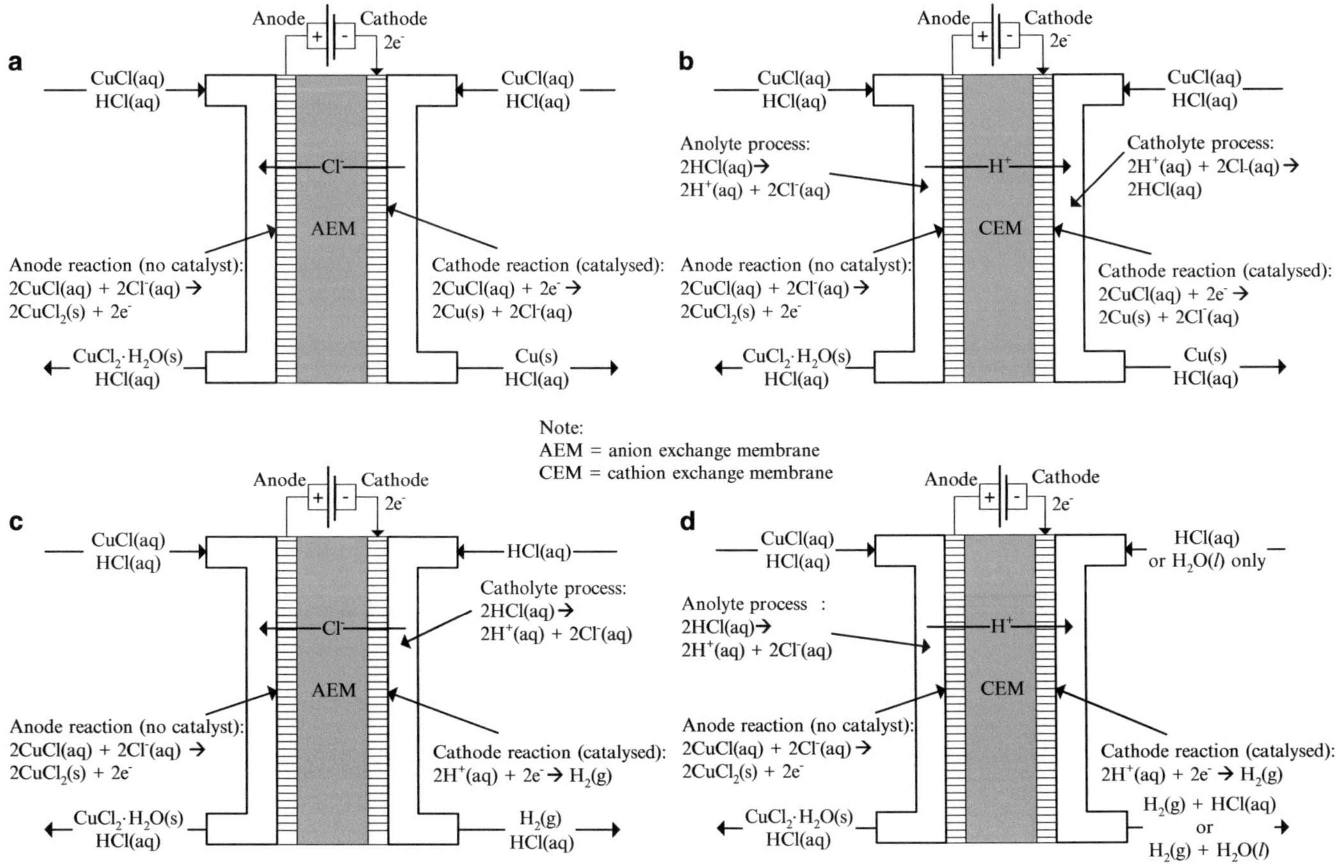

Fig. 6.15 Electrochemical cell variants encountered in copper–chlorine cycles. (**a**) Cell ID: Cu-Cl^-, (**b**) Cell ID: Cu-H^+, (**c**) Cell ID: H_2-Cl^-, (**d**) Cell ID: H_2-H^+

- H_2-Cl^-, see Fig. 6.15c which produces hydrogen gas at the cathode and it has an AEM permeable to chlorine ions.
- H_2-H^+, see Fig. 6.15d which produces hydrogen gas at the cathode and it has a CEM permeable to protons.

Note that the cells Cu-Cl^- and Cu-H^+ conduct the disproportionation process #1 while the H_2-Cl^- and H_2-H^+ cells conduct the HCl/CuCl electrolysis process #7 (see Table 6.2). It was found that the anodic reaction does not require a catalyst but rather an electrode with a large exposed surface. The cathodic side however requires catalysts, in general platinum based. The membrane poses important technical challenges; the main problem is the undesired crossover of ionic species.

The anodic reaction—common for all cells—is a non-catalytic electrochemical process in which aqueous cuprous chlorine combines at the electrode surface with a chlorine ion, forms cupric chloride, and releases two electrons to the anode. The cupric chloride forms the complexation product $CuCl_2 \cdot nH_2O(s)$ which is retrieved from the anolyte exit stream.

The cathodic reaction for the Cu-Cl^- cell consumes aqueous cuprous chloride and electrons (provided at the cathode surface) and generates metallic copper particles (with a dendritic structure) and chlorine ions. Chlorine ions cross the membrane and become consumed in the anodic reaction. The cell Cu-H^+ requires two additional processes which occur spontaneously in the anolyte and catholyte, respectively. Due to the high hydrochloric acid concentration, both the anolyte and catholyte are used as a source of protons and chlorine ions. In the anode side, the anolyte gives the chlorine ions that are required for the cathodic reaction, while the protons cross through the membrane to the cathodic side where they combine spontaneously in the catholyte to form aqueous hydrochloric acid. Overall, in the Cu-H^+ cell, hydrochloric acid is consumed at the anode side and produced at the cathode side.

For both copper-producing electrochemical cells—Cu-Cl^- and Cu-H^+—there is a substantial challenge with the handling of metallic Cu generated at the cathode. As discussed in Ranganathan and Easton (2010a), electrowinning of copper from a solution of chloride is plagued by the copper growth as dendritic structures can physically tear the membrane leading to catastrophic cell failure. It is also complicated to remove the precipitated metallic copper from the cathode side.

The process in cell H_2-Cl^- uses the catholyte as a source of chlorine and protons which result from a spontaneous dissociation of HCl molecules. Chlorine ions cross from the cathode side to the anode side where they are consumed in the reaction of cupric chloride formation, while the protons are reduced at the cathode surface. In cell H_2-H^+, the process of proton generation takes place in the anolyte—sourced from HCl. Protons cross the membrane and reach the cathode side where they are reduced at the electrode surface. At the same time, chlorine ions from anolyte enter into the anodic reaction. Thus, hydrochloric acid is consumed at the anodic side of H_2-H^+ cell, while for the H_2-Cl^- cell, it is consumed at the cathode side.

Although with cells H_2-Cl^- and H_2-H^+ the problems caused by copper electrowinning are eliminated, these cells pose a major technological challenge due to occurrence of a copper crossover process. Copper crosses through membranes whether as ions (if a CEM is used) or as a complex in the aqueous solution. Due to the copper complexation process at the anode side, copper complexes in the aqueous phase exist. Such complexes can cross the AEM. The occurrence of ionic copper or copper complexes to the cathode side leads to the unavoidable formation of metallic copper, which greatly reduces the activity of catalysts and leads to a significant decrease of cell performance. This is therefore a topic of intense research to find a suitable membrane that impedes copper crossover in the hydrogen-producing electrochemical cell (see Naterer et al. 2011a).

6.4.2 Electrode Development for Anodic and Cathodic Reactions

The anodic side half-reaction is essentially the same for all possible cell configurations. It has been extensively studied by Easton et al. at UOIT (see Ranganathan and Easton 2010a, b) and at AECL (Stolberg et al. 2008). Half-cell experiments with various electrodes were performed at UOIT using a standard three-electrode cell arrangement comprising a saturated calomel reference electrode (SCE), a platinum wire counter electrode, and the anode. The aim of the experiment was to determine a better structure for the anode electrode characterized by high current densities and reduced over-potentials. Electro-catalysts are not required; this has been confirmed by Suppiah et al. (2009) at AECL. Data for stirred and quiescent electrolytes were obtained at 298 K. The anodic half-reaction occurs as

$$2\mathrm{CuCl(aq)} + 2\mathrm{Cl^-(aq)} \rightarrow 2\mathrm{CuCl_2(aq)} + 2\mathrm{e^-}. \qquad (6.14)$$

AECL measured the equilibrium potential for reaction (6.14) which was reported in Stolberg et al. (2008) to be 0.237 V. The most promising results obtained at UOIT were with ceramic carbon electrodes (CCE) based on poly-aminopropyl siloxane (PAPS), as described in Ranganathan and Easton (2010b) by using a base-catalyzed sol-gel method. A precursor 3-aminopropyl trimethoxy silane (APTMS) was combined with carbon black (type Vulcan XC72) in a methanol–water solution with ammonium hydroxide addition, which catalyzed the polymerization process of APTMS. A gel was obtained and coated on a strip of 1 cm^2 of carbon fiber paper (CFP) type TPGH-090 (from BASF), which was subsequently dried at room temperature for 6 h, followed by drying in an oven at 80 °C for 12 h. Two types of electrodes were prepared in this way: single-face coated and double-face coated. The coating deposition was of 2.7 mg/cm^2. The weight percent of PAPS in CCE was determined by thermogravimetric analysis and 36 %. Figure 6.16 shows a photo and microscopy photo of the electrode and its coated surface. The surface is visualized through scanning electron microscopy (SEM). The figure shows that the CCE layer is a uniform three-dimensional structure with high porosity and large surface area.

Fig. 6.16 CCE-PAPS anodes obtained at UOIT (courtesy of Prof. B. Easton)

Ranganathan and Easton (2010a) determined the current density vs. voltage diagrams for the half-cell reaction with respect to SCE. Three types of experiments were performed: with uncovered CFP electrode, with single-side coated CFP and double-side coated CFP. The measurements were performed in solutions of 3 and 6 mol/L HCl concentration and 1, 10, and 100 mmol/L CuCl concentration. At 100 mmol/L CuCl, the current saturation has been observed, which indicates mass transport limitation. It was also observed that the cell performance decreased at higher concentrations of HCl (the current density at saturation decreased by ~20 %). The impact of HCl concentration on reaction kinetics is explained by the influence of dissolved CuCl species. The following main species are generated when cuprous chloride is dissolved in an aqueous solution of hydrochloric acid: CuCl(aq), $CuCl_2^{-1}$ (aq), $CuCl_3^{-2}$(aq), and $CuCl_4^{-3}$(aq). In addition, at higher HCl concentrations, the following species may be present: Cu_2Cl^-(aq), Cu_2Cl_2(aq), and $Cu_2Cl_3^{-1}$(aq). Therefore, in addition to the oxidation reaction (6.14), the following concurrent reaction may occur (see Naterer et al. 2011a):

$$CuCl_4{}^{-3}(aq) \rightarrow CuCl^+(aq) + 3Cl^- + e^- \quad \text{at } 6\,\text{M HCl}$$

$$CuCl_4{}^{-3}(aq) \rightarrow CuCl_2(aq) + 2Cl^-(aq) + e^- \quad \text{at } 11\,\text{M HCl}.$$

Since Cu(II) occurs at higher concentrations of HCl, the ionic species have a larger size, they are less mobile, and the convection limitation of kinetic current occurs faster than for the case when HCl concentration is reduced. From experimental data, for similar experimental conditions, the catholyte copper species concentration decreases with increasing HCl concentration. In 6 mol/L HCl, the $CuCl^+$ that is formed can be transported across the membrane by diffusion and migration. In 11 mol/L HCl solution, the $CuCl_2$ and HCl are transported across the membrane by diffusion only. Nonetheless, in highly concentrated solutions of HCl, the neutral and cationic species of copper mono and bivalent are lower, thereby facilitating a reduced crossover rate.

Polarization curves for the anodic reaction were obtained by Ranganathan and Easton (2010a, b) with CCE electrodes as well as bare CFP electrodes. The results

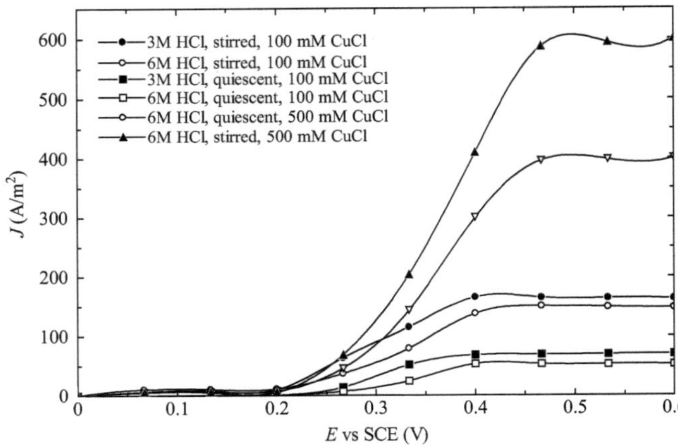

Fig. 6.17 Polarization curve for anodic reaction with a solution of 100 mmol/L CuCl using CCE electrodes prepared at UOIT [data from Ranganathan and Easton (2010a); cubic spline data smoothing is applied]

show the net improvement of CCE electrodes. Figure 6.17 presents the polarization curves for double-side coated electrodes (CCE) having a total exposed area (two sides) of 2 cm^2. The experiments were performed in a solution of 100 mmol/L CuCl. Data was obtained for 3 mol/L HCl and 6 mol/L HCl in stirred and quiescent solution. The results show that the half-cell potential corresponding to the current saturation is around 0.4–0.5 V. The maximum current density is around 600 A/m^2 with the corresponding over-potential for the half-cell reaction at the anode ~0.2 V.

At around 6 mol/L HCl, the electric conductivity of the aqueous HCl solution reaches a maximum. Under quiescent conditions for 6 mol/L HCl solution, the following saturation current is obtained experimentally by Ranganathan and Easton (2010a): 0.15, 0.8, 11.2, and 80 mA for CuCl concentrations of 1, 10, 100, and 500 mmol/L, respectively. If the electrolyte is stirred, the following saturation currents are reported: 0.34, 2.7, 30, and 120 mA for the mentioned CuCl concentrations, respectively. The analysis presented graphically in Fig. 6.18 demonstrates that the saturation current density increases quasi-linearly with molarity of cuprous chloride. If experimental data is extrapolated to 1 mol/L CuCl, it appears that the current density with a stirred anolyte surpasses 1,000 A/m^2, which is promising.

According to Stolberg et al. (2008), AECL performed full-cell experiments which confirm that a current density higher than 1,000 A/m^2 is achievable when 6 mol/L HCL anolyte is used with 1 mol/L CuCl molarity. The half-cell tests at AECL were performed with a catalyst-free graphite separator plate (GSP), coated GSP with Pt/XC-72R catalyst mixture, catalyst-free CFP type EC-TP1-060, and platinized CFP and coated CFP with Pt/XC-72R catalyst. The analyte was stirred and a solution of 1 mol/L CuCl/6 mol/L HCl has been used. The results show that a catalyst-free GSP performs the best with a saturation current density of 1,200 A/m^2 at 0.5 V polarization, which is a 0.263 V anodic over-potential.

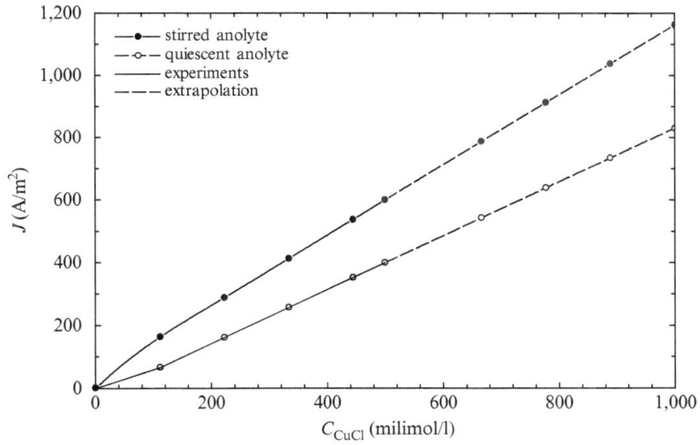

Fig. 6.18 Dependence of anodic saturation current density on CuCl molarity in a 6 mol/L HCl electrolyte based on experiments by Ranganathan and Easton (2010a) and extrapolation

AECL focused mainly on the development of the electrochemical cell for CuCl/HCl electrolysis to generate hydrogen and hydrated cupric chloride (Suppiah et al. 2009). An essential part of the research involves development of advanced electrodes for the cathode. The cathodic half-cell reactions were studied in a three-electrode electrochemical cell. The reference electrode of the half-cell was the SCE while the counter electrode was a graphite rod.

The tested electrodes for the cathode were fabricated by deposition of various platinized catalysts on a CFP type EC-TP1-060 Toray. The application of platinized catalysts was done by painting or direct spraying. The cathodic reaction was carried out in HCl (20 % by weight). The following coatings were applied:

- Electrode #1: Pt/XC-72R catalyst mixture with Pt loading of 19 % painted on graphite
- Electrode #2: Pt deposited on CFP type TGP-H-060 (Toray)
- Electrode #3: Pt/CV-72R catalyst mixture deposited on CFP type EC-TP1-060
- Electrode #4: Catalyst-free graphite
- Electrode #5: Catalyst-free CFP type EC-TP1-060

The polarization curves obtained at AECL with these electrodes #1 to #5 as cathodes are presented in Fig. 6.19. The best cathode is electrode #1 which shows 1,000 A/m^2 at 0.33 V polarization vs. SCE. Catalyst-free electrodes cannot be used for the cathode side reaction. Copper deposits on the cathode surface. AECL observed visually the copper deposits on the cathode after dismantling their single cell which had a cathode with 25 cm^2 area.

The metallic copper deposits can be observed in Fig. 6.20. A proprietary cell (described subsequently) was developed at AECL. This cell showed no (or very limited) copper crossover. A specimen has been extracted from the electrode of the cell after experimental tests. The Pt-based electro-catalyst structure was analyzed

Fig. 6.19 Polarization curves for cathode obtained at AECL with various electrodes [data from Suppiah et al. (2009)]

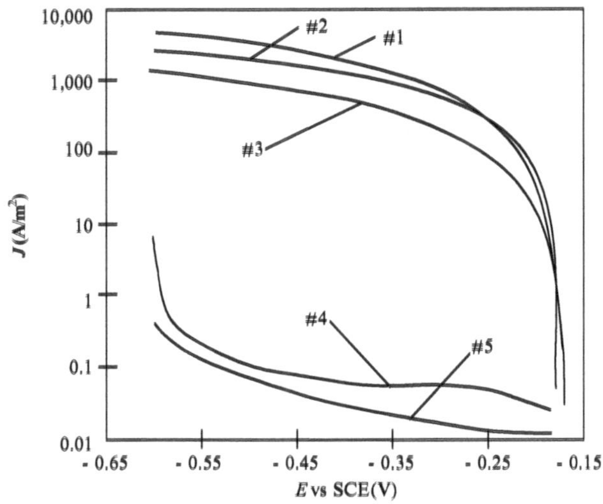

Fig. 6.20 Copper deposits on a 25 cm^2 cathode (courtesy of AECL)

with the help of a SEM and energy-dispersive X-ray spectroscopy (EDX). The technique of EDX characterizes the atomic composition of interactions between electromagnetic radiation and matter, by analyzing X-rays emitted by the sample in response to impact by charged particles. The results regarding the spectrum of atomic percentages are presented in Table 6.12, while the image obtained by the SEM is shown in Fig. 6.21. The figure shows the locations for each of the spectra listed in Table 6.12. Copper was not found in the crystal-free regions of the electrocatalyst.

Balashov et al. (2011) showed that it is not necessary that the cathodic reaction is conducted with a high concentration of hydrochloric acid. It can be conducted simply in water which can maintain the proton transport. These researchers reported encouraging experimental results obtained with water as the catholyte. The results were obtained in a full-cell water electrolysis system which used a proprietary membrane–electrode assembly (MEA) of which cathode materials were undisclosed. The MEA used by Balashov et al. (2011) was a commercial type HYDRion 115 and HYDRion 117 produced by Electrochem Inc. Another conceptual system

Table 6.12 Spectrum of atomic percentages at three locations (see Fig. 6.21)

Element	Spectrum number (atomic %)		
	#1	#2	#3
C	88.93	90.99	90.65
O	3.11	1.62	N/A
F	7.09	5.20	8.89
Al	N/A	0.09	N/A
Si	0.03	N/A	N/A
S	0.30	0.70	0.22
Cl	0.19	0.66	N/A
Cu	0.05	0.10	N/A
Pt	0.30	0.64	0.15

Source: Naterer et al. (2011a)

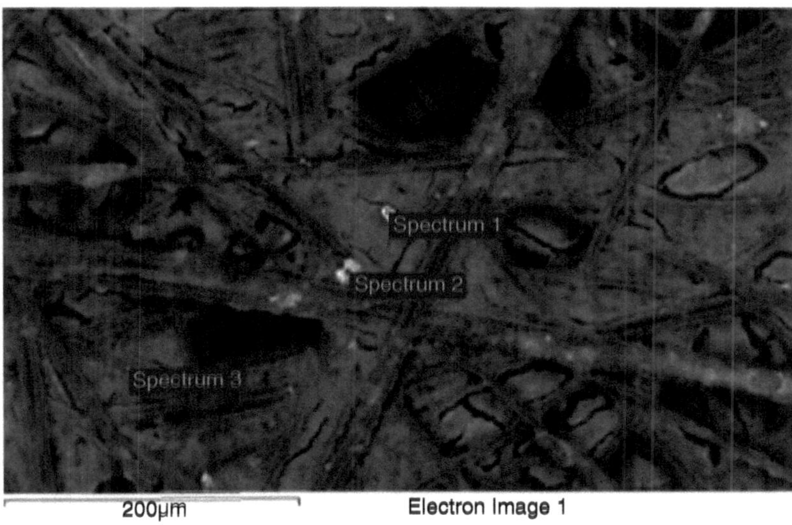

Fig. 6.21 SEM/EDX image of the Pt electro-catalyst for cathode used by AECL (courtesy of AECL)

which uses a nonaqueous catholyte made of acetonitrile or dimethylformamide solvents was proposed by Zamfirescu et al. (2012).

The cathode process in electrochemical cells with Cu-Cl^- and Cu-H_2 generates metallic copper at the cathode according to the half-reaction as

$$CuCl(aq) + e^- \rightarrow Cu(s) + Cl^-(aq). \tag{6.15}$$

Electrowinning technology recommends the use of carbon electrodes at the cathode because copper precipitates as dendritic powder. The anolyte with aqueous cuprous chloride is fed at the bottom of the cathode compartment and the product stream is extracted from the top side. This type of process has been studied in past

Fig. 6.22 Permeability experimental apparatus at Argonne National Laboratory: (**a**) permeability cell and (**b**) arrangement of the cell in a thermostated furnace (courtesy of ANL)

seminal work by Lewis et al. (2003) which reported the formation of dendritic copper powder at the cathode with sizes from 3 to 100 μm. It appears that the development of copper-producing electrochemical cells for the copper–chlorine cycle has been set aside by most of the researchers because the alternative hydrogen-producing electrochemical process appears to be more competitive.

6.4.3 Membrane Development

As depicted in Fig. 6.15, two types of membranes have been studied for electrochemical cell development, namely, an AEM and CEM. Although CEMs appear attractive because they can impede cationic copper species to migrate to the cathode side, there has been no research reported for their use within the Cu–Cl cycle. The focus of researchers has been toward the development of AEMs which are resistant to copper crossover (Naterer et al. 2011a).

The ANL, GTI (De Plaines, IL), Pennsylvania State University (PSU), and AECL collaborated in finding and testing proton-selective membranes with lower rate of diffusion/migration for copper species. ANL developed a diffusion cell system for testing membrane permeability. The cells comprise a solute-rich compartment and a solvent-rich compartment between which the membrane is interposed. The cell used by ANL is shown in Fig. 6.22a. During the tests, the cell was placed in a thermostated furnace where the temperature was kept constant at an imposed value. Because the cell emits gaseous hydrochloric acid during the experiments, it must be safely vented. Therefore, the cell was placed in a secondary enclosure with argon purge and venting lines. The arrangement of a cell in the furnace is shown in Fig. 6.22b.

Table 6.13 Membranes tested at various institutions

Investigator	Membrane	Remarks
ANL/AECL	NAFION 117	This is a reference membrane trademark of Du Pont Du Nemours.
PSU/ANL/AECL	S-Radel	Polysulfonate membrane based on the commercially available polymer from Solvay Advanced Polymers Radel NT 5500.
	CEM IEC 2.0	Cross-linked polysulfone membrane. IEC = 2.5 was also tested.
GTI/ANL	CM2-base	Consists of a double layer of Nafion. The membranes were treated to improve wettability and remove skin effects by an etching process.
	GC2hydromod	Based on commercially available membranes. No heat treatment before testing.

Table 6.14 Measured permeability to $CuCl_2$ and proton conductivity of selected membranes

Membrane	δ (cm)	σ (S/cm)	K (cm^2/s)	σ/K (S s/cm^3)
Nafion 117	0.020	0.0840	4.6×10^{-7}	1.8×10^5
CG2	0.003	0.0054	9.4×10^{-10}	57×10^5
CEM (IEC = 2.0)	0.01	0.038	2.1×10^{-8}	18×10^5
CM2-base	0.035	0.083	5.8×10^{-9}	143×10^5

Note: δ membrane thickness, σ membrane proton conductivity, K membrane permeability to $CuCl_2$, σ/K is called membrane selectivity, namely, the ratio of conductivity and permeability
Data from Naterer et al. (2011a)

A buffer solution comprising 1 mol/L $CuCl_2$ dissolved in 10 mol/L HCl(aq) was placed in the solute side compartment of the permeability cell (see Fig. 6.22a). In the other cell—solvent side—a solution of 10 mol/L HCl(aq) exists. The diffusion process allows various species to pass through the membrane and consequently the solvent side concentrates in $CuCl_2$. ANL determined the $CuCl_2$ concentration at the solvent side by UV/visible light spectroscopy analyses. The $CuCl_2$ concentration in the buffer solution at the solute side has been assumed constant in time. The $CuCl_2$ concentration measurements were performed at several time instances such that the permeability could be determined as a function of time. Two sets of measurements were performed at ANL, namely, one at ambient temperature and the other at 353 K. It was observed that the permeability at 353 K is higher than at ambient temperature in all tests. The measurements at ambient temperature included a test of a commercial NAFION 117 membrane for which the permeability has been found to be $K = 1.3 \times 10^{-8}\ cm^2/s$. The permeability of NAFION 117 to $CuCl_2$ has been proposed by ANL as a reference value and a target was set to develop proton-selective membranes with a permeability to $CuCl_2$ reduced to one-tenth of NAFION 117; thus the permeability target is $K = 1.3 \times 10^{-9}\ cm^2/s$ (Naterer et al. 2011a).

Table 6.13 shows membranes tested at ANL, AECL, PSU, and GTI for $CuCl_2$ permeability. The main results regarding the permeability are listed in Table 6.14.

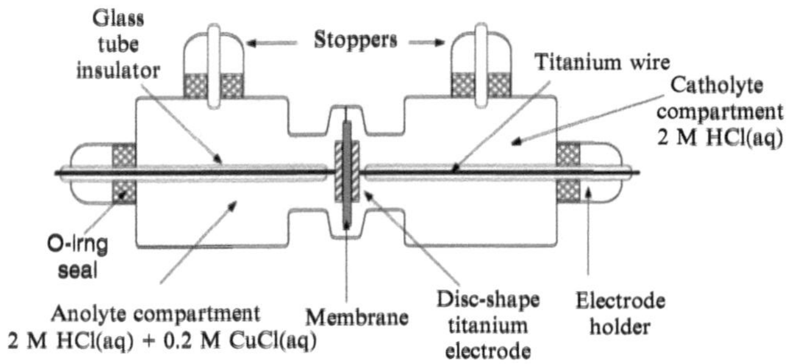

Fig. 6.23 Schematic of the custom-made "clip"-type two-probe conductivity cell used in proton conductivity tests of membranes performed at PSU (modified from Balashov et al. 2011)

Some of the membranes were tested separately in a custom-made electrochemical cell at AECL. As reported by Naterer et al. (2011a), Radel membranes from PSU showed higher copper diffusion than Nafion when tested in an electrochemical cell of AECL, although the test made with an ANL permeability cell showed that these membranes have about a ten times lower diffusivity of $CuCl_2$ than NAFION 117.

The conductivity to protons of various membranes was characterized at PSU with the aid of a custom-made through-plane conductivity cell of which a schematic is presented in Fig. 6.23. The cell comprises two compartments separated by a membrane which is sandwiched between two titanium disks that serve the role of electrodes. The compartments were filled with solutions that simulate the anolyte and catholyte of an actual electrochemical cell. An electrochemical impedance spectroscopy (EIS) method was applied to eventually determine the membrane conductivity in chloride media. The method is described in detail in Balashov et al. (2011), which performed tests with a NAFION 115 membrane. Additional results of the same group were reported by Naterer et al. (2011a) and summarized in Table 6.14. All conductivity measurements were made at ambient temperature. In the same table, the ratio between conductivity and permeability is also presented for each membrane. This ratio is a measure of the membrane selectivity to protons. The membrane with the highest selectivity should be the best choice. The data from Table 6.14 indicates that the membrane CM2-base developed at PSU has the best selectivity.

Experimental observations made at AECL with their custom-made electrochemical cell revealed that copper(II) diffusion through Nafion membranes shows a strong dependence on membrane pretreatment conditions. This may explain why modified Nafion membranes do not perform as well as non-treated membranes.

6.4.4 Copper-Chloride Complexation Studies

The presence of copper-chlorides (CuCl, $CuCl_2$) in anolyte solution leads to the occurrence of various complexation reactions of Cu(I) and Cu(II). It is indicated in

Table 6.6 that copper has two oxidation states (+1 or +2), while chlorine has four oxidation states (−1, +1, +5, +7). Therefore, copper and chlorine may form several molecular and ionic species. The copper (II) cation forms a pale-blue octahedral complex with six water molecules $[Cu(H_2O)_6]^{2+}$.

In chloride media, Cu(I) can exist as several different anionic species (for example, $CuCl_2^-$, $CuCl_3^{2-}$). The abundance of each species is dependent on the concentration of chlorine ions. The copper complex reacts with chlorine ions in aqueous solutions according to the following reactions at equilibrium that describe the complexation process (see Trevani et al. 2010):

$$[Cu(H_2O)_6]^{2+} + Cl^- \rightleftarrows [CuCl(H_2O)_5]^+ + H_2O \tag{6.16}$$

$$[Cu(H_2O)_6]^{2+} + 2Cl^- \rightleftarrows CuCl_2 \cdot 4H_2O + 2H_2O \tag{6.17}$$

$$[Cu(H_2O)_6]^{2+} + 3Cl^- \rightleftarrows [CuCl_3(H_2O)_3]^- + 3H_2O \tag{6.18}$$

$$[Cu(H_2O)_6]^{2+} + 4Cl^- \rightleftarrows [CuCl_4(H_2O)_2]^{2-} + 4H_2O. \tag{6.19}$$

It can be observed that both copper(I) species and copper(II) species can be formed in the electrolyte of the electrochemical cell. Furthermore, molecular cupric chloride is formed according to (6.17) as a hydrated complex. Chemical equilibrium data of copper complexation processes is essential for modeling speciation in both the anolyte and catholyte of the electrochemical cell and copper transport processes. Moreover, it is essential to determine equilibrium constant data and chemical activities at high temperatures and pressures, because it is expected that in an actual process, the electrochemical cell for hydrogen generation will operate at ~15 MPa and temperatures up to 423 K (see Trevani et al. 2010).

Aiming to determine equilibrium constants and other relevant parameters related to copper-chloride complexation within the electrochemical cell, Trevani et al. (2010) performed a series of experiments at high pressure and temperature. The experimental data was further processed by various modeling methods. The testing pressure was 9 MPa for two operating temperatures, namely, 373 and 423 K. An experimental arrangement comprising a high-pressure titanium cell of 0.34 cm^3, having a sapphire window (1.72 cm optical path length), was coupled to a custom-made automatized dosing system for chemicals (e.g., copper-chloride). The concentration measurements were performed with a UV-visible spectrometer which was used to collect spectral data in the range 200–1,100 nm with a browsing step of 1 nm.

The experimental data at high pressure and previously published data at low pressure have been further used to calculate the association constants for the Helgeson–Kirkham–Flowers (HKF) equation of state, which models the behavior of any species in water. The HKF is specifically used to predict the equilibrium constants of various reactions in aqueous solutions. Based on their experiments and modeling, Trevani et al. (2010) predicted the equilibrium constant for equilibrium

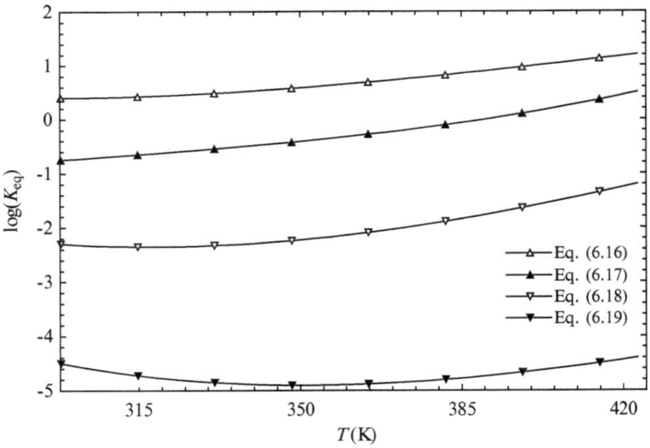

Fig. 6.24 Equilibrium constant of complexation reactions of the copper(II) cation with chlorine anion in aqueous solutions [data from Trevani et al. (2010)]

reactions in (6.16)–(6.19). The results are shown graphically in Fig. 6.24. It is interesting to note that thermodynamics favors reaction (6.17) at high temperature and intermediate chlorine concentrations, which means that the dominant species in these conditions will be the hydrated cupric chloride, $CuCl_2 \cdot 4H_2O$, with over 90 % species abundance (see Trevani et al. 2010).

6.4.5 Full-Cell Experiments

Full-cell experiments for process #7 (Table 6.2), namely, CuCl/HCl electrolysis (or electrochemical chlorination), were performed at AECL and PSU in a cell configuration that corresponds to the Cu-H^+ system described in Fig. 6.15. The essential difference between AECL and PSU cells is that the first uses an aqueous hydrochloric acid solution at the cathode while the other uses water as a proton transport medium through the membrane.

AECL has successfully demonstrated the production of hydrogen over a period of several days with a CuCl/HCl electrolysis cell. The initial experiments were performed with a commercial water electrolysis cell from Electrochem Inc. under low-flow test conditions. The cathode and anode were custom-made, produced from CFP (type Toray EC-TP1-060) painted on one side with a Pt/XC-72R catalyst mixture. Subsequent experiments were performed at higher flow rates using a custom-made electrolysis cell with a catalyst-free anode (graphite separator plate). At the cathode, a catalytic layer was applied by direct spraying of Pt/XC-72R catalyst mixture on the GSP. Figure 6.25 shows photos of the two electrolysis cells at AECL.

The cells were installed into a test loop of which a diagram is shown in Fig. 6.26a and a photograph in Fig. 6.26b. As described by Suppiah et al. (2009), the catholyte

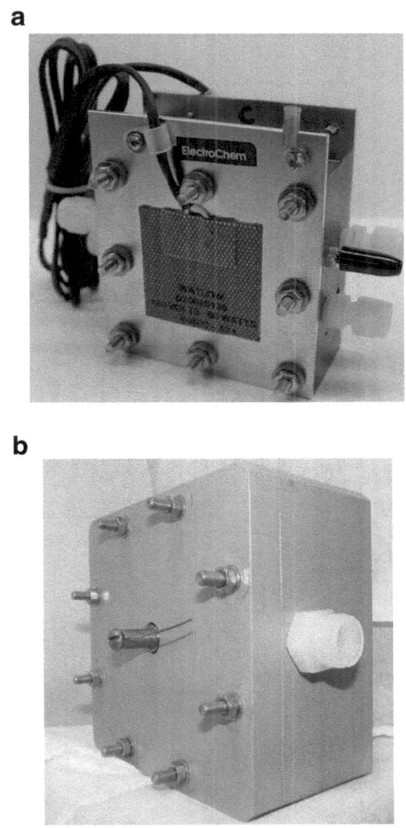

Fig. 6.25 Single cells for electrochemical step of copper–chlorine hybrid cycle for experimental investigations at AECL (courtesy of AECL). (**a**) Commercial cell with custom electrodes, (**b**) Custom-made cell

is kept in a 4 L catholyte reservoir #1 which is circulated with a peristaltic pump and directed to a column comprising anion exchange resin (AER) (see figure) containing a DOWEXTM 21 K XLT AER. This removes complexes of copper-II/ chloride. Further, the catholyte is directed into a column containing a chelating ion exchange resin CR (see figure) of type DOWEXTM M4195 which removes uncomplexed copper (II) cations. After the electrochemical process at the cathode, the hydrogen-rich catholyte is returned to catholyte reservoir #1.

The anolyte is kept in the anolyte reservoir #2 which is circulated with a peristaltic pump through the anode compartment where Cu(I) oxidizes to Cu(II). Further, the catholyte passes through vessels #3 where copper coils are inserted. A reverse reaction occurs to recycle the anolyte, namely, Cu(II) ions from the anolyte react with metallic copper and form two moles of Cu(I) for each mole of Cu(II). Therefore the concentration of Cu(I) ions tends to increase in the anolyte reservoir. In order to compensate, an HCl vessel is installed (see figure) from which concentrated HCl is injected into the anolyte reservoir with a pump. This maintains the concentration of Cu(I) ions relatively constant for the experiment. The anolyte

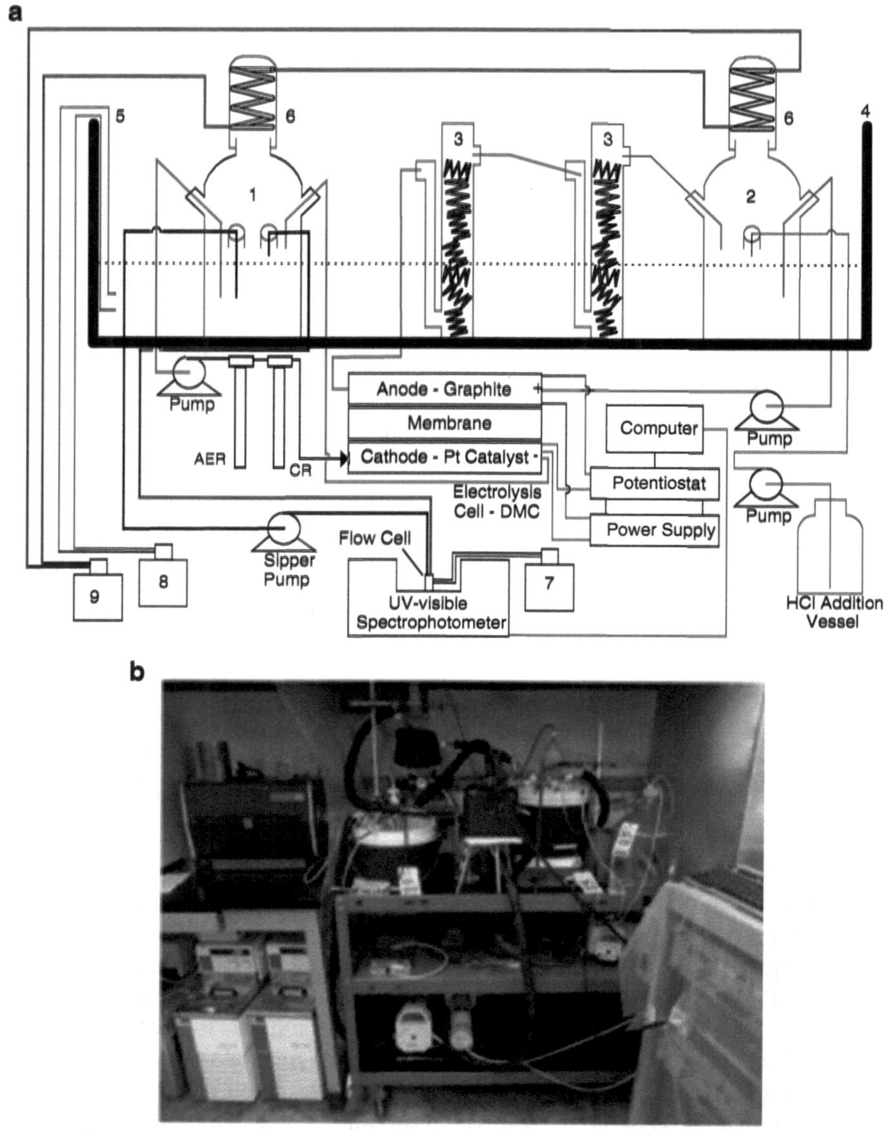

Fig. 6.26 Experimental facility at AECL for testing a single-cell electrolyzer for the CuCl/HCl solution: (**a**) system diagram (modified from Naterer et al. 2011a), (**b**) photograph (courtesy of S. Suppiah)

and catholyte solutions are maintained at constant temperature by immersion of the anolyte and catholyte reservoir as well as a copper coil vessel in a thermostated water bath #4, which is coupled through lines #5 to a temperature-controlling unit #8. In addition, condenser coils #6 are used at the top of the anolyte and catholyte

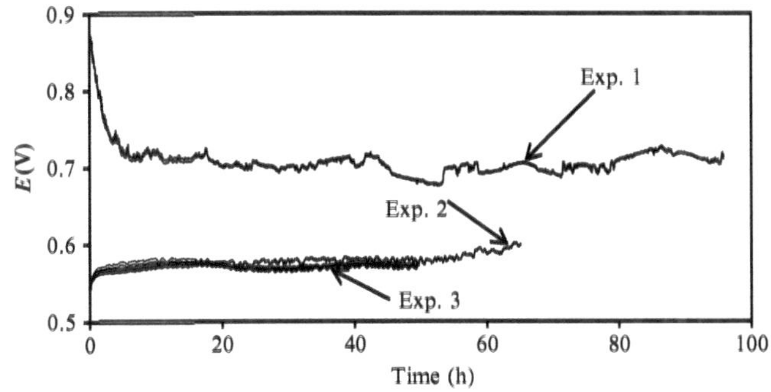

Fig. 6.27 AECL results of cell voltage for CuCl/HCl electrolysis process (courtesy of AECL)

reservoirs which are coupled to a chilled water bath #9. The condensers are used to condense gaseous species (e.g., HCl) in the anolyte and catholyte reservoirs.

An UV/visible spectrophotometer is coupled to the catholyte reservoir through a circuit including a pump. The spectrometer requires cooling via the bath #7; it is used to monitor the species concentration in the catholyte reservoir. The electrochemical cell is coupled to a power supply and a computer-controlled potentiostat.

The experiments were performed over a range of HCl concentrations in the anolyte and catholyte solutions for a domain of CuCl(aq) concentrations in the anolyte. The copper (I) oxidation reaction is presently conducted in a 0.5 mol/L CuCl(aq)/11 mol/L HCl(aq) anolyte and 11 mol/L HCl(aq) catholyte. The experiments confirmed that copper deposition is absent at such a high catholyte concentration in hydrochloric acid, even at 1,000 A/m^2.

The catholyte flow rate was 1.5 L/min and the anolyte flow rate was 1.25 L/min. A zero-gap configuration was used. The operating temperature was ~325 K. The temporal records of cell voltage (E) are presented in Fig. 6.27 for three experimental tests. The experimental data (Exp. 1) was obtained during preliminary experiments with a cell with conventional design; the anolyte was 0.5 mol/L CuCl(aq) in 10 mol/L HCl(aq) and the catholyte was 10 mol/L HCl(aq). The cell voltage was quasi-stable at 0.7 V. As mentioned by Naterer et al. (2011a), stable cell performance at voltages down to 0.477 V has been observed for 10 h using an electrolysis cell of conventional design.

The data sets (Exp. 2) and (Exp. 3) were obtained using an electrolysis cell having a proprietary (nonconventional) design that is capable of suppressing copper species crossover into the catholyte. The proprietary cell of AECL uses a Pt-based electrocatalyst at the cathode and a CEM. No catholyte copper was observed when the electrolysis was conducted using the new electrolysis cell. The results—produced for 0.5 mol/L CuCl(aq) in 11 mol/L HCl(aq) at the cathode and 11 mol/L HCl(aq) at the anode—show very promising stable performance at a cell voltage of about 0.57 V over a period of about 50 h.

A polarization curve with the modified cell of AECL is presented by Suppiah et al. (2009) for 1 mol/L CuCl(aq)/6 mol/L HCl(aq) anolyte and 6 mol/L HCl(aq)

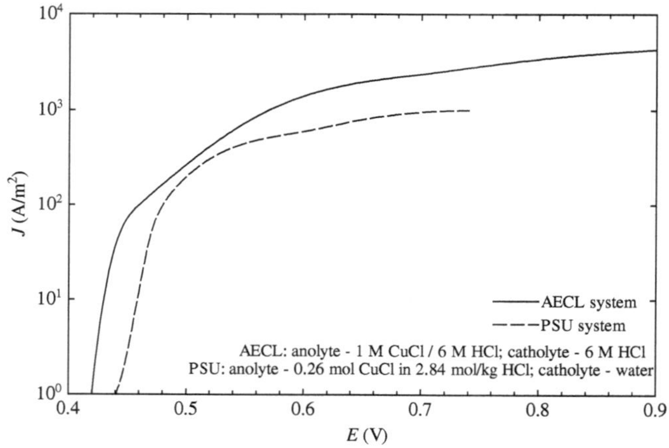

Fig. 6.28 Polarization curves obtained with full-cell systems [data from Suppiah et al. (2009) for AECL, and Balashov et al. (2011) for PSU]

catholyte. It shows that the current density is 1,000, 2,380, 3,420, and 4,290 A/m^2 for 0.57, 0.7, 0.8, and 0.9 V, respectively. Figure 6.28 shows the polarization curve for these experiments. These results show that it is possible to obtain high current densities with the cell at reasonably low cell voltages.

Researchers at PSU reported an experimental investigation of a CuCl/HCl electrochemical process performed with a commercial cell for water electrolysis. The cell is supplied by Electrochem Inc. and has a proton exchange membrane of 5 cm^2, which is the same as the effective area of the electrodes. NAFION membranes were used with two thicknesses, namely, 127 and 178 μm. Two carbon paper diffusion layers (AA10, Ion Power Inc.) were applied to each side of the membrane after which graphite bipolar plates were attached. There is proprietary undisclosed information about the electro-catalysts. The experimental setup at PSU is different than developed at AECL mainly because it does not recycle the spent anolyte after its passage through the cell. There is a large reservoir of anolyte which supplies a solution with CuCl(aq) concentrations in the range of 0.1–0.3 mol/kg in 1.5–4 mol/kg HCl(aq). The catholyte was deionized water. The anolyte flow rate was varied in a range of 30–68 mL/min.

The polarization results for PSU obtained with an anolyte consisting of 0.26 mol CuCl in 2.84 mol/kg HCl at 26 °C are superimposed in Fig. 6.28 for comparison purposes. The current density is 100 and 1,000 A/m^2 for 0.48 and 0.74 V, respectively.

6.4.6 Electrochemical Cell Modeling

Predictive formulations can be obtained by thermodynamic and kinetic modeling of the electrochemical cell. The cell model can be used for parametric analysis purposes aimed to provide new insight for the effective design and improvement of the electrolysis process. Electrochemical chlorination processes of CuCl (process #7, Table 6.2) have been modeled thermodynamically and kinetically at UOIT by Odukoya and Naterer (2011). The reversible cell is derived from thermodynamics, based on the Gibbs free energy of the reaction, assuming the following half-reactions:

$$\left.\begin{array}{l}\text{Anode: } 2\mathrm{CuCl} + 2\mathrm{HCl} \rightarrow 2\mathrm{CuCl_2} + 2\mathrm{H^+} + 2\mathrm{e^-} \\ \text{Cathode: } 2\mathrm{H^+} + 2\mathrm{e^-} \rightarrow \mathrm{H_2(g)}.\end{array}\right\} \tag{6.20}$$

Because two electrons are transferred for one molecule of hydrogen, the reversible open-circuit voltage is $\mathrm{ROCV} = -\Delta G/2F$, where F is Faraday constant, and ΔG represents the Gibbs free energy for the overall cell reaction: $2\mathrm{CuCl} + 2\mathrm{HCl} \rightarrow 2\mathrm{CuCl_2} + \mathrm{H_2(g)}$. The actual cell potential is given by the sum of ohmic, activation, and concentration potentials according to

$$\mathbb{E} = \mathrm{ROCV} + V_{\mathrm{ohm}} + V_{\mathrm{act}} + V_{\mathrm{conc}}. \tag{6.21}$$

The concentration over-potential is calculated based on the Butler–Volmer and Tafel equations for a given current density (j) and under the assumption that the charge transfer coefficient a at the anode and cathode are similar. Therefore,

$$V_{\mathrm{conc}} = \frac{RT}{2(1-a)F} \ln\left(\frac{j}{j_0}\right) \tag{6.22}$$

where j_0 is the exchange current density. The ohmic over-potential is a summation of over-potentials due to effective electric charge transport through the electrolyte (r_e), and the electrical resistances of the anode (r_a) and cathode (r_c). These electrical resistances can be written per unit of surface area of the electrode; using area-specific resistances, the total ohmic over-potential is

$$V_{\mathrm{ohm}} = j \times (r_e + r_a + r_c). \tag{6.23}$$

In the particular case of the CuCl/HCl electrolysis cell, one assumes that the electrolyte is a proton exchange membrane. Therefore the apparent electrical resistance of the electrolyte is obtained from the local electrical conductivity (σ_e) through a membrane via integration as follows:

$$r_e = \int_0^{\delta} \frac{\mathrm{d}x}{\sigma_e(x)}, \quad \text{where } \sigma_e(x) = \sigma_{e0} \times \frac{C_{\mathrm{CuCl}}(x)}{C_{\mathrm{H^+}}}. \tag{6.24}$$

Equation (6.24) was proposed by Odukoya and Naterer (2011) and assumes that σ_{e0} is the membrane conductivity for a molar concentration of cuprous chloride of unity. Moreover, the molar concentration of CuCl varies across the membrane as $C_{\mathrm{CuCl}}(x)$ and the proton concentration $C_{\mathrm{CuCl}}(x)$ is assumed constant. Determination of the concentration profile $C_{\mathrm{CuCl}}(x)$ can be made based on Fick's law of diffusion which can be integrated over the membrane thickness, namely:

$$C_{\mathrm{CuCl,a}} = C_{\mathrm{CuCl,c}} - \frac{n_{\mathrm{d}} \times j}{D \times F} \times \delta, \tag{6.25}$$

where subscripts a and c refer to the anode and cathode, respectively, D is the diffusion coefficient, δ is the membrane thickness, and n_{d} is the electro-osmotic drag coefficient. Since the diffusive flux is proportional to the molar concentration, Ohm's law requires that the molar concentration increases linearly across the membrane. Therefore, from (6.25) one can obtain the following:

$$C_{\mathrm{CuCl}}(x) = C_{\mathrm{CuCl,c}} \times \left[1 + \left(\frac{x}{\delta} - 1\right) \times \frac{j}{j_{\mathrm{L}}}\right], \quad \text{where } j_{\mathrm{L}} = \frac{FDC_{\mathrm{CuCl,c}}}{n_{\mathrm{d}}\delta} \tag{6.26}$$

where the limiting current density is introduced as j_{L}, because when $j = j_{\mathrm{L}}$, there is no presence of CuCl at the cathode–electrolyte interface $C_{\mathrm{CuCl}}(x = 0) = C_{\mathrm{CuCl,c}} = 0$. Substituting (6.26) into (6.24), one obtains

$$r_{\mathrm{e}} = \int_0^{\delta} \left[\left(\frac{C_{\mathrm{H^+}}}{C_{\mathrm{CuCl,c}}}\right) \Big/ \left(1 + \left(\frac{x}{\delta} - 1\right) \times \frac{j}{j_{\mathrm{L}}}\right)\right] \mathrm{d}x. \tag{6.27}$$

Assuming that the electrical resistance of the membrane is the dominant term in (6.23), then consequently V_{ohmic} can be approximated with $V_{\mathrm{ohmic}} = j \times r_{\mathrm{e}}$. Regarding the activation over-potential, this is made based on the Nernst equation; the activation over-potential is

$$V_{\mathrm{act}} = \frac{RT}{2aF} \times \ln\left(\frac{j}{j_0}\right). \tag{6.28}$$

Entropy generation in the electrolytic cell is a measure of irreversibility and voltage loss within the cell. In other words, voltage losses are a direct result of entropy production within the cell. Hence, entropy generation is a robust way to analyze the irreversibilities within the cell. There are other sources of irreversibility within the system, which could impede the overall performance of the cell. These include frictional losses and losses within other devices connected to the cell. The over-potential is a measure of irreversibility. According to (6.21), the total potential that must be applied to the electrodes is a summation of the reversible open-circuit voltage and various over-potentials. Each over-potential has an irreversibility. The sum of all over-potentials, $\mathrm{OVP} = V_{\mathrm{ohm}} + V_{\mathrm{act}} + V_{\mathrm{conc}}$, quantifies the total irreversibilities. Therefore, the generated entropy is $S_{\mathrm{gen}} = 2F \times \mathrm{OVP}/T$, where T is the operating temperature in K. Based on the above equations, one obtains the following expression for generated entropy:

$$S_{\mathrm{gen}} = \frac{2F}{T} \times \left[j \times r_{\mathrm{e}} + \frac{RT}{2aF} \times \ln\left(\frac{j}{j_0}\right)\right]. \tag{6.29}$$

Table 6.15 Operating parameters for electrochemical cell modeling

Operating parameters	Value
Temperature of the cell, T (K)	373.15
Exchange current density, j_o (A/m^2)	0.7
Activation potential, η (mV)	0.515
Electrons transferred in the reaction, Z	2
Charge transfer coefficient, a	0.5
Molar weight of mixture, M (kg/kmol)	0.17048
Universal gas constant, R (J/mol K)	8.314
Faraday constant, F (C/mol)	96,500
Concentration of CuCl (mol)	1.5
Concentration of HCl (mol)	6
Over-potential, E_0 (mV)	0.515
Reversible open circuit voltage, ROCV (mV)	−1.009

Source: Odukoya and Naterer (2011)

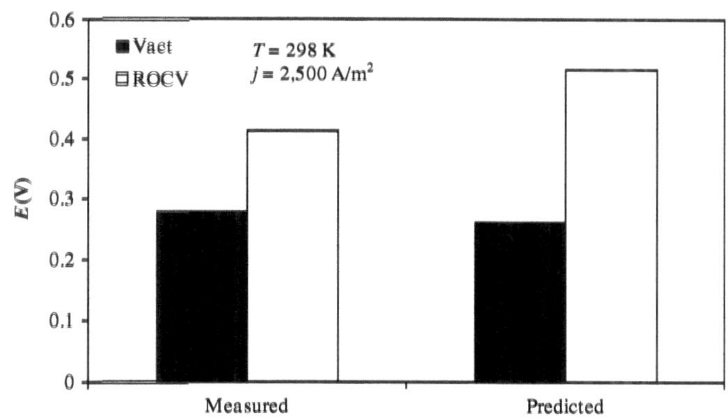

Fig. 6.29 Comparison between predicted and measured data. Predictions are based on modeling by Odukoya and Naterer (2011). Measured data is taken from Stolberg et al. (2008)

Unlike other parameters such as over-potential or polarization, entropy production can provide a more robust metric, from which all irreversible losses within an electrochemical system can be characterized. For example, entropy production encompasses both irreversible voltage losses of concentration polarization within the electrode, as well as viscous losses induced by frictional irreversibilities of fluid motion within the electrochemical cell. The operating parameters in the modeling are listed in Table 6.15; they are intended to replicate conditions of experimental studies conducted by AECL for the CuCl/HCl electrolysis cell.

Figure 6.29 shows a comparison between measured current densities and the predictions. The comparison shows that the magnitudes of the predicted over-potential in the new formulation show close agreement with experimental half-cell data reported previously by Stolberg et al. (2008) for cuprous chloride electrolysis. It can be observed that the activation potential shows only about a 6 % difference between the experimental activation potential and the predicted results. Note that

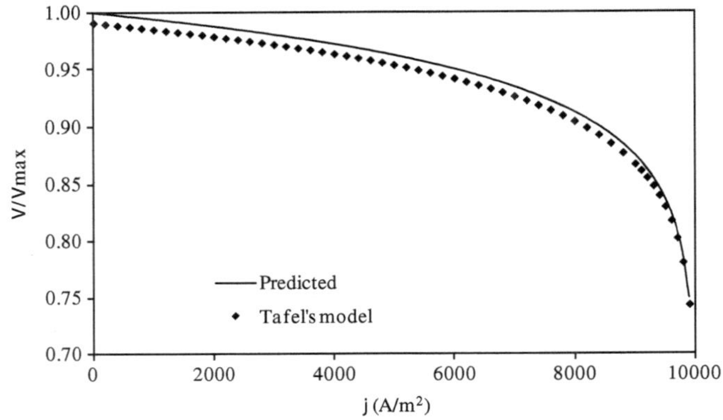

Fig. 6.30 Model validation through comparison with Tafel's reference model [data from Odukoya and Naterer (2011)]

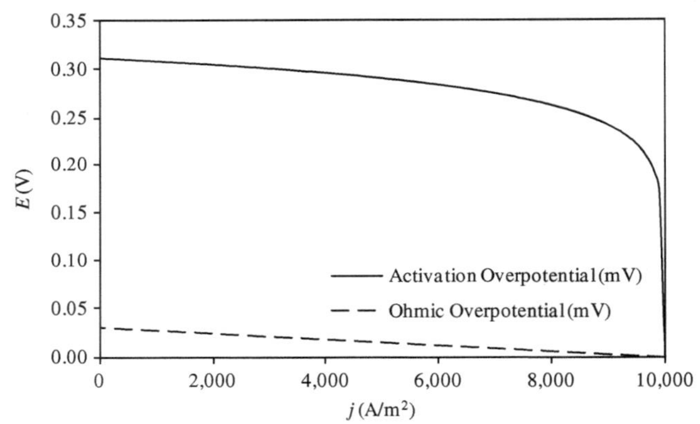

Fig. 6.31 Variation of ohmic and activation over-potentials with current density [data from Odukoya and Naterer (2011)]

the comparison is made for a single operating condition of 298 K and current density is 2,500 A/m^2. Additional model validation is presented in Fig. 6.30 where the Tafel model for over-potential prediction (see Chap. 4) is used as a comparison base. The predictions of the new formulation by Odukoya and Naterer (2011) were compared to the Tafel model. Close agreement between the results provides a useful validation of the current formulation, as the maximum difference between both methods is only about 0.08 %. This comparison serves as a useful verification of the model for predicting entropy production during the oxidation of cuprous chloride.

Based on the parametric analyses with the validated model, it shows that the activation over-potential is much greater than the ohmic over-potential. This is indicated in Fig. 6.31. The change in current density significantly affects the

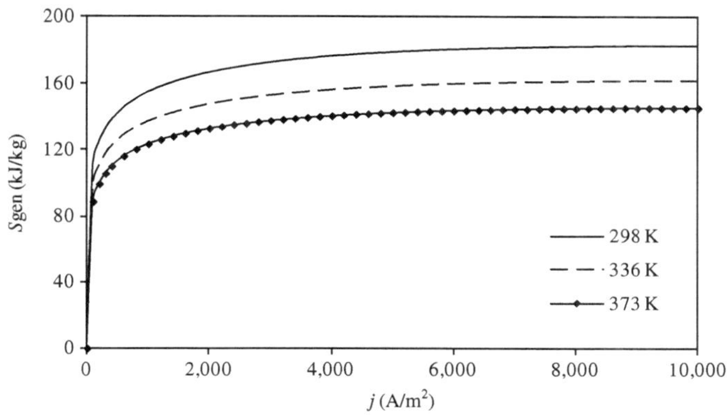

Fig. 6.32 Variation of generated entropy with current density and operating temperature [data from Odukoya and Naterer (2011)]

activation over-potential. This increase in over-potential is about 81 %, when compared with its value at the lowest current density. The ohmic over-potential can be neglected at very high current densities, but it has a significant effect at low and intermediate current densities (0–7,000 A/m^2). The ohmic resistance increases linearly with current density, as expected by the relationship of Ohm's law.

The operating temperature of the electrochemical cell has a significant effect on the entropy generation in Fig. 6.32. The entropy generation for the three temperatures is nearly equal until about 100 J/kg K. Beyond a current density of 2,000 A/m^2, the rate of change of entropy production is not significant for each temperature under consideration. It can be inferred that some level of equilibrium has been obtained for the electrochemical reaction. When the temperature of the cell varies from 298 to 373 K, the entropy generation decreases by about 25 %. Thus, increasing the cell temperature reduces the energy required by the electrochemical cell. For a cell temperature of 298 K, a percentage increase of 58 % was observed between a current density of 100 and 4,000 A/m^2, while the increase beyond this range is only about 3.6 %.

The new model developed by Odukoya and Naterer (2011) predicts the entropy production in the electrochemical cell, particularly diffusive irreversibilities of cuprous chloride electrolysis. The ohmic resistance appears to have minimal effect on the electrolytic cell performance at high current densities. A significant effect of current density on over-potential was observed at low current densities (0–250 mA/cm^2). From the entropy production results, it is preferable to operate the electrochemical cell at higher temperatures to reduce the irreversible losses. However, there are other limitations at high temperatures due to the corrosive nature of hydrochloric acid. The current density has a significant effect on entropy production between 0 and 400 mA/cm^2. The exchange current density has been shown to have little effect on the cell over-potential. The activation over-potential has the most significant effect on the cell performance. Although the ohmic resistance has little effect at high current densities, the effects become significant

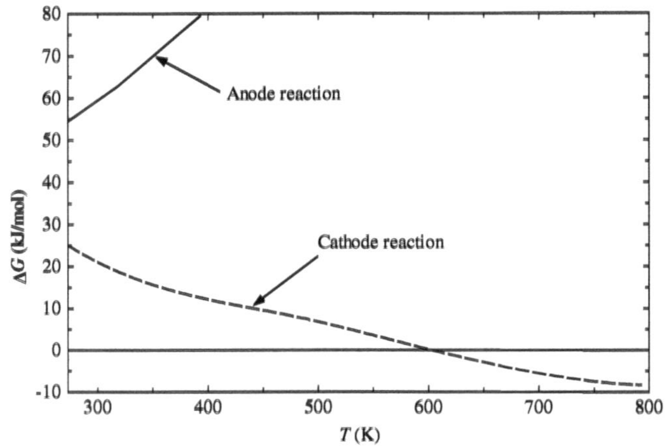

Fig. 6.33 Gibbs free energy for half-reactions of Cu–Cl copper disproportionation process [data from Lewis et al. (2003)]

at lower current densities. Also, high charge transfer coefficients are undesirable, as they increase the entropy production in the system. At a high charge transfer coefficient, the energy required for cell operation nearly doubles, which makes the system less commercially viable.

Other modeling results obtained at UOIT relate to thermodynamic analyses through energy and exergy balance equations for the electrolytic copper production step via disproportionation process #1 (see Table 6.2). Based on Table 6.11, the half-reactions for the disproportionation process can be written as (see Lewis et al. 2003)

$$\begin{cases} 4CuCl + 4Cl^- \rightarrow 4CuCl_2{}^- \text{ at anode} \\ 4CuCl_2{}^- \rightarrow 2CuCl_2 + 2Cu + 4Cl^- \text{at cathode.} \end{cases} \tag{6.30}$$

Using the process representation in (6.31), the Gibbs free energy of the electrochemical process can be calculated as a function of temperature for anodic and cathodic reactions. This data can be further used in thermodynamic analyses through energy and exergy. Figure 6.33 shows the variation of Gibbs free energy. This data reveals that the cathode half-reaction is facilitated by higher temperature. In other words, the reversible cell potential at the cathode can be very low, or reversed if the cathode is maintained at high temperature.

A thermodynamic modeling study of the CuCl disproportionation process has been reported in Orhan et al. (2008a). Several factors affect the energy balances for systems involving chemical reactions. Some of these apply generally, without regard for whether a reaction occurs. For example, it is necessary to determine work and heat transfer, as well as the importance of kinetic and potential energy. Other considerations relate to the occurrence of combustion or other reactions.

For example, it is important to have the states and conditions of the reactants before a reaction occurs, and if fluids are premixed. The state of the products also must be assessed, e.g., if water formed in the reaction has condensed.

The exergy efficiency can be formulated for the reacting system. At steady state, the rate at which exergy enters the reacting system equals the rate at which exergy exits plus the rate at which exergy is destroyed within the system. It is assumed that the reactor is well insulated, so there is no heat transfer and thus no accompanying exergy transfer. The exergy efficiency is defined usually with $\psi = ex_{out}/ex_{in}$. In the analysis, a preliminary analysis assumes that the reaction is complete. Under this assumption, one can obtain the upper bound of exergy efficiency and its correlation with reaction temperature and the temperature of the environment. The results reported in Orhan et al. (2008a) show that the exergy efficiency is higher at lower reaction temperatures. The relationship of exergy efficiency with reference-environment temperature is obtained for a constant reaction temperature of 353 K and with a reaction temperature evaluated at a constant reference-environment temperature of 298 K.

6.5 Hydrolysis Reaction Step

6.5.1 *Thermodynamics of Hydrolysis Reaction*

Thermodynamic analysis is a key step in the study of any thermochemical reaction. The aim is to determine the equilibrium yields and identify possible side reactions. Processes #2 to #5 listed in Table 6.2 are hydrolysis processes encountered in various versions of the copper–chlorine hybrid cycle. Processes #2 and #3 which form copper oxychloride have been actively studied. Process #8, oxychlorination, is a reaction between cupric chloride and steam. The reaction is endothermic and non-catalytic and operates between 625 and 675 K as described in Table 6.2.

The Gibbs free energy and enthalpy of the reaction at 673 K was estimated by Lewis et al. (2003) as ~40 kJ/mol H_2O and ~117 kJ/mol H_2O, respectively, based on the assumption that the molar enthalpy and entropy of copper oxychloride ($CuO{\cdot}CuCl_2$) can be represented as a summation of enthalpies of cupric oxide (CuO) and cupric chloride ($CuCl_2$). Although small, the Gibbs free energy of the reaction is positive. Therefore the reaction is not spontaneous and one may expect relatively low yields.

Besides the main reaction—(6.2), the desired reaction—several side reactions may occur. Thus, hydrolysis is considered by researchers the most challenging reaction of the cycle—see Ferrandon et al. (2008). Table 6.16 shows some of the main reactions that may occur when cupric chloride is in the presence of steam at high temperature. The main products in the hydrolysis reactor are copper oxychloride and hydrochloric acid. However, the temperature in the hydrolysis reactor is sufficiently high for copper oxychloride to decompose partially (reaction #2) and

Table 6.16 List of desired and possible side reactions in a $CuCl_2$ hydrolysis reactor

ID	Reaction	Remarks
1	$CuCl_2(s) + H_2O(g) \rightarrow CuO \cdot CuCl_2(s) + HCl(g)$	Desired reaction, hydrolysis, (6.2)
2	$CuO \cdot CuCl_2(s) \rightarrow 0.5O_2(g) + 2CuCl(s, l, g)$	Thermolysis of copper oxychloride, (6.5)
3	$CuCl_2(s) \rightarrow CuCl(s, l, g) + 0.5Cl_2(g)$	Thermolysis of $CuCl_2$ with CuCl formation
4	$3CuCl(g) \rightarrow Cu_3Cl_3(g)$	Formation of tricopper trichloride
5	$Cl_2(g) + H_2O(g) \rightleftarrows 0.5O_2(g) + HCl(g)$	Reverse Deacon and Deacon reactions

possibly release gaseous oxygen and cuprous chloride. The CuCl may be present as solid, liquid, and/or gas. If the temperature is lower, below the normal boiling point of CuCl, sublimation of solid CuCl may occur—see Ferrandon et al. (2008). For a temperature over the melting point, liquid and vapor CuCl are produced. The most important side reaction is the cupric chloride decomposition (process #3 in the table) which generates cuprous chloride and chlorine.

It is also possible that tricopper trichloride (a gaseous compound, Cu_3Cl_3) is formed from gaseous cuprous chloride directly. Cuprous chloride formed in the hydrolysis reactor is not necessarily problematic because this is the desired product of copper oxychloride decomposition within the cycle. The main problem is the formation of molecular chlorine which must be recycled in the system. Formation of significant quantities of chlorine negatively impacts the cycle of chemical reactions within the thermochemical cycle. Therefore, the process parameters and reactor must be selected and designed such that the chlorine yield is minimized or avoided. In addition, there may be traces of tricopper trichloride generated as side reactions; this must be avoided too, because Cu_3Cl_3 is stable since it has a circular molecule.

Preliminary experimental results from ANL—published by Lewis et al. (2003)—show that the maximum conversion of cupric chloride occurs when steam is supplied by about 20 times more than the stoichiometric quantity. Ferrandon et al. (2010a, b) show that in a spray atomizer operated under vacuum, the conversion of $CuCl_2$ is facilitated better than a reactor operating at a pressure higher or equal than atmospheric. Because the conversion of $CuCl_2$ is not complete, it becomes possible that $CuCl_2$ enters the next reactor—the thermal decomposition reactor for copper oxychloride. In this case, due to a higher operating temperature, thermolysis of cupric chloride occurs and releases chlorine gas. This changes the partial pressure of gases in the reactor and affects the purity of produced oxygen, while at the same time disturbs the balance of chemicals within the plant.

The variation of Gibbs free energy for each of the reactions from Table 6.16 is presented in Fig. 6.34 which has been obtained with EES software. At temperatures over 500 K, the reaction of copper oxychloride decomposition (reaction #2) becomes a spontaneous endothermic reaction. On the contrary, the reaction of interest (#1 in the table) has a positive Gibbs free energy (although small) at the temperatures of interest. This is an indication that the reaction produces lower yields. Yields can be improved if excess steam is supplied; this implies further separation of steam from other products. The cupric chloride direct

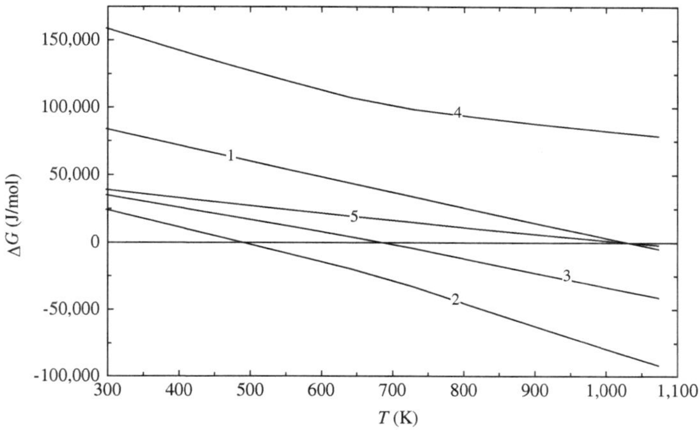

Fig. 6.34 Variation of Gibbs free energy with reaction temperature (reaction numbers correspond to ID from Table 6.16)

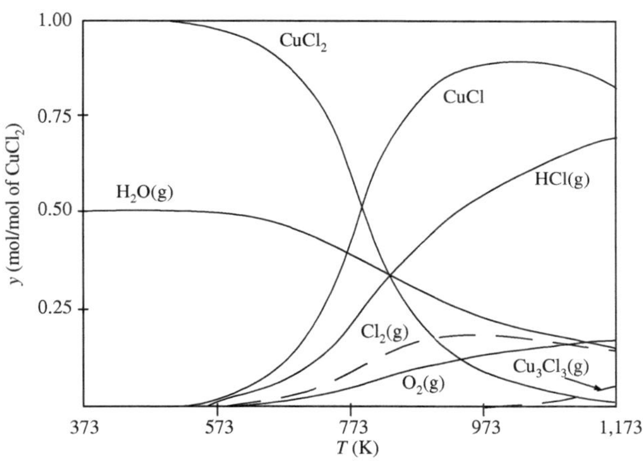

Fig. 6.35 Equilibrium yields of a $2CuCl_2$ + H_2O reaction (*dashed lines* refer to the undesired side products) [data from Lewis et al. (2009a)]

thermolysis—reaction #3—becomes spontaneous at temperatures over 700 K. Therefore its yield may be high. In order to reduce chlorine formation, the temperature in the hydrolysis reactor must be kept below 700 K. Reaction #4 also has a positive Gibbs free energy, which is high, so therefore this side reaction may generate relatively small yields.

The forward reaction #5 is the reverse Deacon reaction that occurs in the gas phase. This reaction consumes the side product Cl_2(g) and generates hydrochloric acid and oxygen. However, it has a positive Gibbs energy and thus is not favored thermodynamically. The backward reaction #5 is the Deacon reaction. This is favored thermodynamically as it will tend to generate chlorine once hydrochloric

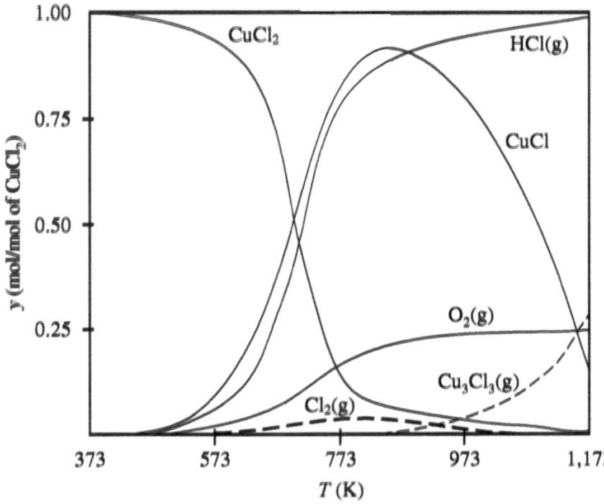

Fig. 6.36 Equilibrium yields of a $2CuCl_2 + 10H_2O$ reaction (*dashed lines* refer to the undesired side products) [data from Lewis et al. (2009a)]

acid has reacted with oxygen. Oxygen is not desired in the hydrolysis reactor in order to minimize/avoid chlorine release.

Figure 6.35 shows the calculated equilibrium yields as a function of temperature in a reaction vessel where 2 mol of cupric chloride and 1 mol of steam are supplied. It can be observed that at 573 K, there is no chlorine generation; however, $CuCl_2$ conversion is very low (below 5 %). At around 773 K, the conversion reaches around 50 %, but the chlorine yield is too high (around 0.5 moles per mole of $CuCl_2$). At 973 K, the chlorine production reaches an apparent maximum of 20 %. Over 973 K, gaseous tricupric trichloride starts to be produced.

This analysis demonstrates that molecular chlorine is the only relevant side product that occurs at temperatures of interest (623–723 K). If excess steam is used then the chlorine generation decreases. Figure 6.36 presents the equilibrium yields of cupric chloride conversion at ten times more steam than required stoichiometrically. The conversion becomes higher than 80 % at temperatures over 773 K when the generated yield of chlorine is ~0.05 moles per mole of $CuCl_2$. If the reaction is conducted at 973 K, the conversion is higher than 95 % while the molar percent of generated chlorine is near or below 1 %. However, around 0.05 mol of undesired side product, Cu_3Cl_3, may be generated per mole of cupric chloride. In the study by Lewis et al. (2009a) it is mentioned that the appropriate engineering design should maintain the temperature in the hydrolysis reactor in the range of about 600–700 K and separate the solids, liquids, and gases to eventually extract steam and hydrochloric acid before the remaining products are fed to the next reactor where thermal decomposition of copper oxychloride occurs predominantly at a slightly higher temperature.

Four hydrolysis processes are listed in Table 6.2. Process #4 has been studied in the past by Dokyia and Kotera (1976). Hydrolysis without the formation of copper oxychloride occurs according to the following simplified reaction:

$$H_2O + 2CuCl_2 \rightarrow 0.5O_2(g) + 2CuCl(l) + 2HCl(g). \quad (6.31)$$

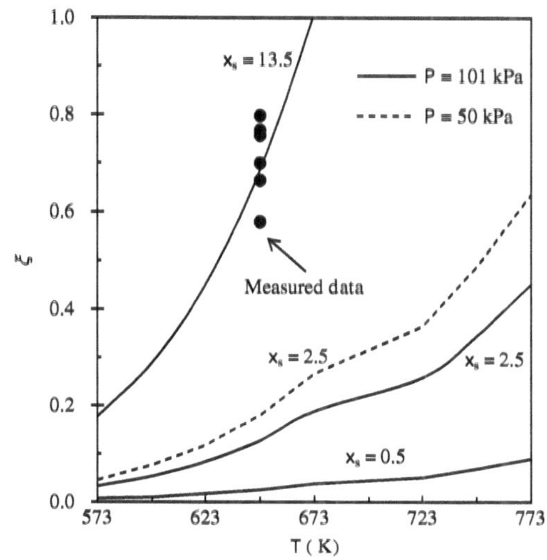

Fig. 6.37 Variation of the extent of reaction (ξ) for the hydrolysis reaction (6.2) at various steam to $CuCl_2$ molar ratios (x_s) [data from Daggupati et al. (2009)]

As mentioned by Wang et al. (2008), the reactor for process #4 must perform simultaneously a spray drying process for $CuCl_2$ dehydration and the thermochemical reaction of hydrolysis, (6.31). As indicated by Wang et al. (2008), around 27.2 mol of steam are necessary for hydrolysis of 2 mol of cupric chloride when the process is conducted according to (6.31). For this estimate, it was assumed that superheated steam is supplied at 873 K and 1 bar and its sensible heat is used until the steam temperature drops to 773 K. It is also assumed that the heat delivered by the steam is used to provide the necessary reaction heat and preheat the reactants at the required temperature.

Process #2 (Table 6.2) is hydrolysis performed with cupric chloride in hydrated form. Wang et al. (2008) reported for this case that 26.7 mol of steam are needed for hydrolysis of two moles of cupric chloride. The process #3 (see Table 6.2) assumes an additional dehydrating step within the cycle to obtain a dehydrated copper oxychloride solid. In this case, the quantity of required steam is substantially reduced, namely, one requires about 8.2 mol of steam for 2 mol of cupric chloride fed to the reactor. An exergy analysis of this reaction step was performed by Orhan et al. (2009a). It shows that the exergy efficiency of hydrolysis step—assuming that the reaction is complete—is around 75 %.

The extent of the hydrolysis reaction—(6.2) and #1 in Table 6.16—has been analyzed by Daggupati et al. (2009), as indicated in Fig. 6.37. The predicted results compare well to the measurements from Ferrandon et al. (2008) at 650 K. As derived by Daggupati et al. (2009), the extent of reaction for (6.2) can be expressed as

$$\xi = \sqrt{0.25 K_{eq} x_s^2 / (K_{eq} + 4P)},$$

where P is the operating pressure and x_s is the molar ratio of steam vs. cupric chloride supplied to the reaction. If the reaction is stoichiometric, then $x_s = 0.5$.

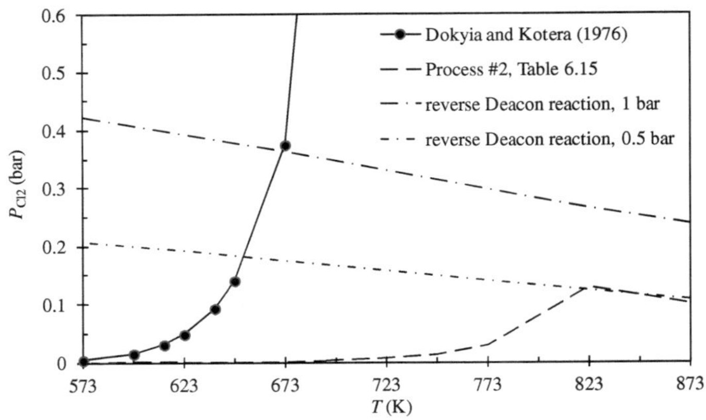

Fig. 6.38 Partial pressure of chlorine in equilibrium with the $CuCl_2$ decomposition reaction and reverse Deacon reaction (for Cl_2 consumption) [data from Daggupati et al. (2009)]

The equilibrium constant is calculated from the Gibbs energy of the reaction at the specified temperature:

$$K_{eq} = -\Delta G^0(T)/(RT).$$

From Fig. 6.37, with excess steam of 13.5 more than stoichiometric (27 mol of steam per 2 mol of cupric chloride), the reaction goes to completion at 673 K; it has reasonable equilibrium yield at 650 K. Figure 6.38 shows the variation of equilibrium partial pressure of chlorine with temperature. The criterion used to identify the operating conditions for complete consumption of chlorine is such that the equilibrium partial pressure of chlorine in the decomposition reaction should be lower than the equilibrium partial pressure of the chlorine reverse reaction for stable operation. Based on this criterion, it yields—according to Fig. 6.38—that the reactor operation can be stable for temperatures in the range of approximately 573–800 K.

6.5.2 *Overview of Experimental Reactor Developments*

A number of experiments and reactor developments for hydrolysis within the Cu–Cl cycle can be observed in the literature back to 2003. Various reactors were tested such as packed bed, fluidized bed, and spray reactors. The reactor sizes range from a small bench, operating with $CuCl_2$ samples of tens of grams, to larger engineering-scale reactors at UOIT. Most of the experimental reactors were a batch type; some semi-batch (in which only the fluids are supplied and extracted continuously and the reacting solids remain in the reactor until the experiment ends); or flow reactors. In this section, the experimental research on reactor development is examined.

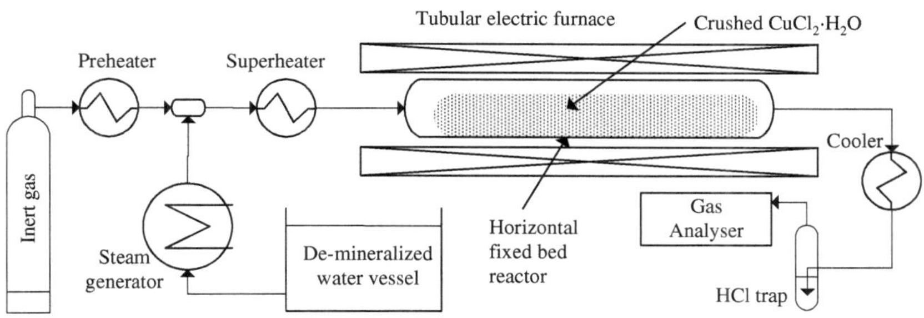

Fig. 6.39 Configuration of hydrolysis experiments in a fixed bed horizontal reactor

Basic thermochemistry experiments aimed to characterize the hydrolysis reaction were performed in a fixed bed reactor in an arrangement illustrated in Fig. 6.39. The reactor typically consists of a quartz tube in which crushed reactant $CuCl_2{\cdot}H_2O(s)$ is deposited in a fixed bed. The reactor is placed in a tubular horizontal furnace. The probe weight is measured before and after hydrolysis. During the experiments, a flow of steam is injected from one side of the reactor. The steam is prepared in an auxiliary unit comprising a demineralized water vessel and steam generator. During the experiment, inert gas is used to adjust various parameters or other procedures. For example, the partial pressure of steam may be reduced by adding an inert gas. In addition, inert gas may be used for preheating purposes, or during the shutoff period. On the opposite side of the reactor, gaseous products are collected. These cases normally comprise HCl, H_2O, and Cl_2. Hot gases may be characterized by various techniques including UV/visible spectroscopy (see Croizé et al. 2010). After cooling the product gases, HCl can be trapped in an alkaline solution (typically NaOH). The pH measurements on the HCl trap may be used to determine the hydrochloric acid concentration. Other gas analyzers may be used such as the chromatograph. Before and after each experiment, the solid sample may be characterized by X-ray diffraction (XRD). The scale of the hydrolysis process at the lab-scale experiments is characterized by the mass of the solid $CuCl_2$ sample.

There have been reported experiments in large vertical fluidized bed reactors. Such reactors consist of a quartz tube placed vertically in an electrical furnace. The sample, consisting of granular matter, is placed on a porous media distributor inside the reactor. Steam and inert gas are injected from the bottom and the bed is fluidized. Typically the steam/copper ratio is much larger than the stoichiometric amount. Figure 6.40 presents a typical configuration of a vertical fluidized bed reactor and auxiliary devices. Vertical fluidized bed reactors were tested with 1.35 g $CuCl_2$ samples at ANL and 2,250 g at UOIT.

Fluidized bed reactors provide the opportunity for better contact between the reactants—steam and particles of $CuCl_2$. However, there are disadvantages: large quantities of steam must be (re)circulated in order to be able to fluidize the bed, to transfer the necessary reaction heat, and to supply the required steam flow rate to the reaction. Large steam requirements may result in an energy penalty to the cycle.

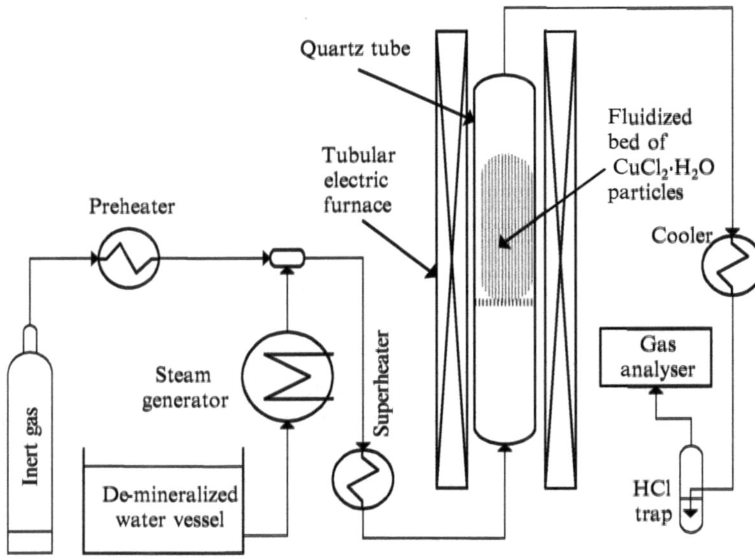

Fig. 6.40 Vertical fluidized bed reactor for hydrolysis experiments and auxiliary systems

An alternative option is a spray reactor. As opposed to a fluidized bed reactor, in a spray reactor, atomized particles formed from an aqueous solution of hydrated $CuCl_2$ are sprayed in a large vessel filled with steam at the reaction temperature. The residence time of droplets is designed to be sufficient to allow particles to reach the reaction temperature and react. The pressure in the reaction vessel can be at vacuum, which facilitates the reaction thermodynamically. This type of design was tested at ANL (Ferrandon et al. 2010a, b).

The solubility limit of $CuCl_2$ in water is about 750 g/l; therefore a solution with dissolved $CuCl_2$ can be pumped at high pressure and injected from the top of a spray reactor. The general configuration of the experimental setup with a spray reactor is presented in Fig. 6.41. The diagram shows that steam together or without a carrier gas (argon or nitrogen) is introduced into the spray reactor at the top, in a co-current configuration with sprayed particles. It is also possible to introduce steam in a countercurrent configuration. An injection nozzle is placed at the topmost position of the reaction vessel. While falling down, the atomized particles react with steam. At the bottom of the reactor, the reacted particles are collected and extracted.

The main experimental reactors developed at various institutions are listed in Table 6.17. Most of the experimental studies were performed at lab and benchtop scales. ANL developed a bench-scale spray reactor which can operate with ca. 300 g cupric chloride di-hydrate $CuCl_2{\cdot}2H_2O$ in a continuous flow system with solid products accumulated at the bottom of the reactor. As mentioned previously, UOIT developed a semi-batch fluidized bed reactor with a load of 2,250 g of solids which is the largest scale reactor developed up to this time.

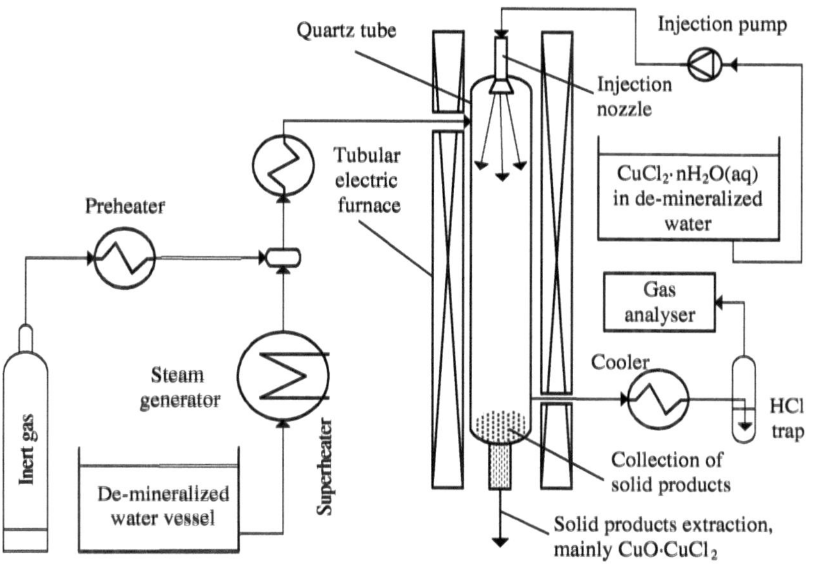

Fig. 6.41 Principle of a spray reactor system for $CuCl_2$ hydrolysis

Table 6.17 Reactor characteristics of main hydrolysis experiments

Research group	Reactor type	Reactor details	References
CEA	Fixed bed horizontal reactor	Horizontal glass reactor, 10 mm i.d. Sample of 860 mg of crushed $CuCl_2{\cdot}2H_2O$. Tubular furnace. H_2O/Cu ratio of 15. Analysis of gaseous HCl and Cl_2 at 663 K.	Ferrandon et al. (2010a)
ANL	Fixed bed horizontal reactor	70 cm long glass tube, 12.7 mm o.d. Placed in a tubular electric furnace; 40 mg of $CuCl_2$ sample; experiments performed in a temperature range of 573–623 K.	Lewis et al. (2005a, b)
	Fluidized bed vertical reactor	25.4 mm i.d., 35.6 cm height quartz tube with quartz frit in the middle. Reaction sample of 1.35 g; molar ratio of steam to $CuCl_2$ of 16:1.	Lewis et al. (2005a, b)
	Fixed bed vertical reactor	12.7 mm o.d. quartz tube with a basket and quartz frit at the bottom. Sample was ~300 mg.	Lewis et al. (2009a, b)
	Continuous spray reactor	Borosilicate glass, 1.3 m height, 11.4 cm i.d., 14 L volume; central port atomizer at top; two ports for gas injection/extraction with co-countercurrent configuration; vertical furnace with three independent heating zones. Injection of $CuCl_2{\cdot}2H_2O$ at 1–5 mL/h in hot steam at 673 K under vacuum, 0.25 bar. Argon sweep gas at 300 mL/min.	Ferrandon et al. (2010a)
UOIT	Vertical fluidized bed reactor	15 cm i.d., 1.1 m height; 2.25 kg bed weight. Flow distributor for steam injection made in ceramic porous media. Split tubular electric furnace of 15 kW with three heating zones.	Naterer et al. (2011a)

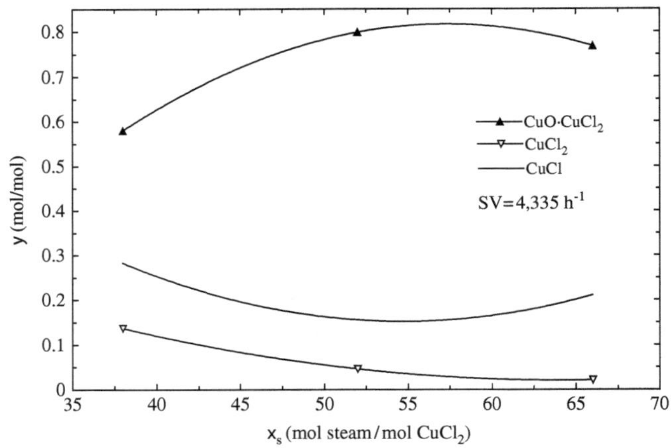

Fig. 6.42 Molar fraction of products of $CuCl_2$ hydrolysis in fixed bed horizontal micro-reactor [data from Ferrandon et al. (2008)]

6.5.3 Benchtop Experiments and Results

The first set of comprehensive experiments for hydrolysis was reported by Lewis et al. (2005b). Two types of experiments were preformed: (1) a fixed bed horizontal reactor (40 mg sample) and (2) a vertical fluidized bed reactor (1,350 mg sample). It was found that the reaction has reasonable kinetics above 573 K, although as shown in Fig. 6.34, the Gibbs free energy of the reaction is positive. The experiments confirmed that the major product of the reaction is copper oxychloride, although some amounts of cuprous chloride are also formed. Excess steam and a lower reaction temperature disfavor the formation of CuCl. However, the observed rate of reaction at lower temperatures is reduced; according to Lewis et al. (2005b) it is about ~2 times slower at 573 K vs. 623 K. In the fluidized bed reactor, the reaction temperature was kept at ~650 K with a molar steam vs. $CuCl_2$ ratio of 16:1.

A subsequent series of hydrolysis experiments was reported by ANL in a fixed bed horizontal reactor of 70 cm length and 12.5 mm o.d. placed in a tubular electric furnace. The results were presented in Ferrandon et al. (2008) for a test temperature of 650 K. The yield of copper oxychloride depends on the steam vs. cupric chloride molar ratio. The trend in Fig. 6.42 was recorded. A maximum yield of 80 % occurs at an ~57.5:1 steam vs. cupric chloride molar ratio. The CuCl by-product yield is minimized for a 55:1 steam-to-$CuCl_2$ molar ratio. No chlorine production was observed.

In addition, it was observed that the CuCl yield is lower at lower reaction temperatures; also, the yield decreases with the size of $CuCl_2$ particles. Another test performed in the same reactor at a fixed steam-to-$CuCl_2$ molar ratio of 17:1 was reported by Lewis et al. (2009b). It was observed—see Fig. 6.43—that the copper oxychloride yield increases at higher gas hourly space velocities (GHSV). Moreover, the influence of GHSV on the desired product yield is more important than the influence of the steam-to-$CuCl_2$ ratio.

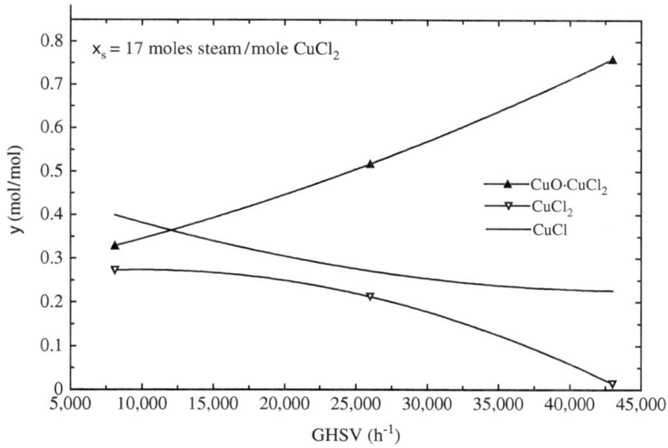

Fig. 6.43 The influence of gas hourly space velocity (GHSV) on yields of the hydrolysis reaction for a fixed molar steam-to-$CuCl_2$ molar fraction [data from Lewis et al. (2009b)]

A horizontal fixed bed reactor has been used at CEA to study hydrolysis reactions during collaborative research with ANL (Ferrandon et al. 2010a). The concentration of gaseous products was determined by Fourier transform infrared (FTIR) spectroscopy in conjunction with the use of a conductivity probe for HCl(g) and Cl_2(g) detection. During hydrolysis at 663 K, no chlorine evolution has been observed. However, when the furnace temperature was increased over 673 K, chlorine presence was detected. The chlorine concentration increases with temperature. It is determined that at 733 K, the chlorine concentration reached a maximum of 1.34 mmol/L; when the temperature was increased to 773 K, chlorine gas was not detected, which indicated that it was "consumed" in the chemical reactions.

Studies on thermochemical decomposition of cupric chloride have been reported by various researchers including Serban et al. (2004), and Ferrandon et al. (2010a). These studies are important for understanding the occurrence of chlorine as a possible side product during hydrolysis. Most studies were performed using thermogravimetric analyzers. As indicated by Serban et al. (2004), the reaction kinetics for $CuCl_2$ thermal decomposition is represented by a zero-order rate law with a pre-exponential factor $\mathcal{A} = 2.9 \times 10^7\,\text{min}^{-1}$ and an activation energy of $E = 102.2\,\text{kJ/mol}$. Therefore, based on the Arrhenius equation, the rate constant is $k = \mathcal{A} \times \exp(-E/RT)$.

6.5.4 *Spray-Type Mini-Reactor for Hydrolysis (ANL)*

A spray reactor has been developed at ANL as reported in several publications by Ferrandon et al. (2010a, b). The principle of the spray reactor system is illustrated in Fig. 6.41 and the main reactor characteristics are listed in Table 6.17. Because flow rates were relatively small, e.g., cupric chloride injection of 1–5 mL/h and liquid water at 0.5 mL/h, syringe pumps were used within the experimental setup.

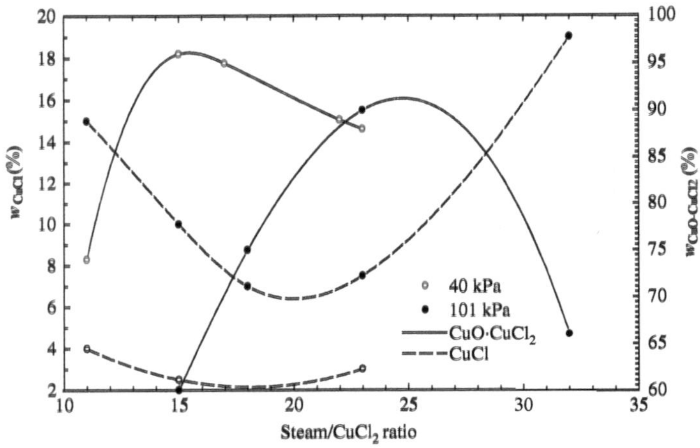

Fig. 6.44 Characterization of solid products of the hydrolysis reaction in terms of $CuO{\cdot}CuCl_2$ and CuCl concentrations by weight (w) [data from Ferrandon et al. (2010b)]

The temperature regulation system is capable of maintaining a constant temperature at the outer wall of the reactor. During the operation, due to continuous flow, the temperature varies axially from about 623 K at the upper heated side to 713 K at the lower side. The resulting solid products are collected at the reactor bottom and maintained at 423 K with electrical heating tape. After the experiments end, the solids are weighted and analyzed with specific methods (e.g., XRD). Two types of particle atomizers were used to spray the $CuCl_2$ from the top of the reaction vessel: a pneumatic nebulizer (which requires the use of a sweep gas) and an ultrasonic nozzle (which consumes electrical energy).

It was reported that an ultrasonic nozzle performs better than the nebulizer. The droplets formed by atomization with the ultrasonic nozzle have around a 25 μm average diameter. As observed in the experiments (by the use of SEM), during the reaction, the particles were broken down to 2–3 μm. The formation of a CuCl side product is mainly influenced by the steam-to-$CuCl_2$ ratio in the input stream; lower ratios are not good as they facilitate formation of CuCl. A ratio of 20:1 minimizes the CuCl formation to about 4 % by weight in the output stream. Furthermore, reduction of the operating pressure leads to an increased yield of copper oxychloride while it reduces the yield of the CuCl side product. This result—reported by Ferrandon et al. (2010b)—is illustrated graphically in the plot in Fig. 6.44.

At a reduced pressure, the maximum yield of copper oxychloride is obtained at a much lower steam/copper ratio (around 40 % steam reduction if one operates at 40 kPa with respect to 101 kPa). Ferrandon et al. (2010b) showed that the residence time of the particle in the reactor decreases when reducing the pressure. With a reduction of residence time, thermolysis of cupric chloride has a lower rate; therefore, less cuprous chloride is observed. The residence time decreases with a lower pressure because natural convection effects due to heat transfer within the reactor are diminished at lower pressure (Grash of number is lower at lower pressures).

6.5.5 Scaled-Up Hydrolysis Reactor and Test Loop (UOIT)

A hydrolysis step testing loop with a production capacity of about 1,000 times larger than lab tests performed for proof-of-concept studies has been built and tested at UOIT (see schematic diagram in Fig. 6.45). The test loop comprises a hydrolysis reactor, nitrogen flow line with a heater, high and low flow superheated steam lines, condensers for cooling the exit gas stream, gas–liquid separator, and a scrubber for noncondensable gases in the downstream region of the hydrolysis reactor. Nitrogen gas from cylinders is needed to preheat the cupric chloride during the experiment startup. Dehydrated cupric chloride particles are enclosed in the hydrolysis reactor (HR). The hydrolysis lines are heated electrically with an electric heater (EH) equipped with temperature controllers.

During the startup operation, nitrogen from the cylinder N_2 is supplied through line #1 and valve V_1 with a flow rate measured and regulated precisely with the flow-meter battery formed by the fine flow meter F_1 for low flow and the flow meter for a high flow rate F_2. A pressure and temperature station is placed in order to determine the density of nitrogen just after the flow rate measurement point. Nitrogen is heated in the electric heater up to the reaction temperature which is reached by stream #4. A valve V_2 is used to close or open the hot nitrogen supply to the reactor. The main flow of steam is generated in electric steam generator ESG_1 and superheated in the electric superheater (ESH). Fresh water is supplied at #6 and saturated vapor is delivered in #7. If needed, for flow regulation or various procedures (e.g., start-up, shutdown), a part of the saturated steam is expelled via valve V_3. A safety valve SV_1 is installed on the steam generator. The steam flow rate can be regulated via valve V_4 and measured by flow meter F_3. Superheated steam is obtained at #13 at a temperature corresponding to the hydrolysis reaction. A secondary generator of saturated steam is also used for the purpose of tuning the experimental parameters. The steam is produced in electric steam generator ESG_2 which is supplied from the deionized water vessel DSV using the water pump (WP).

The flow rate of saturated steam in #17 is regulated by the valve V_6. The hydrolysis reactor is supplied with nitrogen, steam, or superheated steam through line #19, valve V_8, and line #20. This reactor is equipped with a safety valve SV_3 and a line for hot gas exhaust #21 on which the valve V_9 is placed. The product gases of the reaction are collected in #23 and directed toward a condensation battery formed from water-cooled condensers C_1, C_2, and C_3. The booster condenser is C_1, which lowers the reaction gas mixture temperature from 673 to 523 K. The booster condenser is followed by C_2 and C_3 coupled in parallel; they are used to condense the reaction gas mixture at maximum steam flow conditions. In line #29, the condensate comprises water, hydrochloric acid in the aqueous phase, and noncondensable gases. The gases are separated in the LGS at line #33 and passed to the scrubber SCRB. In the scrubber, the remaining traces of hydrochloric acid are absorbed in water and the remaining noncondensable gases are expelled out in #34. The liquid collected at the bottom of LGS in #30 accumulates in the liquid collection vessel (LCV) where the acid is neutralized by an alkaline solution. A sample port #32 is operated with the valve V_{12} installed in order to determine the concentration of hydrochloric acid during the experiments. A chiller is used to cool the

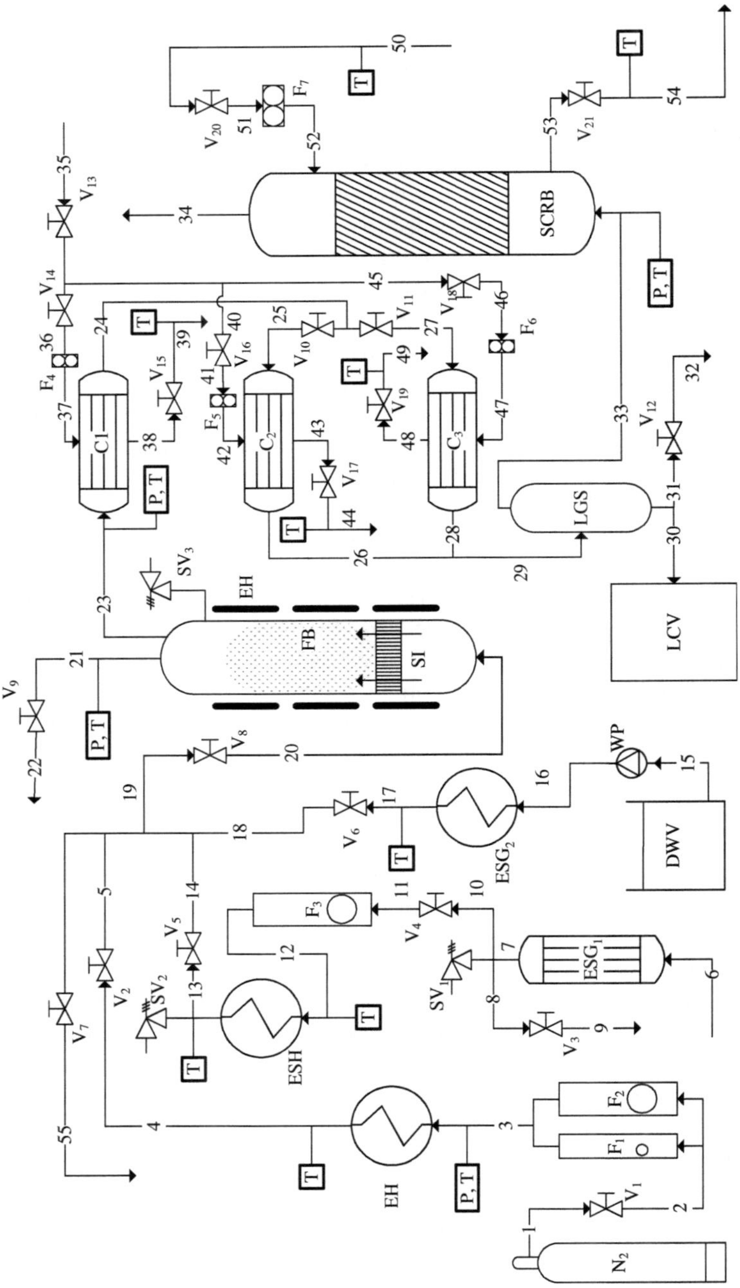

Fig. 6.45 Schematic of the hydrolysis test loop at UOIT

Fig. 6.46 Photograph of the hydrolysis test loop at UOIT

condenser with glycol–water coolant. Cold coolant is provided at #35 and divided in three streams, namely, #36, #40, and #46 toward each of the condensers. For each condenser, the flow rate of coolant and the temperature of coolant before and after the condenser are measured. Valves are installed to regulate the coolant flow. Cold water is supplied in #50 to the scrubber. The water flow rate and temperature are measured. There is also a valve V_7 which allows the exhaust of hot steam and/or nitrogen before its supply to the reactor. This maneuver may be necessary during experiments.

The photograph of the hydrolysis testing loop is presented in Fig. 6.46. Its general characteristics are given as follows.

- The main steam generator consisting of ESG_1 and ESH (see schematics from Fig. 6.45) has a capacity of 45 kW and produces superheated steam at above 773 K with a maximum mass flow rate capacity of 30 kg/h.
- The supplementary saturated steam generator ESG_2 produces 3 kg/h.
- Condenser C_1 is a high-temperature heat exchanger made of quartz glass.
- Condensers C_2 and C_3—made in pyrex glass—have a heat transfer area of 1.2 m^2.

Two types of scaled-up reactors were developed at UOIT: a vertical fluidized bed reactor and a horizontal packed bed reactor. The photograph of the fluidized bed reactor for hydrolysis is presented in Fig. 6.47.

The vertical hydrolysis reactor is made of quartz glass with an internal diameter of 150 mm and height of 1.1 m. The solid cupric chloride reactant is contacted with superheated steam injected from the bottom of the reactor. A steam injection (SI) system (in Fig. 6.45) is placed at the lower side of the reactor. This is made in ceramic porous media with the role of flow distribution. Steam injected from below creates and maintains the fluidized bed at the upper side of the reactor. A tubular electrical furnace encloses the reactor. The furnace is made of ceramic fibers containing three electrical heating elements regulated independently. The power rating is 15 kW.

Fig. 6.47 Scaled-up fluidized bed reactor for $CuCl_2$ hydrolysis at UOIT

An additional test loop was devised for the study of fluidized bed hydrodynamics with the purpose of predicting the optimum fluidization parameters as well as to improve the reactor reliability and safety. The loop has a fluidized bed reactor of 65 cm height which is filled with cupric chloride powder and fluidized with a nitrogen–helium mixture with a density ratio equivalent to steam at high temperatures inside an actual reactor. Pressure probes were installed above the distributor plate and around 10,000 measured data points were collected during the tests. A photograph of the fluidized bed test facility for cold experiments is shown in Fig. 6.48.

Figure 6.49a shows the minimum fluidization velocity using a standard deviation of pressure fluctuations with Puncochar's method. The bed height between the pressure probes was $L = 0.065$ m and around 10,000 data points were collected. Figure 6.49a presents the mean values of pressure drop variation with gas velocity in the bed. The pressure drop values increase with gas velocity in the fixed bed range, after reaching the minimum fluidization velocity. The pressure drop does not increase for a higher gas velocity in the fluidized bed range.

The reaction tests performed with the scaled-up hydrolysis test loop require a start-up sequence. The boiler is cleared by flowing steam through a bypass line to the atmosphere before starting the experiment. Cupric chloride solids are fed into the hydrolysis reactor from the top. The solids are preheated with nitrogen which also fluidizes the bed. After heating the solids to the predetermined test temperature of the reaction, steam is fed into the reactor from the bottom. The reactant superheated steam is generated with the steam boiler and superheater to a required temperature before it is introduced into the reactor. The steam flow can be regulated

Fig. 6.48 Photograph of low-temperature fluidized bed apparatus for the hydrolysis reaction at UOIT

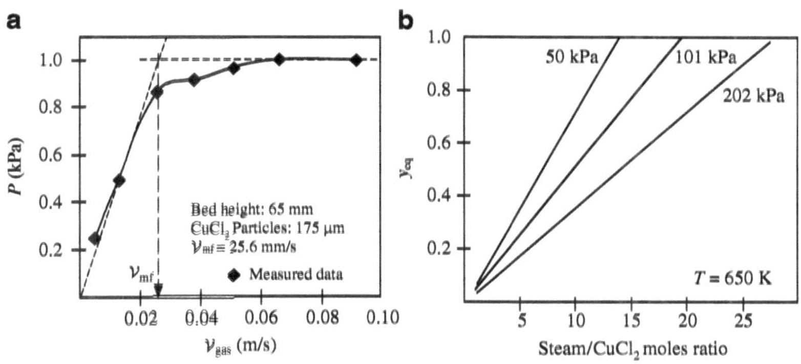

Fig. 6.49 Test results with the fluidized bed reactor at UOIT: minimum fluidization velocity ($\mathcal{V}_{mf}$) of cupric chloride during cold experiments (**a**) and equilibrium conversion (y_{eq}) of cupric chloride (**b**) [data from Naterer et al. (2011a)]

in the measurement range of 3–30 kg/h. Nitrogen gas may be used during the tests to control the reaction.

The major products in the reaction step are copper oxychloride (solid) and HCl (gas). After the reaction, the HCl gas and unreacted steam are condensed using quartz and pyrex condensers for condensation. The noncondensable gases pass through a sodium hydroxide solution in the scrubber to completely remove traces of HCl gas, if any. The liquid separated from the gas–liquid separator is taken into a neutralization tank and neutralized with alkali solution. Reaction gas mixture samples are collected after the quartz condenser C_1 to monitor progress of the reaction. The formation of HCl gas is used to estimate the cupric chloride conversion in the reaction.

Before the tests, the anhydrous $CuCl_2$ powder size distribution and volume mean diameter were quantified with a particle analyzer. The reactant was analyzed for its purity using XRD analysis which determined that the purity is over 98 %.

The tests confirmed the steady production of copper oxychloride from solid/dehydrated cupric chloride. It was found that the cupric chloride conversion increases with a higher steam–copper ratio, temperature, and reduced pressure. This recorded trend is illustrated in Fig. 6.49b. The presence of hydrochloric acid in the input steam limits the conversion of solid. From the experimental studies, a key objective is to further reduce the excess steam requirement. Also, operating parameters need to be optimized to achieve the maximum copper oxychloride yield.

6.5.6 Hydrolysis Process Modeling and Simulations

Although there has been significant experimental research reported in the literature, the hydrolysis process of cupric chloride is not fully understood. From an engineering point of view, it is important to devise a hydrolysis reactor with low steam consumption, high conversion, and no side products. Thus, modeling studies of the process were conducted and additional numerical simulations aimed to help process understanding and design of a more efficient reactor. The research at ANL led to the formulation of a model for a hydrolysis spray reactor which was reported by Ferrandon et al. (2008). The model uses ASPEN Plus simulation software. The related research at UOIT was reported in a series of papers addressing various aspects such as:

- Hydrolysis reaction kinetics (Daggupati et al. 2010b)
- Diffusion of gaseous products through $CuCl_2$ particle surface layer (Daggupati et al. 2010a)
- Conversion of solid cupric chloride reacting particles during hydrolysis (Daggupati et al. 2011a)
- Fluid-particle mass transport modeling in the fluidized bed hydrolysis reactor (Haseli et al. 2009)
- Hydrodynamics of fluidized bed in the hydrolysis reactor (Haseli et al. 2008)

In the modeling results at ANL for the spray reactor, the APEN Plus simulations extended the understanding of spray operation and to better set the operating parameters. Figure 6.50 shows these results which indicate the cupric chloride conversion with temperature for three steam vs. $CuCl_2$ molar ratios. According to these results, the conversion of cupric chloride is very high (near 100 %) at 673 K for a steam-to-$CuCl_2$ ratio of 17. The simulation predicted at the same condition that CuCl side product is generated in amounts as low as 2 % (molar) with respect to the cupric chloride. At temperatures higher than 673 K, the molar fraction of CuCl in the product stream increases fast and reaches ~17 % at 723 K. If the steam-to-$CuCl_2$ ratio (x_s) is set at higher values than 17, then the maximum conversion occurs at lower temperatures, e.g., for $x_s = 30$, the maximum occurs at around 623 K.

Mechanistic models that account for transport phenomena and reaction kinetics are needed for improving the performance of the reactors. Two main reactions must

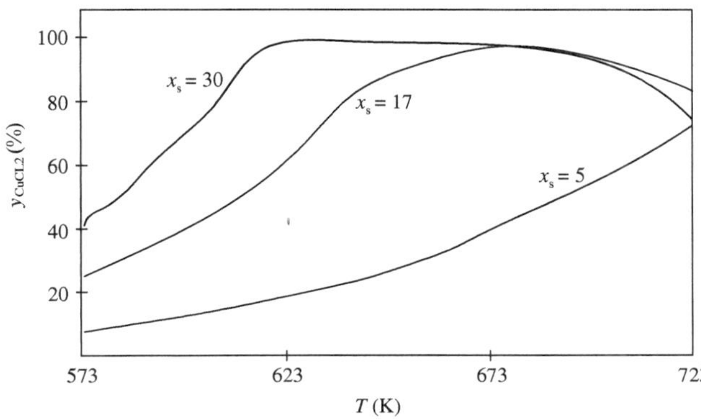

Fig. 6.50 ASPEN Plus simulation results with spray reactor developed at ANL [data from Ferrandon et al. (2008)]

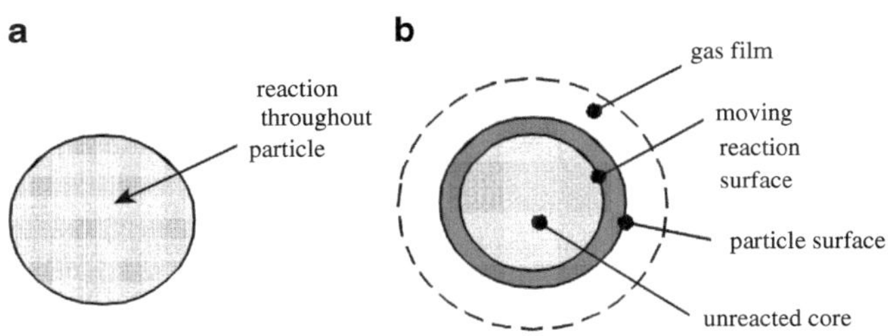

Fig. 6.51 Gas–solid reaction models: (**a**) uniform conversion model (UCM) and (**b**) shrinking core model (SCM)

be considered simultaneously in a model that accounts for gas-to-solid diffusion phenomena and gas–solid chemical reactions for hydrolysis process.

1. The hydrolysis reaction itself—with steam diffusion toward the core of a solid cupric chloride particle, steam consumption in the reaction, hydrochloric acid release, and diffusion toward particle surface.
2. The thermolysis of cupric chloride—with chlorine production and diffusion toward the particle surface; in addition, cuprous chloride is generated; depending on the reaction conditions, this can be solid or liquid (molten salt) or it can sublimate.

Any gas–solid reaction can be modeled based on two well-known models: the uniform conversion model or the shrinking core model. These mechanistic models are explained with the help of Fig. 6.51 where it is shown for gas reactant diffusion of a uniform-reaction model (Fig. 6.51a) and the solid-layer configuration of a reacting particle for a shrinking core model (Fig. 6.51b). A uniform-reaction model is used when the diffusion of gaseous reactant into a particle is much faster than the chemical reaction.

The solid reactant is consumed nearly uniformly throughout the particle. On the other hand, a shrinking core model is used if the diffusion of gaseous reactant is much slower and it restricts the reaction zone to a thin layer that advances from the outer surface into the particle. Diffusion of gaseous reactant occurs through a film surrounding the particle, after which the reactant penetrates and diffuses through a layer of ash to the surface of the unreacted core. The shrinking core model for spherical particles predicts the reaction of gaseous reactant with solid at the particle surface, and diffusion of gaseous products through the ash, back to the exterior surface of the solid. The reaction conditions are assumed isothermal, with a constant pellet size, and pseudo-steady-state conditions that are valid for gas–solid reactions.

The conversion of a particle of $CuCl_2$ in the hydrolysis reactor can be expressed geometrically based on the shrinking core reaction model as the ratio between the outer shell, consisting of copper oxychloride product, and the particle volume:

$$X = 1 - (d_c/d_o)^3, \tag{6.32}$$

where d_c is the diameter of the unreacted core and d_o is the particle radius. As reported by Daggupati et al. (2009), the maximum equilibrium conversion is given by

$$X_{eq} = \left(\frac{K_{eq} x_s^2}{4P + K_{eq}}\right)^{0.5},$$

where K_{eq} is the equilibrium constant calculated based on the Gibbs free energy at the reaction temperature, P is the operating pressure, and x_s is the steam-to-$CuCl_2$ ratio. The reaction rate for water consumption can be expressed in four different ways as indicated by Daggupati et al. (2010b), namely,

$$r_{H_2O} = \begin{cases} k_1\left(P_{H_2O} - \dfrac{P_{HCl}^2}{K_{eq}}\right) \\ k_2 C_{H_2O} \\ D_{e,H_2O}\left(\dfrac{d^2 C_{H_2O}}{dr} + \dfrac{2}{r}\dfrac{dC_{H_2O}}{dr}\right) \\ D_{e,HCl}\left(\dfrac{d^2 C_{HCl}}{dr} + \dfrac{2}{r}\dfrac{dC_{HCl}}{dr}\right), \end{cases} \tag{6.33}$$

where $k_{1,2}$ are the reaction rate constants, which are assumed to follow the Arrhenius law for temperature dependence; C are molar concentrations (moles per volume); D_e is the effective diffusivity; r is the radial coordinate; and P_{H_2O} is the partial pressure of steam. Based on (6.33), the following compact expression for $CuCl_2$ conversion for the kinetic-controlled regime during hydrolysis was reported by Daggupati et al. (2010b):

$$\frac{dX}{dt} = \frac{3(1-X)^{2/3}}{d_o \times \rho CuCl_2} \times k_1 \left[\frac{x_s - X}{x_s + X} P - \frac{1}{K_{eq}}\left(\frac{2X}{x_s + X} P\right)^2\right]. \tag{6.34}$$

Table 6.18 Parameters for kinetic modeling of hydrolysis process

Parameter	Symbol	Value	Units
Rate constant for kinetically controlled regime	k_1	2.2	mmol/m^2 s
Rate constant for first rate reaction	k_2	0.1	mmol/s
Effective diffusivity of chloride at $CuCl_2$ decomposition	D_e	2.4×10^{-9}	m^2/s

Source: Daggupati et al. (2010b)

If one assumes diffusion control for kinetics of the hydrolysis process, then a first-order reaction rate applies. In this case, the $CuCl_2$ conversion is given by

$$\frac{\mathrm{d}X}{\mathrm{d}t} = \frac{3(1-X)^{2/3}}{d_o \times \rho\mathrm{CuCl_2}} k_2 C_{\mathrm{H_2O}}. \tag{6.35}$$

The decomposition is analyzed by diffusion through the product layer to control the reaction. The total gas pressure is assumed to be constant and the pellet is assumed spherical. For the cupric chloride decomposition reaction, the rate of diffusion of chlorine through the product layer, based on the assumption of equimolar counter diffusion, is

$$r_{\mathrm{Cl_2}} = \frac{2\pi d_o d_c D_e}{d_o - d_c}\left(C_{\mathrm{Cl_2,i}} - C_{\mathrm{Cl_2,b}}\right), \tag{6.36}$$

where $C_{\mathrm{Cl_2,i}}$ is the chlorine concentration at the interface and $C_{\mathrm{Cl_2,b}}$ is the chlorine concentration in the bulk phase. The equilibrium constant of the $CuCl_2$ decomposition reaction is related to the partial pressure of chlorine, according to $K_{eq} = P_{\mathrm{Cl_2}}$. The $CuCl_2$ dissociates until the partial pressure of Cl_2 reaches the equilibrium constant, K_{eq}, at a given temperature. Since the rate of chlorine production is equated to the rate of particle shrinking during decomposition, the following equation can be obtained (Daggupati et al. 2010b):

$$3D_e t\left(\frac{K_{eq}}{RT_b} - C_{\mathrm{CuCl_{2b}}}\right) = \rho_{\mathrm{CuCl_2}} d_o^2 \left(1 - 3(1-X_d)^{2/3} + 2(1-X_d)\right), \tag{6.37}$$

where R is the universal gas constant, T_b is the temperature in bulk phase (Kelvin), t is time, and X_d is the conversion during the decomposition reaction.

In the above equations for conversion, (6.34), (6.35), and (6.37), the rate constants k_1 and k_2 and the diffusion coefficient D_e must be specified. Daggupati et al. (2010b) fitted the models to the experimental data of Ferrandon et al. (2008) of cupric chloride hydrolysis at 650 K, for particle sizes of 150–250 μm. The results are summarized in Table 6.18. The comparison between predicted and measured data can be observed in Fig. 6.52.

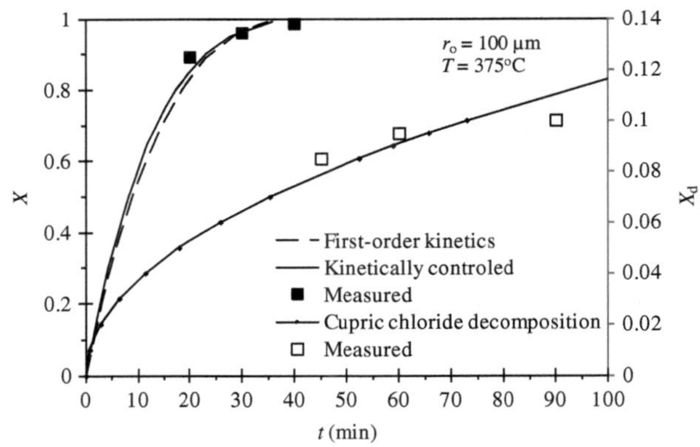

Fig. 6.52 Comparison between predicted and measured conversion rates of $CuCl_2$ hydrolysis and $CuCl_2$ thermolysis [measured data from Ferrandon et al. (2008), predicted data from Daggupati et al. (2010b)]

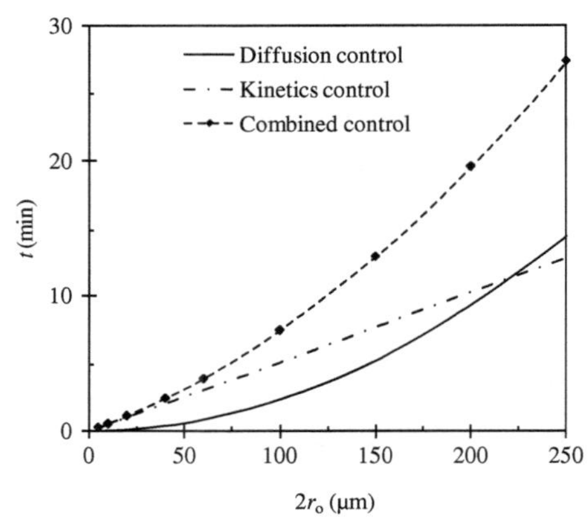

Fig. 6.53 Variation of hydrolysis reaction time with particle size as predicted by a shrinking core model [data from Daggupati et al. (2010a)]

The reaction times for the cases of (1) diffusion control of reaction, (2) reaction kinetics control, and (3) combined control, are presented in Daggupati et al. (2010a). The reaction time is expressed by

$$\begin{cases} \tau_d = \dfrac{\rho d_o^2}{3D_e C_{H_2O}}, & \text{for diffusion controlled reaction} \\ \tau_r = \dfrac{\rho d_o}{kC_{H_2O}}, & \text{for kinetics controlled reaction} \\ \tau_c = \dfrac{\rho d_o}{C_{H_2O}} \times \left(\dfrac{1}{k} + \dfrac{d_o}{3D_e}\right), & \text{for combined control,} \end{cases} \tag{6.38}$$

where k is the intrinsic rate of reaction.

Figure 6.53 illustrates the influence of $CuCl_2$ particle diameter on the reaction time for the three control cases of chemical reaction. The time required for the complete conversion of solids increases with particle size. The diffusion time is smaller than the reaction control time at a smaller particle size, with an increase in

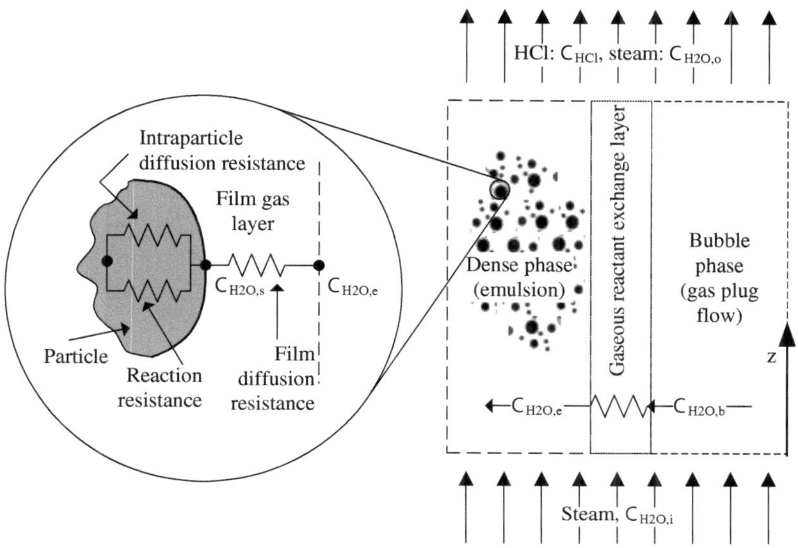

Fig. 6.54 Schematic model of transport processes in a reacting fluidized bed for hydrolysis

particle size, as the diffusion time dominates over the reaction time. With a combined resistance, the required time for complete conversion of solid is much higher than the other two resistances, and it increases with particle size.

A hydrodynamic model of reacting cupric chloride particles in a fluidized bed reactor has been developed by Haseli et al. (2008, 2009). Consider that steam at a concentration $C_{H_2O,i}$ is fed into a fluidized bed reactor, which consists of bubble and emulsion phases. The steam reactant is transferred from the bubble phase (with a concentration bulk concentration $C_{H_2O,b}$) to the emulsion phase (with a lower emulsion-phase concentration $C_{H_2O,e}$) to react with particles of $CuCl_2$ which are fluidized in the bed. The schematic of the process is presented in Fig. 6.54. It was assumed that a solid particle S is made of an active reactant solid particle C, and nonreactive material D. The transport resistances are the bubble to emulsion resistance, external film resistance around the solid particle, and interparticle resistance. The reactor contains particles that have spent different periods inside the bed, and hence they have a distribution of conversion.

The predicted steam and $CuCl_2$ particle conversions at various superficial gas velocities and three typical bed inventories predicted by Haseli et al. (2009) are shown in Fig. 6.55. The solid conversion increases with the fluidizing gas velocity, since the solid particles are more immersed with steam. On the other hand, however, the steam conversion decreases as its superficial velocity increases. For a certain quantity of bed inventory and constant particle feed rate, a higher superficial velocity means a higher gas flow rate, which consequently results in less conversion of steam. Furthermore, from Fig. 6.55, a higher bed inventory improves the conversion of both reactants. At given feed rates of reactants, the mean

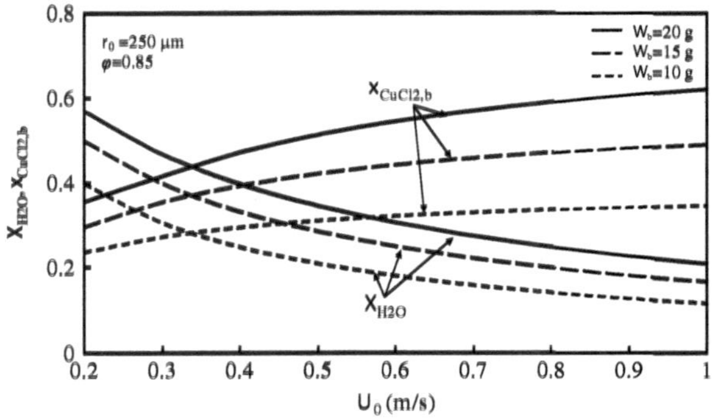

Fig. 6.55 Prediction of conversion rates of gaseous and solid reactants with UCM for three bed inventories W_b in a benchtop fluidized bed reactor of 2.66 cm/16.5 cm bed diameter/height, $d_0 = 250\,\mu m$ particle average diameter with $\varphi = 0.85$ sphericity, and particle mass flow rate at the inlet of $\dot{m}_i = 0.9\,g/s$ [data from Haseli et al. (2009)]

residence time, τ, increases as the bed inventory rises. Thus, the contact opportunity for both reactants increases. In Fig. 6.55, a uniform conversion model (UCM) has been used for the predictions. The steam conversion X_g is defined by

$$X_{H_2O} = 1 - \frac{C_{H_2O,o}}{C_{H_2O,i}}. \tag{6.39}$$

The average conversion of solid particles in the bed $x_{CuCl_2,b}$ depends on the conversion at the inlet conditions $x_{CuCl_2,0}$ and mass fraction of solid reactant in the fed stream $Y_{CuCl_2,0}$ according to

$$x_{CuCl_2,b} = x_{CuCl_2,0} + X_{H_2O} \times a \times (1/Y_{CuCl_2,0} - x_{CuCl_2,0}), \tag{6.40}$$

where a is a dimensionless parameter characterizing the fluidized bed, defined by Haseli et al. (2008), through the following relationship:

$$a \times \dot{m}_i = 2 \times \mathcal{V}_0 \times A \times M \times C_{H_2O,i}, \tag{6.41}$$

where M is molecular mass of $CuCl_2$, A is the cross-sectional area of the bed, $\mathcal{V}_0$ is the superficial gas velocity, and $\dot{m}_i$ is the inlet flow rate of solids.

The mechanism of heat transfer between the gas and a particle, and/or bed of particles, was analyzed by Daggupati et al. (2011a, b). The heat transfer in a gas–solid two-phase flow undergoing a chemical reaction within a fluidized bed is determined by simultaneously solving the mass and energy balance equations for the bed. The important process parameters are the steam-to-cupric chloride mole ratio, superficial velocity of steam through the fluidized bed at a given mass flow rate of solid, and the reaction/bed temperature. In order to maintain the hydrolysis reaction at a required temperature, the amount of heat required for the endothermic reaction at a given operating condition should be supplied externally to maintain isothermal conditions. In practice, removal/input of reaction heat to/from a fluidized bed is usually accomplished by an immersion surface, such as heat transfer tubes that carry cooling/heating fluids.

An alternative approach is to provide the required heat of reaction into the fluidized bed reactor by supplying hot fluidizing gas with an inlet temperature higher than the bed temperature. Due to direct contact of steam with solid particles, this method of heat transfer can be more efficient than utilizing a submerged surface. The steam temperature must be maintained below 423 °C to avoid melting of small quantities of cuprous chloride (CuCl) formed due to the decomposition of cupric chloride during the hydrolysis process. At temperatures above the melting point of cuprous chloride, the solids agglomerate and reduce the bed performance.

Relevant modeling parameters for the hydrolysis reactor of the fluidized bed configuration are listed and defined in Table 6.19. A modeling case study is considered by Daggupati et al. (2011a, b) which refers to the operating conditions of the scaled-up hydrolysis reactor at UOIT. The parameters and operating conditions for the hydrolysis reaction for the case study are listed in Table 6.20.

Based on the predicted results, the required steam inlet temperature to maintain the reactor at a desired temperature under the conditions of maintaining an endothermic reaction is indicated in Fig. 6.56. The steam inlet temperature depends on the melting or the decomposition temperature of solid components of the reaction mixture in the bed. The melting point of the solid reactant, $CuCl_2$, is ~771 K. The solid hydrolysis reaction product, namely, the copper oxychloride Cu_2OCl_2, decomposes to CuO and $CuCl_2$ at ~673 K. If the steam inlet temperature is higher than the melting point of its decomposition product, CuCl (696 K), the solid reactant may agglomerate or coalesce into a larger combined mass, thereby resulting in poor fluidization behavior. The inlet temperature is very important to maintain the required superficial velocities inside the reactor for good fluidization.

Figure 6.57 shows the time required to raise the temperature of a single cupric chloride particle from room temperature to the fluidized bed temperature, T_b. This is determined based on an energy balance equation that leads to the following equation:

$$t = \frac{mc_p}{hA} \ln\left((T_b - T_p)/(T_b - T_{pb})\right), \tag{6.42}$$

where T_p is the temperature of the cold particle and T_{pb} is the temperature of the particle in the bed. As observed in the figure, the time required to raise the solid temperature decreases with fluidization velocity. It is much lower than the time estimated based on the overall bed heat transfer coefficient.

Figure 6.58 shows the variation of conversion of cupric chloride solid and steam with superficial velocity, solid flow rate, and temperature. From this figure, it appears that a higher temperature and steam flow rate increase the solid conversion, and they decrease at a higher solid rate. The conversion of steam is constant for a given temperature, irrespective of the mole ratio of steam, and it only increases with reaction temperature. As shown in Daggupati et al. (2011a, b) the cupric chloride

Table 6.19 Modeling parameters for fluidized bed reactor for $CuCl_2$ hydrolysis

Parameter	Correlation	Units
Minimum fluidization velocity	$\mathcal{V}_{\mathrm{mf}} = \left[33.7^2 + 0.0408\frac{d_\mathrm{p}^3\rho_\mathrm{g}g(\rho_\mathrm{p}-\rho_\mathrm{g})}{\mu_g^2}\right]\frac{\mu_\mathrm{g}}{d_\mathrm{p}\rho_\mathrm{g}}$	m/s
Bubble relative velocity	$\mathcal{V}_{\mathrm{br}} = 0.711 \times (gd_\mathrm{b})^{1/2}$	m/s
Bubble velocity	$\mathcal{V}_\mathrm{b} = 1.6\left[(\mathcal{V}_\mathrm{o} - \mathcal{V}_{\mathrm{mf}}) + 1.13d_\mathrm{b}^{1.35}\right]d_\mathrm{t}^{1.35} + 0.711(gd_\mathrm{b})^{0.5}$	m/s
Bubble diameter	$d_\mathrm{b} = d_{\mathrm{bm}} - (d_{\mathrm{bm}} - d_{\mathrm{bo}})\mathrm{e}^{-0.3L_\mathrm{f}/D_\mathrm{t}}$ $d_{\mathrm{bm}} = \min\left\{0.0065\left[\frac{\pi}{4}d_\mathrm{t}^2(\mathcal{V}_\mathrm{o} - \mathcal{V}_{\mathrm{mf}})\right]^{0.4}, d_\mathrm{t}\right\}$ $d_{\mathrm{bo}} = \frac{0.0278}{g}(\mathcal{V}_\mathrm{o} - \mathcal{V}_{\mathrm{mf}})^2$ for $d_{\mathrm{bo}} > l_{\mathrm{or}}$ $d_{\mathrm{bo}} = \frac{0.082}{g^{0.2}}\left[\frac{\mathcal{V}_\mathrm{o}-\mathcal{V}_{\mathrm{mf}}}{N_{\mathrm{or}}}\right]^{0.4}$ for $d_{\mathrm{bo}} \le l_{\mathrm{or}}$ $d_{\mathrm{b,av}} = \frac{1}{L_\mathrm{f}}\int_0^{L_\mathrm{f}} d_\mathrm{b}\,\mathrm{d}z$	m
Bubble fraction in the bed	$\varepsilon_\mathrm{b} = \frac{\mathcal{V}_\mathrm{o}-\mathcal{V}_{\mathrm{mf}}}{\mathcal{V}_\mathrm{b}-\mathcal{V}_{\mathrm{mf}}}$, for fast bubbles	–
Interchange coefficient between bubble and emulsion	$\mathcal{K}_{\mathrm{be}} = \frac{2\mathcal{V}_{\mathrm{mf}}}{d_\mathrm{b}} + 5.7(D_g\varepsilon_{\mathrm{mf}})^{0.5}g^{0.25}/d_\mathrm{b}^{1.25}$, freely bubbling beds	s^{-1}
Bed height	$L_\mathrm{f} = \frac{L_{\mathrm{mf}}}{(1-\varepsilon_\mathrm{b})}$	m
Number of transfer units	$\mathrm{NTU} = \frac{K_{\mathrm{be}}\varepsilon_\mathrm{b}L_\mathrm{f}}{\mathcal{V}_\mathrm{o}}$	–
Excess gas flow ratio	$\beta = \frac{\mathcal{V}_\mathrm{o}-\mathcal{V}_{\mathrm{mf}}}{\mathcal{V}_\mathrm{o}}$	–
Concentration efficiency	$N_\mathrm{a} = \frac{C_\mathrm{i}-C_\mathrm{o}}{C_\mathrm{i}-C_\mathrm{e}} = 1 - \beta\exp\left(-\frac{\mathrm{NTU}}{\beta}\right)$	–
Dimensionless parameter, reactor level	$\alpha = 2U_\mathrm{o}A_{\mathrm{C,bed}}C_\mathrm{i}M_\mathrm{p}/\dot{m}_\mathrm{i}$	–
Gas concentration, emulsion phase	$C_\mathrm{e} = C_\mathrm{i}(1 - X_\mathrm{g}/N_\mathrm{a})$	mol/m^3
Kinetic coefficient	$K_{\mathrm{re}} = b\frac{M_\mathrm{p}}{\rho_\mathrm{p}}k_\mathrm{o}C_\mathrm{e}^n$	s^{-1}
Damkohler number at reactor scale	$D_{\mathrm{as}} = \frac{K_{\mathrm{re}}W_\mathrm{b}}{\dot{m}_\mathrm{o}} = K_{\mathrm{re}}\bar{t}$	–
Dimensionless factor	$\lambda = \frac{D_{\mathrm{as}}}{1-D_{\mathrm{as}}}$ or $\frac{K_{\mathrm{re}w_\mathrm{b}}}{F_\mathrm{o}-r_{\mathrm{c,b}}}$	–
Overall rate of reaction in the bed	$r_{\mathrm{c,b}} = \dot{m}_\mathrm{o}\lambda\left[\left(1-\frac{\lambda}{3}\right)^2 + \left(\frac{\lambda}{3}\right)^2 - 2\left(\frac{\lambda}{3}\right)^2\exp\left(-\frac{3}{\lambda}\right)\right]$	kg/s
Equilibrium conversion of gas	$X_{\mathrm{e,g}} = 3 \times 10^{-7}\,\mathrm{e}^{0.0175T}$, $T = 573 - 673\,\mathrm{K}$, at 101 kPa	–
Gas conversion	$X_\mathrm{g} = \min\left\{\frac{1}{\alpha}\left(1 - \frac{D_{\mathrm{as}}}{\lambda}\right), X_{\mathrm{e,g}}\right\}$, $X_\mathrm{g} \le X_{\mathrm{e,g}}$	–
Solid conversion	$X_\mathrm{c} = \alpha X_\mathrm{g}$	–

Source: Daggupati et al. (2011a).

solid conversion is limited to 26 % at a 20 mole ratio of steam at 573 K. However, at 673 K, the required mole ratio is 13.5. A three-step calculation procedure is followed for the conversion of steam and cupric chloride solid. The procedure involves (1) calculation of the gas conversion, (2) calculation of the solid conversion, and (3) equating the gas conversion with the solid conversion, taking fluid flow factors of the fluidized bed into consideration.

Table 6.20 Parameters and operating conditions for a case study with the UOIT fluidized bed reactor for cupric chloride hydrolysis

Parameter	Symbol	Range	Units
Particle diameter	d_p	200	μm
Reactor diameter	d_t	0.15	M
Weight of bed material	W_b	2.24	Kg
Mass flow rate of solid	$\dot{m}$	1.5–5.0	kg/s
Minimum fluidization velocity	$\mathcal{V}_{mf}$	0.0343	m/s
Gas velocity	$\mathcal{V}_0$	0.0343–0.4	m/s
Steam initial concentration	C_i	0.0178	kmol/m^3
Order of reaction for steam	n	1	–
Density of cupric chloride solid	ρ_p	3,386	kg/m^3
Reaction temperature	T	573–673	K
Reaction pressure	P	101	kPa

Source: Daggupati et al. (2011a, b, c)

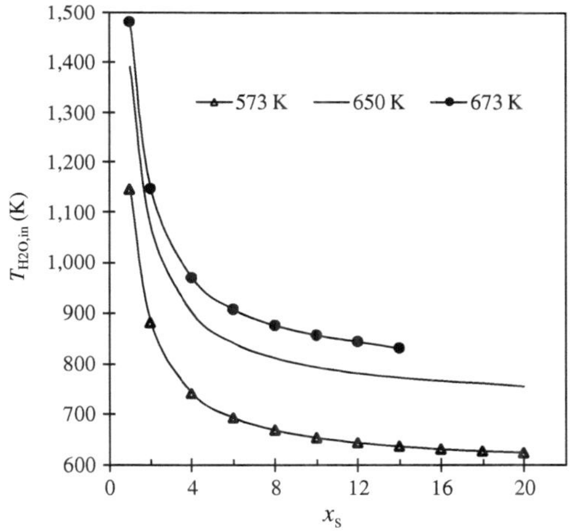

Fig. 6.56 Variation of steam inlet temperature with mole ratio of steam and mass flow rate [data from Daggupati et al. (2011a)]

6.6 Thermolysis Reaction Step

6.6.1 Reaction Kinetics Thermodynamics

The thermolysis process—thermal decomposition of copper oxychloride—is an endothermic solid conversion process which generates oxygen gas and molten cuprous chloride. Consequently, in a thermolysis reactor for copper oxychloride, three phases coexist: solid, liquid, and gas. Copper oxychloride particles decompose into oxygen gas bubbles and molten CuCl. The reactant particles absorb heat from the surrounding molten CuCl salt. Gaseous products leaving the oxygen reactor include oxygen gas and CuCl vapor, and potentially products from side reactions, such as Cl_2 gas, HCl gas, and H_2O vapor (in trace amounts). The other

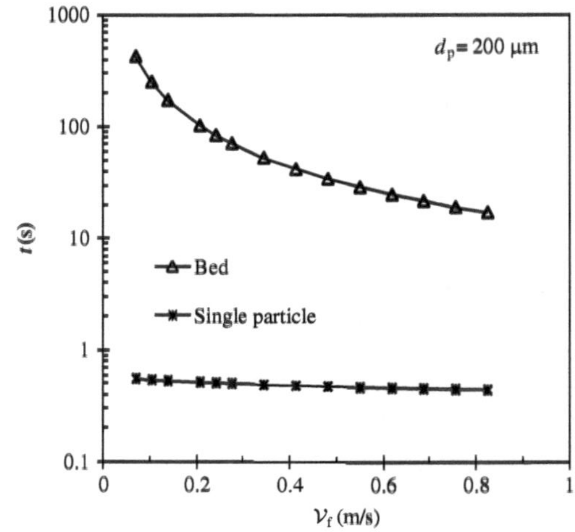

Fig. 6.57 Time required to raise the solid particle temperature from 298 K to the bed temperature of 673 K [data from Daggupati et al. (2011a)]

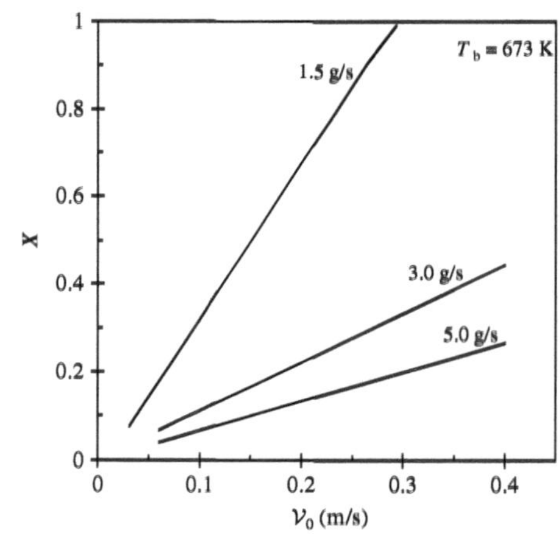

Fig. 6.58 Conversion of solid cupric chloride particles vs. superficial velocity [data from Daggupati et al. (2011a, b, c)]

liquid/solid products leaving the reactor are molten CuCl, unreacted solid $CuCl_2$ from the upstream hydrolysis reaction, as well as reactant particles entrained by the flow of molten CuCl.

The decomposition mechanism is due to a number of elementary reactions as indicated in Table 6.21. Copper oxychloride decomposes thermally after 673 K and one of the released products is cupric oxide. One first pattern of decomposition occurs according to reaction #1 listed in the table: 2 mol of copper oxychloride generate 2 mol of cupric oxide solid and 1 mole of cupric chloride plus 1 mol of cuprous chloride (solid or molten, depending on the temperature), plus 1/2 mol of chlorine gas. This process is identified in Nixon et al. (2011) to occur immediately above 673 K and it may be observed up to 753 K. At higher temperatures, two other competing reactions occur in the range of 753–823 K: one desired reaction of cupric oxide chlorination (#3 or #4) and an undesired reaction of cupric chloride thermolysis (#2).

Table 6.21 Elementary reactions of copper oxychloride thermal decomposition

ID	Process	Elementary reaction	Temperature range (K)
1	Decomposition	$2CuO \cdot CuCl_2(s) \rightarrow 2CuO(s) + CuCl_2(s) + CuCl(s, l) + 0.5Cl_2(g)$	673–753
2	Thermolysis	$CuCl_2(s) \rightarrow CuCl(l) + 0.5Cl_2(g)$	753–823
3	Chlorination	$CuO(s) + 0.5Cl_2(g) \rightarrow CuCl(l) + 0.5O_2(g)$	753–823
4	Chlorination	$CuO(s) + Cl_2(g) \rightarrow CuCl_2(l) + 0.5O_2(g)$	>753
5	Oxidation	$CuCl(l) + 0.5O_2(g) \rightarrow CuO(s) + 0.5Cl_2(g)$	>823
6	Decomposition	$CuO \cdot CuCl_2(s) \rightarrow CuO(s) + CuCl_2(s)$	673–753
7	Oxidation	$2CuCl(l) + 0.5O_2(g) \rightarrow CuO \cdot CuCl_2(s)$	>823
8	Decomposition	$CuO \cdot CuCl_2(s) \rightarrow CuO(s) + CuCl(s, l) + 0.5Cl_2(g)$	>673

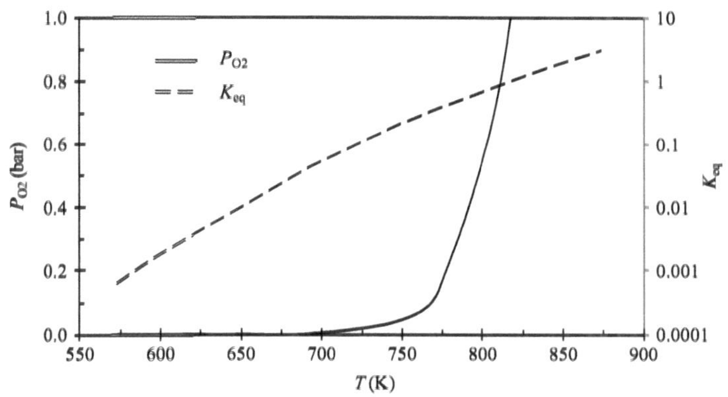

Fig. 6.59 Equilibrium constant and partial pressure of oxygen released by the reaction of copper oxychloride thermolysis

Chlorination consumes chlorine and releases the oxygen product. However, the competing reaction of thermolysis generates additional chlorine as an undesired side product. Over 823 K, the kinetics of the oxidation reaction (#5) of cuprous chloride becomes favorable which consumes oxygen and generates chlorine. Another possible reaction of thermal decomposition is mentioned by Serban et al. (2004); it consists of reaction #6 which generates cupric oxide and cupric chloride at temperatures above 673 K. The generated cupric chloride decomposes thermally according to process #2 and produces chlorine, which in a stoichiometric amount with cupric oxides is consumed in chlorination process #3. Marin et al. (2011a, b) studied also the reverse reaction which can occur in the thermolysis reactor according to process #7. This process is a chemical reaction with a negative Gibbs free energy over the range of operating temperatures of interest. Therefore, it must be disfavored by appropriately setting the reaction parameters. Marin et al. (2011a) mentioned another decomposition pathway, process #8.

The variation of equilibrium constant, $K_{eq}(T) = -\Delta G^0(T)/RT$, of the thermal decomposition reaction, $CuO \cdot CuCl_2(s) \rightarrow 2CuCl(s, l) + 0.5O_2(g)$, is presented in Fig. 6.59. Based on reaction stoichiometry, $K_{eq} \times P = P_{O_2}$, where P is the operating pressure and P_{O_2} is the partial pressure of oxygen. If the operating pressure is 1 bar,

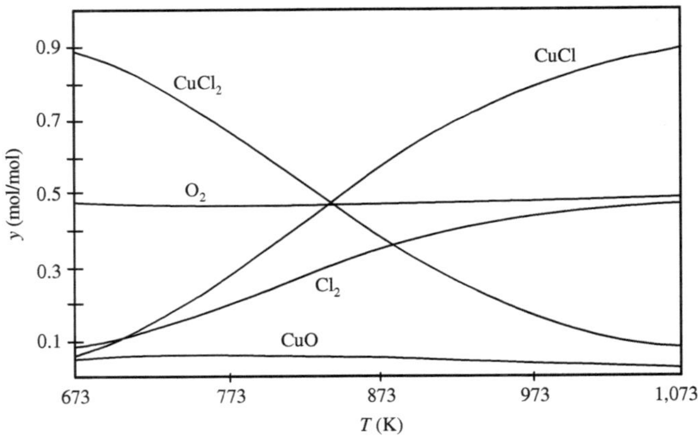

Fig. 6.60 Equilibrium molar fractions of the equimolar CuO/Cl_2 system [data from Nixon et al. (2011)]

it yields the partial pressure of oxygen and its variation with temperature as indicated on the graph in Fig. 6.59. It can be observed that the release of oxygen at equilibrium becomes significant only after 700 K.

Thermogravimetric experiments reported by Nixon et al. (2011) revealed that the decomposition of copper oxychloride becomes important at temperatures higher than 753 K, which is relatively close to the temperature at which cupric chloride starts to decompose. It can be concluded that cupric oxide and cupric chloride must be formed in the first phase of copper oxychloride decomposition. This confirms the assumption made by Serban et al. (2004) that the decomposition starts with processes #6 and #2 (Table 6.21) simultaneously. Chlorine is rapidly consumed in a subsequent reaction via processes #3 and #4 as mentioned above. Chlorine is consumed as per the experiments of Serban et al. (2004) and Nixon et al. (2011). In addition, equilibrium calculations of equimolar CuO and Cl_2 system were performed. As indicated in Fig. 6.60, at lower temperatures, formation of $CuCl_2$ with oxygen release is favored thermodynamically with respect to the formation of CuCl.

Consumption of chlorine in the $CuO(s) + Cl_2(g) \rightarrow CuCl_2(l) + 0.5O_2(g)$ reaction has been studied by Serban et al. (2004). The rate of this reaction is expressed by

$$r = k \times \left(P_{Cl_2}^{0.25} - P_{O_2}/K_{eq}\right), \tag{6.43}$$

where the rate of reaction k is determined by Serban et al. (2004) as ~10 mmol/g_{CuO} h bar$^{0.25}$.

6.6.2 Scaled-Up Thermolysis Reactor (UOIT)

A large reactor and experimental test loop have been developed at UOIT to conduct thermolysis experiments of copper oxychloride at a scale 1,000 times larger than

Fig. 6.61 Scaled-up thermolysis reactor at UOIT, charged with copper oxychloride

prior proof-of-principle experiments (Serban et al. 2004, Lewis et al. 2005a, b, Nixon et al. 2011). The reactor consists of a vertical cylindrical vessel made of stainless steel 316 and hastelloy C. It contains a crucible fabricated in ceramic where the sample of copper oxychloride is placed. There is also a secondary containment to the sample that allows for channeling of the gases outside via a liquid seal that maintains a backpressure within the chamber slightly above atmospheric pressure. This method is used to prevent gases from outside to enter the reactor during tests. A photograph of the reactor is shown in Fig. 6.61. It is placed in an electrical furnace with three heating zones able to control the temperature in the range of 300–1,273 K. The furnace can operate within 1 °C of the target temperature. The schematic of the thermolysis test loop in which the reactor is installed is shown in Fig. 6.62 and a photograph in Fig. 6.63.

Copper oxychloride powder is placed in a cleaned crucible prior to the experiments. Nitrogen is injected into the reactor at a desired flow rate which is continuously measured with a mass flow meter with ±0.8 % accuracy and ±0.2 % repeatability. Thermolysis reaction produces gases which are directed to an ABB-AO202 chlorine gas analyzer that provides readings in a continuous manner, with accuracy to within 1 %, and repeatability of 0.5 %. A separate stream of gas products is directed via the continuous oxygen analyzer AMI-201RSP with five selected ranges of operation, a minimum detection of 0.01 % and repeatability of 0.2 %. Cuprous chloride vapor escapes with the gaseous products of decomposition during the tests. Therefore, a condenser installed on the discharge of the reactor prevents condensate and retains the CuCl. Temperatures of streams in/out of the reactor are monitored via type K thermocouples with an accuracy of ±0.2 %. Solid reactants and products are measured via a Mettler Toledo ML3002E weight scale with a repeatability of 0.01 g and linearity at 0.02 g.

Prior to the start of experiments, the solid reactants are placed in the furnace for a minimum of 24 h at 423 K. The sample is placed inside the reactor, then the reactor is closed, and nitrogen is used to purge and eliminate any traces of oxygen. When the oxygen content is below detectable levels, the furnace begins to heat the sample.

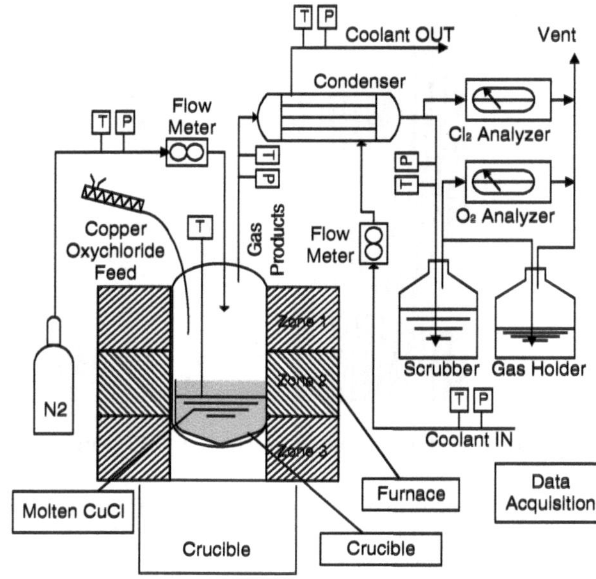

Fig. 6.62 Schematic of copper oxychloride reactor and its test loop at UOIT

At the sample temperature of 723 K, further addition of reactants is initiated depending on the goal of the experiments (see Marin et al. 2011b).

- For experiments focused on the decomposition of $CuCl_2$, a sample of 200 mmol is used and decomposed at various temperatures.
- For experiments focused on the decomposition of the mixture of CuO and $CuCl_2$, solids have been added at a continuous rate to the reactor.

The addition of solids is performed through a special opening of the reactor. The test temperature is maintained after the addition of solids has finished and until no oxygen evolution is detected. Experiments with continuous addition into molten CuCl resulted in an approximately linear release of oxygen with respect to the addition of solids. The total mass of oxygen is then determined from numerical integration of the oxygen generation curve. The solid products of reaction were analyzed by XRD with a Philips PW1830HT PANalytical X'pert diffractometer where the diffraction angle 2-Theta varied between 5° and 60°. Subsequent quantification used X-pert PANalytical software.

The experimental equipment was first tested with a salt that dissociates at temperatures similar to those required by copper oxychloride, in order to provide a pattern of reference and obtain a reactor performance characterization. Potassium chlorate was used in the process of reactor performance characterization. Potassium chlorate is a solid at normal ambient temperatures and it melts at 629 K. It boils and separates into potassium chloride (KCl) and oxygen when exposed to temperatures beyond 673 K.

The average oxygen production rate is approximately 0.04 mol/min from a sample of 0.082 mol of $KClO_3$, when extrapolated to the capacity of the reactor.

Fig. 6.63 Photograph of copper oxychloride decomposition test loop at UOIT

This suggests an oxygen generation capacity beyond 1.13 N m^3/h that corresponds to an H_2 production rate above 4 kg/day, which is 10,000 times larger than past proof-of-principle experimental tests.

In the oxygen reactor, copper oxychloride particles decompose into molten salt and oxygen. The reactant particles absorb heat from the surrounding molten bath. Once the temperature of 773 K is reached, the sample decomposes quickly with oxygen released at the peak rate of 0.006 mol/min. A total of 0.107 mol of oxygen were released with a yield of 89 %. The yield was calculated by comparing the actual oxygen generated with the amount of oxygen that could have been generated if the reaction was completed to 100 %. According to the results presented in Fig. 6.64, the entire decomposition of 0.236 mol of $CuO + CuCl_2$ occurred within 35 min of the initial 10 min, showing a marked oxygen evolution while the balance of time shows the reaction rate slowly decaying to negligible oxygen release.

The solid products of oxygen production from copper oxychloride appear as a black hard sintered material, as shown in Fig. 6.65. The solids are mildly hygroscopic, and they can be pulverized with some mechanical effort. This means that an

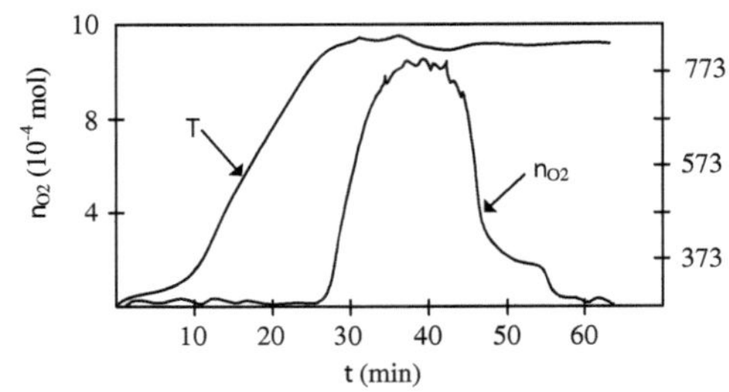

Fig. 6.64 Oxygen production from the reactant copper oxychloride [data from Marin et al. (2010)]

Fig. 6.65 Sample of products of $CuO{\cdot}CuCl_2$ decomposition

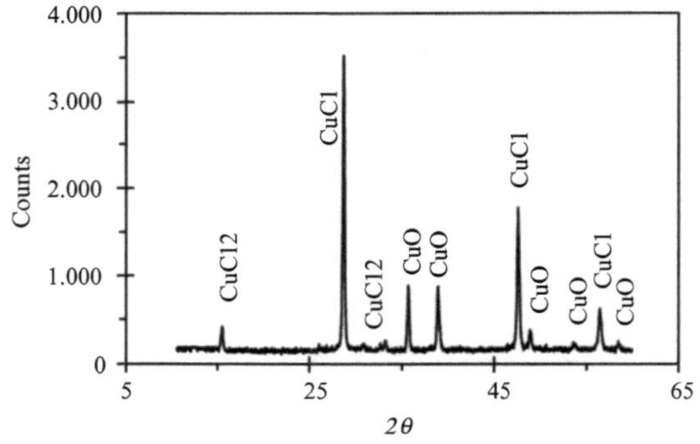

Fig. 6.66 Measured XRD results of copper oxychloride decomposition product [data from Naterer et al. (2011a)]

additional process must be introduced for the cycle integration before the downstream electrolysis reaction. The chunks of solid CuCl must be mechanically removed from the oxygen production reactor and then pulverized. In this condition, the solids are dissolved in water for the preparation of the aqueous solution needed by the downstream electrolytic hydrogen production cell. Since the CuCl must be first solidified, thermal energy recuperated from the phase change of molten salts from the decomposition reactor can be used elsewhere in the Cu–Cl cycle for thermal energy integration.

The analysis based on XRD powder techniques of the composition of the solid product of the oxygen production resulted in patterns shown in Fig. 6.66. The spectrum revealed a mixture of solids where copper oxide, cupric chloride, and cuprous chloride were present. The major product component is CuCl with mixed

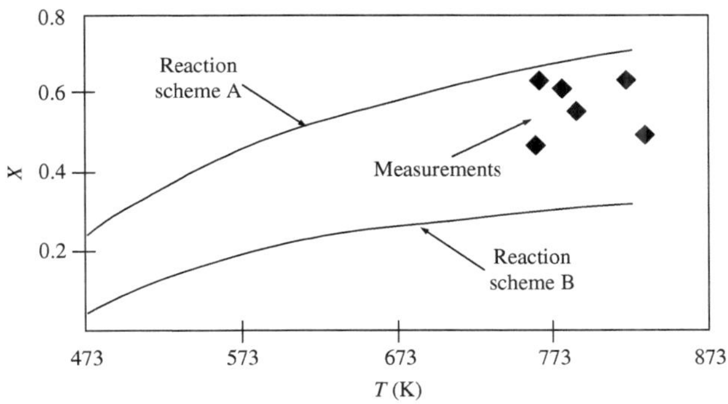

Fig. 6.67 Predicted vs. measured conversion of copper oxychloride at various temperatures and atmospheric pressure [data from Marin et al. (2011b)]

quantities of $CuCl_2$ and CuO. The analysis of the relative amounts of XRD peaks with reference to standard samples indicated the composition of solids.

Figure 6.66 shows that the unreacted $CuCl_2$ in the hydrolysis may not be fully decomposed in the downstream oxygen production reaction. Therefore, $CuCl_2$ may be further carried by CuCl to the downstream cell for the electrolytic hydrogen production. This same situation occurs with CuO present in the output stream from the oxygen reactor. The figure also indicates the presence of unreacted $CuCl_2$ and cupric oxide in addition to cuprous chloride with the solid products of decomposition. A longer residence time will lead to a full decomposition of cupric chloride into cuprous chloride and chlorine.

The theoretical and measured yields of copper oxychloride decomposition were compared in Marin et al. (2011b), as shown in Fig. 6.67, where the predictions represent the predicted equilibrium yield. At equilibrium, the predictions show that the conversion of the mixture of $CuO{\cdot}CuCl_2$ is between 50 and 80 % at 800 K. The experimental decomposition of copper oxychloride resulted in conversions between 60 and 80 % when operated at atmospheric pressure. Two reaction schemes are considered in the models developed by Marin et al. (2011b):

- Scheme A, defined by the reaction sequence:

$$\mathrm{CuO \cdot CuCl_2(s) \rightarrow CuO(s) + CuCl(s,l) + 0.5Cl_2(g)} \tag{6.44}$$

$$\mathrm{CuO(s) + 0.5Cl_2(g) \rightarrow CuCl(l) + 0.5O_2(g)} \tag{6.45}$$

- Scheme B, defined by the reaction sequence:

$$\mathrm{CuO \cdot CuCl_2(s) \rightarrow CuO(s) + CuCl_2(s)} \tag{6.46}$$

$$\mathrm{CuCl_2(s) \rightarrow CuCl(l) + 0.5Cl_2(g)} \tag{6.47}$$

$$\mathrm{CuO(s) + 0.5Cl_2(g) \rightarrow CuCl(l) + 0.5O_2(g)} \tag{6.48}$$

The decomposition rate falls close to the predictions of the reaction scheme A and B. In the analysis, the partial pressures of chlorine in the reactor rise up to four orders of magnitude larger than the partial pressure of CuCl vapor at temperatures between 723 and 833 K. Ongoing research is being conducted to improve the yields of oxygen gas and reduce side reactions and by-products, among which chlorine side product is the most significant, although some traces of cuprous chloride vapor were also encountered.

6.7 Dehydration Step of Cupric Chloride

6.7.1 Spray Drying Experiments (UOIT)

Adjusting the molar fraction of H_2O and $CuCl_2$ is important for optimization of the hydrolysis reaction which eventually generates HCl and $CuO{\cdot}CuCl_2$. The electrolytic step produces a stream of hydrated cupric chloride diluted in aqueous solution of hydrochloric acid: 1 mol of $CuCl_2$ is dissolved in about 8 mol of water. In view of conducting the hydrolysis process, the molar ratio of steam and cupric chloride must be adjusted to about 2:1, namely, $CuCl_2{\cdot}2H_2O$ is the desired reactant for hydrolysis. Alternatively, anhydrous cupric chloride $CuCl_2(s)$ may be used as a reactant for the hydrolysis process—see Naterer et al. (2008b). The hydrated halide complex $CuCl_2{\cdot}2H_2O$ forms crystals with copper atoms surrounded by two chlorine atoms and two water molecules placed approximately in the corners of a square; therefore the crystals appear as an agglomeration of planar square cells.

The solubility limit of $CuCl_2$ in water—which influences the concentration limit of cupric chloride molecules in aqueous solutions—ranges from 0.1075 mol $CuCl_2$ per mole of water at 303 K to 0.1333 mol $CuCl_2$ per mole of water at 353 K. If the solution is concentrated over the solubility limit, a slurry containing solid particles of hydrated cupric chloride in water is formed. Further water removal—obtained by various methods of drying (evaporation)—leads to generation of solid powder of $CuCl_2$ in its hydrated form. If more heating is applied, water is fully eliminated and anhydrous cupric chlorine is formed. The following drying methods were analyzed for cupric dehydration:

- Spray draying of aqueous solution—An evaporative heat and mass transfer process to remove water molecules when spraying the droplets of aqueous $CuCl_2(aq)$ in hot air maintained in a large vessel at 353 K.
- Spray draying of cupric chloride slurry—An evaporative heat and mass transfer process to remove water molecules when spraying a slurry containing about 55 % solids per volume with a ratio of 3.5 mol of water per mole of cupric chloride.
- Flash draying—Evaporation of moisture takes place when the surrounding liquid conditions suddenly change and become lower than the saturation point,

such as conditions leading to a large pressure drop that converts an initial subcooled liquid to superheated liquid.

- Combined flashing and spray drying—An aqueous solution or slurry is sprayed into a chamber where hot air is circulated.
- Crystallization—Formation of solid crystals precipitating from an aqueous solution by cooling and/or concentration changes that bring the solution below the solidification point.

Within the Cu–Cl cycle, approximately 116 kJ of heat is released internally within the thermochemical cycle (by exothermic reactions, solidification, and cooling) and 277 kJ of heat is required (for endothermic reactions, drying, and heating) for each gram of hydrogen produced with the precipitate slurry drying method. If internal heat is ideally recycled, then a net heat input of about 161 kJ/g H_2 would be needed, of which ~82 kJ/g H_2 represents the low-grade heat for spray drying, vaporizing water, and heating of low-temperature solid particles and gas in the Cu–Cl cycle.

Two experimental facilities were used at UOIT for dehydration tests of cupric chloride: (1) low-temperature spray drying unit, and (2) high-temperature spray drying unit. The low-temperature spray drying test unit has been designed and built at Clean Energy Research Laboratory (CERL) at UOIT. This unit—called dryer 1 unit—is based on a Yamato dryer model D-41 which was modified to avoid corrosion caused by the corrosive cupric chloride solution. A titanium two-fluid nozzle was used and stainless steel parts downstream of the nozzle were coated with a thin layer of Viton. The test system comprises the following elements:

- Cylindrical glass drying chamber of 0.45 m diameter with 1.0 m cylindrical height.
- Glass-made cyclone and particle collection vessel.
- One two-fluid atomization nozzle of 0.7 mm diameter made of titanium.
- Air is cleaned via a scrubber.
- Peristaltic pump for solution to spray.
- Flow meter for measurement of the flow rate of air for atomization.
- Evaporation capacity of 3 L/h.
- Temperature sensors mounted in the drying air at the inlet and outlet.
- Digital instruments for barometric pressure and ambient air temperature and humidity.

A schematic diagram of the experimental setup for low-grade heat utilization—called dryer 2—is shown in Fig. 6.68 and a photograph is shown in Fig. 6.69. The high-temperature spray drying unit is based on a Niro mobile minor dryer and has the following characteristics:

- Cylindrical drying chamber: 0.8 m diameter, 0.6 m height, and a 60° conical bottom.
- Two-fluid nozzle of 0.8 mm diameter.
- Drying gas is introduced from the top.
- Cyclone separator used to separate air and powder.
- Comprises an air cleaning unit.

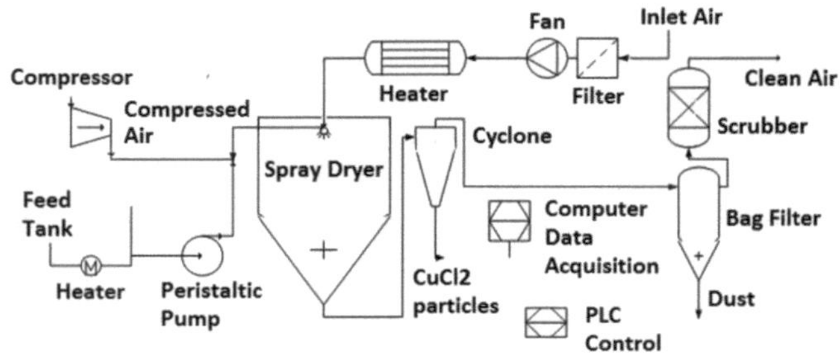

Fig. 6.68 Experimental facility for low-temperature spray drying of hydrated cupric chloride at UOIT

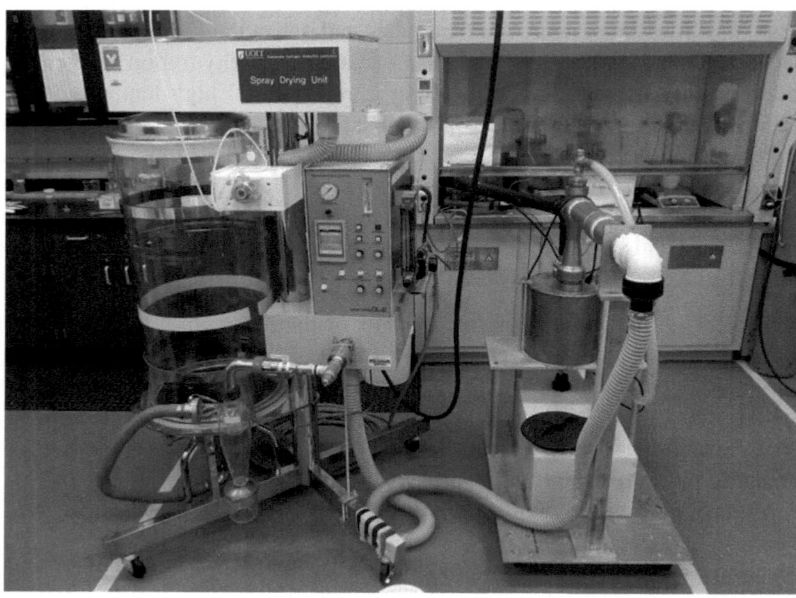

Fig. 6.69 Photograph of the experimental spray drying unit for $CuCl_2 \cdot nH_2O$ dehydration at UOIT

Aqueous solutions of sodium bicarbonate, sodium carbonate, and cupric chloride were used for the study. Experimental data has been collected and analyzed to identify the main effects of operating variables on powder quality, and utilization of low-grade heat input. The main operating variables are the inlet air temperature, atomization liquid feed rate, and atomization air pressure. The dependent variables are the outlet air temperature, moisture content in the product, and powder quality. A larger atomization air pressure increases the atomization airflow rate. The change in atomization airflow with the type of solution and its solid concentration is negligible.

The test parameters for low- and high-temperature experiments are listed in Table 6.22. The experimental data was used to determine the values of the Hausner ratio which indicates the flow behavior of powders from cohesive non-flowing to free-flowing regimes. The Hausner ratio is defined as the ratio of tapped density (ρ_{tap}) to aerated bulk density (ρ_{bulk}) of a powder, as follows:

$$HR = \rho_{tap}/\rho_{bulk}. \tag{6.49}$$

Bulk density of a powder is measured by allowing powder to settle in a measuring vessel under the influence of gravity, whereas tapped density of a powder is measured by tapping the measuring vessel. An Electrolab's microcontroller-based tap density tester (Model ETD-1020) was used for the measurement of tap density and Hausner ratio. In this method, the cylinder is raised and dropped under its own weight by a fixed drop height of 3 mm ±10 % at a nominal rate of 250 drops per minute. The cylinder is initially tapped for 500 times, and then the measured volume to the nearest graduated unit is obtained.

After the measurement, it continued tapping for 750 times and then again it measured the volume to the nearest graduated unit. If the difference between the two volumes is less than 2 %, the final tapped volume is considered. If not, it continues further until the difference between the measurements is less than 2 %. A Microtrac S3500 unit was used for particle size analysis. It works on light scattering technology. The particle range for this model is 0.024–2,800 μm. Each sample was taken more than two times for accuracy and repeatability. The particle analyzer gives the particle volume mean diameter, number mean diameter, and area mean diameter, and also the particle size distribution. The Sauter mean diameter of the solid particle is estimated in this study as

$$d_p = \left(\frac{\sum d^3}{\sum d^2}\right). \tag{6.50}$$

For a distribution analysis of particle size, the span is calculated using the following diameters: $d_{10\%}$—the diameter for which 10 % of the sample is smaller; $d_{50\%}$—the diameter for which 50 % of the sample is smaller; and $d_{90\%}$—the diameter for which 90 % of the sample is smaller. The span is calculated as follows:

$$SP = (d_{90\%} - d_{10\%})/d_{50\%}. \tag{6.51}$$

Product moisture content is represented either on a dry or a wet basis. A drying method for determination of moisture content depends on the nature of the tested material. Sodium bicarbonate decomposes at 373 K. This must be dried under vacuum for moisture removal. Sodium carbonate powder and cupric chloride powders were dried by taking 4–5 g of a sample and heating at 375–378 K. Samples were checked at every hour until there was no change in the subsequent values. The experimental results are summarized in Table 6.23. The data has been compared with reference results published in past literature. As indicated in Fig. 6.70, the

Table 6.22 Experimental parameters for drying tests of $CuCl_2$ at UOIT

		Atomization liquid			Atomization air		Drying air		
Dryer	Experiment ID	$\dot{m}$ (kg/h)	Solid fraction	T (K)	$\dot{m}$ (kg/h)	P (bar)	$\dot{m}$ (kg/h)	T_{in} (K)	T_{out} (K)
1	1	0.58–1.33	4–8 % $NaHCO_3$	293–303	2.36–4.68	2–4	57.6–86.4	323–373	303–318
	2	0.58	8 % $NaCO_3$	293	4.68	4	82	323	303
	3	0.58	10 % $CuCl_2$	293	3.84	4	72	333–393	303–328
2	4	0.464	44.83 % $CuCl_2$	293	10.4	2	82	473	383
	5	0.864	44.83 % $CuCl_2$	293	7.3	1.5	82	443	373

Source: Daggupati et al. (2011c)

Table 6.23 Experimental results of drying tests at UOIT

Experiment ID	Moisture (wt. %)	ρ_{tap} (kg/m^3)	ρ_{bulk} (kg/m^3)	d_p (μm)	SP	HR
1	21.3	547	325	29.67	1.01	1.68
2	20.89	605	385	24.6	1.36	1.57
3	20.88	652	411	21.56	1.58	1.58
4	5.29	752	631	23.97	3.26	1.19
5	2.7	701	590	21.31	3.21	1.19

Source: Daggupati et al. (2011c)

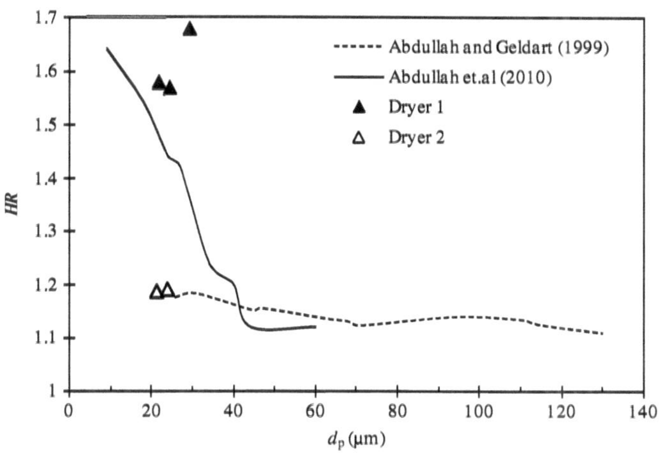

Fig. 6.70 Drying experiment validation against published reference results [data from Daggupati et al. (2011c)]

measured data agrees well with other published works, in terms of Hausner number correlated to the particle size. The Hausner ratio for the cupric chloride powder in high-temperature drying (dryer 2) is 1.18, which is less than 1.25. This indicates a free-flowing character of the powder, whereas the Hausner ratio in low-temperature drying (dryer 1) is above 1.4, which indicates a non-flowing and cohesive nature of powder. This occurs because the powder cohesiveness increased due to an increase in moisture content for low-temperature drying, compared to high-temperature drying. At high-temperature drying (dryer 2), a lower moisture content in the powder allows free flowing behavior, even though there is not much change in the Sauter mean diameter of the particles.

The product recovery rate for these runs was above 80 %. The overall recovery rate for these runs was calculated by taking the powder collected in the cyclone separator, conical section, in the drying chamber, as well as trapped powder in the connecting hose pipe between the drying chamber and cyclone separator. The difference in recovery rate may be due to low feed sampling rates and loss of powder in the conical section and inside the hose connector. There was no coating

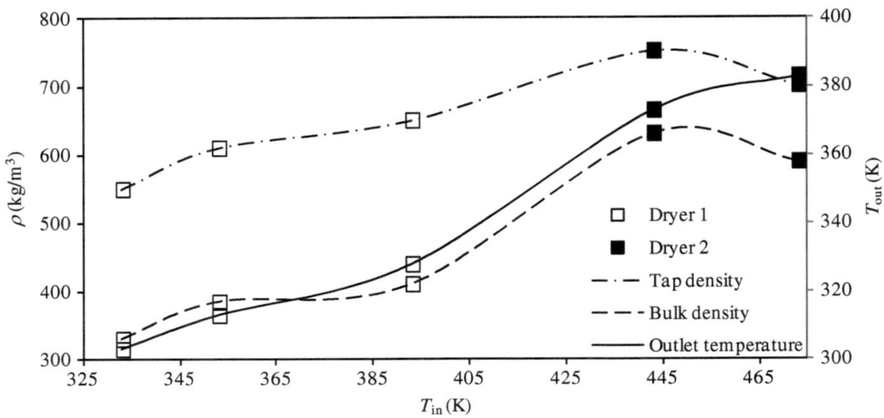

Fig. 6.71 Variation of tapped and bulk densities of the particles obtained by drying with the inlet temperature of the drying air [data from Daggupati et al. (2011c)]

of powders observed on the drying chamber. The experimental tests were stable without fluctuations in the atomization airflow and pressure for cupric chloride solution drying.

The variation of tapped and bulk densities with inlet temperature of drying air is presented in Fig. 6.71. Note that the tapped density is a measure of the flow ability of powders. The figure shows that the tapped and bulk density of the powder produced by drying shows maxima at an air inlet temperature of around 450 K when the temperature at outlet (air) is ~375 K. The tapped bulk density also decreases with increasing moisture content, and in turn reduces the flow ability of solids.

The outlet air temperature depends on the inlet air temperature, drying airflow rate, and atomization liquid flow rate. An increase in inlet air temperature and airflow rate increases the outlet air temperature and it decreases with an increase of atomization liquid rate. The inlet air temperature for cupric chloride solid varied from 333 to 393 K for low-temperature drying using the Yamato spray dryer. The corresponding air outlet temperatures varied from 303 to 328 K. Similar phenomena were also observed with sodium carbonate, and bicarbonate solutions. It shows that low-grade heat can be used for drying of cupric chloride solution to solid. The moisture content in the product depends on the air outlet temperature. The spray-dried product at low temperatures below 373 K contains moisture, which is removed in a downstream fluidized bed in two stages: one for removal of moisture, and the other for the hydrolysis reaction.

Figure 6.72 shows the particle size and size distribution for cupric chloride and sodium bicarbonate powders. The mean particle size for cupric chloride solid was around 31 μm in dryer 1, and 18.5 μm in dryer 2, with a broad size distribution. As indicated in Table 6.23, the span for particles varied from 1.1 to 1.4 for low-temperature drying and it was ~3.2 at high-temperature drying. These results may be attributed to the size of the two-fluid nozzle used in dryer 1. Moreover, the

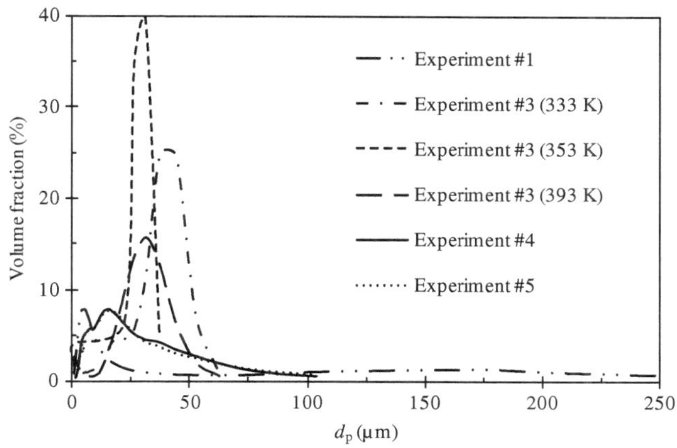

Fig. 6.72 Volumetric distribution of particle mean size [data from Daggupati et al. (2011c)]

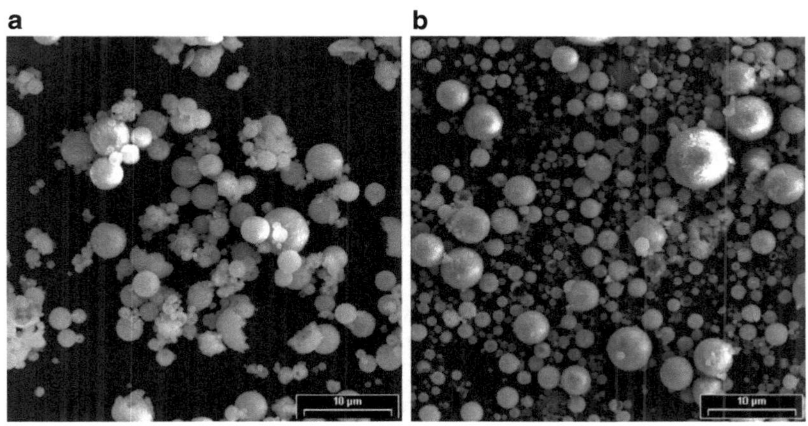

Fig. 6.73 SEM images of $CuCl_2$ particles formed by spray drying at UOIT. (**a**) High-temperature experiments, (**b**) low-temperature experiments

moisture content in the particle might have caused agglomeration of solids for a large particle size and distribution.

Powder characteristics and properties can be controlled and maintained uniformly throughout a drying operation. A sample SEM image of $CuCl_2$ particles obtained by spray drying is shown in Fig. 6.73. The particle morphology was investigated with a JEOL JSM-6400 scanning electron microscope. Since the SEM requires a vacuum, the procedure required the complete dehydration of the sample before placing it into the SEM vacuum chamber. Therefore, the crystal structure observed under the SEM was the anhydrous cupric chloride.

Cubic crystals were obtained by vacuum drying at room temperature, whereas long prisms were formed under normal drying conditions.

As the drying temperature increases, the solid becomes amorphous, because there was less time for crystals to grow before contacting another crystal growing in the droplet. Many nucleation points give rise to many small crystals—see Fig. 6.73a. Because of the increased number of crystals, there is less precipitate per crystal, and the crystals are much smaller in size. As the temperature is reduced, the number of nucleations is more orderly. It was also observed that part of the spheroid is thicker on one side than the other, particularly at lower temperatures, due to the remaining liquid inside the center of the droplet drifting away from the direction of flow causing the wall to be thicker on the side opposite to the direction of flow. At low drying temperatures (333 K), the particles did not form hollow spheroids, indicating that there was sufficient time for only one crystal to form for each droplet—see Fig. 6.73b.

In the process of spray drying the aqueous cupric chloride, volatile compounds (water and hydrochloric acid) evaporate from the droplets generated by the atomization nozzle. Because they are evaporated at the surface of the droplet, nucleation points occur first on the surface. This causes crystals to form and grow internally into the droplet. Based on the SEM images, it was observed that a higher temperature increases the number of nucleation points.

At temperatures of 393 K, the color of the dried solids was light green, indicating that the hydrate was forming first, and then later shrinking to form the anhydrous form. The formation of the hydrate has significant implications on the drying rate, since the first drying rate is determined by the hydrate, and later the drying rate is affected by the anhydrous form.

Ongoing research is examining the crystal structure of the $CuCl_2$ product, particularly to determine if there are interstitial water molecules inside a crystal lattice of $CuCl_2$. Other issues under investigation include whether there are water molecules bonded to the edges of the $CuCl_2$ crystal, whether there are interstitial Cu and Cl atoms inside a lattice of the di-hydrate, and gaining a deeper understanding of the precise mechanism of drying. The molecular structures of both the di-hydrate and anhydrous forms are planar. Bonds are formed between adjacent planes at specific angles. In the hydrate form, removal of water leads to collapse of the crystals.

The low-temperature experiments (dryer 1) showed that the produced powder shows cohesive and non-flowing behavior due to high moisture content. The moisture content in the powder also depends on the drying air inlet, as well as outlet temperatures. However, drying of cupric chloride and sodium bicarbonate solutions at low temperature demonstrates the potential for low-temperature spray drying of cupric chloride solution, for utilizing low-grade heat from nuclear, solar, and other industrial sources. Additional studies are recommended to identify the critical particle size for low- and high-temperature drying to improve the flow ability of cupric chloride powder. Experimental spray drying studies for cupric chloride solid at high temperature (dryer 2) gave acceptable particle size and flow ability of powder for high-temperature drying.

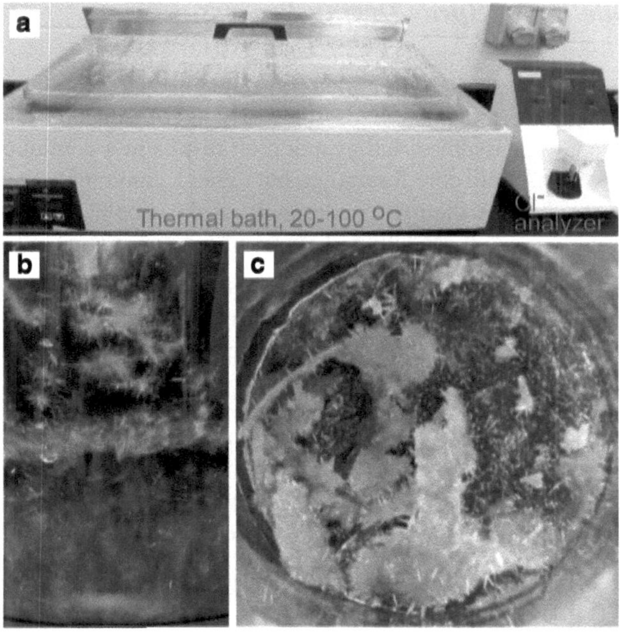

Fig. 6.74 Images of $CuCl_2$ particles formed by crystallization at UOIT

6.7.2 Crystallization Experiments (UOIT)

Due to the large-size equipment required for industrial scale drying of aqueous $CuCl_2$ in a future commercial hydrogen plant, alternative methods of water removal from the $CuCl_2$ were also investigated to improve the economic viability of the processes. In particular, methods of crystallization were examined by cooling the aqueous solution of cupric chloride. Crystallization is a process of formation of solid crystals precipitating from the aqueous solution. It is a physicochemical solid–liquid separation technique, whereby mass transfer of a solute occurs from the liquid to solid crystalline phase. The crystallization occurs by precipitation, through a variation of the solubility conditions of the solute in the solvent, rather than precipitation due to chemical reaction.

Crystallization experiments for the binary system of $CuCl_2 + H_2O$ were performed at saturation conditions using a thermal bath and a chlorine ion analyzer (Fig. 6.74a). It was found that 20 % of the $CuCl_2$ can be crystallized upon cooling from 80 to 20 °C. Figure 6.74b shows a sample of grown crystals during an experiment. The dimensions of the dendritic crystals were in the range of 0.1–0.6 cm. After the clear solution is removed, the crystals appear in Fig. 6.74c. Ongoing research is being performed to increase the portion of solids formed and reduce the temperature requirements of drying. This would enable the Cu–Cl cycle to utilize waste heat from thermal power plants or other industrial sources for the spray drying heat input requirements.

6.7.3 Modeling of Spray Drying Process

Additional numerical studies—reported in Naterer et al. (2008b)—have been conducted to analyze the process of spray drying of aqueous cupric chloride droplets. The analysis assumed that evaporation results in a continuous shrinkage of the droplet with no solid core formation, but rather a continuous growth of $CuCl_2$ crystals, with the average thermal conductivity of the droplet remaining constant. Additional assumptions for modeling of spray drying process are listed as follows:

- Mass and momentum changes of dispersed droplets do not contribute to corresponding changes of the drying medium (air).
- Kinetic energy changes in the drying air stream due to droplet movement are negligible.
- Drying air remains at constant temperature and humidity conditions.
- Evaporation produces a continuous shrinkage of the droplet with no solid core formation.
- Evaporation produces a continuous growth of $CuCl_2$ crystals.
- The average thermal conductivity of the droplet remains constant.

The main correlations used for modeling of the mass transport in spray drying are listed in Table 6.24. If one denotes with α the droplet thermal diffusivity, $\mathfrak{r}$ the radial coordinate of droplet assumed spherical, and h the convective heat transfer coefficient at the droplet surface, and consider symmetry of heat transfer throughout the droplet $(\partial T/\partial \mathfrak{r} = 0, \mathfrak{r} = 0)$, heat flux continuity at the droplet surface $-k_{\mathrm{d}}(\partial T/\partial \mathfrak{r}) = h_{\mathrm{g}}(T_{\mathrm{s}} - T_{\infty}), \mathfrak{r} = d_{\mathrm{d}}/2$, and that at the initial moment the droplet temperature is $T(t=0) = T_{\mathrm{i}}$, then the transient heat conduction equation $r^2(\partial T/\partial t) = \alpha \partial[\mathfrak{r}^2(\partial T/\partial \mathfrak{r})]/\partial \mathfrak{r}$ can be integrated. In addition, the differential mass transfer equation can be formulated—as indicated by Naterer et al. (2008b)—in the following way:

$$\frac{\mathrm{d}(d_{\mathrm{d}})}{\mathrm{d}t} = \frac{2DM_{\mathrm{w}}}{\rho_{\mathrm{w}} d_{\mathrm{d}}}(1 + 0.6Re^{0.5}Sc^{0.33}) \times (C_{\mathrm{s}} - C_{\mathrm{g}}), \tag{6.52}$$

where M_{w} is the molecular weight of the water, ρ_{w} is the density of saturated water in thermodynamic equilibrium with the droplet surface, D is the diffusion coefficient, and C_{s} and C_{g} are the vapor concentrations at the droplet surface and bulk flow, respectively. The variation of droplet diameter in time, due to the evaporation process, can be eventually expressed with

$$\frac{\mathrm{d}(d_{\mathrm{d}})}{\mathrm{d}t} = \frac{2k_{\mathrm{d}}(1+B)^{-0.7}}{\rho_{\mathrm{w}} d_{\mathrm{d}} \lambda}(2 + 0.6Re^{0.5}Pr^{0.33})(T_{\mathrm{g}} - T_{\mathrm{d}}), \tag{6.53}$$

where T_{d} is the droplet temperature and B and λ are defined in Table 6.24. In order to find the droplet shrinkage rate, it can be assumed that the droplet mass decrease is

Table 6.24 Correlations used for spray drying modeling

Equation	Remarks
$d_S = \sum_d^3 / \sum_d^2$	It expresses the Sauter mean diameter of the droplets; each droplet has a diameter d_d.
$\ln(1 - v) = -0.3 \times (d_d/d_S)^{1.6}$	Rosin–Rammler correlation of droplet size distribution in sprays. Here v is the fraction of total volume of spray contained in droplets smaller than d_d.
$Bi = Nu \times k_g/(2k_d)$	Biot number characterizing heat transfer within a droplet as a function of Nusselt number (Nu) and thermal conductivity of gas phase (k_g) and of droplet (k_d).
$Nu = 2 + 0.6Re^{0.5}\,Pr^{0.33}$	Ranz–Marshall correlation for Nusselt number, where Pr is Prandtl number and Re is Reynolds number.
$Re = (\mathcal{V}_g - \mathcal{V}_d) \times d_d\rho_g/\mu_g$	Definition for Reynolds number, where subscript g represents gas phase, $\mathcal{V}$ is velocity, μ is dynamic viscosity, and ρ is density.
$Sh = 2 + 0.6Re^{0.5}\,Sc^{0.33}$	Mass transfer in the convection process is defined by the Sherwood number (Sh), Reynolds number (Re), and Schmidt number (Sc).
$k_c \times d_d = Sh \times d_g$	Definition of (Sh) number based on mass transfer coefficient (k_c).
$C_d = \frac{24}{Re}(1 + 0.17Re^{0.632} + 10^{-6}Re^{2.25})$	Correlation for drag coefficient of the droplet as a function of Reynolds number.
$\lambda = h_{lg}(T_s) + \pi d_d^2 h(T_g - T_s)/\dot{m}$	Definition of dimensionless droplet evaporation constant where $\dot{m}$ is the mass rate of droplet shrinkage, T is temperature at the gas bulk (g) and droplet surface.
$B = \frac{c_p(T_g - T_s)}{\lambda}$	Definition of Spalding number where c_p is the specific heat of water vapor, T is temperature at the gas bulk (g) and droplet surface.

Source: Naterer et al. (2008b)

proportional to the droplet diameter shrinkage; thereby the following energy balance can be written for the droplet:

$$mc_p\frac{dT_d}{dt} = h\pi d_d(T_g - T_d) + \lambda\dot{m}. \tag{6.54}$$

Furthermore, the aerodynamic momentum balance results in the following expression for the velocity of the droplet—see Naterer et al. (2008b) for the detailed derivation:

$$m_0\mathcal{V}_0 = m(t) \times \mathcal{V}_d(t) + \int_0^t \left[0.5\pi\rho_g C_d \mathcal{V}_d^2(t) d_d^2\right] dt, \tag{6.55}$$

where C_d is the drag coefficient defined in Table 6.24.

Model validation was performed by comparing the model predictions with relevant published experimental data from past literature. Figure 6.75 shows a validation example where predictions of the particle diameter distribution are compared to experimental results from Cleary et al. (2007). For the results shown

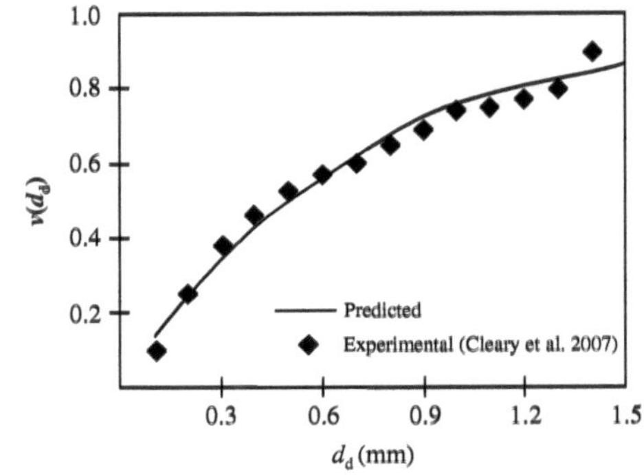

Fig. 6.75 Validation of spray drying model against published experimental data—fraction of the total volume of spray contained in the droplets smaller than d_d vs. droplet diameter [data from Naterer et al. (2008b)]

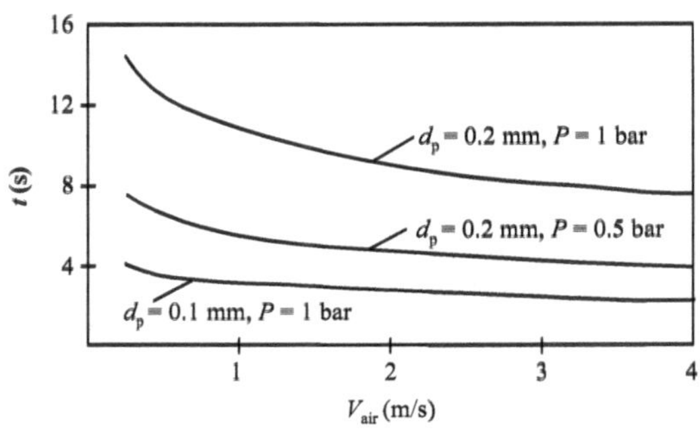

Fig. 6.76 Predicted drying time for inlet air at 303 K with 0.01 kg moisture per kg dry air for various droplet diameters and pressures [data from Naterer et al. (2008b)]

in the figure, it is assumed that the release pressure of the droplets is 24 bar and the Sauter diameter is 0.344 mm. The model results compare well against the experiments. Moreover, it can be observed that the amount of liquid evaporated as a consequence of flashing is relatively small, compared to the amount of rained-out liquid with droplet diameter. Furthermore, the Sauter mean diameter strongly depends on the nozzle characteristics and also the downward movement of spray flow, as this may increase the potential for coalescence and droplet size. This result makes the flashing effect minor in spray/flash drying, in comparison to the spray drying process of evaporation.

Figure 6.76 illustrates the variation of drying time with air velocity, temperature, and operating pressure. It was observed that the drying time depends strongly on the inlet air humidity, temperature, particle size, and operating pressure. At a low humidity of 0.0025 kg water/kg dry air, the drying time is less than 6 s for droplet sizes less than 200 μm, at 308 K. At a humidity of 0.01 kg water/kg dry air, the drying time is less than 6 s for droplet sizes less than 100 μm, at 308 K, and droplet sizes less than 150 μm, at 343 K and 1 bar operating pressure. At low operating pressures of 0.5 bar, the drying time is less than 8 s for droplet sizes less than 200 μm, at 308 K. The results predict that evaporative drying is possible down to temperatures as low as 308 K, although such low-temperature drying may limit the product quality and throughput.

6.8 Thermochemical Chlorination Step of Copper Particles

6.8.1 Reaction Kinetics and Thermodynamics

The reaction of copper chlorination with HCl which evolves hydrogen within some versions of Cu–Cl cycle has been intensively studied experimentally at ANL in 2003–2005; see Lewis et al. (2003), Lewis et al. (2004), Serban et al. (2004), and Lewis et al. (2005b). A numerical model for kinetics of this reaction was presented by Zamfirescu et al. (2010b). The reaction equilibrium and reaction extent were analyzed by Daggupati et al. (2009). This reaction is listed as process #6 in Table 6.2 and it is exothermic.

The equilibrium constant of the hydrogen production reaction can be expressed either in terms of Gibbs free energy or in terms of fugacities of reacting species, according to

$$K_{eq}(T) = -\frac{\Delta G(T)}{RT} = f_{H_2} f_{HCl}^2 \left(\frac{f_{CuCl}}{f_{CuCl}^0}\right)^2 \left(\frac{f_{Cu}}{f_{Cu}^0}\right)^{-2}, \tag{6.56}$$

where f^0 are the fugacities at 1 bar. Assuming that the pressure is low enough for ideal gas conditions at equilibrium, the equilibrium constant from (6.56) can be rewritten as

$$K_{eq} = \frac{1}{P} \times \frac{y_{H_2}}{y_{HCl}^2}, \tag{6.57}$$

where y_{H_2} and y_{HCl} are the mole fractions of hydrogen and hydrochloric acid in the gas phase.

The mole fraction of gaseous components can be expressed in terms of the extent of reaction, ξ. In stoichiometric conditions, 2 mol of copper are reacted with 2 mol of HCl. In order to decrease the equilibrium constant—for higher conversion—hydrochloric acid must be supplied in excess; one denotes $N \geq 2$ the number of moles of HCl supplied for each two moles of copper. As detailed by Daggupati et al. (2009), one can then express (6.56) and (6.57) in the following form:

$$-\frac{\Delta G(T)}{RT} = \frac{1}{P}\left(\frac{\xi}{N-\xi}\right) \bigg/ \left(\frac{N-2\xi}{N-\xi}\right)^2. \tag{6.58}$$

The second degree equation (6.58) gives a direct solution for ξ, with two roots. The positive root, less than unity, can be used for predicting the extent of the reaction. Figure 6.77 illustrates the effects of operating pressure, temperature, and excess reactant on the extent of the reaction for the hydrogen production

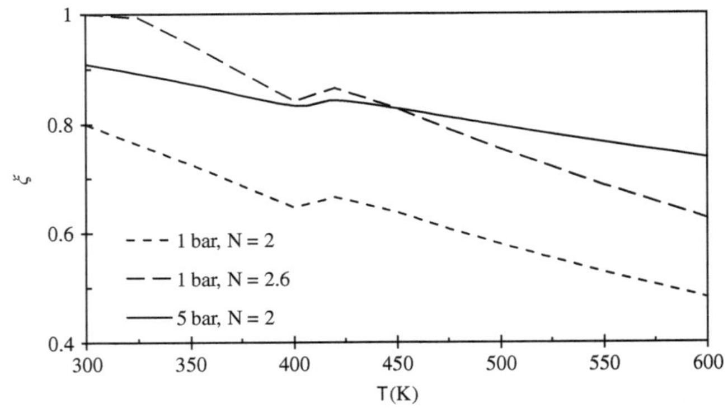

Fig. 6.77 Variation of reaction extent of copper chlorination reaction with reaction temperature, pressure, and excess HCl reactant [data from Daggupati et al. (2009)]

reaction. An increase in temperature reduces the equilibrium conversion. The increase in the operating pressure and excess reactant hydrogen chloride raises the equilibrium conversion values. At the operating pressure of 1 bar, the equilibrium conversion values are above 90 %, with 40 % excess reactant (hydrogen chloride) and 55 % excess reactant for complete conversion of the copper solid. The gas–solid reaction may be conducted either by increasing the operating pressure or excess reactant below 400°°C. Above 800 K, the reaction system becomes three phase, with gas–liquid–solid phases, because of the melting of cuprous chloride at about 703 K.

Serban et al. (2004) determined the reaction rate in a fixed bed reactor made of a quartz tube of 12.7 mm i.d., 35.6 cm height placed in a vertical electric furnace, and connected to a mass spectrometer calibrated to detect HCl, H_2, and Cl_2 gases. At a temperature where cuprous chloride starts melting (703 K), the reaction accelerates considerably. The gas–solid mass transfer is controlled by chemical reactions at the interface. The forward reaction is order 0.5 and the backward reaction is order one. The reaction is reversible and the reaction constant varies in the range 0.85–2.5 atm$^{0.5}$ when the temperature falls from 745 to 673 K. Serban et al. (2004) fit the experimental results on a "ln K vs. $1/T$" curve and determined the Arrhenius equation for the reaction rate constant. The kinetic experiments were performed with commercially available Cu particles of 3 μm. The reaction rate is expressed by

$$r = k_f P_{HCl}^{0.5} - k_b P_{H_2} = k_f (P_{HCl}^{0.5} - P_{H_2}/K_{eq}), \tag{6.59}$$

where k is the reaction constant and the index "f" stands for forward and the index "b" refers to backward reactions, and $K_{eq} = k_f/k_b$. Based on the data published by Serban et al. (2004) and Zamfirescu et al. (2010a, b, c), the rate

Fig. 6.78 Hydrogen yields with a fixed bed reactor [data from Serban et al. (2004)]

constant of the forward reaction can be estimated with the following Arrhenius-type equation:

$$k_f = 1,737,573 \exp(-7,515.5/T), \tag{6.60}$$

where k_f is given for 1 atm, 1 g of copper, and 1 µmol of HCl and it has units of h^{-1}.

The size and shape of electrolytic copper particles were affected by the operating parameters of the cell. To directly measure the kinetics of the reaction between HCl and Cu, the reaction rates were measured at four different temperatures (673, 700, 723, and 750 K). The hydrogen yields were determined by comparing the measured amount of hydrogen produced vs. the stoichiometric amount of hydrogen that would be formed if all Cu was oxidized to CuCl. Hydrogen yields decreased with larger Cu particle sizes, indicating that high surface areas of contact between HCl and Cu are necessary for high hydrogen yields. Figure 6.78 shows SEM images of the spheroidal and dendritic Cu particles used in the hydrogen generation reaction. No gaseous products other than hydrogen and HCl were observed in the effluent stream. XRD examination of the solid product resulting from the reaction showed patterns only for CuCl and Cu. No secondary reactions were favored in the temperature range studied.

6.8.2 Hydrodynamic Cold Tests with a Scaled-Up Packed Bed Reactor (UOIT)

Scale-up studies were performed on the process to generate larger capacities of hydrogen production (about 3 kg/day in a larger reactor, see Naterer et al. 2009). A packed bed reactor was developed and tested at UOIT for this purpose. The packed bed has been tested at cold operation conditions with the aim to study the flow hydrodynamics of various particle sizes and fillings. Also, the purpose of the tests

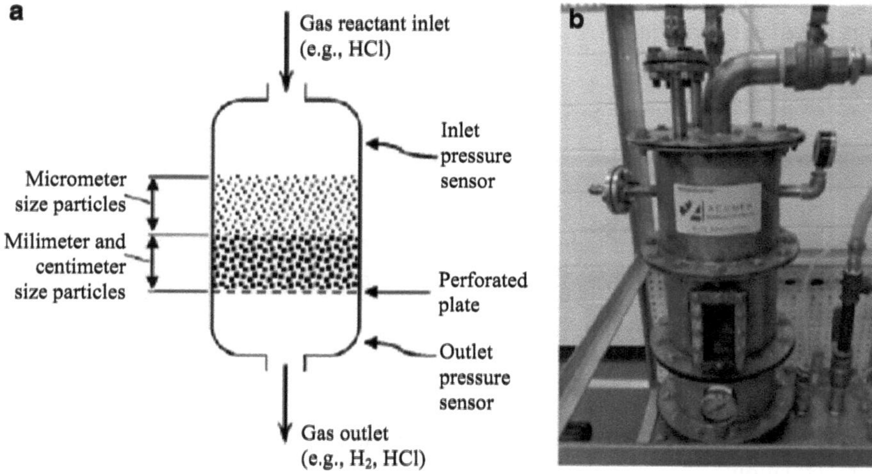

Fig. 6.79 (**a**) Schematic and (**b**) photograph of the packed bed reactor for thermochemical chlorination step

was to obtain hydrodynamic data for designing two of the reactors needed for the copper–chlorine cycle: the chlorination reactor (namely, the reaction step discussed in this section) and hydrolysis reactions (the process presented in Sect. 6.5). This section summarizes the past research results obtained during cold tests with the scaled-up packed bed reactor for conducting the thermochemical chlorination of copper particles.

The test reactor consists of a vertical pressure vessel 21 cm in diameter with a support mesh installed inside to hold the packed solids. A schematic and a photograph of the reactor are illustrated in Fig. 6.79. The system is connected to a compressed air station which is adjusted to provide pressurized air at 2 atm (gauge) at the reactor inlet (top side in the figure). The gas pressure is measured at the inlet and outlet (bottom) with pressure transducers (PX142-001D5V, 22 Pa uncertainty). The assembly photograph of the test loop is presented in Fig. 6.80. On the air line, a Chemline FT0025 flow meter is installed (uncertainty 0.27 N m^3/h).

The system calibration has been made with an empty reactor to determine a reference pressure drop. The reference pressure drop has been subtracted from pressure drop measurements of a filled reactor to determine the dynamic effects of the packing.

Figure 6.81 shows photographs of packing materials used for three sets of experiments. Table 6.25 lists the characteristics of the packing material and the associated experimental conditions. A Microtrac particle was analyzed to determine the particle characteristics. The 4 mm diameter particles are used as a support bed for the 450-mm diameter particles. The pressure drop measurements from the 4 mm particles are subtracted from the combined 4 and 450 mm experiments, to obtain data for the 450-mm particles.

Fig. 6.80 Scaled-up reactor for thermochemical chlorination step (UOIT)

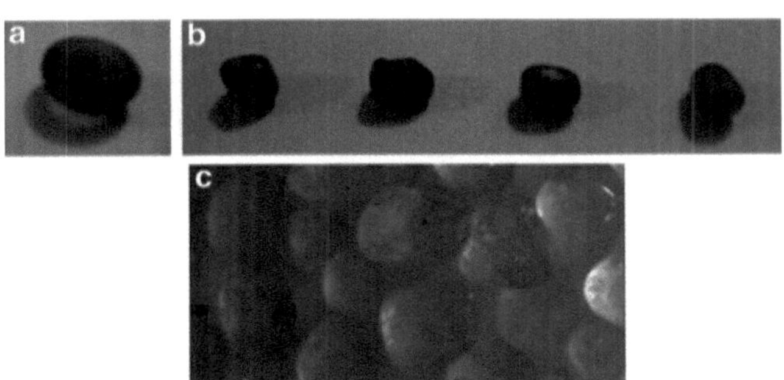

Fig. 6.81 Packing materials used in hydrodynamic tests with the scaled-up chlorination reactor: (a) 1 cm, (b) 4 mm, (c) 450 μm

The calibration data for the experimental tests is presented in Fig. 6.82 in terms of the pressure drop over the reactor against the airflow rate. Two measurements were performed: one with the reactor empty, and one with the perforated plate installed. A net pressure drop increase of around 10 Pa is due to the perforated plate.

Table 6.25 Packed bed parameters used in the hydrodynamic experiments

Material	d_p (mm)	Φ	ε	L (cm)	Re_p
Glass beads	10	0.88	0.38	10, 20, 30	15–150
Peas	4.0	0.98	0.33	10, 15	5.4–54
Sand	0.436	0.75	0.17	10, 15, 20	0.47–4.7

Note: d_p particle diameter, Φ sphericity, ε void fraction, L bed height, Re_p bed Reynolds number
Airflow rate range of 490–4,900 cm^3/s
Source: Pope et al. (2011)

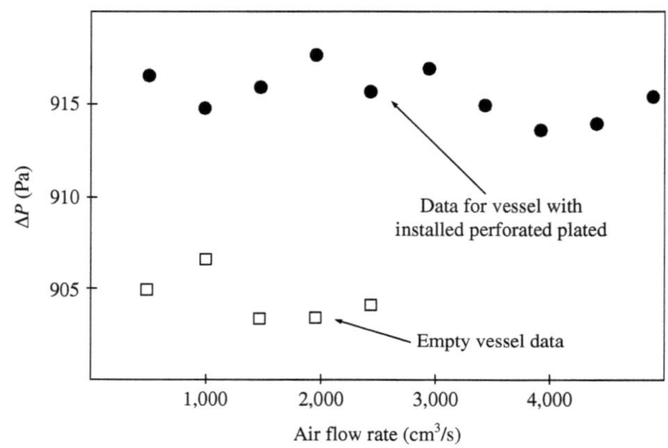

Fig. 6.82 Experiment calibration data for packed bed chlorination reactor [data from Pope et al. (2011)]

Table 6.26 Experimental correlation of f_p with Re_p for packed bed reactor

d_p (mm)												
0.45	Re_p	1	1.25	1.5	2.	2.5	3	4	5	N/A	N/A
	f_p	11	6	4.5	4	3	2.8	2.5	2	N/A	N/A
4	Re_p	5	10	16	22	26	34	38	43	48	54
	f_p	3.7	4	7.5	8.2	6	5.5	5.3	5.2	5.3	5.1
10	Re_p	14.5	30	44	58	74	87	104	118	132	147
	f_p	50	33	10	1	6.5	4	3.9	5.8	4.5	3

Source: Pope et al. (2011)

The collected experimental data are shown graphically in Table 6.26 where the packed friction factor (f_p) is correlated with the particle Reynolds number (Re_p). Here the Reynolds number for the bed is defined by the following equation:

$$Re_p = \frac{\rho \times \mathcal{V} \times d}{\mu \times (1 - \varepsilon)}, \tag{6.61}$$

where ρ is gas density, V is the superficial velocity of the gas phase, d is diameter, μ is dynamic viscosity of the gas, and ε is the void fraction. The friction factor of the bed can be estimated—as detailed in Pope et al. (2011)—by

$$f_{\mathrm{p}} = \frac{150}{\Phi \times Re_{\mathrm{p}}} + 1.75. \tag{6.62}$$

6.8.3 Modeling of Chlorination Reactor

Thermodynamic, kinetic, and hydrodynamic modeling of the chlorination reactor which conducts the solid–gas reaction of copper with hydrochloric acid, $2Cu(s) + 2HCl(g) \rightarrow H_2(g) + 2CuCl(l)$, which is exothermal, is a key step for reactor analysis and design. Orhan et al. (2008b) reported a thermodynamic analysis of the chlorination reaction step, based on energy and exergy methods. Zamfirescu et al. (2010b) reported a reactor modeling study based on reaction kinetic analysis. Pope et al. (2011) modeled the flow hydrodynamics through a scaled-up packed bed reactor for chlorination. Subsequently, the main analysis and modeling results for system kinetics and hydrodynamics were summarized.

The modeling of the chlorination reactor is also complicated because three phases are involved: gaseous, liquid (molten salt), and solid. Some multiphase reactor concepts and their auxiliary systems were reported by Wang et al. (2008). Using packed bed reactors is one of the options. In a multiphase packed bed reactor for copper chlorination, the solid and gaseous reactants are fed and a molten salt product is produced together with hydrogen gas. Heat is released by the reaction which must be removed via specific means of heat transfer. One of the key challenges is that three phases are present in the chemical reactor—molten (liquid) CuCl; gaseous HCl, H_2, and possibly CuCl vapor; and solid copper.

Packed bed reactors expose a large total surface area of solid particles to a fluid. The efficiency of the packed bed is an important factor in the operating and product costs. Investigating the pressure drop through the packed bed is important for maximizing the efficiency of the hydrolysis reaction for a variety of reasons. The use of auxiliary pumps and other parasitic losses are affected by a large pressure drop, since they need to supply capacity and input power.

The flow paths taken by the fluid through a packed bed are very complex and irregular. This makes it difficult to obtain exact solutions and precise representations of the fluid flow. The radial fluid velocity is significant because the irregularity of the packing material causes highly variable flow paths and large differences in local and average velocities. As specified in Pope et al. (2011), numerical predictions of fluid flow through a variety of structured packing arrangements showed good agreement with the Ergun equation, when the flow conditions were represented by Reynolds numbers (Re_{p}) less than 130.

This equation correlates the friction factor of the packed bed to the pressure drop and relevant geometric characteristics of the bed, according to the following mathematical expression:

$$f_p = \frac{\Phi d}{L\rho V^2}\left(\frac{\varepsilon^2}{1-\varepsilon}\right)\Delta P, \tag{6.63}$$

where ΔP is the pressure drop across the bed and all other quantities are specified in Table 6.25.

When the flow conditions are represented by a very low Reynolds number, the Stokes law can be applied. Low Reynolds flow ($Re_p < 5$) represents conditions where conduits through the packing material are extremely small (due to the microscale particles), so viscous effects between the fluid and packing material become important. A friction factor for the bed based on the Stokes law has been derived in Pope et al. (2011) as follows:

$$f_p = \frac{24\mu\varepsilon}{\rho\Phi dV}. \tag{6.64}$$

For the hydrodynamic modeling of the packed bed, (6.64) is applied when $Re_p < 5$ and (6.63) is applied when $Re_p > 22$. A composite solution that predicts the friction factor in the transition, creeping flow, and Ergun regions was proposed in Pope et al. (2011). The equation, valid for $5 \leq Re_p \leq 22$, is expressed by

$$f_{p,c} = f_{p,S} + \left[f_{p,T}^{-n} + f_{p,E}^{-n}\right]^{-1/n}. \tag{6.65}$$

In (6.65) $f_{p,S}$ represents the packed bed friction factor in the Stokes flow regime, defined by (6.64); $f_{p,E}$ represents the friction factor for Ergun flow regime, predicted with (6.63); n is an empirical coefficient of correction; and $f_{p,T}$ represents a transitional friction coefficient defined by

$$f_{p,T} = f_{p,E} \times (Re_p/20)^m, \tag{6.66}$$

where m is a correction coefficient determined as 1.3, based on experimental data.

The composite solution, (6.66), exhibits close agreement throughout the entire range of Re_p, for which experimental measurements were obtained. Four different values of the coefficient n for (6.65) are tested: 0.5, 1, 5, and 50. There is a low sensitivity exhibited between the predicted friction factor and the parameter n. However, when n is between 1 and 5, this provides the best agreement. A comparison of modeling results and experimental determinations is shown in Fig. 6.83.

The results of the hydrodynamic modeling suggest that predictions of friction factor through a packed bed, for micrometer-sized packing material, exhibit altered

Fig. 6.83 Modeling versus experimental results for packed bed reactor [data from Pope et al. (2011)]

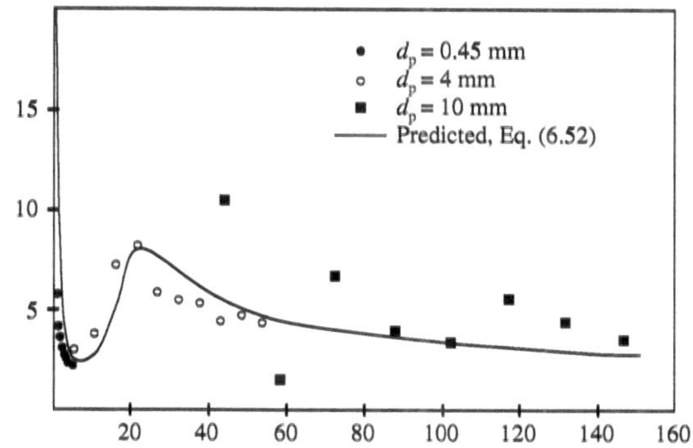

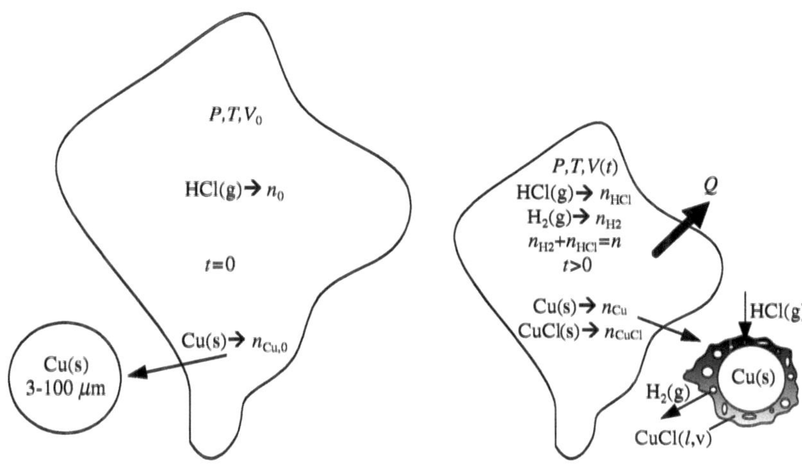

Fig. 6.84 Model of chlorination reaction process at constant pressure [modified from Zamfirescu et al. (2010b)]

flow phenomena than when the particles have larger diameters. These provide useful information and measured data to improve packed bed designs by limiting the adverse impact of undesirable pressure drops during operation.

The chemical reaction is modeled kinetically, and a parametric study is performed to determine the influences of particle size, temperature, and molar ratios on the hydrogen conversion aspects. For modeling purposes, it is assumed that chemical reactions and heat transfer are decoupled. Since the reaction is exothermic, it transfers reaction heat in the amount of about ~46 kJ/mol H_2 at 723 K.

The relevant transport phenomena at the gas/solid interface are heat and mass transfer and the chemical reaction. Figure 6.84 shows a model of the reactor. The reactor is a constant-pressure, isothermal batch type and it can be viewed as a thermodynamic system of variable volume $V(t)$, where $t \geq 0$ is time. At the initial moment, $t = 0$, in the reactor, one finds only n_{Cu} moles of copper and $n_{HCl} = n_0$

moles of gaseous hydrochloric gas. At any later moment, in addition to the two reactants in the reactor, there are n_{H_2} moles of hydrogen and n_{CuCl} moles of cupric chloride (assumed here in the solid state). The figure suggests that the gas volume around a reacted particle shrinks, because 1 mol of hydrochloric acid gas converts to a half mole of hydrogen. The fractional conversion X of hydrochloric acid gas is the ratio of converted reactant to the initial molar content, n_0. Since the converted HCl at the current time t is $n_0 - n_{HCl}$, the fractional conversion is

$$X = \frac{n_0 - n_{HCl}}{n_0}. \tag{6.67}$$

Based on stoichiometry and fractional conversion of HCl, the number of moles of gases participating to the chemical reaction (hydrogen and hydrochloric acid) within the reactor is $n = n_0 \times (1 - 0.5X)$. Initially, copper particles present a bare surface exposed to HCl reactant. During the reaction, at the surface of the particle, the CuCl product is formed (see Fig. 6.84). The reaction heat will increase the particle temperature up to the melting point of CuCl. As saturated liquid, cuprous chloride partially evaporates. Its vapor pressure is about 10 Pa close to the normal melting point. Through the two-phase CuCl shell, gaseous HCl diffuses toward the copper surface and generated gaseous hydrogen diffuses toward the exterior.

This depicted situation suggests that the outer shell that embeds the copper core is unstable and thus facilitates good mass transfer. For gas–solid non-catalytic reactions, the mass transfer mechanism may be controlled by diffusion through the gas film, diffusion through the outer layer, or the chemical reaction. In the last case, which is assumed to represent the present situation, the mass transfer depends on the concentrations in the bulk flow (HCl) and at the copper surface (H_2). At the same time, the molar consumption rate of copper is the same as the rate of production of hydrochloric gas:

$$\dot{n}_{Cu} = \dot{n}_{HCl} = r_{HCl}. \tag{6.68}$$

In (6.68), the rate of HCl consumption (r_{HCl}) depends on partial pressures of HCl and H_2 gases in the reactor space. The reaction rate is defined by (6.59) and (6.60). Therefore, one has

$$d^3 - d_0^3 = \frac{6M_{Cu}}{\pi \rho_{Cu}} \int_0^t \dot{n}_{HCl}\, dt. \tag{6.69}$$

The right side of (6.69) is negative (HCl is consumed) and the radius of the copper particle diminishes until it vanishes. On the right side, by setting $R = 0$, one may determine the time to complete the reaction, t_c. As detailed in Zamfirescu et al. (2010b), the conversion rate may be expressed by

$$\frac{dX}{dt} = k_f \left[\left(\frac{1 - X}{1 - 0.5X} \right)^{0.5} - \frac{1}{K_{eq}} \left(\frac{0.5X}{1 - 0.5X} \right) \right]. \tag{6.70}$$

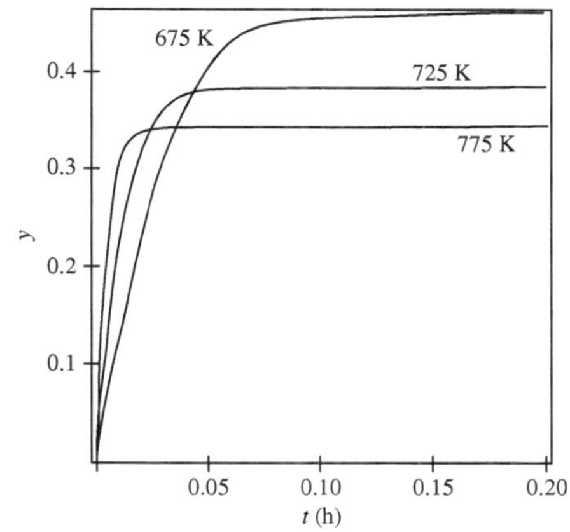

Fig. 6.85 Hydrogen yield variation with time for three reaction temperatures [data from Zamfirescu et al. (2010b)]

Equation (6.70) can be integrated numerically for a given operating temperature at which the constants are known. The fractional HCl conversion at equilibrium can also be computed by setting $dX/dt = 0$ and solving for X. It results in the time to reach chemical equilibrium in the batch reactor. The solution $X(t)$ of (6.70) is essential in the reactor design because it allows for calculation of other quantities, e.g., $\dot{n}_{HCl}$ as a function of time. This leads to a particle diameter variation $d(t)$ with (6.69), but also allows for the calculation of heat generation at the particle surface based on the heat of reaction, $\Delta H_r(T)$:

$$q''(t) = -\frac{\Delta H_r(T)}{\pi\, d(t)^2}\dot{n}(t). \tag{6.71}$$

The hydrogen yield (y) represents the number of moles of hydrogen generated per mole of HCl reactant. Hydrogen yield can thus be calculated based on fractional HCl conversion, by $y = 0.5X$. In Fig. 6.85, the hydrogen yield variation against the time is shown, whereas the time origin is the moment when copper particles enter the reaction. The hydrogen yield is obtained by integration of (6.55); this integration has been made for three reaction temperatures. It is observed from the figure that the magnitude of the hydrogen production is faster at higher temperatures, although the equilibrium yield is lower. The time to equilibrium t_{eq} varies from 0.06 to 0.14 h.

Figure 6.86 presents the variation with reaction temperature of the hydrogen yield at equilibrium (y_{eq}), hydrochloric acid conversion at equilibrium (X), and equilibrium hydrogen concentration (molar fraction) in the gas phase (C_{eq}). Obtaining good conversions (over 80 %) is possible at the lowest temperatures in the range of interest, but this leads to larger equilibrium times, or in other words, larger residence times of particles in the reactor.

Denote n as the excess number of moles of hydrochloric acid, defined as the moles of HCl supplied to the reactor per mole of copper. In mathematical terms, this can be written as $n_{Cu} \times n = n_0$, where n_0 is the number of moles in HCl in the feed

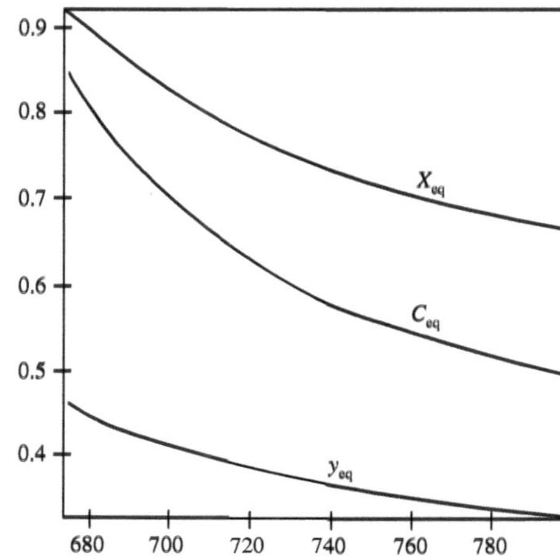

Fig. 6.86 Variation with reaction temperature of the H_2 yield at equilibrium (y_{eq}), HCl conversion at equilibrium (X), and equilibrium hydrogen concentration in the gas phase (C_{eq}) [data from Zamfirescu et al. (2010b)]

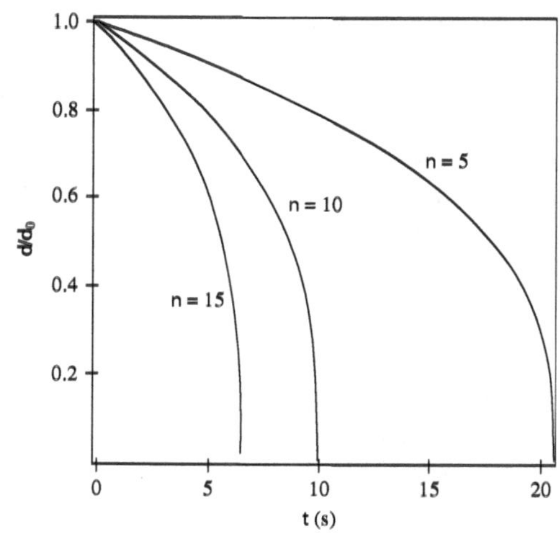

Fig. 6.87 Copper particle shrinkage during the reaction for various excess numbers (n) of hydrochloric acid in the feed stream [data from Zamfirescu et al. (2010b)]

stream. Then, as detailed in Zamfirescu et al. (2010b), the copper particle shrinkage with time is governed by the equation

$$\mathrm{d}(t) = d_0 \times \left[1 - n\int_0^t \left(\frac{\mathrm{d}X}{\mathrm{d}T}\right)\mathrm{d}t\right]^{1/3}. \tag{6.72}$$

Equation (6.72) can be integrated numerically to observe the particle shrinkage in time, during the reaction process. For stoichiometry, $n = 1$, so for 1 mol of copper there is 1 mol of HCl. In a reactor, the HCl must be supplied in excess in order to shift the reaction equilibrium to the forward direction. Figure 6.87 shows the decrease during the reaction time of the particle diameter relative to its initial diameter for excess hydrochloric acid, $n = 5$, 10, and 15. The particle shrinks in an accelerated manner especially after the diameter decreases below 20 % of the initial value. The solid copper particle completely disappears after a few tens of seconds. It is found that the residence time of copper particles varies between 10 and 100 s, depending on the operating conditions. The hydrogen conversion at equilibrium varies between 55 and 85 %, depending on the reaction temperature.

6.9 Heat Recovery and Reuse Within a Cu–Cl Plant

6.9.1 *Pinch Analysis*

The Cu–Cl cycle is expected to be driven in an environmentally benign manner using nuclear energy. Since each step occurs at a different temperature, the product of a step needs to be cooled or heated to the next step's temperature. Thus, there are many heat recovery opportunities within the cycle. The recovered energy as well as the energy released from the exothermic reactions could be reused in the cycle. Heat is transferred between various endothermic and exothermic reactions in the Cu–Cl cycle through heat exchangers that supply or recover heat from individual processes. There is also opportunity to integrate heat pumps and heat engines with the cycle.

Effective thermal management within the Cu–Cl cycle is crucial for achieving high efficiency. The cycle's efficiency is improved drastically when all heat released by the products of reactions is recycled internally. Therefore, the energy matching and hence heat exchangers within the cycle have a significant role in terms of overall effectiveness of the cycle.

As a first step of the pinch analysis of the cycle, the heat fluxes crossing the thermodynamic boundary of the water-splitting cycle are identified and listed in terms of their temperature level and enthalpy. In order to have uniform energy units, the enthalpy is expressed in kJ/mol of hydrogen produced.

The five-step cycle Cu–Cl-5 has the most opportunities of heat recovery and reuse because it comprises four thermal reactors. Other heat fluxes are required to preheat the reactants at the desired level of temperature. Some other fluxes are associated with heat recovery from the hot products. The inventory of heat transfer for the five-step Cu–Cl cycle is presented in Table 6.27. One observes that copper particles must be heated from 333 to 723 K, which is the temperature of the hydrogen production reactor. For the purpose of the thermodynamic analysis, the Cu particle heating is divided into three processes, namely: process $1 \rightarrow 2$ discussed above, where the enthalpy required to heat Cu and $CuCl_2$ and water between 333 and 373 K is 14 kJ/mol H_2; process $3 \rightarrow 4$ along which three different substance streams are heated from 373 to 673 K, i.e., 2 mol of Cu, 1 mol of steam, and 2 mol of dry $CuCl_2$ with an enthalpy input of 75 kJ/mol H_2; and lastly the 2 mol of Cu particles are heated from 673 to 723 K within process $5 \rightarrow 6$ that consumes

Table 6.27 Energy inputs and outputs for Cu–Cl-5 cycle

Process	ΔH (kJ/mol H_2)	T (K)	Remarks
1 → 2	+14	333	Electrolytic Cu(s) production step: $4CuCl(s) \rightarrow 2CuCl_2(aq) + 2Cu(s)$
2 → 3	+75	333 → 373	Heating: $2Cu(s) + 2CuCl_2(aq) + 1.248H_2O(l)$
3 → 4	+87	373 → 673	Heating: $H_2O(g) + 2Cu(s)$; $2CuCl_2(s)$
4 → 5	+132	673	Hydrolysis reactor: $2CuCl_2(s) + H_2O(g) \rightarrow CuO{*}CuCl_2(s) + 2HCl(g)$
5 → 6	+3	673 → 723	Copper particles heating: 2 mol Cu(s) per mol H_2
7 → 8	+146	823	Oxygen production reactor: $CuO{*}CuCl_2(s) \rightarrow 2CuCl(l) + 0.5O_2(g)$
9 → 10	−10	823 → 723	Cooling: $0.5O_2(g) + 2CuCl(l)$
10 → 11	−47	723	Hydrogen production reactor: $2Cu(s) + 2HCl(g) \rightarrow 2CuCl(l) + H_2(g)$
11 → 12	−162	723 → 373	Cooling: $H_2(g) + 0.5O_2(g) + 4CuCl(l \rightarrow s)$
12 → 13	−75	373 → 333	Cooling: $H_2(g) + 0.5O_2(g) + 4CuCl(s) + 1.248H_2O(g \rightarrow l)$
W_{EL}	+110	–	Electrical power needed to drive Cu(s) production step (1 → 2)
Q_H	+458	–	Total heat input into the cycle
Q_C	−294	–	Total heat rejected by the cycle
Q_{Loss}	−21	–	Assumed heat losses, amounting to 5 %

Sign convention: (+) = heat or work input, (−) = heat output

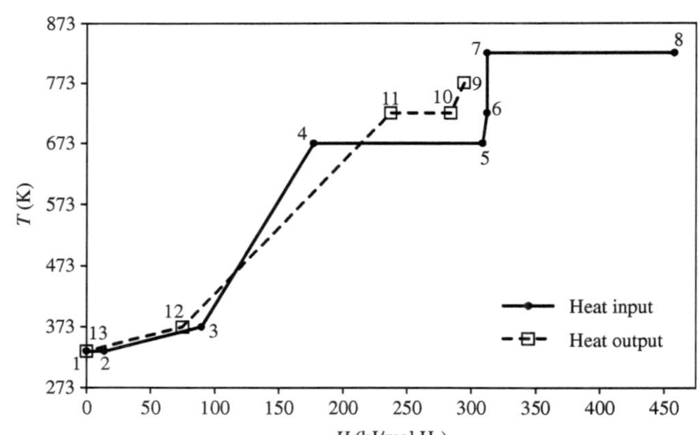

Fig. 6.88 Pinch diagram for the five-step copper–chlorine cycle

3 kJ/mol H_2. A similar approach is pursued to identify all processes listed in Table 6.1, where the energy efficiency of the cycle is also indicated.

Starting from the process data in Table 6.27, it is possible to construct a temperature–enthalpy diagram (pinch diagram) as shown in Fig. 6.88. This indicates the required heat inputs and outputs and their associated temperature levels. In the diagram, the continuous line refers to heat input streams, while the dashed line refers to heat outputs. The processes illustrated in the figure have labels corresponding to those indicated in the table. Thus the streams requiring heat are identified with the line 1–2–3–4–5–6–7–8, and those releasing heat are 9–10–11–12–13. The construction of the pinch diagram imposes a convention for "zero" enthalpy. The enthalpy of the state at the lowest temperature is set to zero. This is state 1—the isothermal process in the electrolytic bath that produces copper from cuprous chloride at 333 K.

Note that the value of 333 K is adopted here from the perspective of coupling the process to the moderator of a nuclear power plant. However, in general, the temperature associated with the Cu production step can be closer to the ambient. For the heat releasing process, zero enthalpy is assumed for the cooled product streams of H_2, O_2, CuCl(s), and $H_2O(l)$—the last two chemicals are recycled. This temperature is 333 K, the lowest in the cycle, even though in an actual implementation, the streams of hydrogen and oxygen should be cooled more, i.e., to the ambient temperature or below, to facilitate compression and storage.

Considering the temperature–enthalpy profiles from Fig. 6.88, one observes that for $T < 373$ K, the temperature of streams releasing and absorbing heat are almost superimposed. This suggests that relatively straightforward flow arrangements can be devised for heat recovery. Orhan et al. (2011) report an exergy and energy analysis of heat exchangers within the copper–chlorine cycle aimed to enhance the heat recovery and reuse options for improved efficiency.

The analysis of heat exchangers involves calculations of the number of heat transfer units (NTU) and heat exchanger effectiveness (ε). These parameters are defined by known relationships, depending on the specific heat (c_h, c_c) of the hot and cold streams, respectively, overall heat transfer coefficient (U), heat exchanger area (A), and mass flow rate $\dot{m}$; the definition relationships are given below:

$$\varepsilon = \frac{c_c \times (T_{c,out} - T_{c,in})}{c_{min} \times (T_{h,in} - T_{h,out})}, \qquad \text{NTU} = UA/(\dot{m} \times c)_{min}, \tag{6.73}$$

where $c_{min} = \min\{c_c, c_h\}$ and $(\dot{m} \times c)_{min}$ is the minimum of the products of the mass flow rate and specific heat for hot and cold fluids.

The effect of the heat exchanger effectiveness on thermal management in the Cu–Cl cycle is illustrated in Fig. 6.89. The amount of recovered heat within the cycle can be increased by using heat exchangers with higher effectiveness. Then, the total heat requirement and consequently the net heat requirement for the cycle are decreased. At $\varepsilon = 0.85$, the total heat demand is equal to the recovered heat and thus the net heat requirement is zero. Beyond that point $(\varepsilon = 0.85)$, the net heat requirement is negative since the recovered heat is more than the total heat demand in the cycle. However, the heat requirement for chemical reactions and the electrical energy needed for electrolysis, pumping, and compression are not included, as the analysis is only considering the heat exchangers.

The variation of the overall efficiencies of the Cu–Cl cycle with heat exchanger effectiveness is—as shown in Orhan et al. (2011)—approximately linear. Both energy and exergy efficiencies of the cycle increase with more effective heat

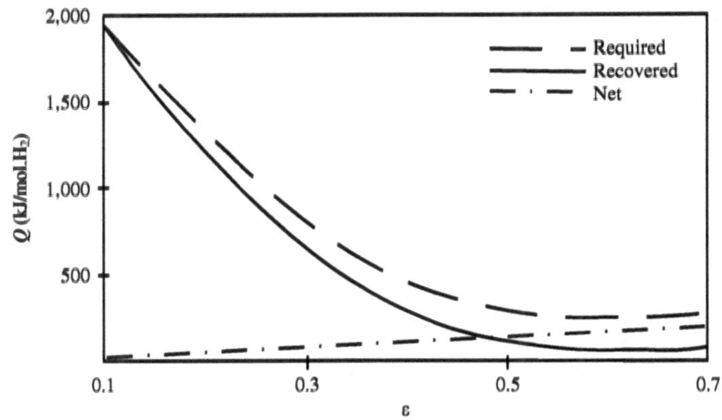

Fig. 6.89 Effect of heat exchanger effectiveness on heat recovery in the Cu–Cl cycle [data from Orhan et al. (2011)]

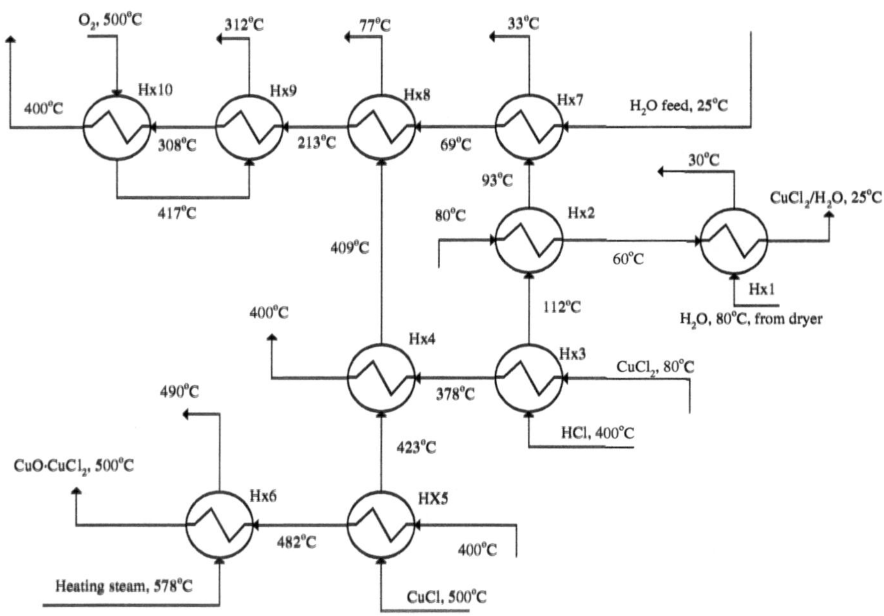

Fig. 6.90 Heat exchanger network of heat recovery and reuse within the Cu–Cl cycle [modified from Orhan (2011)]

exchangers in the cycle. This is because heat exchangers with a higher effectiveness result in more heat recovery within the cycle and thus decrease the external heat requirements for the cycle. Based on a pinch analysis, integrated heat exchanger networks of enhanced heat recovery and reuse can be used. A heat exchanger network was presented by Orhan et al. (2011) as indicated in Fig. 6.90.

6.9.2 Heat Recovery from Molten CuCl

The most challenging heat recovery process within the copper–chlorine cycle is from molten cuprous chloride droplets that solidify and must be cooled to a temperature corresponding to the electrochemical step (about 353 K). Recovering heat during the solidification of molten CuCl and efficient utilization of this recovered heat are crucial to the efficient performance of the copper–chlorine cycle. Several studies reported theoretical results of heat recovery process and modeling—Jaber et al. (2010a, b) and Ghandehariun et al. (2011)—and some yet unreported experimental work is under way at UOIT.

The process of cuprous chloride droplet cooling is a type of multiphase flow, similar to droplet/particle flows in spray columns, packed beds, and other devices involving the interaction between particles or droplets with gas streams. Heat recovery from molten CuCl droplets has similarities with the dynamics of droplets with high viscosity ratios like solid particles, as with liquid metal droplets. Heat transfer from the continuous phase to the droplet involves convective heat transfer, and complex interactions of velocity, droplet breakup, elongation, and so forth.

Various methods for recovering heat from molten CuCl—suggested by Ghandehariun et al. (2011)—are presented in this section. Table 6.28 presents various heat recovery options that were considered in cooling of cuprous chloride droplets. Several lab experiments for cuprous chloride solidification were performed at UOIT. Figure 6.91 presents two experimental units used for heat recovery experiments from molten CuCl. Figure 6.92 illustrates some samples obtained during the solidification experiments. A scaled up tilt furnace apparatus is used at UOIT for large-scale heat recovery from molten cuprous chloride. A photograph is shown in Fig. 6.93.

Direct heat and mass transfer from sprayed droplets of CuCl to a gaseous coolant has been modeled by Jaber et al. (2010a, b). Fluid flow around droplets can be categorized into low Reynolds number flows, and high Reynolds (*Re*) number flows. Low Reynolds number flows can be creeping flow if $Re \ll 1$ for which the inertia of the fluid can be neglected. The flow around the droplets is asymmetric only for small Reynolds numbers. For flows with $Re > 1$, the pattern becomes distorted, and a wake is generated at the back of a droplet. At very high Reynolds numbers, the wake becomes unsteady and very strong, with shedding of vortices.

The following flow regimes are encountered for high Reynolds numbers of solid spheres.

- *Attached flow* ($1 < Re < 20$): The flow remains attached to the droplet surface and separation does not occur in this flow regime. The generated wake behind the droplet is weak and steady.
- *Steady-state wake* ($20 < Re < 130$): At the beginning of this flow regime, a weak circulation region is attached to the back of the droplet. As *Re* increases, the wake becomes stronger, its width and length increase, and its point of attachment moves forward on the surface of the droplet.

Table 6.28 Heat recovery options from molten cuprous chloride within the Cu–Cl cycle

Type and description	Diagram
Drum flaker • Solidification onto the outer surface of a rotating cylinder • Inner surface is cooled by a coolant • Knife or blade removes the solids • 80 % of the circumference in contact with material • Drum surface temperature lower than the melting point • Molten CuCl solidifies at the outside wall forming a solid layer	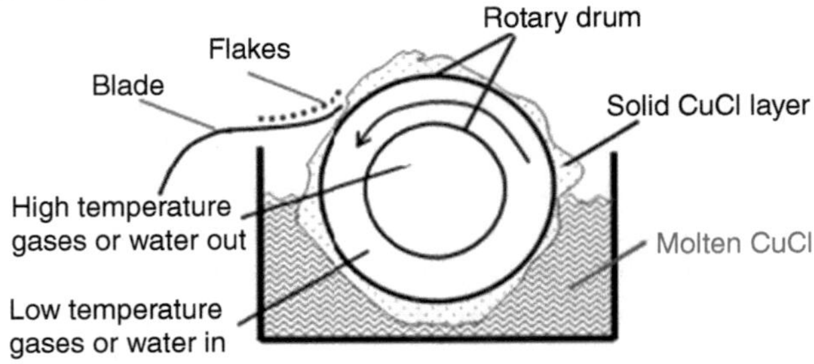
Casting extruder • Tube in a tube heat exchanger configuration is used • Water supplied to the bottom at the outer shell is the coolant • Steam is generated via heat recovery • Molten CuCl enters at ~800 K • Solids exit at ~323 K • Extruder removes solids to prevent plugging of channel • Method involves thorough selection of materials	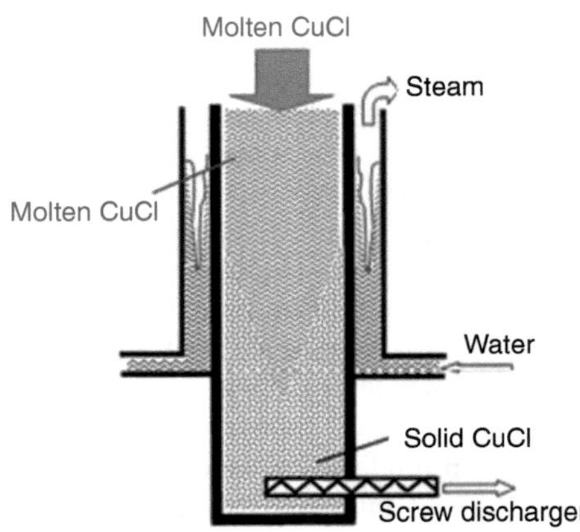
Spinning atomizer • Nozzle introduces the molten CuCl • Spinning disk or rotary cup generates centrifugal force • Droplets disperse around and fall as a spray • Air or inert gas is distributed through a perforated plate • Heat from CuCl is transferred to the coolant • Care must be taken to retain all droplets inside the system • Solid particles are removed at the bottom	

(continued)

Table 6.28 (continued)

Type and description	Diagram
• The vessel must be relatively large • care in selecting materials for rotary cup must be taken	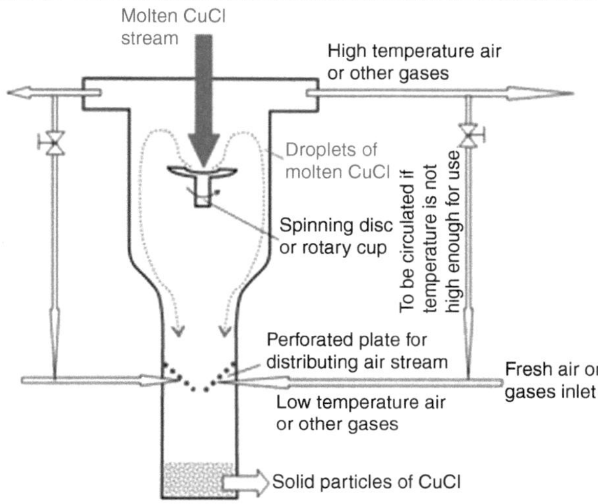
Inert gas atomizer • CuCl is granulated by atomization with inert gas or air • Possible to atomize a slurry of water and CuCl • System is largely extended horizontally • Second air/gas gun is needed to extent particle jet • Design must avoid the buildup of particles on walls • Particles must solidify before colliding with walls • cyclone may be placed to separate the entrained solids	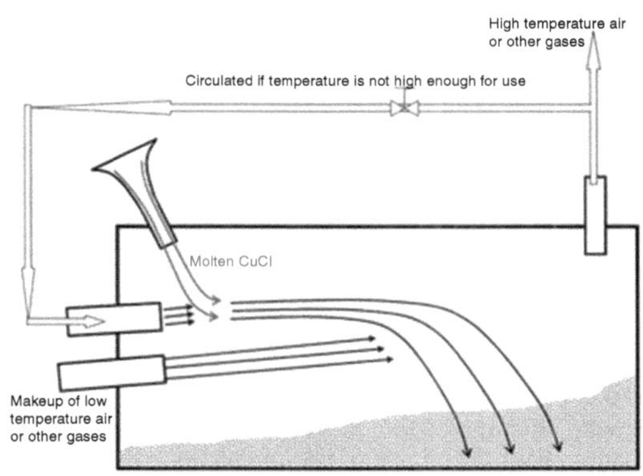
Falling-through atomizer • Molten CuCl sprayed by a sieve shaker • Spraying may be difficult due to corrosive effects of molten salt • If upstream process is carried out at high temperature, it facilitates	

(continued)

Table 6.28 (continued)

Type and description	Diagram
spraying • Droplets solidify before reaching the bottom of the vessel • Countercurrent airflow used to recover heat • A fluidized bed can be also used at the column base to remove additional heat • Resulting hot air from recovery is at ~473 K	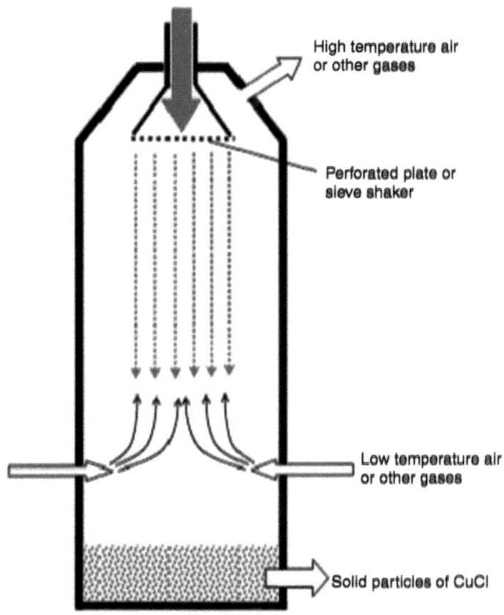
Quenching atomizer • Molten salt must be pumped through a spray nozzle at the top of the vessel • Injection temperature must be ~800 K for good atomization • Atomized salt then falls into a quenching water bath • CuCl further transfers its residual heat to the upward flowing steam • Not all droplets are solidified during descent; some solidify in the quenching bath • Heat received from quenching bath generates steam • Steam is superheated in the region above the bath • No atomization gases are required • CuCl can be easily separated from water and directed toward the next process	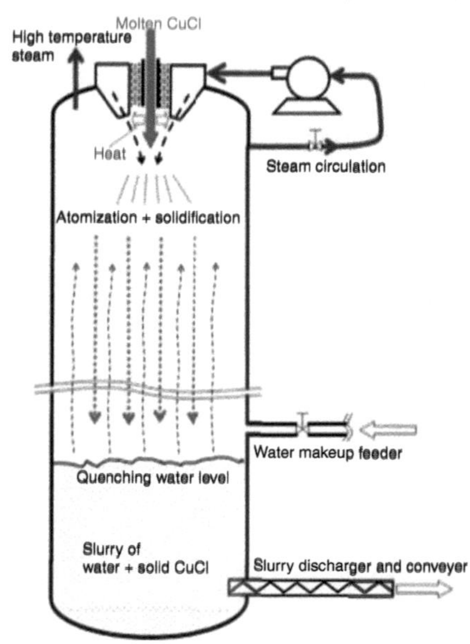

(continued)

Table 6.28 (continued)

Type and description	Diagram
Quenching atomizer with slurry vessel • No quench bath is used, rather all droplets are solidified in the column • Solid particles agglomerate at the bottom of the main (granulation) vessel • Particles are conveyed to a water bath where they release their sensible heat • Saturated steam flows in the granulation vessel • Steam is superheated in the granulation vessel by heat transfer from droplets • Control of droplet dimensions is not essential because the risk of steam explosion is eliminated • Corrosion protective coatings are needed inside the vessel	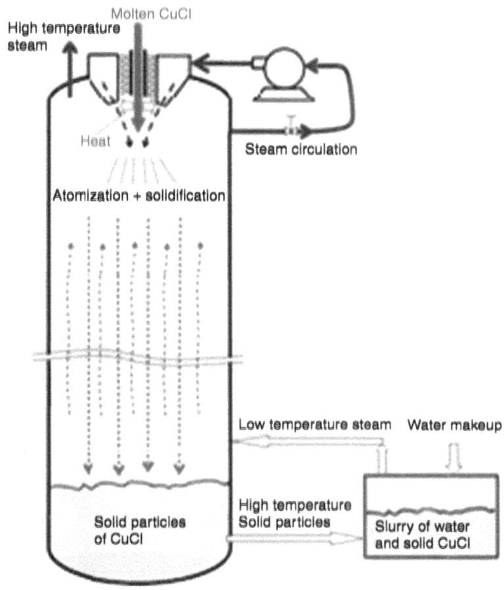

Source: Ghandehariun et al. (2011)

Fig. 6.91 Heat exchanger (**a**) and buffer tank for steam (**b**) used for heat recovery experiments from molten CuCl at UOIT

- *Unsteady wake in laminar flow* (*130 < Re < 270*): at the beginning of this flow regime, a weak, long-period oscillation at the tip of the wake occurs. Oscillations increase with increasing *Re*, but the wake remains attached to the droplet. In this flow regime, the viscous effects have a small role on the value of the drag coefficient. Most of the generated drag is due to the pressure difference between two sides of the droplet. The flow outside of the wake remains laminar throughout this flow regime.

Fig. 6.92 Samples with solidified cuprous chloride

Fig. 6.93 Tilt furnace apparatus for large-scale molten salt heat recovery at UOIT

- *Vortex shedding (270 < Re < 6,000)*: Pockets of vorticity are shed away from the tip of the droplet in this flow regime. This influences the flow velocity and its fluctuations at far fields. The separation point of the flow moves forward on the droplet's surface, reaching a point on the droplet that corresponds to an angle measured from the forward axis of the sphere of 83°.
- *Unsteady boundary layer separation (6,000 < Re < 300,000)*: At the beginning of this flow regime, which is usually called the "lower critical Reynolds number," the separation point at the droplet surface starts rotating around the droplet in a frequency equal to the vortex shedding frequency. The wake behind the droplet does not reach a turbulent stage, and the value of the drag coefficient reaches an almost constant value throughout this flow regime.
- *Transition and supercritical flow (Re > 300,000)*: The wake becomes turbulent in this flow regime. The transition to a turbulent wake starts at $Re = 2 \times 10^5$ and it is completed at $Re = 3.7 \times 10^5$. However, the critical point of transition is usually at around $Re = 3 \times 10^5$. The point of separation moves downstream of the droplet, from an angle of 83°–120°, and it remains constant thereafter. The free shear layer in the flow becomes turbulent and attached to the surface of the droplet, which causes a sharp reduction in the value of the drag coefficient.

Flow around liquid droplets has the same characteristics as solid particles. However, at high Reynolds numbers, the shear stresses acting on the droplet's surface lead to a significant shape distortion that might lead to breakup of the droplet. There are several differences between flow around liquid droplets and solid particles. Firstly, internal circulations within liquid droplets delay the formation of the wakes by delaying the boundary layer separation. They also cause the length of the wake to be slightly shorter. Secondly, two symmetric internal weak wakes appear at the back of liquid droplets at high Reynolds numbers. These cause a flow reversal at the rear of the droplet, so the velocity vector points in the positive direction for a short distance. Liquid droplets exhibit a lower drag coefficient than that of solid droplets. A high Reynolds number causes the shear stresses to deform liquid droplets and change their shape from perfect spheres to spheroids. Internal flow fields within liquid droplets also affect the external flow field, which can significantly lower the vorticity compared to solid particles. These characteristics and features of fluid flow past droplets and spheres will be used for physical development and understanding of the droplet flow modeling and results. In the next section, predictive formulations will be developed for the fluid flow and heat transfer from the molten CuCl droplets.

In a falling-through heat recovery system configuration as indicated in Table 6.28, the droplets fall under gravity, and transfer heat to the upward flowing surrounding gas stream. The flow is assumed to be one-dimensional countercurrent symmetrical flow. Droplets enter the heat exchanger from the top, initially at a near-zero velocity, and then increase as they fall down the heat exchanger under the action of different forces.

The heat transfer phenomena between the droplets and air is assumed as one-dimensional transient heat transfer, with heat transfer at the surface of the droplet to the air by forced convection. The temperature distribution within a droplet is assumed to be spatially uniform. The lumped capacitance assumption is made for the droplet so that internal temperature gradients within a droplet can be neglected. To determine the temperature distribution, the convective heat transfer rate can be found as

$$\dot{Q}_{\mathrm{c}} = 0.5\bar{h}A_{\mathrm{d}}(T_{\mathrm{d,i}} + T_{\mathrm{d,e}} - T_{\mathrm{f,i}} - T_{\mathrm{f,e}}), \tag{6.74}$$

where A_{d} is the surface area of the droplet corresponding to the droplet mass flow rate; $T_{\mathrm{d,i}}$ and $T_{\mathrm{d,e}}$ are the temperatures of the droplet at the inlet and exit of each element, respectively; and $T_{\mathrm{f,i}}$ and $T_{\mathrm{f,e}}$ are the temperatures of the fluid at the inlet and exit of each element, respectively; and $\bar{h}$ is the average heat transfer coefficient. The heat transfer rate from the droplet during sensible cooling can be written as

$$\dot{Q}_{\mathrm{d}} = \dot{m}_{\mathrm{d}}c_{p,\mathrm{d}}(T_{\mathrm{d,i}} - T_{\mathrm{d,e}}), \tag{6.75}$$

where $\dot{m}_{\mathrm{d}}$ is the droplet mass flow rate, and $c_{p,\mathrm{d}}$ is the specific heat of the droplet at constant pressure. The heat transfer rate to the fluid may be expressed as

$$\dot{Q}_{\mathrm{f}} = \dot{m}_{\mathrm{f}}c_{p,\mathrm{f}}(T_{\mathrm{f,e}} - T_{\mathrm{f,i}}). \tag{6.76}$$

The heat transfer rates are then equated for the sensible heating/cooling portions of the heat exchange trajectory of the droplets, namely, $\dot{Q}_c = \dot{Q}_d = \dot{Q}_f$. Therefore, from (6.74), (6.75), and (6.76), one obtains the exit temperature of the fluid as reported by Jaber et al. (2010a):

$$T_{f,e} = \left[\frac{\bar{h}A_d}{2\dot{m}_f c_{p,f}} \times T_{d,i} + \frac{\bar{h}A_d}{2\dot{m}_f c_{p,f}} \times T_{d,e} + \left(1 - \frac{\bar{h}A_d}{2\dot{m}_f c_{p,f}}\right) \times T_{f,i}\right] \Big/ \left(1 + \frac{\bar{h}A_d}{2\dot{m}_f c_{p,f}}\right). \tag{6.77}$$

Equation (6.77) does not depend on the CuCl properties; it only requires the droplet temperature at the inlet and exit. There exist two cases of heat transfer due to phase change of the droplet. In the first case, the temperature of the droplet is constant during the phase change process, so $T_{d,i}$ and $T_{d,e}$ are equal in (6.77). In the second case, the temperature of the droplet changes between the inlet and exit due to sensible heat loss.

The increase in heat transfer does not take long because the first-phase transformation process differs with the second-phase transformation process by 11 °C. In the first-phase transformation process, the heat transfer rate decreases again and reaches close to zero, because the temperature difference reaches a value of 1 °C. Finally, the heat transfer rate increases again in the final sensible heat transfer process.

Because there is no available experimental data on spraying of molten CuCl, the validation of the model by Jaber et al. (2010a) is performed against reference data of a problem of heat transfer from a droplet to a stream of air. It is found that predictions from Jaber et al. (2010a) have a 0.4 % relative error for the Nusselt number, 1.7 % for convective heat transfer coefficient, and 1.7 % for the heat transfer rate.

With a validated model, two case studies were considered by Jaber et al. (2010a, b) with the parameters indicated in Table 6.29. The variation of the heat transfer rate throughout the height of the heat exchanger is presented in Fig. 6.94. It is evident from the results that the larger droplet diameter has a higher heat transfer rate. This occurs because of the area ratio between two droplets. A droplet with a diameter of 1 mm has four times the surface area of a droplet with a 0.5 mm diameter. The heat transfer rate increases initially at the bottom of the heat exchanger because of the increasing temperature difference between both streams in this region, which yields a higher potential for heat transfer.

The heat transfer rate reaches a maximum value when the second-phase transformation ends. After this point, it starts decreasing due to the decreasing temperature difference between both fluids in the phase transformation process. When the second sensible heat transfer process is reached, the heat transfer rate starts increasing again.

Figure 6.95 illustrates a case with a CuCl flow rate of 3.4 g/s, which is equivalent to the flow from an oxygen production reactor for a Cu–Cl plant that generates 3 kg of hydrogen per day. The airflow rate is 3 g/s, which is slightly higher than the

Table 6.29 Simulation parameters of CuCl spray heat exchanger for two case studies

Parameter	d_d 1 mm	0.5 mm
CuCl mass flow rate (g/s)	3.4	3.4
Air mass flow rate (g/s)	3	3
CuCl inlet temperature (°C)	530	530
CuCl exit temperature (°C)	70	70
Air inlet temperature (°C)	25	25
Air exit temperature (°C)	493	493
Number of CuCl droplets	1,779	14,226
CuCl droplet inlet velocity (m/s)	0.02	0.07
Air inlet velocity (m/s)	0.2	0.2
Droplet travel time (s)	0.36	0.40
Number of droplet injectors	100	100
Heat exchanger height (m)	0.6	0.8
Heat exchanger diameter (m)	0.13	0.13

d_d droplet diameter
Source: Jaber et al. (2010a, b)

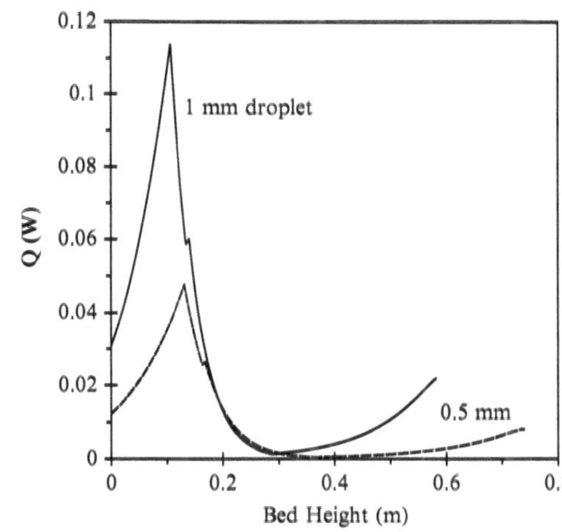

Fig. 6.94 Heat transfer rate variation with bed height [data from Jaber et al. (2010a)]

minimum airflow required to recover all heat from a stream of CuCl droplets at a temperature of 800 K. Because of the high average temperature difference between the CuCl droplets and airstreams, the heat transfer rate is high. The heat transferred to the airstream is more than three times the amount of heat transferred in the latent heat transfer process. This implies that the average required length to transfer one unit of thermal sensible energy was about 0.02 m, compared to 0.03 m to transfer the same amount of heat in a latent heat transfer process, with the same inlet temperature difference. Once the CuCl droplet stream reaches a temperature of 695 K, a second-phase transformation process occurs. Here CuCl droplets have a constant temperature solid–solid phase change. The air temperature still increases, but with a decreasing rate of increase. When the air temperature increases, the temperature difference between both streams decreases and causes the heat transfer

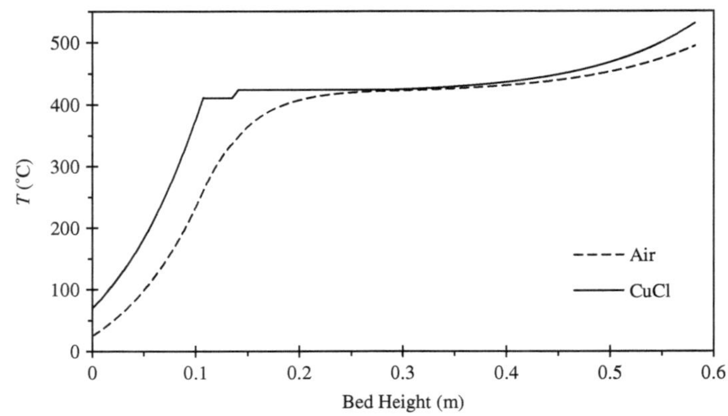

Fig. 6.95 Temperature distribution in air and CuCl for sprayed droplets of 1 mm [data from Jaber et al. (2010a)]

rate to decrease. This causes the required length of interaction between both streams to recover one unit of thermal energy to increase. The rate at which the air temperature increases keeps decreasing until it nearly reaches a value of zero. This occurs at the end of the process, where the air temperature reaches a value of 675 K. This point has the lowest heat transfer rate throughout the heat exchanger.

The final sensible heat transfer process starts when all latent heat energy is absorbed by the airstream from the first-phase change process. In this final process, the CuCl droplets and air temperatures start increasing at a higher rate, similar to the temperature increase in the first sensible heat transfer process. The temperature difference between both streams also increases. The process ends when the CuCl droplets reach a temperature equal to 803 K and the air reaches a temperature of about 766 K. The required dimensions of the heat exchanger for this example case study to complete all of the heat transfer processes are a height of 0.6 m and a diameter of 0.13 m.

6.9.3 Heat Pump Integration in a Copper–Chlorine Plant

As discussed previously, heat from hotter streams can be recovered and used to heat other streams within the Cu–Cl cycle. With respect to the pinch diagram in Fig. 6.88, for $T < 373$ K, the temperature of streams releasing and absorbing heat are almost superimposed. This suggests that straightforward flow arrangements can be devised for heat recovery between them. For the other streams, it is desirable to use heat recovery as much as possible and upgrade the temperature of the rejected heat such that it reaches the level needed for heat input. The heat released by stream 9–10–11–12 should be received by a heat pump, which at the expense of additional power input upgrades the temperature and enthalpy content and delivers a hot stream to the Cu–Cl cycle that matches the profile 3–4–5–6–7–8.

Fig. 6.96 General layout of an integrated copper–chlorine cycle and heat pump system

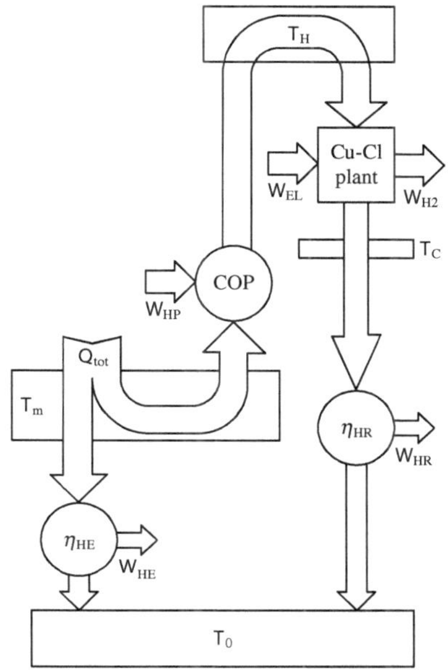

The work required to operate such a heat pump can come from a sustainable heat source available at an intermediate temperature T_m, which drives a heat engine operating between T_m and the ambient at T_0. This heat engine also delivers power to the Cu–Cl cycle. However, because the ambient temperature is lower than 333 K (assumed to be the minimum temperature within the system), it is better from a thermodynamic point of view to use the heat rejected to drive a heat engine that discharges heat to the environment while producing useful work output. For example, if the ambient is a lake at 288 K and the average temperature rejected by the Cu–Cl cycle is 573 K, then the Carnot efficiency of the heat engine is ~50 %.

A thermodynamic model for coupling the Cu–Cl cycle to a sustainable heat source is proposed in Fig. 6.96. This model comprises four components, namely, the Cu–Cl plant, a heat pump "COP," and two heat engines, η_{HE} and η_{HR}. The model assumes four temperature reservoirs as follows:

- Ambient temperature reservoir at T_0
- Sustainable heat source reservoir at $T_m > T_0$
- High-temperature reservoir that delivers useful heat to the Cu–Cl cycle at $T_H > T_m$
- Cold-temperature reservoir at which the Cu–Cl cycle rejects heat at T_C, with $T_H > T_C > T_0$

The plot from Fig. 6.97 shows the variation of hydrogen production efficiency of the Cu–Cl cycle, with idealized constant-temperature heat reservoirs, as a function of the abatement of heat engines/pumps from the Carnot efficiency. For 50 % abatement, one can expect a hydrogen production efficiency range from 4 to 39 % if the temperature of the sustainable energy source T_m varies from 333 to 623 K. The

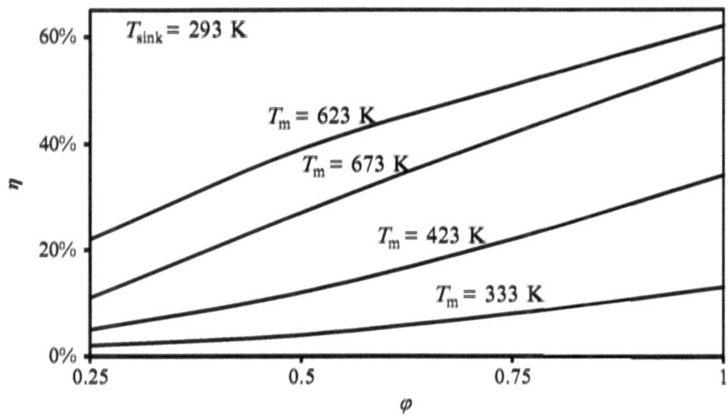

Fig. 6.97 Upper bound energy efficiency variation with abatement φ from the Carnot efficiency for water-splitting system integrated with a heat pump

Carnot abatement factor—denoted with φ—is a parameter ranging from 0 to 1 that expresses the abatement of the COP of a heat pump or efficiency of a heat engine from the Carnot value. For $\varphi = 0$, the COP takes its minimal value of 1, while for $\varphi = 1$, the COP and the efficiency are maximum and equal to the Carnot limit.

Using the Carnot abatement efficiency, the following set of relationships express the energy efficiency and COP of heat engines and heat pumps of the system in Fig. 6.96:

$$\left.\begin{array}{l} \mathrm{COP} = 1 + \varphi[Q_{\mathrm{H}}/(W_{\mathrm{HP}} - W_{\mathrm{HR}}) - 1] \\ \eta_{\mathrm{HE}} = \varphi(1 - T_0/T_{\mathrm{m}}) \\ \eta_{\mathrm{HR}} = \varphi W_{\mathrm{HR}}/Q_{\mathrm{C}}. \end{array}\right\} \tag{6.78}$$

The pinch diagram of the process (see Fig. 6.88) can be transposed in *T*–*S* coordinates, where *S* is the stream entropy expressed in kJ/kmol H_2. The streams are thermodynamic systems that interact with the surroundings by heat transfer only. Consequently, for any stream, $\delta Q = \mathrm{d}H$; thus, $\mathrm{d}S = \int (1/T)\mathrm{d}H$.

For processes evolving at constant temperature (like 4 → 5, 7 → 8, 10 → 11), one obtains $\Delta S = \Delta H/T$. If the temperature is not constant, assume a linear variation of the enthalpy with temperature; then

$$S - S_{\mathrm{a}} = \frac{H_{\mathrm{b}} - H_{\mathrm{a}}}{T_{\mathrm{b}} - T_{\mathrm{a}}} \times \ln\left(\frac{T}{T_{\mathrm{a}}}\right), \tag{6.79}$$

where a is the initial state, b is the final state, and *T* and *S* correspond to any intermediate state of the process.

The area embraced by the lines 1–2–3–4–5–6–7–8 and that indicated by "Heat Source" in Fig. 6.98 represent the ideal/minimum work needed by the heat pump to

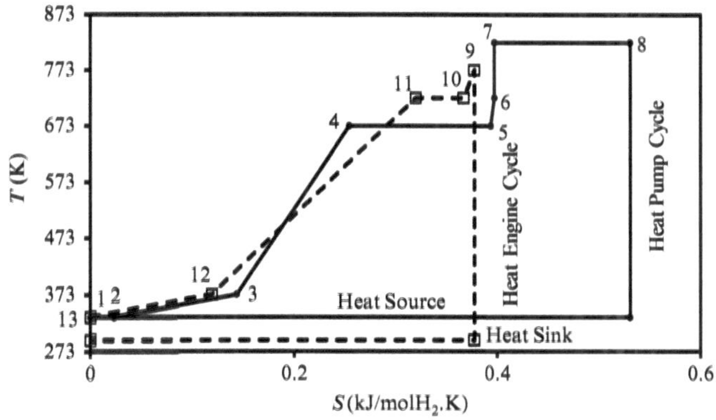

Fig. 6.98 Ideal power and heat pump cycles integrated to a Cu–Cl plant, represented in a *T–S* diagram

upgrade the temperature of the heat source and deliver useful heat to the Cu–Cl cycle. The area encompassed by lines 9–10–11–12–13 and that indicated with "Heat Sink" in Fig. 6.98 represents the ideal/maximum work that can be obtained from the heat recovered from the Cu–Cl cycle. The temperature of the "Heat Source" shown in Fig. 6.98 is taken as 333 K; however, it can be any value in the range above for T_m. The temperature of the "Heat Sink" represented in the figure is taken as 293 K. The engineering challenge regarding the integration of the Cu–Cl plant with the thermal energy source T_m consists of finding suitable working fluids and processes that allow for the construction of a heat pump and heat engine cycle operating according to that presented in Fig. 6.98.

The upper bound efficiency of hydrogen production with the ideal system varies between 6 and 23 % for heat source temperatures in the range of 353–523 K if the Carnot abatement of the integrated heat engines and heat pump is taken as 0.5. Various options of heat pumps—chemical processes based on salt/ammonia- and magnesium-oxide reactions—for upgrading the temperature of thermal energy are presented by Naterer (2008); these systems can be integrated with the copper–chlorine cycle. Integration of a chemical heat pump based on the reversible steam–methane reaction with a copper–chlorine plant linked to a nuclear reactor is presented by Granovskii et al. (2008). Zamfirescu et al. (2009, 2010c, 2011) presented a number of vapor compression heat pumps applicable to a copper–chlorine plant.

Figure 6.99 shows a scheme of cascaded heat pump systems integrated with a copper–chlorine cycle and a cascaded heat engine system. In this concept, the main power plant (e.g., nuclear power plant) generates net power with an efficiency η_1 (the heat source is assumed at a level of ~623 K, while the sink is at 373 K). Further heat recovery (e.g., from the moderator of the nuclear reactor) allows for the operation of a low-temperature heat engine as a sink at the environment at 298 K. This heat engine generates sustainable power ($W_{sustainable}$), which is used to drive cascaded heat pumps that upgrade the temperature of recovered heat up to the level

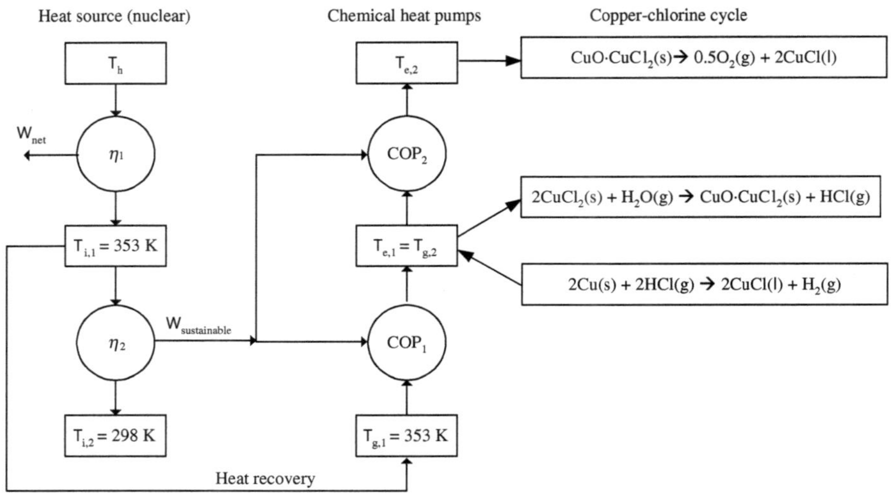

Fig. 6.99 Layout of a copper–chlorine plant integrated with two heat engines and two heat pumps [modified from Naterer (2008)]

required for driving chemical reactions within the Cu–Cl cycle. The exothermic reaction within the cycle is also used at a heat source for further upgrading of the temperature to the level required by the thermolysis reaction (~800 K). Two chemical heat pumps are proposed by Naterer (2008) for the integrated system shown in Fig. 6.99. These are salt/ammonia and magnesium oxide based heat pumps. The chemical reactions with these pumps are presented in Table 6.30.

The lower temperature heat pump is the salt ammonia pump which requires a heat source at 353 K and delivers heat at an upgraded temperature level of 532 K. For the operation, an intermediary heat source is needed at 413 K and intermediary temperature level at 393 K. The magnesium oxide heat pump operates at higher temperatures. It requires 532 K at the source and delivers heat at 783 K at the sink. There is a need for an intermediate heat source and sink of temperature levels of 573 and 583 K for the heat storage mode of operation. The results presented in Naterer (2008) show that the cascaded heat pump system can reach an overall COP in the range of 2–3 which appears promising for linkage with the Cu–Cl cycle.

Granovskii et al. (2008) have shown that it is possible to integrate a steam Rankine power plant—coupled to a supercritical water reactor—with a chemical heat pump based on the following reactions:

- Endothermic, steam methane reaction: $CH_4 + H_2O + Q_{inp} \rightarrow CO + 3H_2$
- Endothermic, reverse water gas shift reaction: $CO_2 + H_2 + Q_{inp} \rightarrow CO + H_2O$
- Exothermic, water gas shift reaction: $CO + H_2O \rightarrow CO_2 + H_2 + Q_{out}$
- Exothermic, methane synthesis reaction: $CO + 3H_2 \rightarrow CH_4 + H_2O + Q_{out}$

It has been shown that the integrated system allows for improved energy utilization efficiency by at least 2 % without jeopardizing the reactor safety.

Table 6.30 Chemical reactions and temperatures of two chemical heat pumps

	Operation mode	T (K)	Reactions
A	Heat storage mode	413	$NiCl_2 \cdot 6NH_3 + Q_{in} \rightarrow NiCl_2 \cdot 2NH_3 + 4NH_3$
		393	$MnSO_4 \cdot 2NH_3 + 4NH_3 \rightarrow MnSO_4 \cdot 6NH_3 + Q_{rej}$
	Heat supply mode	353	$MnSO_4 \cdot 6NH_3 + Q_{waste} \rightarrow MnSO_4 \cdot 2NH_3 + 4NH_3, \quad Q_{waste} = +57.6\,kJ/mol\ NH_3$
		532	$NiCl_2 \cdot 2NH_3 + 4NH_3 \rightarrow NiCl_2 \cdot 6NH_3 + Q_{out}, \quad Q_{out} = -55.3\,kJ/mol\ NH_3$
B	Heat storage mode	583	$Mg(OH)_2 + Q_{in} \rightarrow MgO(s) + H_2O(g), \quad Q_{in} = 81.02\,kJ/mol$
		573	$H_2O(g) \rightarrow H_2O(l) + Q_{rej}$
	Heat supply mode	783	$MgO(s) + H_2O(g) \rightarrow Mg(OH)_2, \quad Q_{out} = 81.02\,kJ/mol$
		532	$H_2O(l) + Q_{waste} \rightarrow H_2O(g)$

Note: *A* salt ammonia heat pump, *B* magnesium oxide heat pump
Source: Naterer (2008)

The manner in which the heat pump is integrated with a nuclear steam cycle and copper–chlorine cycle is presented in the next chapter of the book.

Regarding the use of vapor compression heat pumps in connection with the copper–chlorine cycle, there are at least two major technical constraints: finding a suitable working fluid, and obtaining a coefficient of performance (COP) high enough to justify the use of a heat pump. If the COP is close to 1, then one can use electric heating instead of a heat pump. Zamfirescu et al. (2009) analyzed six organic and titanium-based working fluids for mechanical heat pumps such as biphenyl, biphenyl-methane, naphthalene, isoquinoline, titanium tetrabromide, and titanium tetraiodide. The organic fluids allow heat pump operation using a heat sink with a temperature in the range of 483–533 K, while the heat sink temperature is in the range of 573–753 K. Titanium-based working fluids can operate in heat pumps with a source between 500 and 630 K and heat sink temperature in the range of 573–823 K. Another option is TI_4 with a COP higher than 2. A cascaded heat pump with vapor compression operating with cyclohexane as a bottoming cycle and biphenyl as a topping cycle was reported in Zamfirescu et al. (2010c). The topping cycle is supercritical and delivers heat at variable temperatures (sensible heat). During the cooling process of the working fluid at the sink side heat exchanger of the heat pump, the temperature varies between 873 and 523 K. Under the assumption that both compressors have an isentropic efficiency of 0.75 and the heat source temperature is 333 K, the COP of the cascaded heat pump becomes higher than 5.

A new method of internal heat recovery and reuse with the five-step copper–chlorine cycle was reported by Zamfirescu et al. (2011). A vapor compression heat pump which uses cuprous chloride as a working fluid is used to couple thermally the exothermic hydrogen production reactor to the thermolysis reactor. The heat and temperature level is upgraded through compressors that raise the enthalpy of cuprous chloride vapors. Cuprous chloride is one of the products of the oxy-decomposer (thermolysis reactor) and the working fluid of the heat pump creates an opportunity of better thermal coupling through direct contact heat exchangers for reduced temperature gaps.

Figure 6.100 shows a system where copper oxychloride is fed at #1 into the thermolysis reactor at 675 K where the particulate $CuO{\cdot}CuCl_2$ is heated by direct contact with molten cuprous chloride injected at state point # 15 at 950 K. The molten salt transfers its sensible heat and cools down to the average reactor temperature which is assumed to be 850 K for the reference case in this study.

The heat delivered by the hot stream #15 at 950 K makes copper oxychloride decompose thermally and release oxygen (expelled in #2) while forming molten cuprous chloride (drained out in #3). One part (#4) of the CuCl stream is sent for further processing within the plant while the remaining (streams #5 and #6) are further processed thermo-mechanically within the heat pump. The stream (#5) is throttled to 0.2 mbar and reaches a two-phase state at #7 at a saturation temperature of 755 K. This stream is further heated to saturation with heat recovered from the chlorination reactor of the copper–chlorine cycle.

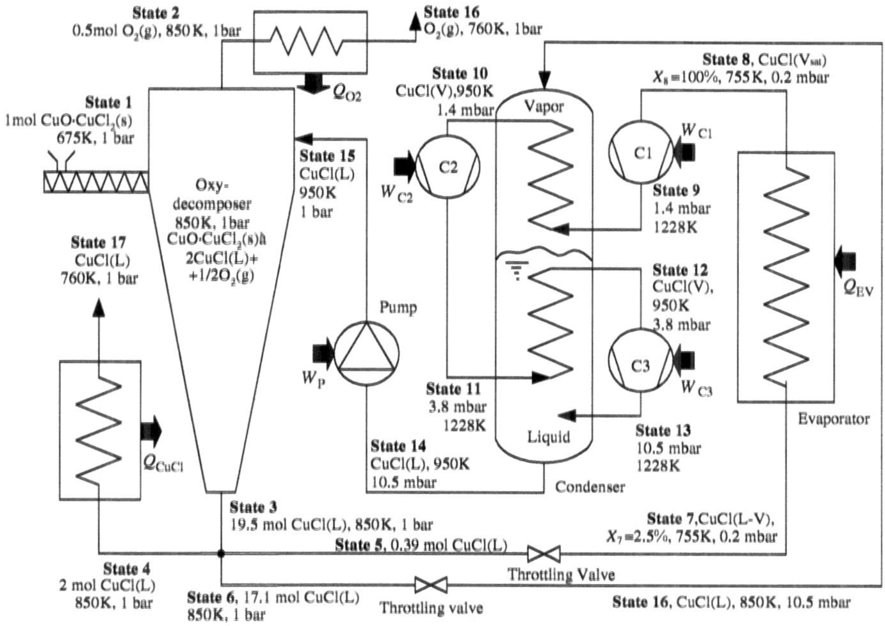

Fig. 6.100 Vapor compression heat pump system with CuCl working fluid, integrated with the five-step copper–chlorine cycle [modified from Zamfirescu et al. (2011)]

The vapor is then compressed in three stages with intercooling. The final pressure is 10.5 mbar, for which the corresponding saturation temperature is 950 K. The condenser in the figure consists of a vessel that incorporates two coils where superheated vapors are cooled after stage 1 and 2 of compression. During the operation, at the bottom of the condenser, saturated CuCl collects. The superheated vapor at the final stage of compression (#13) is injected into this liquid, cooled by direct contact, and condensed. The condensation heat is transferred to molten CuCl which originates from stream #6. The liquid collected in #14 is pumped to atmospheric pressure and injected into the oxy-decomposer.

The thermodynamic cycle in the *T–S* diagram is presented in Fig. 6.101. The reported COP of the heat pump is remarkably high, namely, 6.6 with an assumed isentropic efficiency of 0.85 for all compressors and the pump. The work needed to drive the heat pump associated with production of 1 mol of hydrogen by the Cu–Cl cycle is studied in Fig. 6.102. This work is calculated for a fixed temperature of 850 K in the oxy-decomposer and a range of temperatures at the heat pump sink, T_{15}. Three values were considered for the isentropic efficiency, which is taken to be the same for all compressors and the pump. At a reasonable isentropic efficiency of 0.7, 5 the specific work needed to drive the pump is around 15 kJ/mol of produced hydrogen. Moreover, one observes that further increases in isentropic efficiency do not lead to significant reduction of the specific work.

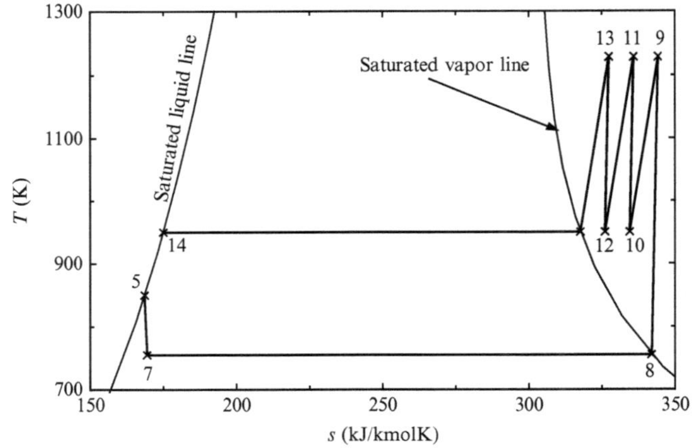

Fig. 6.101 Thermodynamic cycle of CuCl vapor compression heat pump [data from Zamfirescu et al. (2011)]

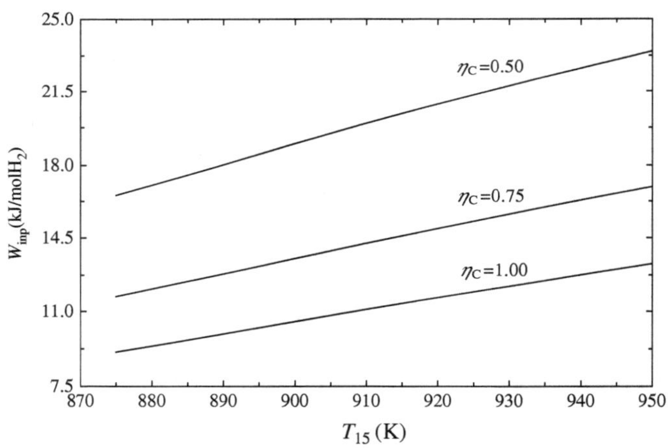

Fig. 6.102 Specific molar work required to drive the CuCl heat pump [data from Zamfirescu et al. (2011)]

6.10 Material Selection and Research

Development of advanced materials for the copper-chlorine thermochemical cycle has major importance. Many components of the cycle work in highly corrosive conditions and high temperatures. One major focus is to develop thermal spray coating technology for thermochemical reactors and other components of the cycle.

A collaborative research program between UOIT and the University of Toronto (UT) has characterized the necessary coatings, as well as test their surface integrity in molten CuCl. Some of the tested materials and procedures are given as follows:

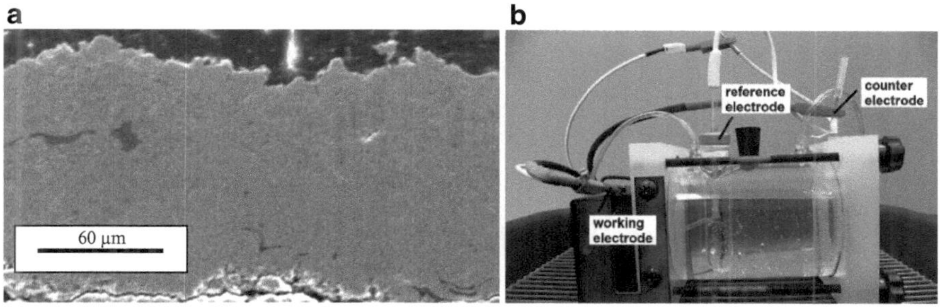

Fig. 6.103 Micrograph of coating cross section deposited from Metco 143 (**a**) (courtesy of Prof. J. Mostaghimi) and electrochemical cell for corrosion testing (**b**)

- Inconel 625 (a nickel-based superalloy) thermally sprayed.
- Austenitic stainless steel AL-6XN with 24 % Ni, 6.3 % Mo thermally sprayed.
- Metal coating with Diamalloy 4006 from Sulzer-Metco ($Ni_{20}Cr_{10}W_9Mo_4Cu_1C_1B_1Fe$).
- Spraying of additional ceramic top layer on metal coatings for various samples. Two ceramic powders were used: zirconia-based Metco 143 ($ZrO_2 \cdot 18TiO_2 \cdot 10Y_2O_3$) of 70 μm thickness, and 170 μm thick alumina-based coating Metco 101NS ($Al_2O_3 \cdot 3TiO_2$).

In the process of coating, a high-pressure mixture of fuel and oxygen ignites in the combustion chamber and the combustion products are accelerated through a converging–diverging nozzle. Particle temperatures are over 2,000 K. As a result, the injected particles attain high velocity at relatively low temperatures. The high kinetic energy of the particles upon impact leads to the formation of a dense and well-adhered coating. The relatively low temperature prevents the particles from excessive oxidation.

Figure 6.103a shows a micrograph of coating cross section deposited from Metco 143 material which was measured by an image analysis technique. The porosity of the coatings was in the order of 1–2 %. The ceramic top coat provides additional protection because of its low thermal conductivity. This insulates the underlying materials, thereby reducing the temperature of the amorphous metallic coating. For example, a 100 μm thick ceramic layer acting as a thermal barrier coating can decrease the temperature of the amorphous coating by about 50 K. Yttria-stabilized zirconia (8 %Y_2O_3-ZrO_2) and Al_2O_3 are two common compositions used for this purpose. They were deposited on amorphous metallic coatings. The metal and ceramic coatings have a dense, low-porosity microstructure that enhances the temperature and corrosion resistance of the multilayer coating.

Selected specimens were tested for corrosion and showed fractured ceramic coatings due to high cooling rates after spraying. The results indicated that the coating integrity can be identified using the electrochemical cell (UOIT) shown in Fig. 6.103b. Electrochemical characterization has been performed on Inconel 625 and AL6XN Stainless Steel samples coated with Diamalloy 4006 metallic layer, and Diamalloy 4006 specimens top coated with either zirconia or alumina.

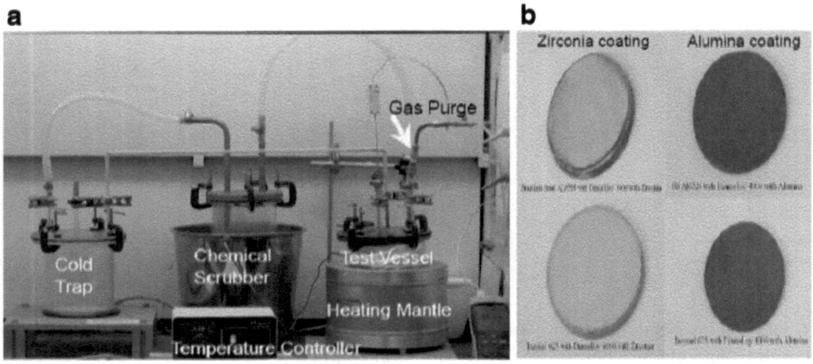

Fig. 6.104 Immersion apparatus for CuCl corrosion tests (**a**) and coating test specimens (**b**)

Salt solutions up to 5 mol/L NaCl were used for characterization at room temperature of the corrosion behavior of coated surfaces with two methods: SEM of the surface and EIS/AC impedance. SEM imaging of the surface was used to identify any physical breakdown, loss of integrity, or restructuring of the surface. Using EIS/AC impedance, the chemical integrity of the film, chemical breakdown, interfacial breakdown, and solution penetration of the specimen were examined. The results of the experiments were consistent with the presence of a passive film.

A CuCl immersion test apparatus was developed for corrosion testing in Fig. 6.104a. Several test immersion were performed with a test duration limited to 8 h for safety concerns. The corrosion specimens remained immersed in the salt as it froze each night. Difficulties in extricating the specimen from the frozen salt have been identified and methods were developed that will enable the removal of solid CuCl from the specimen without damaging or corroding the specimen. Selected surface coating specimens are shown in Fig. 6.104b.

6.11 Scaled-Up Test Loop at UOIT

A scaled-up test loop of the copper–chlorine cycle is under development at UOIT for a projected production of 3 kg hydrogen per day, when all unit operations are connected and integrated together. The test loop consists in four separate main units which are the CuCl/HCl electrolyzer, dehydration reactor, hydrolysis reactor, and oxy-decomposer. In addition, several auxiliary units necessary for system integration are under development. These include the heat recovery system for molten cuprous chloride, apparatus for hydrochloric acid/water separation, among others. A general layout of the scaled-up system is depicted in Fig. 6.105. Each major unit of the test bench is operational. Research studies have been conducted at UOIT on system integration of experimental unit operations of the Cu–Cl cycle, including integration of electrolysis, water separation, and hydrolysis processes, as well as the integration of hydrolysis, oxygen, and hydrogen generation processes.

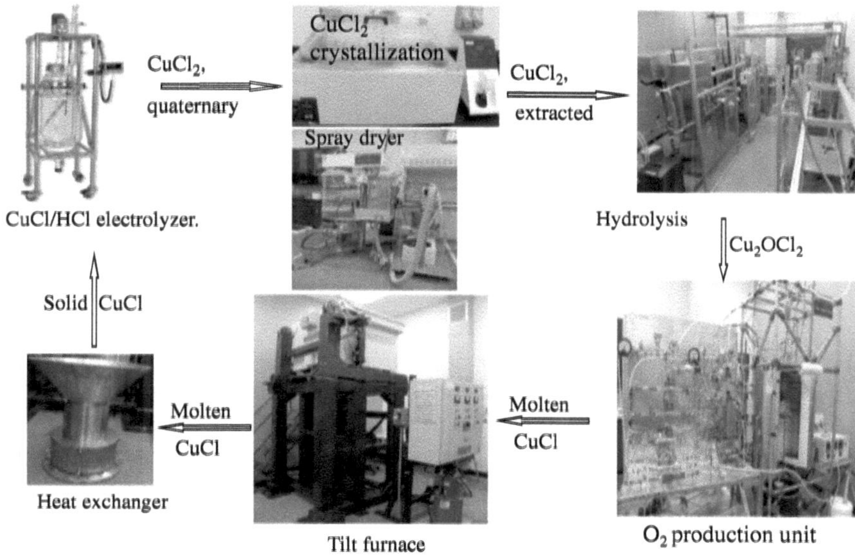

Fig. 6.105 System integration for scaled-up and integrated Cu–Cl cycle at UOIT

The dominant parameter that influences the integration of the electrolysis and hydrolysis processes is the steam quantity required by hydrolysis step. Another parameter that affects the processes of integration is the concentration of cupric chloride exiting the electrolytic cell. The output stream from the electrolytic cell has been simulated using the quaternary system of H_2O, $CuCl_2$, CuCl, and HCl in the experiments with varied concentration of each constituent. Crystallization and spray drying experiments have been performed to concentrate the $CuCl_2$ solution so as to accommodate the variations of electrolysis outputs and deliver $CuCl_2$ of different water contents to the hydrolysis reactor. Crystallization experiments of cupric chloride were also performed by varying the HCl concentration and temperature of the quaternary system. The $CuCl_2$ output from the crystallization and spray drying is delivered separately to the hydrolysis reactor to produce HCl liquid.

With respect to system integration of hydrolysis and oxygen and hydrogen generation processes, studies have also been conducted on the interactions between the Cu_2OCl_2 decomposition and $CuCl_2$ hydrolysis steps. The effects of completion of the hydrolysis reaction on the Cu_2OCl_2 decomposition were studied by varying the fraction of $CuCl_2$ in a mixture of $CuCl_2$ and Cu_2OCl_2 that are fed into the Cu_2OCl_2 decomposition reactor. The primary influences include the conversion of by-products such as Cl_2, partial evaporation of molten CuCl, and heat requirements. The composition of the gaseous products and molten product are analyzed by XRD. The heat input to the oxygen production reactor was characterized by the enthalpy change of the Cu_2OCl_2 and $CuCl_2$ decomposition processes. A mixture of $CuCl_2$ and CuO was also used in the experiments to study the oxygen production reaction so as to accommodate different hydrolysis technologies that may reduce the excess steam and temperature requirements.

Fig. 6.106 Experimental apparatus for HCl/H_2O separation (UOIT)

For separation of HCl and steam downstream of the hydrolysis reactor, the mixture of H_2O and HCl can be concentrated (individually or a combination) by distillation and extractive rectification performed near or below the azeotrope temperature (381 K, at 1 bar; 0.4 HCl mass fraction). An experimental apparatus, shown in Fig. 6.106, was set up at UOIT for investigating water/hydrochloric acid separation. Significant quantities of heat flow are required to condense the high-temperature products, which then must be reheated for other processes. Thus, the downstream design of the hydrolysis reactor can have significant impact on the thermodynamic efficiency of the Cu–Cl cycle. Several techniques to condense the HCl/H_2O mixture are being analyzed and studied experimentally to determine and compare the effectiveness of each system.

Experimental work is also under way on the processing of solidified CuCl and entrained by-products produced from the oxygen production reactor. The processed CuCl will provide the feedstock to the downstream HCl/CuCl electrolyzer. The preparation of the ternary system of CuCl, HCl, and water at various concentrations was for the integration of oxygen and hydrogen production steps. It has been shown that the high conversion of $CuCl_2$ to Cu_2OCl_2 in the hydrolysis step has vital importance to reduce the challenges of the integration of hydrolysis and oxygen and hydrogen production steps.

6.12 Conceptual Designs of Full-Size Plant and Reactors

Experiments have confirmed the feasibility of the copper–chlorine cycle and offer the promise of economic competitiveness of the process to produce hydrogen efficiently in an environmentally benign manner from an inexhaustible resource, water. As discussed in the previous section of this chapter, most experiments related to chemistry were conducted in standard test tube reactors. Larger engineering-scale test reactors were developed at UOIT for a hydrogen production of about 3 kg/day which is 1,000 times higher than initial benchtop proof-of-concept experiments. Reactors of fluidized bed, packed bed, or spray reactors were successfully demonstrated. Much knowledge has been acquired for the process to be scaled up to an industrial plant of much larger capacity and eventually linked to next-generation nuclear reactors.

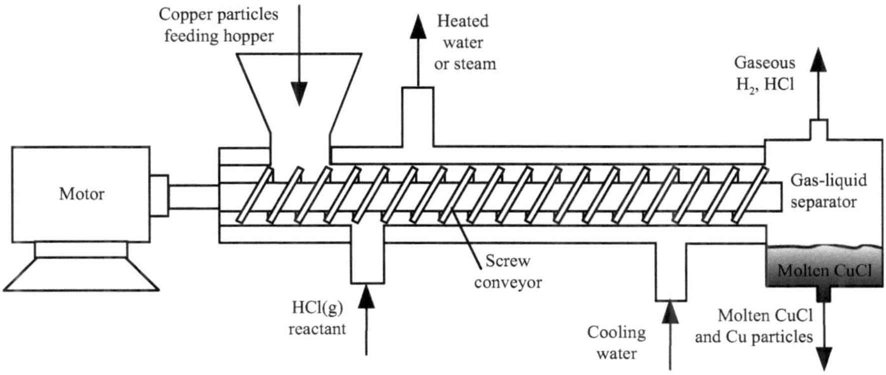

Fig. 6.107 Conceptual design of full-scale chlorination reactor for five-step copper–chlorine cycle [modified from Wang et al. (2008)]

This section presents some design concepts published in past literature for a full-scale copper-chlorine hydrogen plant for nuclear-driven water splitting. Apart from reactors, major equipment such as a sedimentation cell for copper separation in the five-step version of the cycle or a hydrochloric acid gas absorption tower (applicable for all cycle variants) are discussed. Scale-up design issues were examined by Wang et al. (2008) for handling three phases within the molten salt reactor, i.e., solid copper oxychloride particles, liquid (melting salt), and exiting gas (oxygen). It is also important to specify required capacities of feed materials for the multiphase reactors, steam, and heat requirements, as well as other key design parameters for reactor scale-up to a pilot-scale, up to full-plant-scale capacity.

Large-scale implementation of the cycle implies challenges with equipment design especially due to solid handling because solid injection and removal can block equipment operation and generate undesirable side reactions in downstream processes. Corrosion problems are other issues to consider for full plant design because expensive materials may be required. Surfaces exposed to molten CuCl, spray drying of aqueous $CuCl_2$, and high-temperature HCl and O_2 gases can be corroded or damaged during the plant operation unless special measures are taken to impede corrosive attack. Separation of chemicals requires design of special equipment, customized to specific requirements of the cycle. Mechanical equipment such as pumps, blowers, compressors, mixing devices, heat exchangers, and conveyors require special design and thorough material selection.

A possible configuration of the chlorination reactor and its subsystem—for a full-scale plant—was proposed by Wang et al. (2008). The reactor itself is in the form of a kiln with an axial screw conveyor for solid particles. Recall that in the chlorination reaction, 2 mol of copper are reacted to 2 mol of gaseous hydrochloric acid to release hydrogen and form a molten cuprous chloride. As outlined in Sect. 6.8, three phases must be handled in a chlorination reactor for the copper–chlorine cycle: solid, liquid, and gas phases. The design in Fig. 6.107 is one of the possible reactor configurations. During the start-up period, this reactor must be heated up together with the initial

Table 6.31 Flow rates in chlorination reactor for various scales of production

H_2 production				Mass flows (t/h)			Volume flows (m^3/h)			
Scale	Nm^3/day	t/day	MW_{th}	HCl (g)	Cu (s)	CuCl (l)	Cu (s)[a]	CuCl (l)	HCl(g)	V_{sed} (m^3)
Test-bench	36	0.003	>0.01	0.005	0.008	0.0124	0.002	0.003	7.42	0.025
Pilot plant	36,000	3	>10	4.5	7.94	12.4	1.77	3.351	7,420	25.6
Full scale	3,000,000	250	~600	375	662	1,034	148	279	620,000	2,133

[a]Assumed 50 % void fraction; V_{sed} volume of sedimentation vessel
Source: Wang et al. (2008)

charge of copper particles to around 623 K after which the reaction is initiated. It is important to keep the reaction temperature constant. Based on thermodynamic calculations, it is estimated that without external cooling, the adiabatic reaction temperature is ~877 K (Wang et al. 2008).

Another major technical issue is to avoid generation of cuprous chloride vapors in the reactor. At temperatures over 823 K, the saturation pressure of CuCl vapor becomes high. Based on energy balance calculation, at a 723 K reaction temperature, with copper particles fed at 353 K, and eight times HCl excess feed at 673 K, one generates 2 mol of molten copper cuprous chloride and 1 mol of hydrogen per 2 mol of copper. Some unreacted copper remains in the molten CuCl output stream and must be separated and recycled. A sedimentation vessel is needed for this process. To provide enough contact area between HCl reactant gas and solid Cu in the hydrogen reactor, enough volume of copper is accumulated in the sedimentation cell for copper particles. The sedimentation time of copper particles may be several hours, so the sedimentation cell would have about 5-h quantities of solid Cu particles and molten CuCl. Based on this assumption, the volume of the sedimentation vessel can be calculated based on the volume flow rates. The estimated flow parameters and the volume of sedimentation vessel are indicated in Table 6.31 as a function of production scale. Two configurations are proposed for the sedimentation vessel as shown in Fig. 6.108. There is a vertical configuration (Fig. 6.108a) and a horizontal configuration (Fig. 6.108b).

Gaseous hydrochloric acid in the exit stream must be separated from the hydrogen product. The HCl gas can be absorbed by an alkali solution, preferably sodium hydroxide. To absorb efficiently, a bubbling packed tower would be advantageous over a shower hollow tower. In a bubbling packed tower, a gas mixture of HCl and H_2 flows through the NaOH solution. The gas flow path is detoured by packing materials like small plastic rings or cylinders. In a shower hollow tower, the NaOH solution is showered down, while the gas mixture flows upward. Boiling must be avoided within the absorption tower. The neutralization heat of dilute strong acid and dilute strong alkali is 57.3 kJ/mol water; this heat must be removed.

A possible conceptual design of the five-step copper-chlorine plant subsystem which performs the chlorination process is shown in Fig. 6.109. It requires both hydrochloric acid recycling and copper particle recycling. The input into the chlorination subsystem are copper particles from the chlorination step—coming

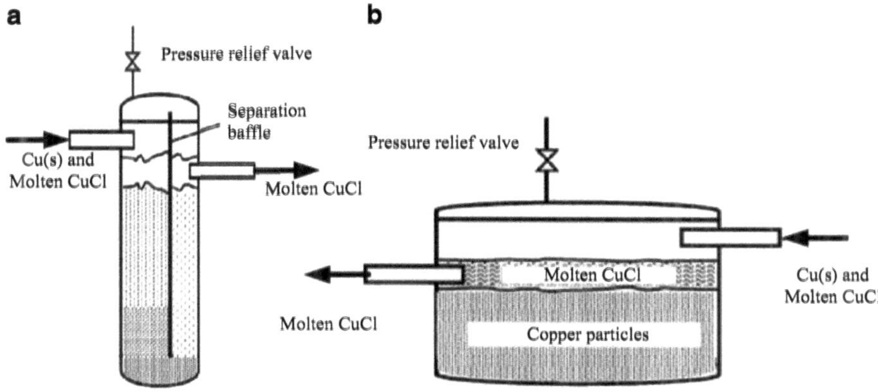

Fig. 6.108 Separation systems of molten cuprous chloride and copper particles: (**a**) vertical configuration, (**b**) horizontal configuration [modified from Wang et al. (2008)]

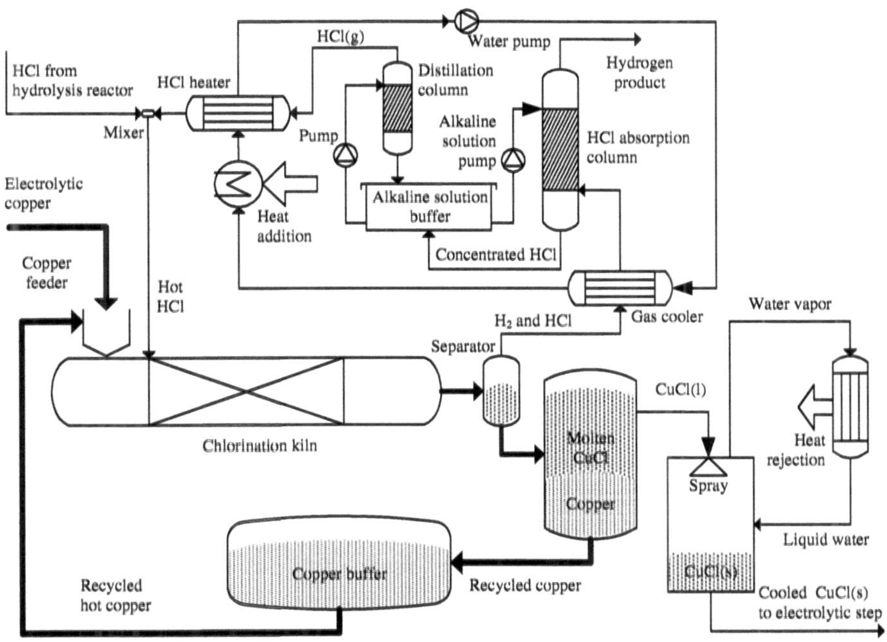

Fig. 6.109 Conceptual design of the chlorination step in Cu–Cl-5 thermochemical cycle [modified from Wang et al. (2008)]

at about 353 K—and hot hydrochloric acid ($T > 600$ K) coming from the hydrolysis subsystem, while the output is a stream of cuprous chloride particles, delivered to the electrolytic process. Unreacted copper particles are first separated by gravitational sedimentation and then recycled at high temperature. The gas phase of the

Table 6.32 Design issues related to the copper oxychloride decomposition step

Category	Technical issues and relevant parameters to consider for design.
Reaction kinetics	Decomposition temperature, activation time and energy of reaction, particle reaction rate/kinetics, residence time (corresponding to each specific particle structure/size, ratios of reactants), heating medium (molten salt/gas/wall/particles).
Flow, phases, mixing	Feeding of reactant particles (hot/cold), size and lifetime of aggregations (particles, molten salt, bubbles). Descent of particles in molten salt. Up-flow of bubbles in molten salt. Velocity, size, volumetric fraction of bubbles at different temperatures. Splashing effect on liquid surface due to the bubbles exiting the chamber. Flooding/overflow of molten salt pushed by bubbles due to sudden pressure decrease. Particle distribution in molten salt. Flow field of molten salt.
Material	Pressure should be lower than 4 atm. Molten CuCl and O_2 are highly corrosive. Vapor CuCl, chlorine gas, and HCl gas may occur in trace amounts.
Oxygen handling	Oxygen removal rate. Entrainment of reactant particles and molten salt droplets. Heat recovery from oxygen stream must be applied: heat transfer rate, thermal efficiency, type of heat exchanger. Composition of effluent gas to be considered (O_2, CuCl vapor, HCl, Cl_2, H_2O)
Molten CuCl handling	Recover heat from molten CuCl by means of direct contact (over a temperature range). Heat transfer rate between water vapor and CuCl particles (either liquid droplets or solid pellets). Thermal efficiency of direct contact. Maximum ratio of water vapor to CuCl. Type/structure of sprayer. Mixing solid particles and liquid water. Recover heat from CuCl by indirect contact. Thermal efficiency of heat exchanger. Type/structure of shell heat exchanger. Composition of molten CuCl. Solid $CuCl_2$ from reaction and side reaction. Reactant particles entrained by molten CuCl. Composition of molten CuCl. Solid $CuCl_2$ from reaction and side reaction. Reactant particles entrained by molten CuCl.

Source: Wang et al. (2008)

reaction product is separated in a gravitational separator. Furthermore, the gases are cooled and hydrochloric acid is absorbed in an alkaline solution using an absorption column. The alkaline buffer is maintained at constant pH by continuous removal of hydrochloric acid in a distillation column. The recycled HCl gas is reheated prior to reinjecting it in the chlorination reactor.

Another major subsystem of the copper–chlorine plant is the thermolysis reactor, which is common to all variants of the cycle. As mentioned in prior sections, in this reactor the copper oxychloride is decomposed and oxygen is produced. Three phases coexist in the reactor, namely, oxygen gas (and possible chlorine gas and CuCl vapors), molten cuprous chloride, and solid copper oxychloride. The major design issues of this reaction steps are listed in Table 6.32.

The thermolysis (oxy-decomposer) subsystem must start by preheating cuprous chloride in the reactor until molten state is achieved. Then oxygen is introduced into the molten CuCl to form bubbles, after which the volume and rate of bubble growth,

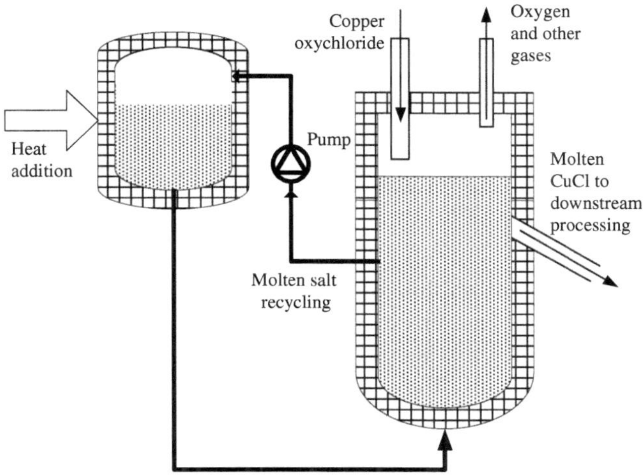

Fig. 6.110 Schematics of the reactor for copper oxychloride decomposition and oxygen generation [modified from Wang et al. (2008)]

as well as the mixing rate of gas and molten salt, can be controlled. Then solid CuCl particles are injected into the reactor to reach the level of overflow after melting, which corresponds to the volume of the reaction zone below the molten salt level. The reactor can then be heated at the reaction temperature and the CuCl melts in the reactor. The reaction starts by feeding copper oxychloride.

For the molten salt reactor, the effective liquid and solid heights should be about ten times the bulk height, since the volumetric ratio of $CuO{\cdot}CuCl_2(s)$ to CuCl is 1/9, excluding the volume occupied by bubbles. The gas bubbles will also occupy a portion of volume. Assuming that the initial size of a gas bubble is about the same as a particle of copper oxychloride, this size will grow when the bubble is ascending to the liquid (molten salt) level.

In the thermolysis reactor, the reactant particles absorb decomposition heat from the surrounding molten bath. While the particles are descending in the molten CuCl bath, oxygen ascends through the molten salt and then leaves the reactor. Simultaneously, another product (liquid CuCl) is formed within the molten bath and exits separately from the reactor. Within the reactor, heat is transferred from liquid (molten CuCl) to solid $CuO*CuCl_2$ (reactant) particles and the molten bath can be sustained by the reaction product itself. The oxygen product will leave the reactant particles immediately due to buoyancy of gas in the molten salt. This fast separation aids to minimize heat transfer resistance to the reactant particles, which then helps make the overall reaction rate closer to the intrinsic reaction rate itself. The design of the reactor requires a high efficiency of heat exchange and separation of reactant from products, as well as product (oxygen) from product (molten salt).

A practical and efficient option is to heat the molten salt from the wall with a double-shell structure as indicated in the reactor schematic in Fig. 6.110. Flowing in

Table 6.33 Flow rates of various streams of the copper oxychloride decomposition reactor

H_2 production		Oxygen at 800 K		$CuO \cdot CuCl_2(s)$		CuCl(l)	
Scale	N m^3/day	kg/day	m^3/h	kg/h	m^3/h	kg/h	m^3/h
Test-bench	36	0.024	2.1	13.4	0.003	12.4	0.003
Pilot plant	36,000	24	2,100	13,375	2.758	12,400	3.352
Full scale	3,000,000	2,000	175,000	1,114,600	230	1,033,333	280

Source: Wang et al. (2008)

the double shell, a secondary flow of molten salt provides heat to the reactor. The secondary flow of molten salt could be provided from a solar energy storage reservoir or nuclear heat exchanger, which uses molten salt as the heat storage and transferring medium. The cuprous chloride itself can act as a heat storage medium. If necessary, agitation could be introduced into the bath. This approach would take advantage of both free and forced convection with heterogeneous mixing that facilitates both heat and mass transfer. The flow rates of significant streams related to thermolysis reactor are listed in Table 6.33 for three capacities of production of the Cu–Cl plant.

In Lewis et al. (2009b), a conceptual design of a three-reaction-step copper–chlorine cycle was described: (1) an electrolytic cell with AEM (H_2-H^+, Fig. 6.15); (2) a crystallizer to dehydrate $CuCl_2 \cdot nH_2O, \quad n > 1$ to the form $CuCl_2 \cdot 2H_2O$; (3) a spray reactor for hydrolysis (Fig. 6.41); and (4) an oxy-decomposer with bayonet heat exchangers. The overall system is presented in Fig. 6.111 in a modified form which includes a molten salt heat exchanger for heat recovery. The electrolyzer operates at high pressure—24 bar—while the spray reactor operates in a slight vacuum or at 1 atm.

As suggested in Fig. 6.111, water is fed into the anolyte buffer where there is a solution of hydrochloric acid, cuprous chloride, and cupric chloride. This solution is pumped to 24 bar and heated to 373 K when the pressurized solution is still in the liquid phase. The anodic reaction occurs between cuprous chloride and chlorine ions, namely, $2CuCl(aq) + 2Cl^-(aq) \rightarrow CuCl_{2(aq)} + 2e^-$. Once formed, cupric chloride complexates with water molecules as $CuCl_2 \cdot nH_2O, \quad n \geq 2$.

The anolyte is cooled down to ~303 K and passed to a crystallizsation unit where it partially dehydrates. Bayonet heat exchangers are used in the crystallization unit to remove heat. Next, cupric chloride in aqueous form is heated such that the aqueous solution can be concentrated in $CuCl_2 \cdot 2H_2O(aq)$. The solution is rapidly expanded in spraying nozzles which form micrometer droplets scattered in steam at 673 K where the hydrolysis reaction occurs with formation of copper oxychloride particles that deposit at the bottom of the reactor. Hydrochloric acid and steam are extracted as gases and cooled to 373 K. A part of the liquid is further condensed and subcooled to ~303 K and further mixed with cooled cuprous chloride particles generated by the molten salt heat exchanger. The particles partially or totally dissolve in the acidic solution. The copper oxychloride solids fall into the oxy-decomposer which is supplied with heat by bayonet heat exchangers sunk into a molten cuprous chloride bath. The decomposition reaction generates oxygen gas

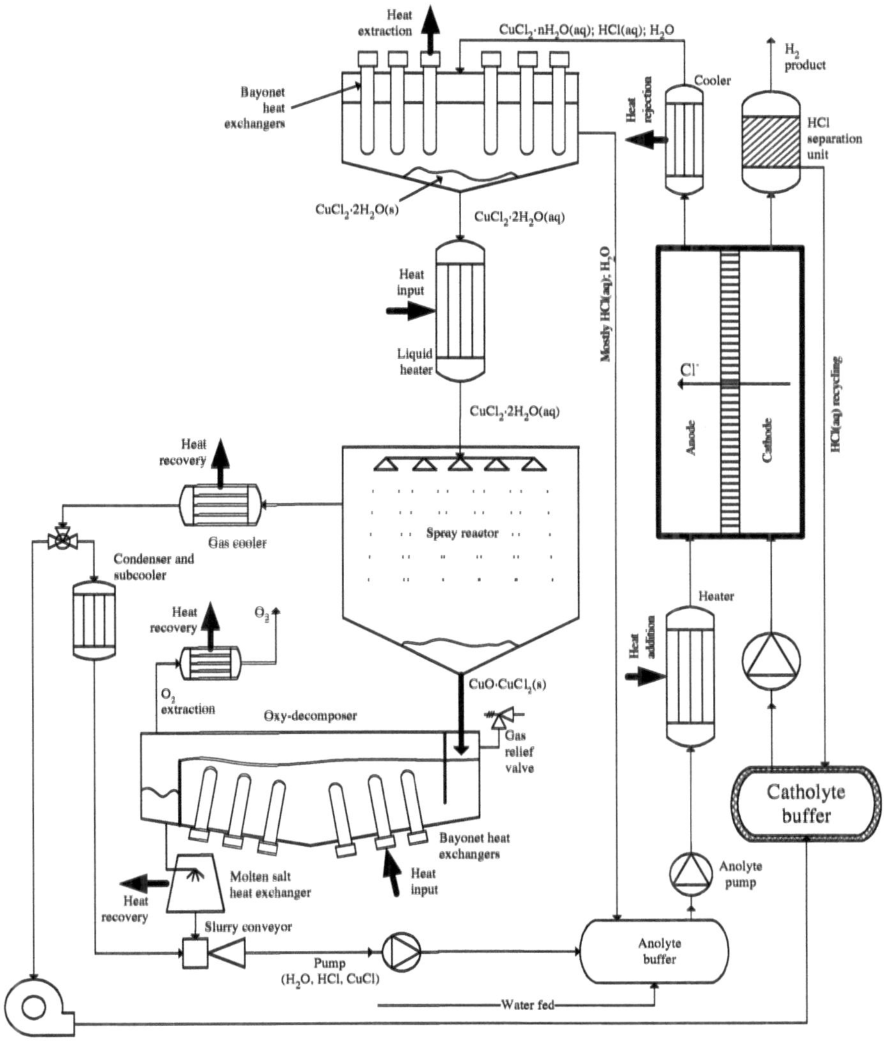

Fig. 6.111 Conceptual design of the Cu–Cl plant with cation exchange membrane electrolytic cell and spray reactor for hydrolysis [modified from Lewis et al. (2009b)]

which is removed from the top of the vessel. The upper layer of molten salt is free of other solid impurities and passes over a baffle into a room from where it is extracted. A molten salt heat exchanger is used for heat recovery and quenching of fine CuCl droplets. With a selected conveyor, CuCl is mixed with water and hydrochloric acid is directed toward the anolyte reservoir. If the spray reactor and oxy-decomposer operate in a slight vacuum, a pump is needed to pressurize the liquid prior to the return in an anolyte buffer. A pump pressurizes the anolyte to 24 bar, the operating pressure of the electrolyzer and crystallization cells. Hot gases

Table 6.34 Stream flow rates of a large-scale copper–chlorine plant with three steps

Stream	H_2O	HCl	CuCl	$CuO{\cdot}CuCl_2$	$CuCl_2{\cdot}2H_2O$	H_3O^+	Cl^-	$CuCl^+$	O_2	H_2
H_2O feed	47	0	0	0	0	0	0	0	0	0
Anode feed	328	3.4	515			40	107	93	0	0
Cathode feed	1,173	0.03	0	0	0	99	184	0	0	0
H_2 product	0	0	0	0	0	0	0	0	0	5.2
Crystallizer feed	168	0.2	0	0	979	414	91	39	0	0
$CuCl_2$ hydrolyzer feed	198	2.5	0	0	0	0	187	519	0	0
H_2O hydrolyzer feed	1,153	0	0	0	0	0	0	0	0	0
Oxydecomposer feed	0	0	0	561	0	0	0	0	0	0
O_2 product	0	0	0	0	0	0	0	0	41.6	0
CuCl recycle	0	0	515	0	0	0	0	0	0	0

Flow rate units: Mt/h
Source: Ferrandon et al. (2008)

of HCl and steam at 373 K are directed toward the catholyte buffer where water condenses and HCl is dissolved. The hot catholyte is pumped to 24 bar and supplied at the cathode.

Hydrochloric acid at high concentration generates protons and chlorine anions according to $HCl \rightarrow H^+ + Cl^-$. The chlorine anions cross the membrane while the protons reduce at the cathode forming hydrogen. A separation unit (not detailed) is used to separate and scrub hydrogen and to return the acidified liquid solution at lower pressure but high temperature to the catholyte buffer.

The calculated data for a copper–chlorine plant of 125,000 kg hydrogen per day operating in a diagram approximately the same as that presented in Fig. 6.111 was reported by Ferrandon et al. (2008). Accordingly, the plant requires 210 MW thermal energy and ~88 MW electric energy. If one assumes that electricity is generated with 40 % efficiency, the total requirement of thermal energy is 430 MW; therefore the energy efficiency is 29 %. In order to generate the same amount of hydrogen with an alkaline electrolysis system of 65 % energy efficiency, one needs ~313 MW of electric power, or an equivalent of ~780 MW electric energy, with an energy overall efficiency of 26 %. The flow rates reported by Ferrandon et al. (2008) are summarized in Table 6.34.

6.13 ASPEN Plus Plant Simulation and Flow Sheets

Aspen Plus simulation software, which predicts the behavior of process reactions and steps using standard engineering relationships, mass and energy balances, as well as phase and chemical equilibrium data, has been used for Cu–Cl plant simulation studies by Serban et al. (2004), Lewis et al. (2005a, b), Ferrandon et al. (2008), Rosen et al. (2010), and Orhan (2011).

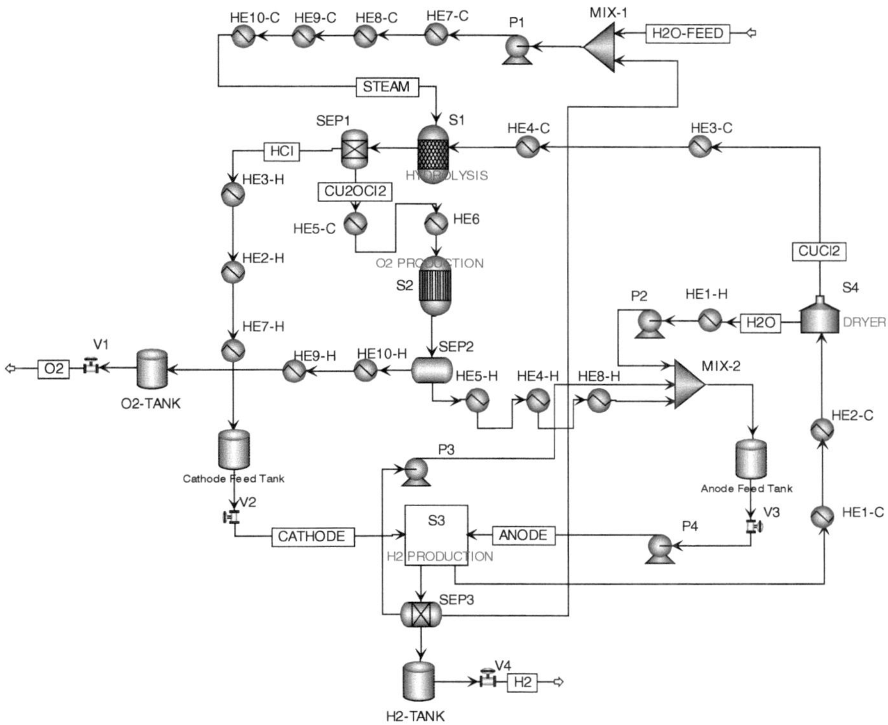

Fig. 6.112 ASPEN Plus flow sheet of a four-step cycle [modified from Naterer et al. (2011b)]

These studies focus mainly on four main unit operations: hydrolysis, oxychloride decomposition reactors, CuCl/HCl electrolyzer, and crystallizer (or spray dryer). An "electrolyte nonrandom two liquid" (ELECNTRL) activity coefficient model was used for modeling of the electrolyte step in the stoichiometry reactor.

The Soave–Redlich–Kwong (SRK) cubic equation of state was used to evaluate component properties and phase equilibria in the steps that involve vapor–liquid phase change. The liquid-phase equilibrium in the unit operation models uses physical property and phase equilibrium calculations. Aspen Plus has the capability of estimating physical property parameters for components that are not present in the Aspen Plus database, based on a combination of experimental data and molecular formulae.

The Aspen Plus simulations revealed that the amount of steam impacts the hydrogen production costs significantly. Handling and condensing the effluent steam from the hydrolysis reactor were investigated by Aspen Plus simulations, in order to reduce the steam consumption requirements. It is inefficient to condense and then re-vaporize the water. Partial condensation of the effluent is another alternative under investigation. Optimization and effective heat matching with

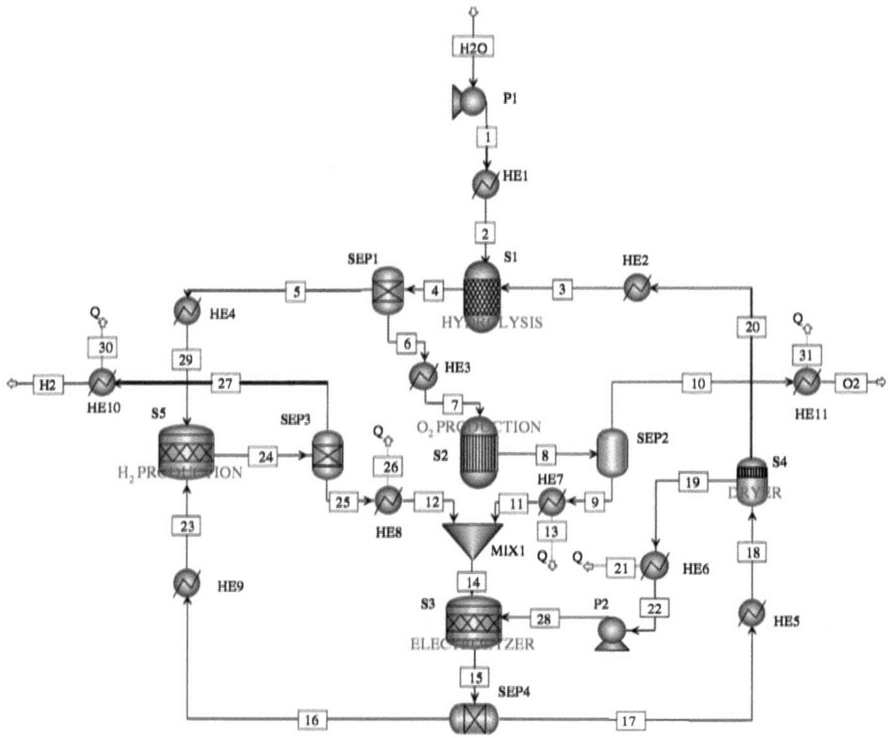

Fig. 6.113 ASPEN Plus flow sheet of a four-step cycle [modified from Orhan (2011)]

pinch analysis and better design of various steps within the Cu–Cl cycle are expected to lead to further efficiency improvements.

Figure 6.112 shows a flow-sheet simulation of a four-step copper–chlorine cycle variant, while the simulation for a five-step cycle is shown in Fig. 6.113. Energy streams resulted from the five-step flow-sheet simulation are listed in Table 6.35.

6.14 Plant Control, Safety, and Reliability

Al-Dabbagh and Lu (2010a, b) reported reliability, control systems, and safety assessments of a nuclear hydrogen plant with the Cu–Cl cycle under various risk scenarios. The studies were developed and analyzed with a dynamic flowgraph methodology (DFM) that provides the capability of modeling the behavior of control systems in the Cu–Cl cycle.

The methodology provides an integrated hardware/software model and also identifies how certain critical events of interest may occur. The DFM model is capable of verifying design requirements, performing failure analysis, and defining

Table 6.35 Energy balance of the five-step Cu–Cl cycle simulation process

Block	Process	ΔH (kJ/mol H_2)	Block	Process	ΔH (kJ/mol H_2)	Block	Process	ΔH (kJ/mol H_2)
S1	T	120.2	S2	T	125.5	S3	E	53.2
S4	M	33.2	S5	T	−41.6	H1	T	80
HE2	T	61.3	HE3	T	20.8	HE4	T	4.0
HE5	T	57.6	HE6	T	−30	HE7	T	−64
HE8	T	−60.6	HE9	T	35.2	HE10	T	−9
HE11	T	−5.6	SEP1	M	0.9	SEP2	M	1.2
SEP3	M	1.8	SEP4	M	2	P1	E	3
P2	E	1.9	MIX1	M	0.4	N/A	N/A	N/S

Processes: *T* thermal; *E* electrical or electrochemical; *M* mechanical.
Source: Orhan (2011)

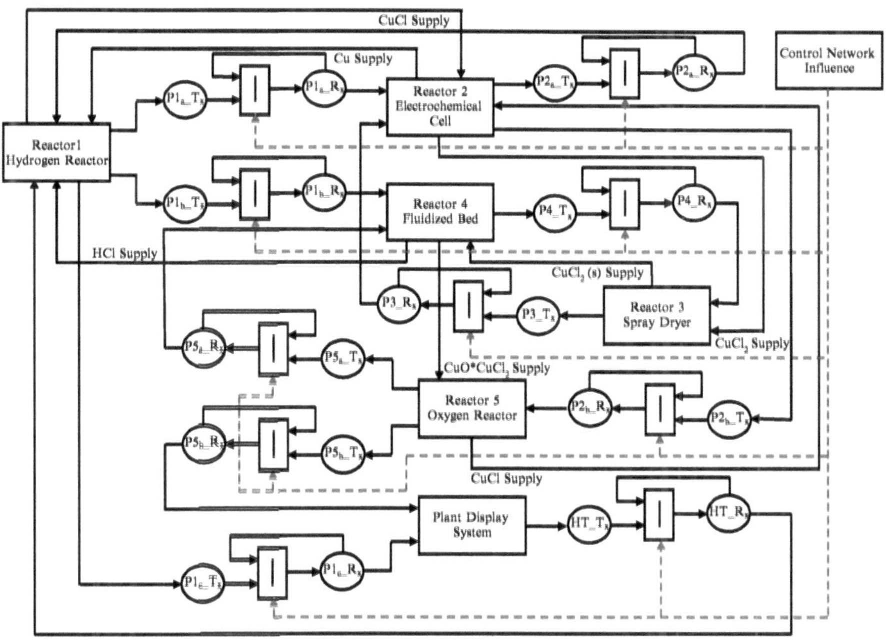

Fig. 6.114 DFM flow sheet of the Cu–Cl cycle [modified from Al-Dabbagh and Lu (2010a)]

test cases. The modeling strategy uses a two-step process, namely, construction and analysis of the model. The construction of the DFM models was performed by using a detailed multistate representation of the cause-and-effect and time-varying relationships within system components.

Figure 6.114 shows a flow sheet of the DFM model of the Cu–Cl cycle. The model is capable of tracing sequences of events inductively and/or deductively to identify paths that result in system events of interest. An inductive analysis was performed by specifying a set of component states and investigating the

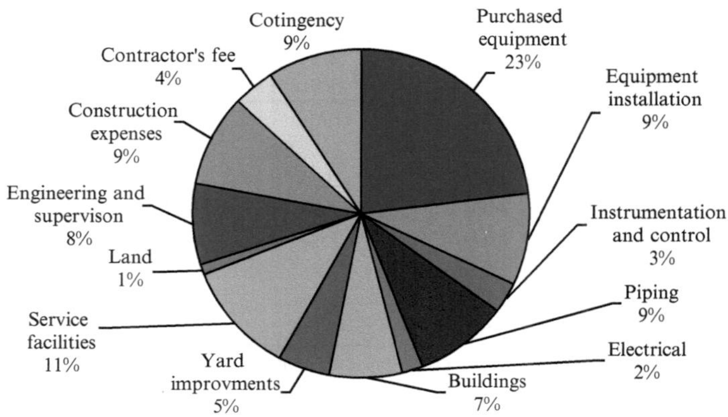

Fig. 6.115 Percentages of cost components for a hybrid Cu–Cl plant [data from Orhan et al. (2008c)]

propagation through the system and the influence on the system state level of interest. Also, a deductive analysis was performed by identifying a state of interest (desirable or undesirable) and finding the combination and sequences of parameters that led to the identified state. The control architecture, communication structure, and configuration of the instrumentation and control systems were developed and the DFM has been successfully applied to the Cu–Cl cycle.

6.15 Cost, Exergo-Economic, and Life Cycle Assessments

A cost analysis of a hybrid Cu–Cl plant was reported by Orhan et al. (2008c) assuming an amortization period of 10 years, 15 % annual return on investment, and an annual capital charge of 20 %. A production cost of $300/kW for the electrolysis cells was assumed, along with storage costs of $800,000/t of hydrogen via tube storage. The working capital comprises:

- Raw materials and supplies carried in stock
- Finished products in stock and semifinished products in the process of being manufactured
- Accounts receivable
- Cash kept on hand for monthly payment of operating expenses such as salaries, wages, and raw material purchases
- Accounts payable
- Taxes payable

Figure 6.115 shows percentages of cost components estimated for a hybrid Cu–Cl plant, as reported by Orhan et al. (2008c). Figure 6.116 shows that the total product cost varies in the range of 1–15 million dollars. The costs increase

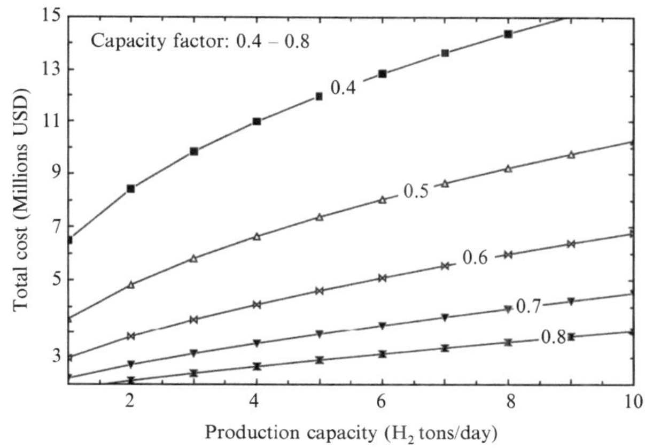

Fig. 6.116 Variation of total product cost with production capacity and capacity factor [data from Orhan et al. (2008c)]

with plant capacity, according to a concave down logarithmic trend. This shows that the cost of unit capacity decreases with production capacity. With an increasing capacity factor, the costs decrease. The curves for capacity factors of 0.6, 0.7, and 0.8 have close similarity, while greater variation exists for capacity factors of 0.5 and 0.4. This suggests that the capacity factor will lie between 0.6 and 0.8.

Exergo-economic analysis using exergy–cost–energy–mass (EXCEM) provides a quantitative relationship between capital costs and thermodynamic losses. Recent EXCEM studies have examined these features of the Cu–Cl cycle, to determine costs for improving the economic viability of the cycle, as reported by Orhan et al. (2010a). The overall energy and exergy efficiencies of the entire cycle were reported for several different cases, i.e., varying reaction temperatures, environment temperatures, pressures, chemical compositions, etc. In addition, the recent studies compared the Cu–Cl cycle costs against other energy options such as fossil fuels (coal and natural gas), renewable energies (solar, wind, hydro, geothermal, and biomass), and nuclear energy.

A thermodynamic analysis using energy and exergy, as well as several parametric studies, was conducted for various configurations of coupled systems to assess and compare their efficiencies (Orhan et al. 2010b). The methodology provides a useful approach for improving the cost-effectiveness of the Cu–Cl cycle. The hydrogen cost was found to be inversely related to the exergo-economic factor, plant capacity, and exergy efficiency.

Life cycle assessments (LCA) were reported in Ozbilen et al. (2011a, b). The environmental emissions and impacts based on standard categories are shown in Figs. 6.116 and 6.117. Table 6.36 defines the environmental impact categories. Emissions from the overall system are the sum of outputs from a nuclear plant and thermochemical hydrogen plant. The LCA of hydrogen production with the thermochemical Cu–Cl cycle was compared against other hydrogen production methods. The LCA was used to evaluate the total environmental impact of the hydrogen production processes and identify environmentally critical phases.

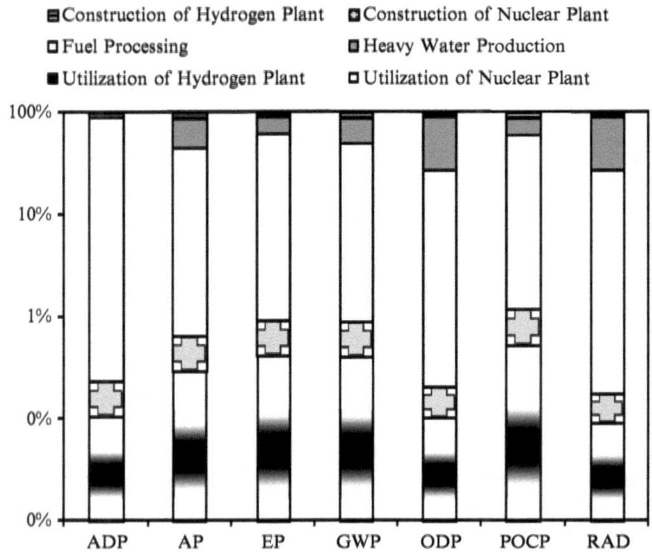

Fig. 6.117 Percentage contributions from different hydrogen generation system processes [data from Ozbilen et al. (2011a)]

Table 6.36 Definitions of environmental impact categories

Environmental impact category	Definition
Abiotic resource depletion potential (ADP) (g extracted element)	Abiotic depletion involves the extraction of nonrenewable raw materials
Global warming potential (GWP) (g CO_2-eq)	Amount of CO_2 in the earth's atmosphere
Ozone depletion potential (ODP) (g CFC-eq)	Depletion of ozone layer leads to an increase in the ultraviolet radiation reaching the earth's surface
Eutrophication potential (EP) (kg phosphate-eq)	Overfertilization or nutrition enrichment at a certain location
Acidification potential (AP) (g SO_2-eq)	Acid depletion on soil and into water may lead to changes in the degree of acidity
Photochemical ozone creation potential (POCP) (kg ethene-eq)	Due to volatile organic compounds in the atmosphere
Radioactive radiation (RAD) (disability-adjusted life years, DALY)	Emission and propagation of energy in the form of rays or waves

The "cradle-to-grave" analysis included production, transportation, installment, operation, and disposal phases of the processes. The LCA analysis had four phases: (1) goal and scope definition (objectives, boundaries); (2) life cycle inventory (LCI) analysis (inventory data on energy and material flows); (3) life cycle impact assessment (LCIA; evaluates environmental impacts of material and energy flows); and (4) improvement analysis (results, conclusions, recommendations, and improvements).

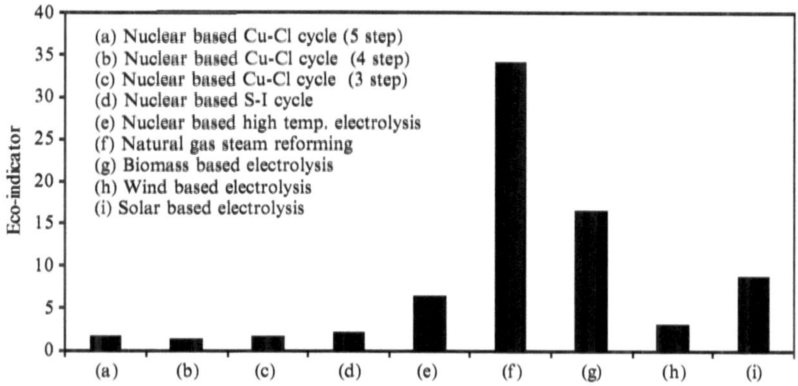

Fig. 6.118 Global warming potential of different hydrogen production methods [data from Ozbilen (2011)]

Table 6.37 Uranium requirements for all stages of the uranium fuel cycle

Fuel cycle stages	Three-step Cu–Cl cycle		Four-step Cu–Cl cycle		Five-step Cu–Cl cycle	
	Input (kg)	Output (kg)	Input (kg)	Output (kg)	Input (kg)	Output (kg)
Mining	6.81×10^{-3}	6.81×10^{-3}	6.00×10^{-3}	6.00×10^{-3}	7.29×10^{-3}	7.29×10^{-3}
Milling	6.81×10^{-3}	3.40×10^{-3}	6.00×10^{-3}	3.00×10^{-3}	7.29×10^{-3}	3.64×10^{-3}
Conversion	3.40×10^{-3}	3.40×10^{-3}	3.00×10^{-3}	3.00×10^{-3}	3.64×10^{-3}	3.64×10^{-3}
Enrichment	3.40×10^{-3}	3.77×10^{-4}	3.00×10^{-3}	3.33×10^{-4}	3.64×10^{-3}	4.04×10^{-4}
Fuel production	3.77×10^{-4}	3.77×10^{-4}	3.33×10^{-3}	3.33×10^{-4}	4.04×10^{-4}	4.04×10^{-4}

Source: Ozbilen et al. (2011a)

Figure 6.118 shows that the four-step Cu–Cl cycle has the lowest environmental impact compared to the three-step and five-step cycles. If the electrical energy output of the nuclear plant is used for all processes, then the global warming potential (GWP) will decrease from 15.8 to 0.56 kg CO_2-equivalent. The primary processes of mining and conversion stages have the highest environmental impact due to the fossil fuel used during those processes. Parametric studies also showed that the increase in plant hydrogen production capacity and lifetime do not have a significant effect on the impact categories if the capacity and lifetime are sufficiently large. It is observed that the nuclear-based S–I cycle and four-step Cu–Cl cycle are the most environmentally benign methods of hydrogen production in terms of GWP.

Table 6.37 shows the uranium requirements for all stages of a fuel cycle for a nuclear hydrogen plant. The assumptions adopted for the reference case have a total lifetime of the system (TLS) of 60 years and a daily hydrogen production capacity of the system (HPC) of 125,000 kg. The heat recovery percentage in the Cu–Cl cycle from exothermic processes is assumed to be 70 %. The functional unit of the system in Table 6.37 is 1 kg of hydrogen production.

6.16 Conclusions

In this chapter, a promising emerging water splitting technology—the hybrid copper–chlorine cycle—is examined and discussed extensively. Seven versions of the cycle were identified and categorized in detail. The three and four-step variants have been actively studied. The cycle versions including a copper electrowinning process had most interest in the period 2003–2008. More recently, AECL demonstrated successfully the hydrogen-evolving electrochemical process and a shift in the research focus toward cycle variants which include this process is observed. UOIT has focused primarily on the development of a four-step process that includes a dehydration system of cupric chloride. ANL has examined a system using a spray reactor with dehydration of copper oxychloride by crystallization. Prediction of thermochemical properties of chemical compounds has significant importance for cycle modeling and design. Electrochemical process and hydrolysis steps are also crucial for cycle development. AECL, UOIT, and PSU developed experimental facilities of electrolysis cells with proton exchange membranes that successfully demonstrate the process. Another crucial reaction step of the cycle is hydrolysis. The parameters were identified based on experiments and scaled-up reactors. It is crucial to set the reaction temperature correctly in order to minimize the side products such as chlorine. At the same time, the amount of excess steam is an important parameter. Dehydration of cupric chlorine has been studied extensively with the aim to reduce the temperature level of the input heat and amount of required heat input. It appears that a combination of crystallization and spray draying can efficiently dehydrate the cupric chloride. Alternatively, ANL developed a method of crystallization followed by a spray reactor for hydrolysis with promising results.

An integrated facility of 3 kg hydrogen per day is currently under development at UOIT to establish the large-scale viability of the Cu–Cl cycle. In addition, multiple studies have been dedicated to process scale-up and simulation at an industrial scale, to cost analysis, LCA, and environmental impact, among others. Based on this knowledge developed over the past decade, the copper–chlorine cycle can be considered as the prime competitor of sulfur–iodine and hybrid-sulfur thermochemical cycles, along with high-temperature electrolysis technology for large-scale hydrogen generation with nuclear energy.

Nomenclature

a	Charge transfer coefficient
a	Fluidized bed parameter
a, b	Regression coefficients
A	Area (m^2)
B	Spalding number

Bi	Biot number
c_p	Specific heat (kJ/kg K)
C	Molar concentration (mol/dm^3)
C_d	Drag coefficient
C_P	Molar specific heat at constant pressure (kJ/mol K)
COP	Coefficient of performance
d	Diameter (m)
d_p	Sauter mean diameter (m)
D	Diffusion coefficient (m^2/s)
D_as	Damkohler number at reactor scale
E	Molar specific energy (kJ/mol)
$\mathbb{E}$	Electric potential (V)
f	Fugacity (Pa)
f_p	Friction factor of the bed
F	Faraday constant (C/mol)
g	Gravitational acceleration (m/s^2)
G	Molar Gibbs free energy (kJ/mol)
GHSV	gas hourly space velocity (h^{-1})
j	Current density (A/m^2)
k	Thermal conductivity (W/m K)
$\mathtt{k}$	Reaction rate constants
K	Equilibrium constant
$\mathcal{K}$	Interchange coefficient
h	Heat transfer coefficient (W/m^2 K)
$\mathtt{h}$	Specific enthalpy (kJ/kg)
H	Molar enthalpy (kJ/mol)
HR	Hausner ratio
L	Fluidized bed height (m)
m	Mass (kg)
$\dot{m}$	Mass flow rate (kg/s)
M	Molecular mass (kg/kmol)
n	Order of reaction
$\dot{n}$	Moles generation rate (mol/s)
n_d	Electro-osmotic drag coefficient
N	Number of atoms
NTU	Number of thermal units
OVP	Total over-potential (V)
P	Pressure (Pa)
Pr	Prandtl number
q	Heat flux (W/m^2)
Q	Heat (kJ)
$\dot{Q}$	Heat transfer rate (W)
r	Superficial specific resistance (Ω/m^2)
$\mathtt{r}$	Reaction rate (mol/s)

$\mathfrak{r}$	Radial coordinate (m)
R	Universal gas constant (kJ/mol K)
ROCV	Reversible open-cell voltage (V)
Re	Reynolds number
S	Molar entropy (kJ/mol K)
Sc	Schmidt number
Sh	Sherwood number
SP	Span
t	Time (s)
T	Temperature (K)
U	Overall heat transfer coefficient (W/m^2 K)
v	Molar specific volume (m^3/mol)
V	Voltage (V)
$\mathcal{V}$	Velocity (m/s)
W	Work (kJ)
W_b	Bed inventory (kg)
x	Transversal coordinate (m)
$\boldsymbol{x}$	Conversion of solid particles
x_s	Molar steam fraction
X	Conversion factor
y	Mole fraction
Y	Mass fraction of solid

Greek Letters

α	Volumetric thermal expansion coefficient (1/K)
δ	Thickness (m)
ε	Bubble fraction in the bed
ε	Heat exchanger effectiveness
η	Efficiency
φ	Carnot abatement factor
Φ	Sphericity
μ	Dynamic viscosity (Pa s)
ψ	Exergy efficiency
θ	Dimensionless temperature
ρ	Density (kg/m^3)
ξ	Extent of reaction
σ	Electric conductivity (S/m)
τ	Reaction time (s)
ω	Acentric factor

Subscripts

0	Exchange current
a	Anode
act	Activation
b	Bulk phase
b	Backward
c	Critical
c	Cathode
c	Cold
conc	Concentration
d	Decomposition
d	Droplet
D	Debye
e	Electrolyte
e	Effective
eq	Equilibrium
E	Einstein
f	Fluidized
f	Forward
g	Vapor
gen	Generated
h	Hot
i	General index
l	Liquid
L	Limiting current density
lg	Latent
m	Melting
ohm	Ohmic
p	Particle
r	Reduced
r	Reaction
sat	Saturation
t	Tube
w	Water

Superscripts

0	Reference state
″	Per unit of surface
‾	Average value (overbar)

References

Abanades S, Charvin P, Flamant G, Neveu P (2006) Screening of water-splitting thermo-chemical cycles potentially attractive for hydrogen production by concentrated solar energy. Energy 31:2469–2486

Al-Dabbagh AW, Lu L (2010a) Design and reliability assessment of control systems for a nuclear-based hydrogen production plant with copper–chlorine thermochemical cycle. Int J Hydrogen Energy 35:966–977

Al-Dabbagh AW, Lu L (2010b) Dynamic flowgraph modeling of process and control systems of a nuclear-based hydrogen production plant. Int J Hydrogen Energy 35:9569–9580

Avsec J, Naterer GF, Predin A (2009) Calculation of thermodynamic properties for hydrochloric and copper compounds in a hydrogen production process. J Energy Technol 2:47–64

Balashov VN, Schatz RS, Chalkova E, Akinfiev NN, Fedkin MV, Lvov SN (2011) CuCl electrolysis for hydrogen production in Cu-Cl thermochemical cycle. J Electrochem Soc 158: B266–B275

Carty RH, Mazumder MM, Schreiber JD, Pangborn JB (1981) Thermochemical production of hydrogen. Final report 30517, vol. 1–4. Institute of Gas Technology, Chicago, IL

Cleary V, Bowen P, Witlox H (2007) Flashing liquid jets and two-phase droplet dispersion I. Experiments for derivation of droplet atomization correlations. J Hazard Mater 142:786–96

Croizé L, Doizi D, Larousse B, Pailloux A, Reaux D, Gallou C, Dauvois V, Roujou JL, Zanella Y, Carles P (2010) Interest of absorption spectroscopy for the control of industrial processes. Application to H2 massive production. Appl Phys B 100:409–415

Daggupati V, Naterer GF, Gabriel K, Gravelsins R, Wang Z (2009) Equilibrium conversion in Cu-Cl cycle multiphase processes of hydrogen production. Thermochem Acta 496:117–123

Daggupati V, Naterer GF, Gabriel K (2010a) Diffusion of gaseous products through a particle surface layer in a fluidized bed reactor. Int J Heat Mass Transf 53:2449–2458

Daggupati VN, Naterer GF, Gabriel KS, Gravelsins RJ, Wang ZL (2010b) Solid particle decomposition and hydrolysis reaction kinetics in Cu-Cl thermochemical hydrogen production. Int J Hydrogen Energy 35:4877–4882

Daggupati VN, Naterer GF, Dincer I (2011a) Convective heat transfer and solid conversion of reacting particles in a copper(II) chloride fluidized bed. Chem Eng Sci 66:460–468

Daggupati VN, Pope K, Naterer GF, Gabriel KS, Wang ZL (2011b) Conversion of solid cupric chloride in a fluidized bed hydrolysis reactor. Proceedings of international conference of hydrogen production ICH2P-11, Thessaloniki, Greece, 19–22 June, paper #112

Daggupati VN, Naterer GF, Gabriel KS, Gravelsins RJ, Wang ZL (2011c) Effects of atomization conditions and flow rates on spray drying for cupric chloride particle formation. Int J Hydrogen Energy 36:11353–11359

Dokyia M, Kotera Y (1976) Hybrid cycle with electrolysis using Cu-Cl system. Int J Hydrogen Energy 1:117–121

Ferrandon MS, Lewis MA, Tatterson DF, Nankanic RV, Kumarc M, Wedgewood LE, Nitsche LC (2008) The hybrid Cu-Cl thermochemical cycle. I. Conceptual process design and H_2A cost analysis. II. Limiting the formation of CuCl during hydrolysis. In: NHA annual hydrogen conference, Sacramento convention center, CA, 30 March to 3 April

Ferrandon MS, Lewis MA, Tatterson DF, Gross A, Doizi D, Croizé L, Dauvois V, Roujou JL, Zanella Y, Carles P (2010a) Hydrogen production by the Cu-Cl thermochemical cycle: Investigation of the key step of hydrolysing $CuCl_2$ to Cu_2OCl_2 and HCl using a spray reactor. Int J Hydrogen Energy 35:992–1000

Ferrandon MS, Lewis MA, Alvarez F, Shafirovich E (2010b) Hydrolysis of $CuCl_2$ in the Cu-Cl thermochemical cycle for hydrogen production: experimental studies using a spray reactor with an ultrasonic atomizer. Int J Hydrogen Energy 35:1895–1904

Ghandehariun S, Rosen MA, Naterer GF, Wang Z (2011) Comparison of molten salt heat recovery options in the Cu-Cl cycle of hydrogen production. Int J Hydrogen Energy 36:11328–11337

Granovskii M, Dincer I, Rosen MA, Pioro I (2008) Performance assessment of a combined system to link a supercritical water-cooled nuclear reactor and a thermochemical water splitting cycle for hydrogen production. Energy Conv Manage 49:1873–1881

Haseli Y, Dincer I, Naterer GF (2008) Hydrodynamic gas-solid model of cupric chloride particles reacting with superheated steam for thermochemical hydrogen production. Chem Eng Sci 63:4596–4604

Haseli Y, Naterer GF, Dincer I (2009) Fluid-particle mass transport of cupric chloride hydrolysis in a fluidized bed. Int J Heat Mass Transf 52:2507–2515

Haynes WM (2012) CRC handbook of chemistry and physics, 92 (Internet Version)th edn. CRC, Boca Raton, FL

Ikeda BM, Kaye MH (2008) Thermodynamic properties in the Cu-Cl-O-H system. Seventh international conference on nuclear and radiochemistry, Budapest, Hungary, August

Jaber O, Naterer GF, Dincer I (2010a) Convective heat transfer from solidified molten salt in a direct contact heat exchanger. Heat Mass Transf 46:999–1012

Jaber O, Naterer GF, Dincer I (2010b) Heat recovery from molten CuCl in the Cu-Cl cycle of hydrogen production. Int J Hydrogen Energy 35:6140–6151

Khan MA, Chen Y (2005) Preliminary process analysis and simulation of the copper-chlorine thermochemical cycle for hydrogen generation. Nuclear hydrogen production, third information exchange meeting, Oarai, Japan, 5–7 October. Nuclear Energy Agency 6122

Lewis MA, Serban M, Basco J (2003) Hydrogen production at 550 °C using a low temperature thermochemical cycle. Proceedings of the OECD/NEA meeting, Argonne National Laboratory

Lewis MA, Serban M, Basco JK (2004) A progress report on the chemistry of the low temperature Cu-Cl thermochemical cycle, Trans Am Nucl Soc 91:113–114

Lewis MA, Petri MC, Masin JG. (2005a) A scoping flowsheet methodology for evaluating alternative thermochemical cycles. Nuclear hydrogen production, third information exchange meeting, Oarai, Japan, 5–7 October. Nuclear Energy Agency 6122

Lewis MA, Masin JG, Vilim RB, Serban M (2005b) Development of the low temperature thermochemical cycle. International congress on advances in nuclear power plants, 15–19 May, Seoul, Korea

Lewis MA, Masin JG (2009) The evaluation of alternative thermochemical cycles—Part II: the down-selection process. Int J Hydrogen Energy 34:4125–4135

Lewis MA, Masin JG, O'Hare PA (2009a) Evaluation of alternative thermochemical cycles, Part I: the methodology. Int J Hydrogen Energy 34:4115–4124

Lewis MA, Ferrandon MS, Tatterson DF, Mathias P (2009b) Evaluation of alternative thermochemical cycles—Part III further development of the Cu-Cl cycle. Int J Hydrogen Energy 34:4136–4145

Marin GD, Wang Z, Naterer GF, Gabriel K (2010) Chemically reacting and particle-laden multiphase flow in a molten salt vessel. Tenth AIAA/ASME joint thermophysics and heat transfer conference, Chicago, Illinois, 28 June to 1 July, Paper #225388

Marin GD, Wang Z, Naterer GF, Gabriel K (2011a) X-ray diffraction study of multiphase reverse reaction with molten CuCl and oxygen. Thermochim Acta 524:109–116

Marin GD, Wang Z, Naterer GF, Gabriel K (2011b) Byproducts and reaction pathways for integration of the CueCl cycle of hydrogen production. Int J Hydrogen Energy 36:13414–13424

McQuillan BW, Brown LC, Besenbruch GE, Tolman R, Cramer T, Russ BE, Vermillion BA, Earl B., Hsieh H-T, Chen Y, Kwan K, Diver R, Siegal N, Weimer A, Perkins C, Lewandowski A (2010) High efficiency generation of hydrogen fuels using solar thermal-chemical splitting of water. General Atomics Project 3022

Miller AI (2005) An update on Canadian activities on hydrogen. Nuclear hydrogen production, 3rd Information Exchange Meeting, Oarai, Japan, 5–7 October. Nuclear Energy Agency 6122

Naterer GF (2008) Second law viability of upgrading waste heat for thermochemical hydrogen production. Int J Hydrogen Energy 33:6037–6045

Naterer GF, Gabriel K, Wang Z, Daggupati V, Gravelsins R (2008a) Thermochemical hydrogen production with a copper-chlorine cycle, I: oxygen release from copper oxychloride decomposition. Int J Hydrogen Energy 33:5439–5450

Naterer GF, Daggupati VN, Marin G, Gabriel KS, Wang ZL (2008b) Thermochemical hydrogen production with a copper–chlorine cycle, II: flashing and drying of aqueous cupric chloride. Int J Hydrogen Energy 33:5451–5459

Naterer GF, Suppiah S, Lewis M, Gabriel K, Dincer I, Rosen MA, Fowler M, Rizvi G, Easton EB, Ikeda BM, Kaye MH, Lu L, Pioro I, Spekkens P, Tremaine P, Mostaghimi J, Avsec A, Jiang J (2009) Recent Canadian advances in nuclear-based hydrogen production and the thermochemical Cu-Cl cycle. Int J Hydrogen Energy 34:2901–2917

Naterer GF, Suppiah S, Stolberg L, Lewis M, Wang Z, Daggupati V, Gabriel K, Dincer I, Rosen MA, Spekkens P, Lvov L, Fowler M, Tremaine P, Mostaghimi J, Easton EB, Trevani L, Rizvi G, Ikeda BM, Kaye MH, Lu L, Pioro I, Smith WR, Secnik E, Jiang J, Avsec J (2010) Canada's program on nuclear hydrogen production and the thermochemical Cu-Cl cycle. Int J Hydrogen Energy 35:10905–10926

Naterer GF, Suppiah S, Stolberg L, Lewis M, Ferrandon M, Wang Z, Dincer I, Gabriel K, Rosen MA, Secnik E, Easton EB, Trevani L, Pioro I, Tremaine P, Lvov S, Jiang J, Rizvi G, Ikeda BM, Luf L, Kaye M, Smith WR, Mostaghimi J, Spekkens P, Fowler M, Avsec J (2011a) Clean hydrogen production with the Cu-Cl cycle—progress of international consortium, I: experimental unit operations. Int J Hydrogen Energy 36:15472–15485

Naterer GF, Suppiah S, Stolberg L, Lewis M, Ferrandon M, Wang Z, Dincer I, Gabriel K, Rosen MA, Secnik E, Easton EB, Trevani L, Pioro I, Tremaine P, Lvov S, Jiang J, Rizvi G, Ikeda BM, Luf L, Kaye M, Smith WR, Mostaghimi J, Spekkens P, Fowler M, Avsec J (2011b) Clean hydrogen production with the Cu-Cl cycle—progress of international consortium, II: simulations, thermochemical data and materials. Int J Hydrogen Energy 36:15486–15501

NIST (2012) Chemistry webbook. National Institute of Standards and Technology. Available from: http://webbook.nist.gov/chemistry/

Nixon A, Ferrandon M, Kaye MH, Trevani L (2011) Thermochemical hydrogen production Synthesis, characterization, and decomposition of copper oxychloride. J Thermal Anal Calor. doi:10.1007/s10973-011-1998-3

Odukoya A, Naterer GF (2011) Electrochemical mass transfer irreversibility of cuprous chloride electrolysis for hydrogen production. Int J Hydrogen Energy 36:11345–11352

Orhan MF (2011) Conceptual design, analysis and optimization of nuclear-based hydrogen production via copper-chloride thermochemical cycles. PhD thesis, University of Ontario Institute of Technology

Orhan MF, Dincer I, Rosen MA (2008a) Thermodynamic analysis of the copper production step in a copper-chlorine cycle for hydrogen production. Thermochim Acta 480:22–29

Orhan MF, Dincer I, Rosen MA (2008b) Energy and exergy assessments of the hydrogen production step of a copper-chlorine thermochemical water splitting cycle driven by nuclear-based heat. Int J Hydrogen Energy 33:6456–6466

Orhan MF, Dincer I, Naterer GF (2008c) Cost analysis of a thermochemical Cu–Cl pilot plant for nuclear-based hydrogen production. Int J Hydrogen Energy 33:6006–6020

Orhan MF, Dincer I, Rosen MA (2009) Energy and exergy analyses of the fluidized bed of a copper-chlorine cycle for nuclear-based hydrogen production via thermochemical water decomposition. Chem Eng Res Des 87:684–694

Orhan MF, Dincer I, Rosen MA (2010a) An exergy-cost-energy-mass analysis of a hybrid copper-chlorine thermochemical cycle for hydrogen production. Int J Hydrogen Energy 35:4831–4838

Orhan MF, Dincer I, Rosen MA (2010b) Exergoeconomic analysis of a thermochemical copper-chlorine cycle for hydrogen production using specific exergy cost (SPECO) method. Thermochim Acta 497:60–66

Orhan MF, Dincer I, Rosen MA (2011) Exergy analysis of heat exchangers in the copper-chlorine thermochemical cycle to enhance thermal effectiveness and cycle efficiency. Int J Low-Carbon Technol 6:156–164

Ozbilen A (2011) Life cycle assessment of nuclear-based hydrogen production via thermochemical water splitting using a copper-chlorine (Cu-Cl) cycle. MSc thesis, University of Ontario Institute of Technology

Ozbilen A, Dincer I, Rosen MA (2011a) Environmental evaluation of hydrogen production via thermochemical water splitting using the Cu-Cl Cycle: a parametric study. Int J Hydrogen Energy 36:9514–9528

Ozbilen A, Dincer I, Rosen MA (2011b) A comparative life cycle analysis of hydrogen production via thermochemical water splitting using a Cu-Cl cycle. Int J Hydrogen Energy 36:11321–11327

Parry T (2008) Thermodynamics and magnetism of copper oxychloride. MSc thesis, Bringham Young University, Provo, UT, USA

Perret R, Chen Y, Besenbruch G, Diver R, Weimer A, Lewandowski A, Miller E (200). High temperature thermochemical solar hydrogen generation research. UNLV Research Foundation. Report to Department of Energy DE-FG36-03GO13062

Pope K, Naterer GF, Wang Z (2011) Pressure drop of packed bed vertical flow for multiphase hydrogen production. Int J Hydrogen Energy 36:11338–11344

Ranganathan S, Easton EB (2010a) High performance ceramic carbon electrode-based anodes for use in the Cu–Cl thermochemical cycle for hydrogen production. Int J Hydrogen Energy 35:1001–1007

Ranganathan S, Easton EB (2010b) Ceramic carbon electrode-based anodes for use in the Cu-Cl thermochemical cycle. Int J Hydrogen Energy 35:4871–4876

Rosen MA, Naterer GF, Chukwu CC, Sadhankar R, Suppiah S (2010) Nuclear-based hydrogen production with a thermochemical copper-chlorine cycle and supercritical water reactor: equipment scale-up and process simulation. Int J Energy Res. doi:10.1002/er.1702

Serban M, Lewis MA, Basco JK (2004) Kinetic study of the hydrogen and oxygen production reactions in the copper-chloride thermochemical cycle. AIChE SPRING NATIONAL MEETING, New Orleans LA, 25–29 April

Sim K-S, Son Y-M, Kim J-W (1993) Some thermochemical cycles composed of copper compounds with three-step reactions. Int J Hydrogen Energy 18:287–290

Stolberg L, Boniface HA, McMahon S, Suppiah S, York S (2008) Electrolysis of the CuCl/HCl aqueous system for the production of nuclear hydrogen. Proceedings of the fourth international topical meeting on high temperature reactor technology HTR-2008. 28 September to 1 October, Washington, DC

Suppiah S, Li J, Sadhankar R, Kutchcoskie KJ, Lewis M (2005) Study of the hybrid Cu-Cl cycle for nuclear hydrogen production. Nuclear hydrogen production, third information exchange meeting, Oarai, Japan, 5–7 October. Nuclear Energy Agency 6122

Suppiah S, Stolberg L, Boniface H, Tan G, McMahon S, York S, Zhang W (2009) Canadian nuclear hydrogen R&D programme: Development of the medium-temperature Cu-Cl cycle and contributions to the high-temperature sulphur-iodine cycle. Forth international exchange meeting on nuclear hydrogen production, Oakbrook IL, USA, 13–16 April. Nuclear Energy Agency and Organization for Economic Co-operation and Development, NEA 6805

Trevani L, Ehlerova J, Sedlbauer J, Tremaine PR (2010) Complexation in the Cu(II)–LiCl–H_2O system at temperatures to 423 K by UV-visible spectroscopy. Int J Hydrogen Energy 35:4893–4900

Trevani L (2011) The copper-chloride cycle: synthesis and characterization of copper oxychloride. Hydrogen and fuel cells international conference and exhibition, Vancouver, BC, 15–18 May

Vilim RB (2004) Commodity inventories and equipment sizes for a low temperature nuclear-chemical plant for production of hydrogen. AIChE spring national meeting, conference proceedings. pp 2706–2712

Wang Z, Naterer GF, Gabriel K (2008) Multiphase reactor scale-up for Cu–Cl thermochemical hydrogen production. Int J Hydrogen Energy 33:6934–6946

Wang ZL, Naterer GF, Gabriel K, Gravelsins R, Daggupati V (2009) Comparison of different copper-chlorine thermochemical cycles for hydrogen production. Int J Hydrogen Energy 34:3267–3276

Wentorf RH, Hanneman RE (1974) Thermochemical hydrogen generation. Science 185:311–319

Williams LO (1980) Hydrogen power. An introduction to hydrogen energy and its applications. Pergamon, New York

Zamfirescu C, Dincer I, Naterer GF (2009) Performance evaluation of organic and titanium based working fluids for high-temperature heat pumps. Thermochim Acta 496:18–25

Zamfirescu C, Dincer I, Naterer GF (2010a) Thermophysical properties of copper compounds in copper-chlorine thermochemical water splitting cycles. Int J Hydrogen Energy 35:4839–4852

Zamfirescu C, Naterer GF, Dincer I (2010b) Kinetics study of the copper/hydrochloric acid reaction for thermochemical hydrogen production. Int J Hydrogen Energy 35:4853–4860

Zamfirescu C, Naterer GF, Dincer I (2010c) Upgrading of waste heat for combined power and hydrogen production with nuclear reactors. J Eng Gas Turbines Power 132:102911

Zamfirescu C, Naterer GF, Dincer I (2011) Vapor compression CuCl heat pump integrated with a thermochemical water splitting cycle. Thermochim Acta 512:40–48

Zamfirescu C, Naterer GF, Dincer I (2012) Photochemical disproportionation of cuprous chloride for electrolytic hydrogen generation. Int J Hydrogen Energy. doi:10.1016/j.ijhydene.2012.01.183

Chapter 7
Integrated Nuclear Hydrogen Production Systems

Abstract In this chapter, various integrated systems for nuclear hydrogen production are presented. Generation IV nuclear reactors and their conceptual processes are presented. Particular attention is given to VHTR and SWCR reactors, which represent two different alternatives for future development of nuclear-based hydrogen production at a large scale. The VHTR integrated system will generate hydrogen by methods adapted to high-temperature heat requirements of over 1,100 K. The SWCR is suitable for processes at intermediate temperatures at around 800 K. Due to its higher temperature, the VHTR (and its variants) can be used with high-temperature electrolysis, as well as the sulfur–iodine and hybrid sulfur thermochemical cycle, and the copper–chlorine cycle as well. The hybrid copper–chlorine cycle is adaptable for the linkage with lower temperature nuclear reactors. Selected integrated systems for nuclear-based conversion of coal and natural gas to hydrogen are also presented. These systems are relevant during the transition period until a future hydrogen economy is implemented. A system that integrates a nuclear reactor with a copper–chlorine cycle and a desalination process is also presented. This type of integration is attractive because it produces fresh water from brackish or sea water which is then used partially as process water for hydrogen production and partially for drinking water.

7.1 Introduction

Integration of a nuclear reactor with a chemical or electrochemical process for hydrogen production is a relatively old idea that originated over 40 years ago. Past literature has been published regarding the linkage options and their specific advantages. In an integrated system, at least three main functional units must work together, namely: (1) nuclear reactor, (2) power plant, and (3) hydrogen production plant.

A common system is a nuclear reactor coupled to a power plant which generates electricity. The electricity is used for alkaline water electrolysis to generate

G.F. Naterer et al., *Hydrogen Production from Nuclear Energy*,
DOI 10.1007/978-1-4471-4938-5_7, © Springer-Verlag London 2013

hydrogen. Other options exist such as nuclear heat to drive thermochemical cycles, linkage of hybrid water splitting cycles with nuclear reactors, or using nuclear energy to convert hydrocarbons to hydrogen.

In the first part of this chapter, the principles of operation of most future nuclear reactors (including Generation IV reactors and others) are discussed. Thereafter, nuclear-based systems for conversion of natural gas or coal to hydrogen are presented. Such systems are relevant for a transition period towards a fully implemented hydrogen economy. Implementing such integrated systems would significantly reduce carbon dioxide emissions. It follows with a section on nuclear-based integrated high-temperature electrolysis systems. Thereafter, systems for integration of the S–I, hybrid sulfur and Cu–Cl cycles with nuclear reactors are presented. A case study is included for an integrated system that uses a nuclear reactor, copper–chlorine plant and a desalination plant.

7.2 Next Generation Nuclear Reactors

A classification of next generation nuclear reactors was provided in Chap. 3 of this book. They include the supercritical water cooled reactor (SWCR), gas cooled fast reactor (GFR), lead-alloy fast reactor (LFR), very high-temperature gas reactor (VHTR), molten salt reactor (MSR), sodium-cooled fast reactor (SFR) and other non-generation IV reactors. Abu-Khader (2009) categorized Generation IV reactors "theoretical designs currently under research," while other categories of reactors are "advanced reactor designs under investigation and development," and established nuclear fission reactors. Based on these categories, Fig. 7.1 illustrates a classification of past, present, and future nuclear reactors for power and/or process heat production. In this section, the main types of future reactors, capable of coproduction of electricity and hydrogen, will be discussed.

The SCWR represents an evolution of older reactor concepts such as BWR, PWR, and PHWR, which use water as a coolant. In an SWCR, the pressure of the water coolant in the reactor is maintained over a critical value of 22.1 MPa such that water does not boil. This simplifies the design, increases the efficiency up to 50 % and addresses the issues related to safety, although more stress occurs on materials. There are two design options of SWCR, either with pressure tubes or a pressure vessel. Variants of SWCR concepts with pressure tubes are underdevelopment in Canada (SCWR-CANDU) and Russia, while variants with pressure vessels are under development in France, Japan, and Korea.

The planned electricity generation range with SWCR reaches up to 1.5 GW electric (no SCWR has been built yet to date). The system uses light water coolant at an operational pressure of 25 MPa. It has design flexibility with an inlet temperature up to ~625 K and outlet up to 900 K, with both thermal (up to 60 GW days per ton of heavy metal) and fast (up to 120 GW days per ton of heavy metal) neutron spectra, three types of fuel (uranium dioxide, thorium and mixed oxide fuel), two types of fuel cycles (once-through or closed) and three

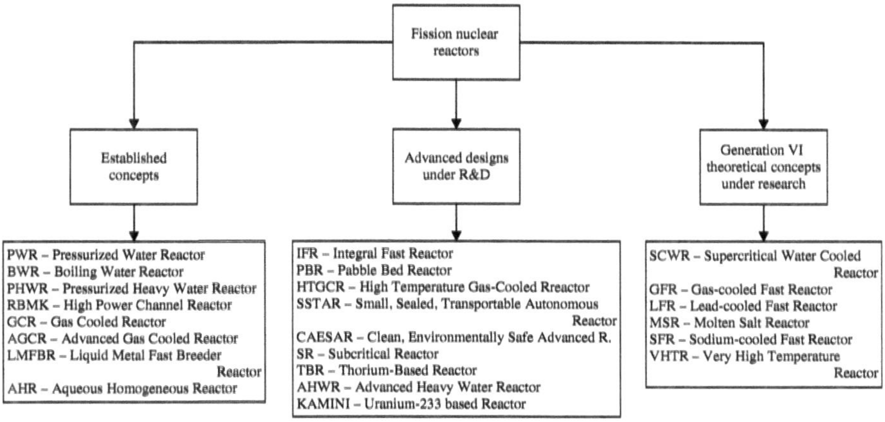

Fig. 7.1 Classification of nuclear reactors

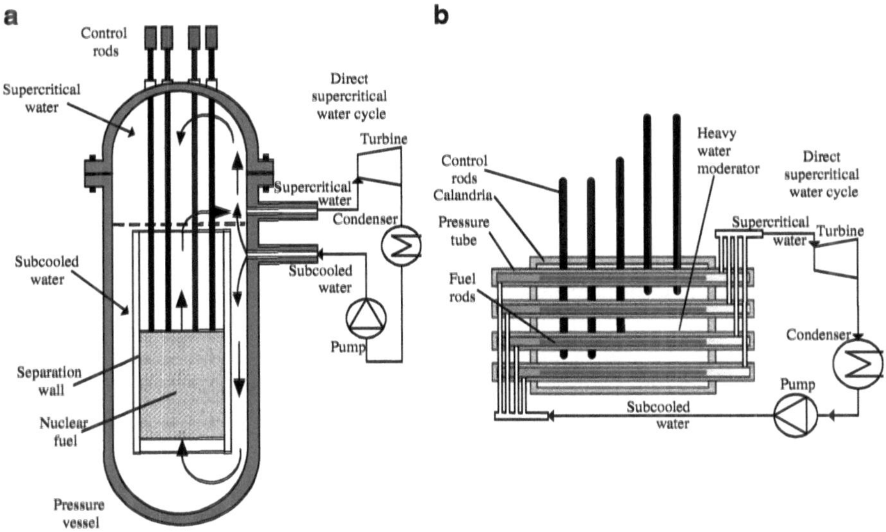

Fig. 7.2 Supercritical water-cooled reactor configurations; (**a**) pressure vessel, (**b**) pressure tube (CANDU-SWCR)

variants of moderators (light water, heavy water, or zirconium hydride solid moderator).

Figure 7.2 illustrates the two main design concepts of SWCR, namely the pressure vessel configuration and the pressure tube configuration. The pressure vessel configuration—illustrated in Fig. 7.2a—represents an evolution of BWR and PWR concepts where the nuclear fuel is placed inside a pressure vessel surrounded by water which has the role of coolant and moderator. Since SWCR

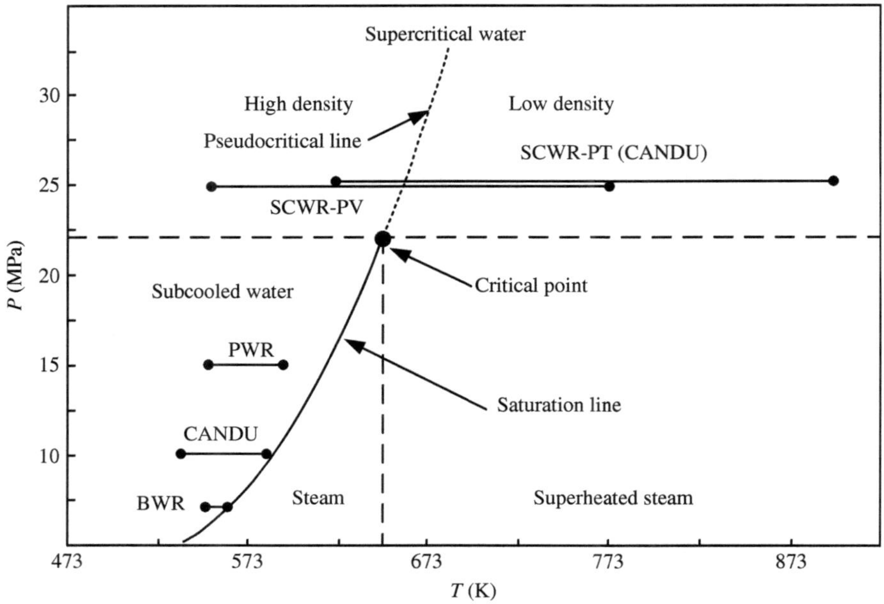

Fig. 7.3 Coolant pressure and temperature variation in various water reactors (*BWR* boiling water reactor, *PWR* pressurized water reactor, *SCWR* supercritical water cooled reactor, *PV* pressure vessel variant, *PT* pressure tube variant) [data from Naidin et al. (2009)]

operates at supercritical pressure, no boiling occurs inside the pressure vessel, but rather water reaches the state of a supercritical fluid, characterized by both the temperature and pressure with values above the critical point. Supercritical water simplifies the design because there is no need for a primary and a secondary circuit for heat transfer as typically used in current subcritical water reactors. Moreover, the need for a pressurizer vessel is also eliminated because water in a supercritical state is compressible.

The CANDU-SWCR configuration uses pressure tubes as illustrated in Fig. 7.2b. The CANDU-SWCR, described in more detail by Khartabil (2009), has a similar geometrical arrangement as traditional heavy water reactors developed by AECL (Atomic Energy of Canada Limited). However, in the supercritical version, light water is circulated in tubes and the nuclear fuel is based on a thorium fuel cycle. The moderator is heavy water which surrounds the pressure tubes in a calandria vessel with enhanced safety functions.

Figure 7.3 illustrates the $P - T$ diagram of water with a superimposed isobaric heating process in traditional and Generation IV water cooled reactors. With traditional reactors, the coolant is kept in a liquid state or—at most—it is found as saturated vapor (see BWR). The next generation of SWCR will likely operate at about 25 MPa. The pressure vessel variant of the reactor under development in Japan and Europe is part of the Generation IV International Forum. In Japan, both

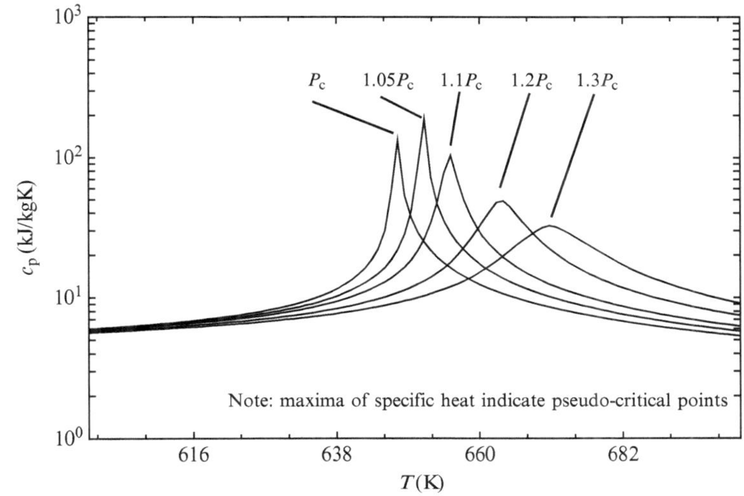

Fig. 7.4 Variation of specific heat in supercritical water as a function of temperature

fast and thermal SCWR development is projected. The pressure tube variant of SCWR is under development in Canada, for an inlet water temperature in the reactor of ~625 K and outlet at ~900 K. Beside the development of the Canadian reactor, which is a GIF adopted concept, there has been a pressure tube version developed in Russia as a non-GIF design, namely by the Research and Development Institute of Power Engineering (RDIPE). The reactor planned by RDIPE is a thermal reactor which operates with water at 25 MP and with 543 K inlet temperature and 818 K outlet temperature. From Fig. 7.3, with supercritical reactors, the process of water heating in the reactor is represented by an isobaric (horizontal) line, whereas the fluid entrance is at a subcritical temperature characterized by a high density (specific to liquids). During the heating process inside the reactor, the temperature of the coolant increases and its specific heat as well. Once the temperature becomes higher than the critical value (T_c), the specific heat increases sharply due to supercritical fluid behavior. The specific heat reaches a maximum value at a certain temperature $T_{pc} > T_c$, denoted by the pseudo-critical temperature (T_{pc}). Thereafter, the specific heat starts to decrease. This behavior is illustrated in Fig. 7.4 while the line that connects the pseudo-critical points is shown in Fig. 7.3.

Another GIF concept is the gas-cooled fast reactor (GFR) with variants in development at Euratom (Joint Research Centre), France (CEA—Commissariat of Atomic Energy), Japan (JAEA—Japan Atomic Energy Agency, ANRE—Agency for Natural Resources and Energy), and Switzerland (PSI—Paul Scherrer Institute) (Fig. 7.5). This is a fast-neutron-spectrum reactor with a closed fuel cycle and helium gas as a coolant. It is envisioned that the GFR will have on-site fuel processing for actinide recycling with a minimal requirement of nuclear material transportation. The helium coolant enters the reactor at ~733 K and leaves at

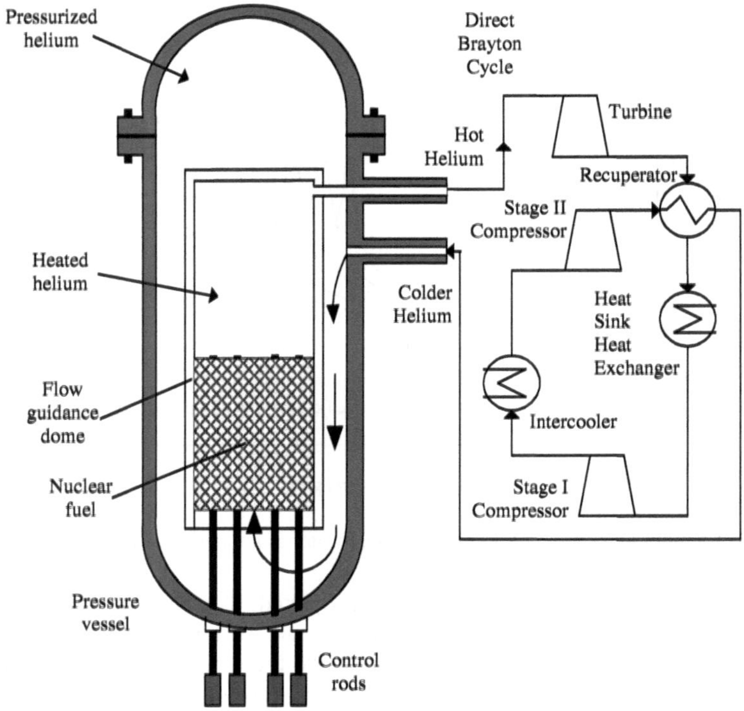

Fig. 7.5 Conceptual schematic of a Generation IV gas-cooled fast reactor

1,023 K and 6 MPa (Van Rooijen 2009). The planned thermal power of this reactor is 2.4 GW. The active core of a nuclear reactor is about 2 m diameter with a height of 6 m and uses a mixed nitride fuel with a breeding ratio of 1.2.

Another Generation IV fast reactor is the Sodium Fast Reactor (SFR), as schematically illustrated in Fig. 7.6. Variants of SFR have been developed by Euratom, France, Japan, China, Korea, Russia, and USA. The reactor features a closed fuel cycle (spent fuel is reprocessed). The concept of the reactor has been verified by systems operated during the period of 1967–1985 on smaller scale reactors in the USA, Japan, India, Germany, France, UK, and Russia.

There are two main variants of SFR, namely a loop type and a pool type. Both systems consist of primary and secondary sodium loops that eventually transfer heat to a steam generator which produces superheated steam at 19.2 MPa and 793 K, while the water inlet is approximately 512 K. The temperature of the secondary sodium loop at the inlet/outlet of the steam generator is 793/608 K; the envisaged thermal power is ~3.5 GW. Figure 7.6 illustrates the conceptual schematic of a loop-type SFR. Note that the pool-type variant, although different, from the point of view of operation, is similar to the loop-type design. The reactor core comprising the nuclear fuel is submerged in molten sodium. The arrangement comprises guiding walls for the flow such that forced convection is promoted and eventually

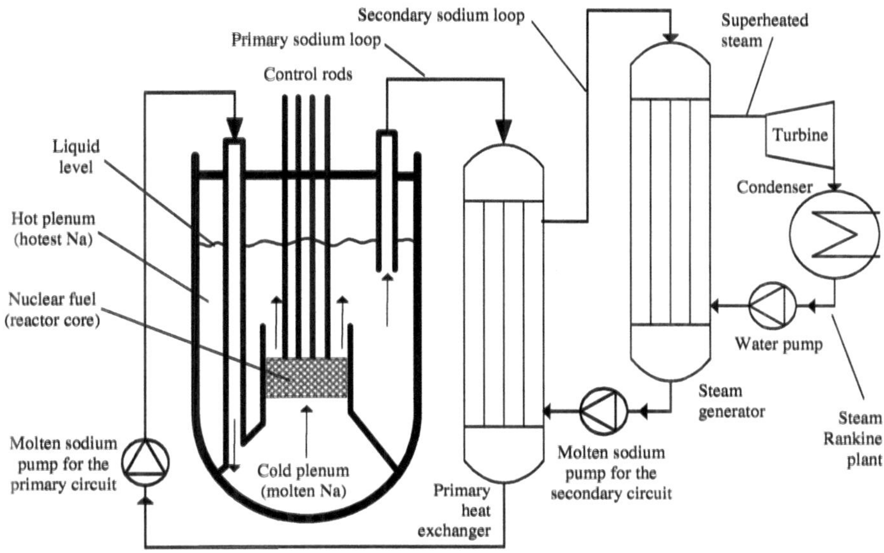

Fig. 7.6 Conceptual schematic of Generation IV sodium fast cooled reactor (loop-type)

the molten sodium is heated and further forced to flow through a primary heat exchanger. The heat is transferred to a secondary loop with molten sodium. A pump is used to circulate the molten sodium through the secondary circuit. Eventually the heat is transferred to the steam generator.

The very high-temperature reactor (VHTR) is planned for a thermal generation capacity of around 600 MW per unit to supply process heat or generate electricity at high efficiency. The reactor uses a thermal neutron spectrum moderated with graphite and cooled with helium. Either pebble bed or prismatic graphite block cores are considered for VHTR. The temperature of helium at the core outlet is 1,273 K. The VHTR technology is built on previous practical experience from many research groups and industries with most relevant examples of test reactors currently under operation in Japan—the HTTR (high-temperature test reactor) with a 30 MW thermal capacity, and in China—the HTR-10 (high-temperature reactor of 10 MW thermal output). Other variants of VHTR concepts are: (1) the Pebble Bed Modular Reactor (PBMR) proposed in South Africa, (2) GTHTR-300C proposed in Japan, (3) ANTARES proposed in France (a reactor design adopted by AREVA), (4) NHDD proposed in Korea (Nuclear Hydrogen Development and Demonstration reactor), (5) GT-MHR proposed in the USA (gas turbine modular helium reactor developed by General Atomics), (6) NGNP proposed in the USA (next generation nuclear reactor for process heat and electricity developed at the Idaho National Laboratory under funding support by the Department of Energy). Past test units and prototypes of high-temperature gas reactors were operated in the USA, Germany, and UK. The test reactor of Japan—the HTTR—uses helium at 4 MPa with 668 K inlet temperature and 1,223 K outlet temperature.

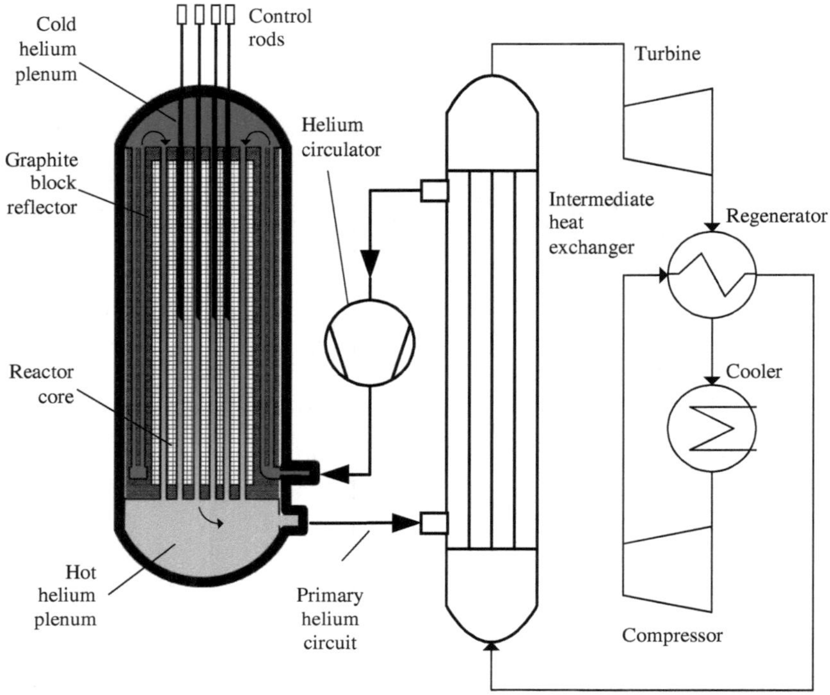

Fig. 7.7 Typical configuration of a VHTR plant

The typical design of a VHTR system is illustrated schematically in Fig. 7.7. The reactor's core consists either as a prismatic block or a pebble bed. Helium gas is circulated through the reactor core in a top-down direction such that hot helium accumulates at a bottom plenum. An intermediate heat exchanger is used to transfer the heat for downstream processing—either for power generation or for use as process heat at high temperature. The test reactor of Japan—the HTTR—uses helium at 4 MPa with a 668 K inlet temperature and 1,223 K outlet temperature. The Chinese HTR-10 test reactor works with helium coolant at 3 MPa and 523 K inlet temperature and 973 K outlet temperature. The GTHTR-300C reactor design for full commercial use is planned to operate at 5–7 MPa helium pressure with an inlet temperature in the range of 860–936 K and outlet temperature of 1,123–1,223 K.

Both Brayton and Rankine power cycles are considered for electricity generation with VHTR systems. Brayton gas turbine cycles can operate either in a direct configuration coupled to the reactor (where hot helium is expanded in a turbine, thereafter is further cooled in a regenerator heat exchanger and then compressed, reheated in the regenerator and then returned at the reactor inlet) or an indirect configuration (which uses an indirect heat exchanger and a secondary helium or helium–nitrogen loop). When a secondary heat exchanger is used, the temperature decreases according to the log-mean temperature difference. At a capacity for full-scale plants (hundreds of MW thermal), it is around 100–150 K. For example, the intermediate temperature heat exchanger in the GTHTR-300C reactor is about 750 K at the inlet and 1,173 K at the outlet. The Chinese full-scale VHTR reactor is projected for ~500 MW thermal output with a pressure of helium gas of 7 MPa, delivered at 1,023 K from an input temperature of 523 K. It generates superheated steam in a secondary loop at ~13 MPa at ~840 K from subcooled water at ~475 K.

The lead-cooled fast reactor (LFR) is another concept of Generation VI systems which uses a fast-spectrum nuclear process with a closed fuel cycle. The core is cooled with a liquid metal consisting of lead or lead-bismuth eutectic. As opposed to all other reactors (such as water reactors, gas reactors), lead-cooled (or liquid metal cooled) reactors do not require a moderator. The LFR design is currently in a development phase in the USA (SSTAR variant—Small Secure Transportable Autonomous Reactor) and at Euratom (ELSY variant—European Lead-cooled System). The planned capacity is about ~20 MW electric for SSTAR and 600 MW for ELSY. The ELSY generates superheated steam at 18 MPa and 723 K from the water inlet at 610 K. The SSTAR delivers supercritical CO_2 at 20 MPa and 825 K. The ELSY is an LFR version adaptable for large-scale nuclear hydrogen production. In its first planned implementation, the system can be coupled to a high-temperature alkaline electrolyzer to split water or a steam-methane reforming plant to extract hydrogen from natural gas with reduced pollution, compared to other methods.

Recent progress in the ELSY design was reported by Alemberti et al. (2011). A typical reactor configuration is illustrated in Fig. 7.8. It includes a cylindrical housing with a hemispherical bottom inside which the liquid metal is recirculated. In the liquid metal, there are eight submerged steam generators with a unit thermal power of 175 MW each.

Future advancement of LFR is under development in the USA after a first phase of testing with the SSTAR reactor, namely a hydrogen generation (or process heat) dedicated LFR will be deployed in the 2020s—denoted STAR-H_2 (Tsvetkova et al., 2003). This will generate 400 MW thermal power at 1,073 K.

The Molten Salt Reactor (MSR) operates with dissolved nuclear fuel in a molten fluoride salt. According to Delpech et al. (2009), the initial MSR concept has been developed in the 1950s at the Oak Ridge National Laboratory. This reactor is suitable for a thorium fuel cycle. Recently anon-moderated thorium molten salt reactor (TMSR-NM) has been developed at CNRS-Grenoble in France. The reactor is designed for a 2.5 MW thermal output and uses a fluoride-based molten salt mixture which comprises uranium and thorium. The operating temperature of the reactor is 900 K (see Delpech et al., 2009). Consequently the MSR can be integrated with the copper–chlorine cycle, and possibly with the hybrid sulfur cycle, but the temperature level of the heat source is insufficient for integration with a sulfur–iodine cycle.

Figure 7.9 shows the temperature profiles in the main reactors. The coolant temperature is plotted against the specific enthalpy (in kJ/kg). For LFR, the profile #2 refers to the Euratom ELSY reactor cooled with water, while profile #3 refers to the US reactor SSTAR which operates with supercritical carbon dioxide. Regarding

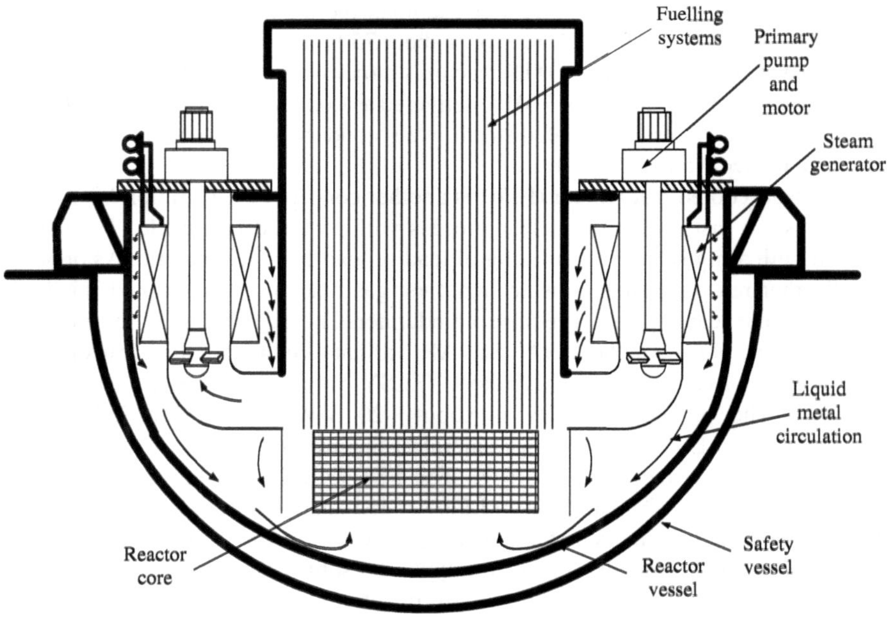

Fig. 7.8 Conceptual schematic of the European ELSY lead fast reactor (LFR)

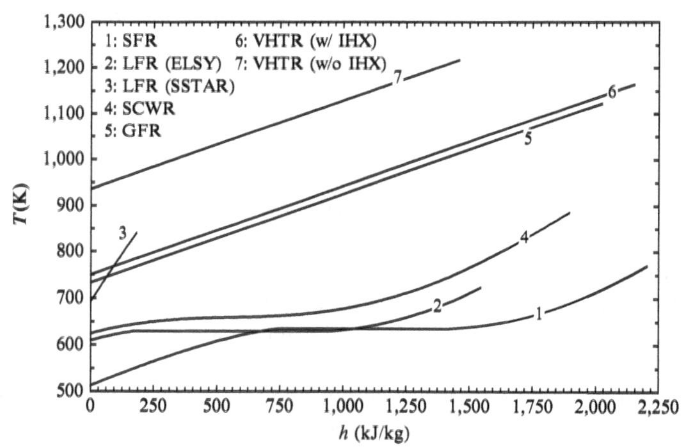

Fig. 7.9 Coolant temperature profiles in Generation IV reactors

the VHTR, the span of coolant temperature is indicated in the diagram for different variants. Profile #6 approximates the case of VHTR systems operating with intermediate heat exchangers (IHX) in which the temperature level of the coolant is lower. The profile #7 represents the case when an intermediate heat exchanger is not used but rather the coolant is used directly for the downstream process.

Table 7.1 Nuclear hydrogen production methods

	Next generation reactors (T_{out})					
Hydrogen production	SFR (723 K)	LFR (723–1,000 K)	SCWR (823 K)	GFR (1,023 K)	VHTR (1,223 K)	MSR (973 K)
Alkaline water electrolysis at 473 K	Yes	Yes	Yes	No	No	Yes
High-temperature steam electrolysis	No	No	No	Yes	Yes	Yes
Sulfur–Iodine cycle	No	No	No	Yes	Yes	No
Hybrid Sulfur cycle	No	No	No	Yes	Yes	No
Copper–Chlorine cycle	No	No	Yes	Yes	Yes	Yes
Natural gas reforming	No	No	No	Yes	Yes	Yes
Coal gasification	No	No	No	Yes	Yes	Yes

Note: Outlet temperature is indicated below the reactor type

Seven major technologies can be envisaged for large-scale production of hydrogen with nuclear energy. These are alkaline water electrolysis at an elevated temperature (473 K), high-temperature steam electrolysis at temperatures over 1,000 K, thermochemical water splitting via sulfur–iodine, hybrid sulfur or copper–chlorine cycles, nuclear assisted natural gas reforming and coal gasification. These technologies can be linked to specific reactors, depending of the required temperature level of the input thermal energy. Table 7.1 indicates several nuclear hydrogen production methods that link reactors to a technology of large-scale hydrogen production. The nuclear reactors which generate heat at intermediate temperatures (below 900 K)—SFR, LFR, SCWR, MSR—can drive large alkaline water electrolysis plants that operate at elevated temperatures and pressures (see Chap. 4). Steam water electrolysis can be linked to intermediate-high and high and very high-temperature nuclear reactors such as MSR, GFR, and VHTR. The sulfur–iodine and hybrid sulfur cycles are only compatible with GFR and VHTR. The copper–chlorine cycle is compatible with roughly 70 % of the proposed next generation reactors including the following reactors adopted by GIF (Generation IV International Forum): SCWR, MSR, GFR, VHTR. Natural gas reforming and coal gasification processes are compatible with MSR, GFR, and VHTR. In the next sections of this chapter, several major schemes of nuclear reactor integration with hydrogen production technologies are introduced.

7.3 Integrated Nuclear Systems for Natural Gas and Coal Conversion to Hydrogen

Coal and natural gas are energy resources that will continue to be used in the near future although it is well accepted that their use leads to pollution of the environment, especially through carbon dioxide emissions. Natural gas reforming involves an endothermic reaction of methane and steam at high temperatures according to the following overall process: $CH_4 + 2H_2O \rightarrow 4H_2 + CO_2$. The process temperature must be higher than 900 K in order to generate a satisfactory yield.

Hydrogen can also be obtained via coal gasification (see Chap. 3) according to the overall endothermic reaction $C + 2H_2O \rightarrow 2H_2 + CO_2$. The use of a nuclear reactor to produce high-temperature heat which is transferred to a coal gasifier or a natural gas reformer represents a possible strategy to significantly reduce pollution of the environment by carbon dioxide emissions characteristic of fossil fuel processes. For example, based on experimental data of Verfondern (2007), at least 30 % more hydrogen is generated for the same amount of CO_2 emission if the heat is derived from a nuclear reactor rather than from combustion of an additional quantity of natural gas (in an autothermal reforming process).

The most technologically advanced developments of nuclear-based natural gas reforming were developed in Japan and Germany where specific research programs were initiated 3–4 decades ago. The process of steam methane reformation leads to emissions of carbon dioxide. In the traditional approach, the emissions are much larger than that corresponding to the stoichiometric amount because the reformation plant must generate enough heat in the process to compensate for the required heat of reaction and, moreover, some natural gas may be used to generate power to maintain pumps, compressors, and other equipment operation. At a nuclear plant, the energy required by the reaction and all auxiliary energy including electricity is derived from the nuclear reactor, thereby leading to significant reductions of CO_2 emissions.

Although GFR, VHTR, and MSR can all be considered as candidates for steam methane reformation. The best suited for application of the technology is the VHTR, because of its superior process temperature which allows for good yields of the process and the use of a well-understood known thermo-catalytic reformation process. The linkage with GFR and MSR will require more progress on the side of catalyst development, and membrane separation for hydrogen extraction at large production scales, in order to become more competitive.

In a VHTR plant, the steam can be heated directly in the IHV (intermediate heat exchanger) and supplied at elevated temperatures to a reformer. Alternatively, a helium loop can be used to deliver heat both to a steam generator and a steam-methane reformation reactor. The most straightforward approach is the system which does not require any modification of the reactor and IHX. Schematically, this system is depicted in Fig. 7.10.

The VHTR facility generates hot helium at around 1,100 K which is directed toward the steam methane reformer. A possible design of the reformer is shown in the figure. The hot helium heats the catalytic bed as it is circulated through a long coil. The colder helium is used to transfer heat to a heat recovery steam generator (HRSG) and recirculated through the IHX of the VHTR. Water is fed to the system (in excess of the stoichiometric amount) and preheated using heat recovered from expelled gases and product hydrogen gas, after which superheated steam is produced. Preheated compressed natural gas (CNG) and superheated steam are fed to the reactor and passed through the hot catalytic bed where successive reactions occur (steam methane reaction and water gas shift reaction). A hydrogen selective

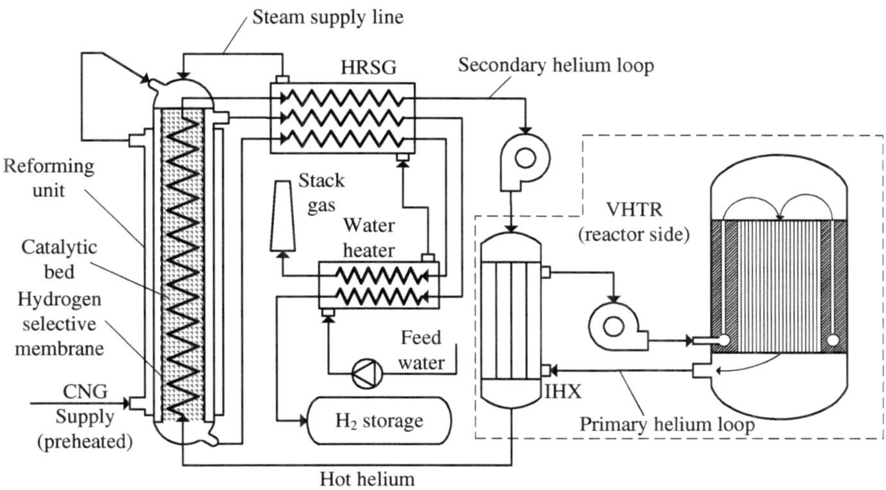

Fig. 7.10 Conceptual system for VHTR-driven natural gas reforming

membrane (palladium based) can be used to extract high purity hydrogen in the vicinity of the reaction sites. The membrane can have a tubular geometry and hydrogen is collected at the outer shell. The remaining (un-reacted) gases are collected at the bottom and used for heat recovery, then further expelled.

Ohashi et al. (2004) reported a stable operation at 4 MPa of an experimental steam reforming unit heated with helium and supplied by compressed natural gas and steam in a mass proportion of 1:3.9 (natural gas vs. steam), while the natural gas vs. heating helium supply rate was 1:7.6. The reported temperature of supplied helium is ~1,150 K.

A detailed description of past German research on nuclear-based natural gas reforming was reported in Verfondern (2007). A helium loop—heated electrically such that it simulates a VHTR system—has been used as a heat source for the reforming process during long-duration experiments at the Jülich Research Centre in Germany. A bayonet type reformer system—named EVA—has been developed with hot helium supplied at the outer shell and reformation processes in the core tube. The heating capacity has been 10 MW with generation of helium at 4 MPa and 1,223 K. The steam reformer was placed in series with a steam generator along the helium circuit. Heat is supplied to the steam reformer as helium cools down from 1,223 to 923 K, while heat is supplied to the steam generator when helium cools down from 923 to 623 K. The process gas is supplied to the reactor in a proportion of 1:4 methane-to-steam at a temperature of 500 K. Due to the heat addition, the gas temperature after leaving the catalytic bed becomes 1,073 K. The heat from the product gas (comprising H_2, CO_2, CO, CH_4) is recovered in the range of 1,073 K to 723 K. The reaction rate is expressed as 1,000 Nm^3 of hydrogen per m^3 of catalyst per hour. The heat transfer coefficient at the process side was reported at around 1,000 W/m^2K, while at the hot helium side, ~500 W/m^2K.

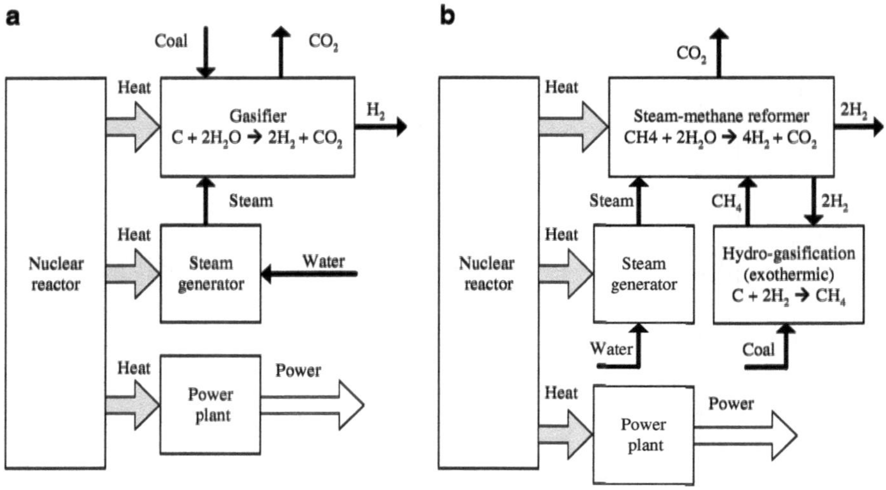

Fig. 7.11 Nuclear coal gasification methods: (**a**) steam-coal gasification; (**b**) hydro-gasification of coal followed by steam methane reformation

From the past German research reported by Verfondern (2007), in order to generate 100,000 Nm^3/h hydrogen using steam reforming of natural gas,there is a need for nuclear reactor heat of 84.82 MW. The recommended reforming unit design comprises a cylindrical shell housing of 3.8 m diameter which accommodates about 300 tubes of 17 m length filled with catalysts deposited on Rasching rings. The flow rate of helium which transfers heat from the nuclear reactor facility must be around 75 kg/s.

Gasification of coal requires also heat at a temperature level higher than 1,000 K for the process to be economically competitive. Because the hydrogen/carbon molar fraction is reduced in coals, the gasification process emits more carbon dioxide as compared to any other reforming process of fossil fuels. One way to reduce the carbon dioxide emissions of coal gasification can be obtained by blending coal with quantities of biomass. When high-temperature heat is provided by VHTR or GFR to drive the process, a substantial reduction of pollution can be obtained.

Kirchhoff et al. (1984) described a pilot plant for hydrogen generation with coal gasification and nuclear energy. In an integrated plant for nuclear coal gasification, the heat source from the nuclear reactor is used for three processes: (1) power generation, (2) steam generation, and (3) providing the required heat of reaction for steam gasification to hydrogen. A general block diagram schematic of an integrated plant for nuclear hydrogen generation is presented in Fig. 7.11a. In Figure 7.12 is presented the diagram of a an integrated system that couples the VHTR with a steam gasifier and uses a secondary helium look and an Rankine steam cycle. The process requires nuclear heat of more than 1,100 K to be effective.

A pioneering study of nuclear coal gasification was published by Schrader et al. (1975) followed by other studies such as Rastoin et al. (1979). In this specific process, a hydro-gasification of coal was proposed rather than direct coal gasification. In a hydro-gasification process, coal was combined with hydrogen to produce synthetic

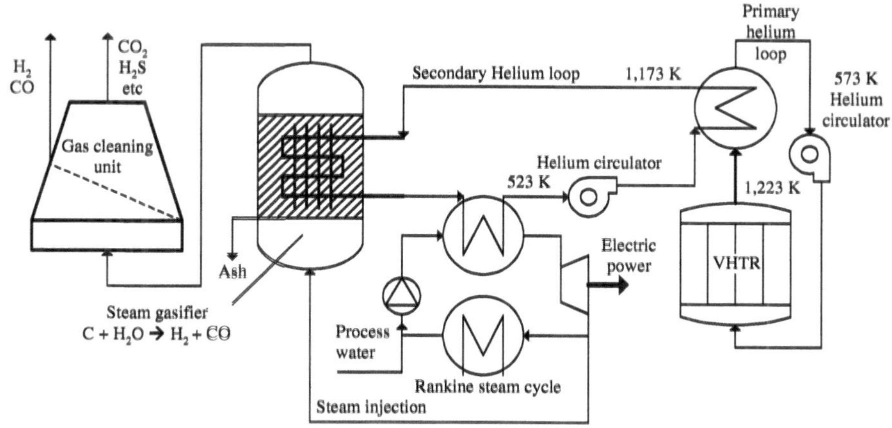

Fig. 7.12 Integrated steam-coal gasification plant for synthesis gas and power generation with nuclear energy [data from Kirchhoff et al. (1984)]

natural gas with an exothermic reaction (which is clean methane). Synthetic natural gas can be used either directly (e.g., as fuel) or it can be reformed to hydrogen onsite. Regardless of the type of the final product (either pure hydrogen or methane), the process requires onsite generation of hydrogen for the hydro-gasification. Therefore, if methane is the desired product, then half of the generated CH_4 is cracked to hydrogen and the other half is delivered as a product. If hydrogen is the desired product—as indicated by the integrated system presented in Fig. 7.11b—then all generated methane is reformed to hydrogen by a steam reforming process.

There are certain advantages of the scheme in Fig. 7.11b with respect to 7.11a, although the last appears more direct. In an integrated hydro-gasification plant, there are five main units (nuclear reactor, power plant, steam generator, hydro-gasifier, and steam-methane reformer), while in an integrated coal gasification system, there are four main units (nuclear reactor, power plant, steam generator, and gasifier). Conversion of steam to methane is more direct than coal gasification because the former involves an exothermic reaction; there is less need of coal preheating; hydrogen is very reactive with coal; the reaction evolves at a reasonable rate; once the synthetic methane is generated, there is an option to maintain it as a final product or convert it (partially or totally) to hydrogen, depending on the actual needs. Also, a well-developed steam reforming process with all reactants and products in the gas phase can be used to generate hydrogen, rather than a more difficult process to handle steam-based coal gasification which involves preheating of coal and a gas–solid reaction of coal and steam (with less reactivity than coal and hydrogen).

The more detailed layout of an integrated plant with gasification of coal and a VHTR reactor is presented. A pilot plant demonstration of this concept was reported by Kirchhoff et al. (1984). The steam gasification process is conducted at 4 MPa in a fluidized bed gasifier to generate synthesis gas according to the

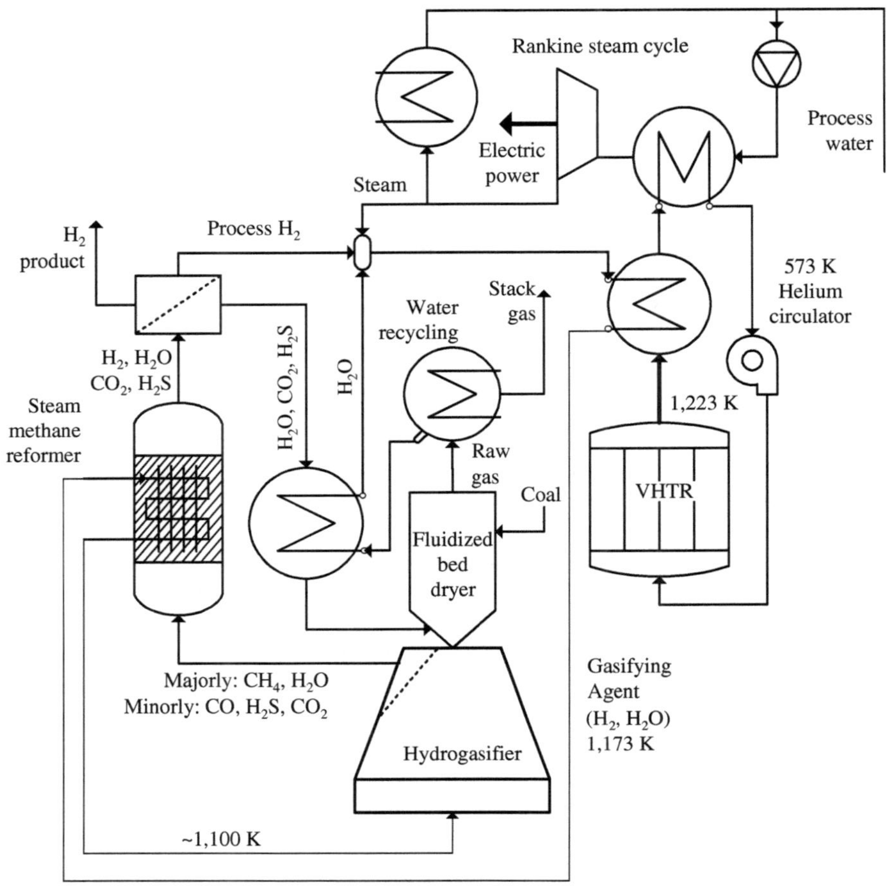

Fig. 7.13 Integrated system for power and hydrogen production by hydro-gasification of coal using nuclear energy

chemical reaction: $C + H_2O \rightarrow H_2 + CO_2$. The synthesis gas can be used further for many purposes including hydrogen generation, a case in which a water–gas shift reactor must be added for downstream processing. Catalysis has been investigated for a reduction of the gasification temperature to as low as ~1,000 K.

The general configuration of an integrated system for nuclear-based power generation and hydrogen generation by hydro-gasification of coal is illustrated in Fig. 7.13. The method offers a key advantage of eliminating the intermediate heat exchanger on the nuclear reactor side. This is possible because the gasification agent can be used as a heat transfer fluid simultaneously. There is only a primary circuit for the reactor where helium is circulated. Thus, the gasification agent comprising steam and hydrogen can be heated up to ~1,173 K and heat is transferred to a steam methane reformer which generates hydrogen. Coal is supplied to a fluidized bed dryer operated with hot stack gases from which water is recovered in a downstream process. The preheated and dried coal is fed to the gasifier where it enters in contact with hot hydrogen reactant and mainly methane is formed. The product gas of the gasifier passes to the catalytic steam methane reformer. A membrane process for hydrogen separation can be used to extract hydrogen product and process hydrogen from the gaseous stream present at the steam-methane reformer exit. Process steam is extracted from the low pressure turbine of the steam Rankine cycle integrated with the plant. This steam is sourced from supply water that is fed to the Rankine cycle pump. Helium from the primary circuit of the nuclear reactor is used to generate superheated steam for the Rankine cycle.

A pilot plant based on hydrogasification of coal has been demonstrated at the "RheinischeBraunkholenwerke AG" in Germany and the results were reported in Schrader et al. (1975) with helium preheated at temperature levels that simulate the situation in an actual VHTR. The system is similar to that suggested in Fig. 7.13 with an essential difference that it does not include any steam-methane reformer. In other words, the pilot plant by Schrader et al. (1975) generates SNG (synthetic natural gas—methane). The pilot plant has been devised for 400 kg raw lignite per hour or 150 kg hard coal per hour. It generates 700 m^3 per hour of raw gas at 4–10 MPa. Based on the experimental demonstration, some full-scale conceptual designs were provided. It was shown that a VHTR of 3 GW thermal can process 2.5 Mt/h lignite and generate 410,000 Nm^3/h of methane and 180 MW electrical power. It also yields 330 t/h residual char. The system efficiency of SNG production is estimated at 72 %. Therefore, accounting for the reaction heat for steam methane reforming, one may expect a hydrogen generation efficiency of over 60 % for the integrated plant of hydrogen production via coal hydrogasification.

In summary, nuclear-based hydrogen production from natural gas reforming and coal gasification (or hydrogasification), which has been studied in the 1970s and 1980s mostly in Germany, France, and Japan, has significant potential as a competitive method for hydrogen production at a large scale.

7.4 Nuclear-Based High Temperature Electrolysis of H_2O

The Department of Energy (DoE) in the USA deployed significant efforts to select the most promising option for the development of its "Next Generation Nuclear Plant" coupled to a hydrogen production technology. An independent, recent study—conducted for DOE—was reported by Varrin et al. (2011). The study aimed at adown selection of the most promising technology for hydrogen production with the high-temperature gas reactor (HTGR).

The down-selection process led to three—most promising—hydrogen production technologies to be linked to the planned HTGR in the USA, namely: (1) the sulfur–iodine (SI) cycle, (2) the hybrid sulfur (HyS) cycle, and (3) high-temperature steam electrolysis (HTSE) process. Among these three technologies, the ranking by Varrin et al. (2011) suggested that HTSE has the highest potential for successful deployment and demonstration of the NGNP. This conclusion is based on the fact that HTSE requires a less complex design with no intermediate chemical recycling

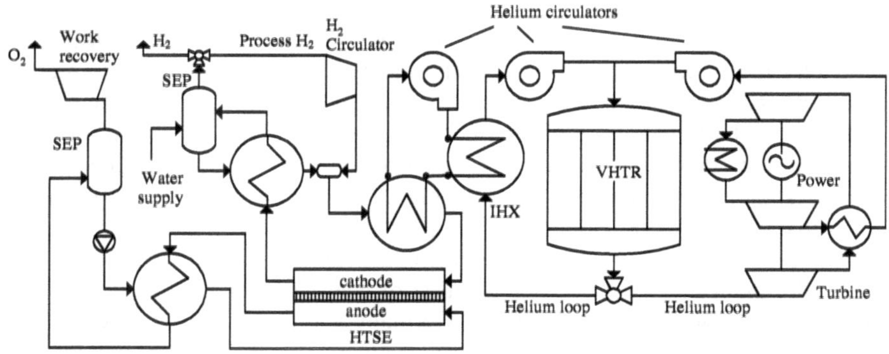

Fig. 7.14 Conceptual layout of an integrated nuclear plant for power and hydrogen generation with high-temperature steam electrolysis

and a hydrogen production efficiency that is less sensitive to the level of operating temperature in the range of 923–1,223 K.

The Idaho National Laboratory (INL) has been engaged in a long-term R&D program of DOE to develop high-temperature electrolysis for hydrogen production from steam. It is envisaged that the overall process efficiency is about 30–34 % although efficiency goals of over 42 % do exist with respect to the LHV of hydrogen (Elder and Allen, 2009). Nevertheless, if instead alkaline electrolysis units are used in conjunction with VHTR at 50% power generation efficiency, then the hydrogen generation efficiency of 34% may be achieved.

The conceptual diagram of an integrated nuclear plant for power and hydrogen generation with high-temperature steam electrolysis is illustrated in Fig. 7.14. The integrated system has three units: (1) VHTR, (2) Brayton power cycle operating with helium as a working fluid and, (3) high-temperature electrolysis. Hot helium generated by the reactor is partially circulated through the Brayton cycle. There it expands first in a turbine, after which it is cooled in a regenerative heat exchanger, then compressed in two stages with an intercooler and then reheated to a temperature level corresponding to the VHTR inlet.

The parallel helium circuit involves the passage of helium through the IHV (intermediary heat exchanger) to transfer heat to a secondary helium loop. The secondary helium loop is used to further transfer heat to the steam generator and then superheated forthe HTSE plant. Liquid water is supplied to the cathodic circuit of the HTSE plant—which is based on a solid oxide electrolysis cell. Water is heated, boiled and superheated using heat recovery from the hot stream exiting the electrolysis unit and partially the heat transferred from the secondary helium loop. A certain amount of hydrogen is maintained (and recycled) in the cathodic loop to promote the electrochemical half reaction at an optimal rate. In the anode feed, some amount of steam is provided as sweep gas for molecular oxygen. The plant has liquid separators for separation of hydrogen and oxygen from water.

The following simulated plant characteristics of the integrated HTSE were presented by Richards et al. (2005):

- VHTR output temperature, 1,223 K
- HTSE process temperature, 1,100 K
- Power generation, 312 MW
- Power supplied to HTSE process, 292 MW
- Thermal energy transferred to HTSE, 68 MW
- Hydrogen production efficiency based on LHV, 46 %

According to Elder and Allen (2009), the process of HTSE with a nuclear reactor is underdevelopment mostly at GA (General Atomics) and based on their design version of VHTR, namely the Modular Helium Reactor (MHR). The planned production capacity is ~270 Mt hydrogen per annum at a delivery pressure of ~5 MPa. The HTSE plant consists of modules of 0.5 MW equivalent electric power consumption. Each module comprises 500 planar cells.

7.4.1 *Integrated VHTR with Sulfur–Iodine Cycle*

Two schemes exist for linkage between the high-temperature reactor and the sulfur–iodine cycle. One of the schemeswasinvestigated by General Atomics and it involves the use of IHX with a secondary helium loop to transfer high-temperature heat at a remote location from the reactor where the SI (sulfur–iodine) plant is located. The transmission distance may be a few hundred meters. As described in Elder and Allen (2009), the GA concept, from the US-DOE NERI (Nuclear Energy Research Initiative), is designed around the modular helium reactor of 600 MW thermal energy output per unit. The integrated hydrogen production site will include four nuclear reactors totaling 2.4 GW thermal power generation. The heat is transferred at 115 m from the reactor via asecondary helium loop which delivers heat at the consumption point at 1,100 K to the S–I cycle. The return flow comes at 838 K at the IHX inlet. Aschematic of the US-NERI integrated plant is presented in Fig. 7.15. The overall hydrogen generation efficiency of the process is 38 % with respect to the LHV of hydrogen.

The second integration scheme of the S–I cycle was developed at JAERI. It links the S–I plants with the GHTTR-300C reactor via an IHX with a helium circulator. The HTTR reactor has a secondary cooling loop at a lower temperature with helium (see figure). According to Sakaba et al. (2007), the temperature of helium at hottest side is 1,223 K and at the coldest side (return), it is 763 K. The hot helium from the primary circuit passes first to the IHX where its temperature decreases to 1,123 K; thereafter, it is expanded in a turbine. The expanded flow reaches about 863 K. Furthermore, this flow is cooled to 300 K and then recompressed, a process after which helium reaches ~500 K. In the regenerative heat exchanger, this flow is reheated to the level of temperature corresponding to the reactor inlet, namely 763 K.

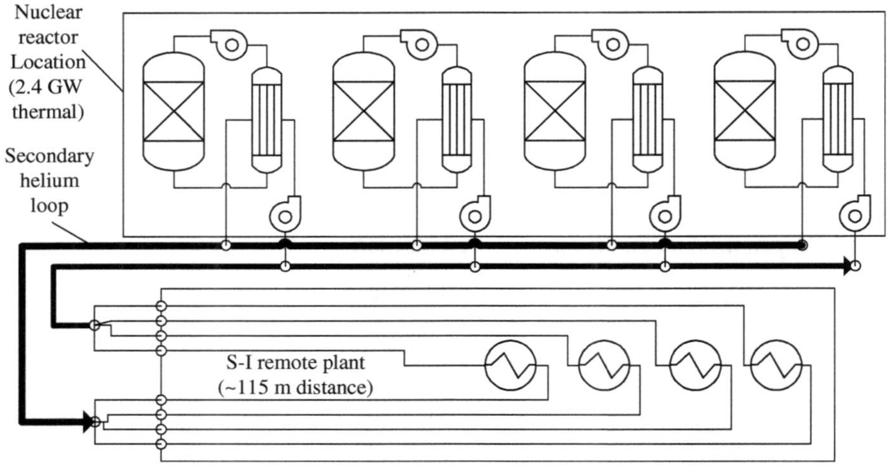

Fig. 7.15 Configuration of the integrated MHR/S–I plant proposed by DOE-NERI and GA

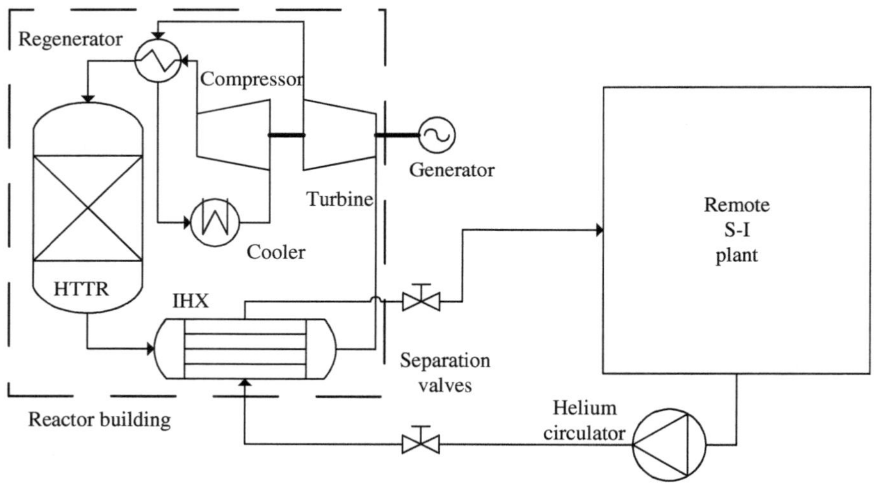

Fig. 7.16 Integrated nuclear reactor/S–I plant developed at JAERI (GHTTR-300C)

In the secondary helium loop, helium leaves the IHX at 1,180 K and it is supplied remotely with the SO_3 decomposed in the S–I plant at a temperature level of 1,150 K (30 K is the estimated temperature loss during fluid transport). Furthermore, hot helium is supplied in series to the sulfuric acid vaporizer and the HI decomposer after which the temperature reaches a value of 750 K. Two versions of integrated plants for hydrogen production are planned at JAEA based on the general layout in Fig. 7.16. These versions are denoted as GTHTR-300C for power and hydrogen cogeneration and GTHTR-300H mainly dedicated to hydrogen production. The characteristics of these plants are listed in Table 7.2.

Table 7.2 JAERI integrated system for nuclear hydrogen production with S–I cycle

Parameter	GTHTR-300C	GTHTR-300H
H_2 production	60 t/day	120 t/day
H_2LHV efficiency	~42 %	~40 %
Power production	175 MW	35 MW
Total heat generation	395 MW th	540 MW th

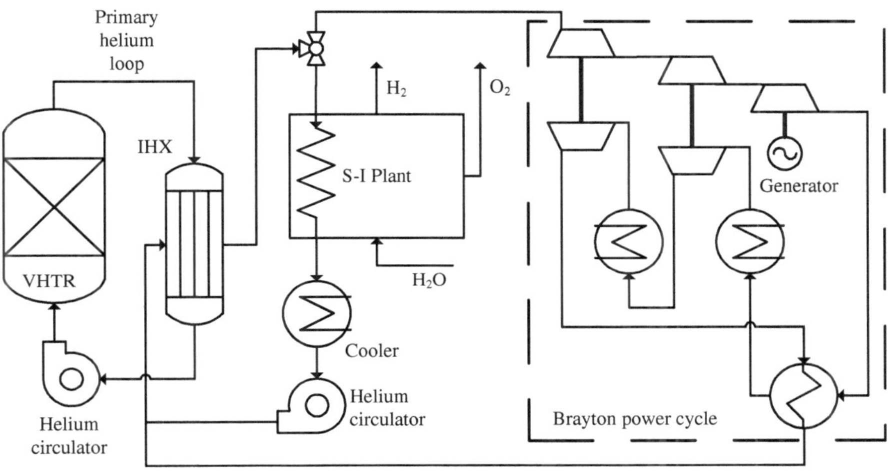

Fig. 7.17 Integrated S–I cycle with VHTR and Brayton cycle according to a HYTHEC/ RAPHAEL coupling scheme

The GTHTR-300 plants use a non-intercooled direct Brayton cycle for power generation with a pressure ratio of 1.5–2 and helium pressure of 5.1 MPa in the core. The efficiency range for power generation is 38–47 %. It is envisaged that a hybrid version of the S–I cycle which includes an electro-dialysis-based HI distillation process will be used for improved efficiency.

The European research programs RAPHAEL and HYTHEC developed an integrated plant of the VHTR and S–I cycle. Its layout is presented in Fig. 7.17. The S–I cycle and the Brayton power cycle are supplied in parallel with heat from the secondary helium loop. The temperature generated by the VHTR is 1,223 K. After the IHX, the helium stream in the primary loop reaches its minimum temperature of 660 K and after the circulator, it is returned to the reactor at 673 K. In the secondary loop, hot helium is delivered at 1,163 K and it returns at 623 K. Heat is transferred to the S–I cycle, a process during which the temperature of helium becomes 695 K. A cooler is placed in series with the S–I cycle, which further cools helium to 600 K before its supply to the circulator of the secondary loop. The Brayton power cycle has three turbines and two compressors. The high and intermediate pressure turbines are sized such that they drive the compressors on the same shaft. The low pressure turbine produces net mechanical power which is

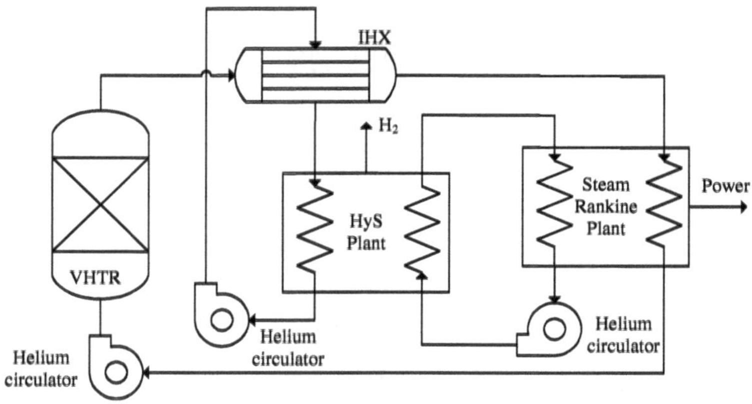

Fig. 7.18 Integrated plant with hybrid sulfur process and PBMR for hydrogen generation

transferred to an electric generator. The design is made so that power generation satisfies all plant needs and there is no power supplied to the grid; the only plant output is hydrogen. A plant of 319 MW is planned, which generates 27 t hydrogen per day.

7.5 Integrated Nuclear Plant for Hybrid Sulfur Process

According to Elder and Allen (2009), a consortium involving the Shaw Group and Westinghouse conceptually developed an integrated plant for hydrogen production with a hybrid-sulfur process and PBMR reactor. The configuration of the integrated plant is depicted in Fig. 7.18. For power generation, the plant uses a standard steam Rankine cycle which uses two heat sources: (1) heat recovered from the hybrid sulfur plant; and (2) heat delivered from heat exchangers coupled in series with the IHX (helium) of the PBMR reactor, at the bottom side.

One single module of the PMBR-based hybrid sulfur (HyS) plant comprises the VHTR, IHX, three helium loops and the HyS unit. There are four linked PBMR/HyS modules to a steam Rankine cycle. The PBMR reactor unit generates 500 MW thermal while heating helium in the primary loop from 700 K to 1,223 K. In the IHX, the temperature of helium from the primary loop is reduced to 1,020 K. The temperature of helium is reduced to ~700 K due to heat transfer to the Rankine cycle. From the 500 MW thermal output, around 195 MW are transferred to the HyS process via a secondary helium loop (see figure) with 1,173 K temperature at the hot end and 970 K at the cold end. Low grade heat (around 100 MW thermal) rejected by the HyS cycle is recovered and transferred to the power plant. As reported by Elder and Allen (2009), each HyS unit uses ~54 MW electrical energy provided by the power plant.

The power plant unit generates ~600 MW electrical energy with an efficiency higher than 38 %. Around 62 % of the total power generated by the plant is

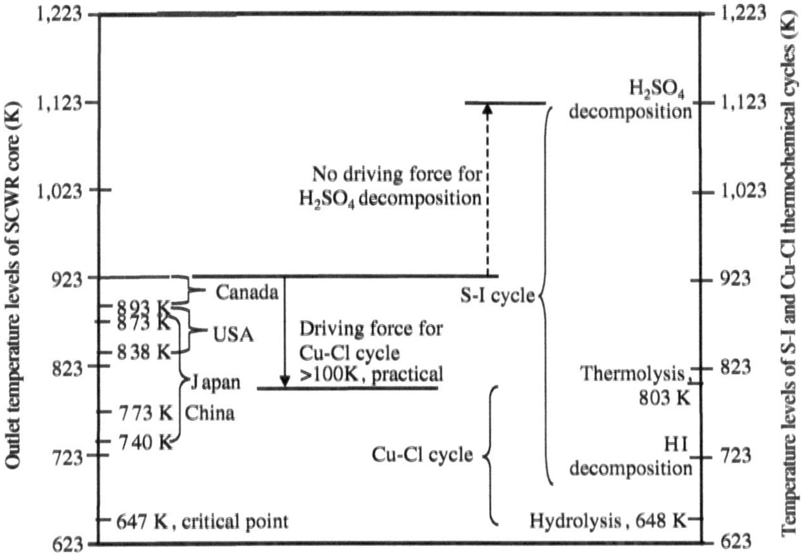

Fig. 7.19 Temperature level of heat in SCWR vs. requirements of Cu–Cl and S–I cycle [data from Wang et al. (2010)]

delivered to the grid while the remainderis used to power the hydrogen production process (including its auxiliary equipment). The total hydrogen production rate of a full plant is ~20 Mt per day. In addition, 380 MW power output is generated. Note that the total heat input generated by the four reactors per plant is 2.4 GW.

7.6 Integration of Copper–Chlorine Cycle with SCWR Reactor

The supercritical water cooled reactor presents a viable linkage to a copper–chlorine thermochemical plant for hydrogen production. Figure 7.19 illustrates the temperature levels of the heat transfer fluid in the SWCR in comparison to the requirements of two major thermochemical cycles: Cu–Cl and S–I cycles. It can be observed that most SCWRs can cover the temperature requirements of all steps in the Cu–Cl cycle, whereas the driving process for the heat exchange provided by CANDU-SCWR of Canada has the largest differential (>100 K).

For the S–I cycle, the sulfuric acid decomposition step requires 1,123 K, which cannot be reached with current technology of SCWRs. Although the temperature of SCWRs can supply theHI decomposition step by providing heat at 723 K, the heat input for this reaction occupies less than 10 % of the total heat required by the S–I cycle. By comparison, the heat quantity required at 1,123 K accounts for about 90 % of the total heat requirement of the cycle. Consequently, the SCWR is not linked to the S–I cycle. Similarly, due to the high temperature requirement, the

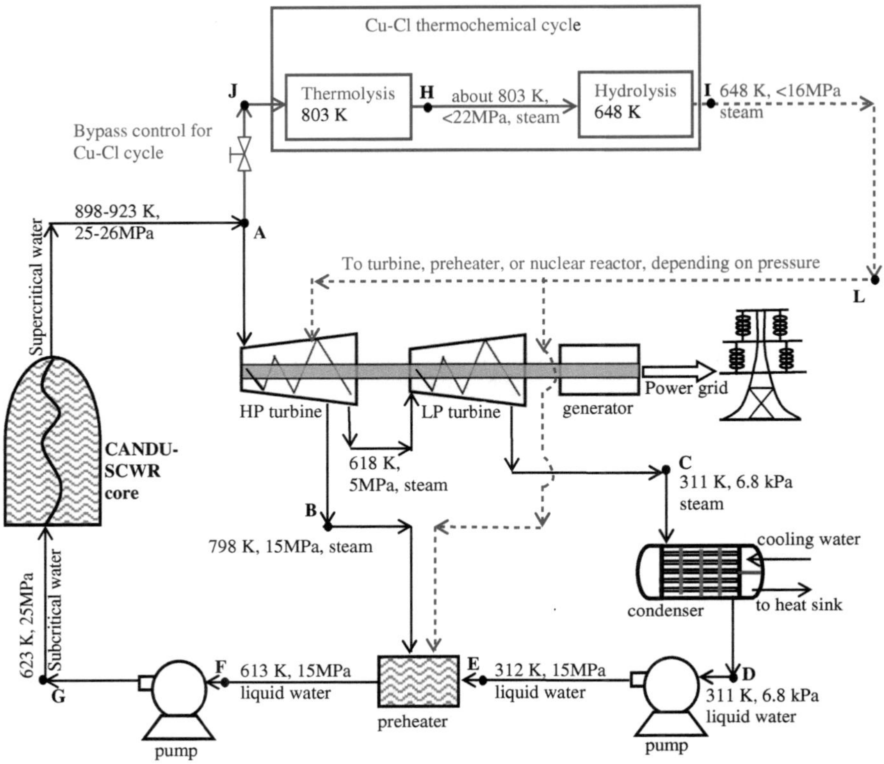

Fig. 7.20 Flowchart of SCWR coolant stream for heat exchange with a Cu–Cl cycle and reheat Rankine cycle [modified from Wang et al. (2010)]

hybrid sulfur process cannot be coupled to the SWCR (see Table 7.1). This explains the research efforts performed in countries that are developing SCWRs (mainly Canada, USA, Japan, China, Korea, France) to advance the development status of the copper–chlorine thermochemical cycle for water splitting.

Because of safety reasons, indirect heat transfer methods with intermediate heat exchangers must be developed in order to link the SWCR to a Cu–Cl plant. The temperature levels of SCWR and the Cu–Cl cycle also have significant influence on the location of the heat extraction from an SCWR. The Cu–Cl cycle will influence the water flow arrangement of the SCWR.

Wang et al. (2010) proposed an integration scheme of a copper–chlorine thermochemical cycle with SWCR according to the flowchart from Fig. 7.20. The power plant is a steam Rankine cycle with a single-reheat system that uses a preheater and two turbines. In the proposed system, a part of the supercritical water stream at the outlet of the SCWR (state A on the diagram) is diverted via a bypass control valve toward the copper–chlorine thermochemical plant. This stream acts as a heat transfer fluid for two reactors of the copper–chlorine cycle:

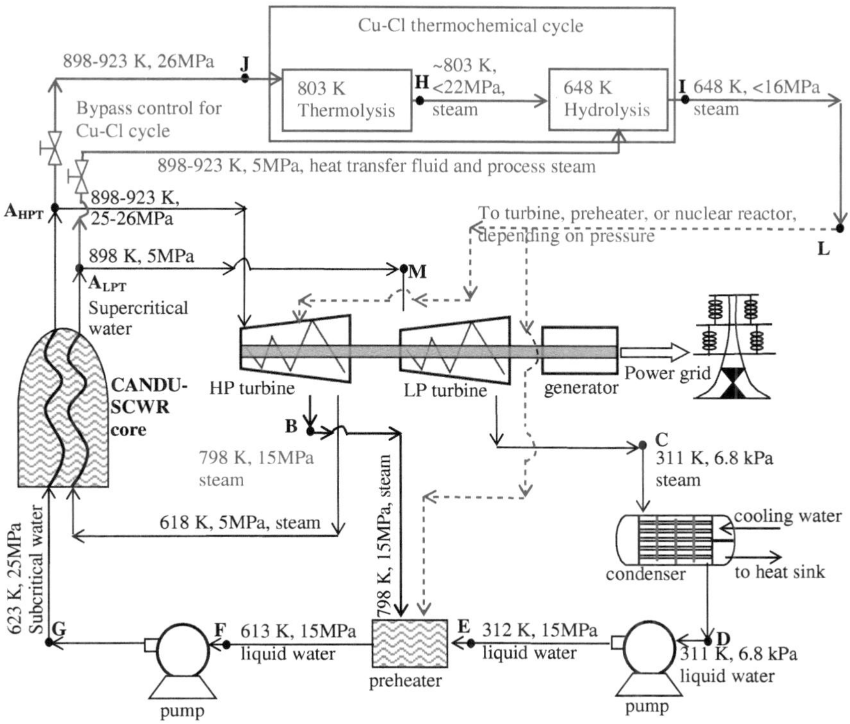

Fig. 7.21 Flowchart of SCWR coolant stream for heat exchange with the Cu–Cl cycle and a single-reheat Rankine cycle [modified from Wang et al. (2010)]

thermolysis and hydrolysis. First, heat is delivered to the thermolysis reactor where the endothermic chemical process is maintained at ~ 800 K. Subsequently the hydrolysis reactor is used for the chemical process at ~640 K. After delivering heat to the copper–chlorine cycle, supercritical water converts into steam (state J in the figure) with the following parameters at the return point: ~16 MPa, 648 K. Steam is further used in the Rankine cycle either by injecting it at the high pressure turbine or at the preheater.

A second linkage scheme between SCWR, Cu–Cl and Rankine plants is illustrated in Fig. 7.21. In this case, a double reheat Rankine plant is considered for more efficient power generation. Supercritical water at 26 MPa from the reactor outlet is diverted partially to a Cu–Cl plant and partially to the high pressure turbine. The stream of supercritical water is mainly used to provide reaction heat to the thermolysis and hydrolysis reactions in a similar manner as the method in Fig. 7.20.

However, an additional stream is diverted from the reactor to the copper–chlorine plant in the case of the flowchart in Fig. 7.21. The superheated steam generated in a reheater is split into two parts: one goes to the Cu–Cl plant and the other to the low pressure turbine. Steam at the low pressure turbine is assumed at

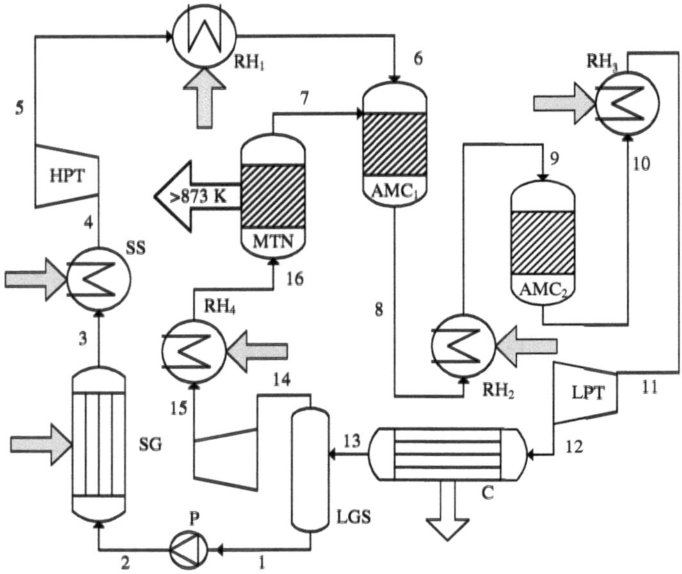

Fig. 7.22 Integratedchemical heat pump—steam Rankine cycle system linked to SCWR and a thermochemical water splitting cycle for hydrogen production [modified from Granovskii et al. (2008)]

5 MPa. In order to use effectively the enthalpy of the superheated intermediate pressure steam, which is diverted to the hydrolysis reactor, a system comprising heat exchangers and a work recovery turbine can be devised such that eventually process steam is provided at ~650 K and ~0.1 MPa. In order to keep the water inventory in the Rankine and reactor circuit, the same amount as the consumed process steam is supplied as fresh water in the preheater vessel (state E in the figure). A scheme similar to Fig. 7.21 was analyzed by Naidin et al. (2009) which concluded that the single-reheat scheme is advantageous with respect to no-reheat because it increases the power generation efficiency to >49 %.

An alternative way to integrate SWCR with a Rankine cycle and a thermochemical cycle is through chemical heat pumps. Chemical heat pumps perform cyclical chemical processes with a net effect of upgrading the temperature level from a heat source. In the context of nuclear-driven thermochemical cycles, it is often required to match the temperature of the reactor coolant to that required by endothermic reactions of the thermochemical water splitting cycle.

An interesting option is discussed next as a case study to integrate SWCR with the Cu–Cl cycle by an intermediate chemical heat pump. This study was published in Granovskii et al. (2008). The working fluid of the Rankine power cycle—steam—can be reacted inside the power plant circuit with methane to perform an endothermic steam-methane reaction. Then at higher temperatures, reverse exothermic reactions can be conducted. The integration scheme is presented in Fig. 7.22.

The system is supplied with heat by heat transfer from supercritical water generated by the SCWR (Fig. 7.22). The system comprises a pump (P) which pressurizes water to 20 MPa. Water is preheated and boiled in a steam generator SG which produces saturated steam. A superheater SS then follows to heat the steam to a superheated state #4 at 873 K. The superheated steam is expanded in the high pressure turbine HPT after which steams reaches the parameters of state #5 at 636 K and 5 MPa. The intermediate pressure steam is reheated via heat transfer from supercritical water from the SWCR in reheater RH_1 it reaches the parameters of state #6, namely 873 K and 5 MPa. Furthermore, the superheated steam is directed toward an autothermal methane conversion process where is reacted catalytically with methane in the reactor AMC_1 according to the endothermic reaction:

$$CH_4 + H_2O \rightarrow CO + 3H_2 - Q \tag{7.1}$$

where Q represents the heat required for the reaction. The heat of reaction in the autothermal process is extracted from the enthalpy of reactants; consequently the output stream in #8 is at a lower temperature according to the absorption of the heat of reaction.

The reaction products after the first stage of steam methane reformation are reheated to 873 K in the reheater RH_2 and directed to the second stage steam-methane converter AMC_2. After the second stage of conversion, the output stream #10 is reheated in RH_3 and then expanded to 0.1 MPa in the low pressure turbine LPT. The excess steam is condensed in the condenser C and water is separated from noncondensable gases in the liquid–gas separator LGS. The gases atstate #14 are then compressed in the compressor COMP up to 5 MP (state #15) and then heated up to 873 K in the reheater RH_4 and supplied to the methanator MTN in state #16. The methanation reaction is exothermic and evolves overall according to:

$$CO + 3H_2 \rightarrow CH_4 + H_2O + Q \tag{7.2}$$

Note that reactions (7.1) and (7.2) proceed simultaneously with the reversible water-gas shift reaction at equilibrium; this is the reaction between carbon monoxide and steam, according to:

$$CO + H_2O \rightleftarrows CO_2 + H_2 \tag{7.3}$$

At equilibrium and with the rapid kinetics, reaction (7.3) does not affect the overall chemical cycle of the integrated heat pump. The methanathor generates useful heat at a temperature higher than 873 K which can be used—via heat transfer—to drive a thermochemical water splitting cycle.

Based on the results of Granovskii et al. (2008), the state points parameters for the integrated system in Fig. 7.22 are given in Table 7.3 whereas the inputs and outputs of mechanical work are listed in Table 7.4. Table 7.5 indicates the composition of the gaseous stream at principal state points of the system. The results indicated in Granovskii et al. (2008) show that at least a 2 % improvement of energy utilization is obtained with the heat pump-based system, with respect to a system with direct heat transfer to the thermochemical cycle such as Fig. 7.21, without jeopardizing the reactor safety.

Table 7.3 State point parameters for the integrated Rankine cycle/chemical heat pump system

State point	1	2	3	4	5	6	7	8
T (K)	298	300	640	873	636	873	>873	<873
P (MPa)	0.1	20	20	20	5	5	5	5
Content	W	W	SaS	SuS	SuS	SuS	Meth	Syn
State point	9	10	11	12	13	14	15	16
T (K)	873	<873	873	NS	298	298	NS	873
P (MPa)	5	5	0.1	0.1	0.1	0.1	5	5
Content	Syn	Syn	Syn	Syn	Syn	Meth	Meth	Meth

W water, *SaS* saturated steam, *SuS* superheated steam, *Meth* reaction products of methanation (CH_4 + unreacted CO and H_2), *Syn* synthesis gas and steam (CO + H_2 + H_2O and unreacted CH_4), NS not specified
Source: Granovskii et al. (2008)

Table 7.4 Heat and work balances for system components

Device	LPT	COMP	P	HPT	MTN	RH_2	RH_3	C	RH_4	SG + SS	RH_1
Energy type	Work	Work	Work	Work	Heat	Heat	Heat	Heat	Heat	Heat	Heat
Energy[a] (kJ)	+437.5	−58.6	−7.1	156.2	94.2	−69.5	−26.0	918.2	−12.6	1228.9	−203.1

[a]Energy is given in kJ per 20 mol of H_2O circulated in the system
Source: Granovskii et al. (2008)

Table 7.5 Molar flows[a] of gaseous streams for integrated system from Fig. 7.22

State	CH_4	H_2O	H_2	CO	CO_2
6	0	20	0	0	0
7	0.89	1.10	0.44	0.015	0.10
8	0.52	20.36	0.44	0.0015	0.10
9	0.52	20.36	1.91	0.010	0.47
10	0.39	20.10	2.43	0.020	0.59
12	0.39	20.10	2.43	0.020	0.59
14	0.39	0.11	2.43	0.020	0.59
16	0.39	0.11	2.43	0.02	0.59

[a]Data is given for 20 mol of water circulated in the main loop
Source: Granovskii et al. (2008)

7.7 Case Study: Integrated Nuclear, Desalination, and Hydrogen Production System

Nuclear-based hydrogen production by the Cu–Cl hybrid thermochemical cycle using sea or brackish water was considered by Orhan et al. (2010). Various configurations for coupling the Cu–Cl cycle with a desalination plant using nuclear or renewable energy were presented and detailed thermodynamic analyses and comparisons of the configurations, as well as relevant parametric studies, were discussed.

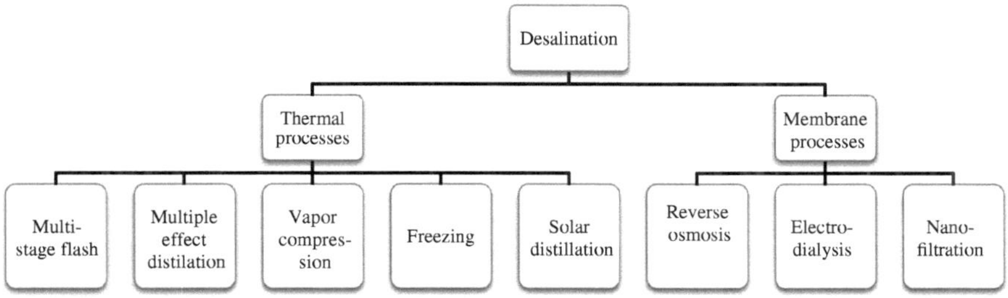

Fig. 7.23 Classification of desalination technologies

The motivation of coupling a nuclear-based thermochemical water splitting plant with a desalination system is motivated by the limited resources of fresh water in many countries. In addition, a thermochemical water splitting cycle must work with a relatively clean water supply in order to generate hydrogen and avoid byproducts or other issues such as scale deposits on process equipment. Fresh water is defined as water containing less than 1,000 mg/L of salts or total dissolved solids (TDS). Above this concentration, properties such as taste, color, corrosion propensity, and odor are negatively affected. Once fresh water is produced by desalination, this consists of a valuable product. It can be partly used for water splitting processes and partly as drinking water.

The feed water salinity for desalination facilities ranges from approximately 1,000 mg/L TDS to 60,000 mg/L TDS, although feed waters are typically labeled as one of two types: seawater or brackish water. Although most seawater sources contain 30,000–45,000 mg/L TDS, seawater reverse osmosis membranes are used to treat water within the TDS range of 10,000–60,000 mg/L. Brackish water reverse osmosis membranes are used to treat water sources (often groundwater) within a range of 1,000–10,000 mg/L TDS. The feed water type can affect several design choices for a treatment plant, including the desalination method, pretreatment steps, waste disposal method and product recovery (fraction of water that becomes product). The desalination technologies can be categorized as indicated in Fig. 7.23.

Five linkage configurations were considered by Orhan et al. (2010) to link the copper–chlorine water splitting cycle to a desalination plant and nuclear reactor. The options are: (1) using waste heat from a nuclear reactor for desalination processes, (2) using recovered energy from the Cu–Cl cycle for desalination processes, (3) using direct (process) energy from a nuclear reactor for desalination processes, (4) using solar photovoltaics for desalination processes and nuclear energy for copper–chlorine processes, and (5) using off-peak electricity from a nuclear reactor for desalination processes.

A flowsheet example of an integrated copper–chlorine and desalination processes is presented in Fig. 7.24. The selected desalination process is humidification–dehumidification (HD). When air flow is in contact with salt water, the air extracts a certain quantity of water vapor at the expense of sensible heat of salt water, inducing cooling. When distilled water is recovered by bringing humid air near or into contact with a cooling surface, this causes condensation of part of the

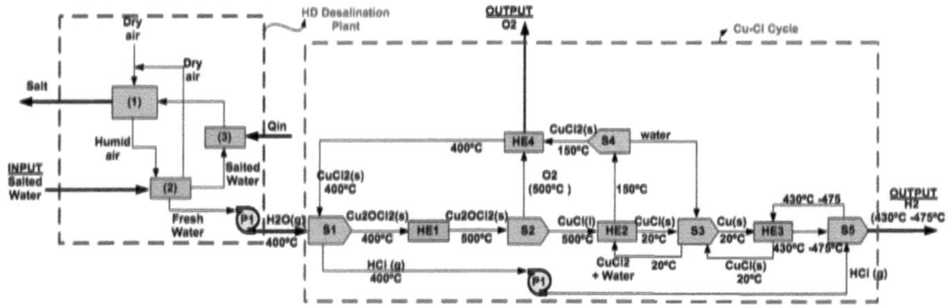

Fig. 7.24 Integrated system of copper–chlorine and humidification–dehumidification (HD) desalination processes [modified from Orhan et al. (2010)]

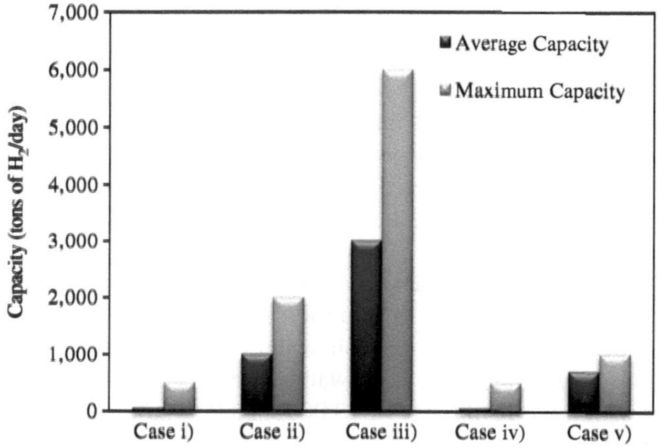

Fig. 7.25 Capacities of desalination plant [data from Orhan et al. (2010)]

water vapor mixed with air. Generally the condensation occurs in another heat exchanger in which salt water is preheated through latent heat recovery. An external heat contribution is thus necessary to compensate for the sensible heat loss. The heat exchangers are often constructed of tubes placed on perforated plates.

In Fig. 7.24, the unit #1 is an air humidifier, unit #2 is an air dehumidifier, and unit #3 is a brine heater. The capacities for each case (see case descriptions above) in the analysis are given in Fig. 7.25, while the specific energy consumptions are shown in Fig. 7.26. This unit cost includes capital costs of both the Cu–Cl cycle and the desalination plant. As shown in the figure, it varies with production capacity but not with the type of desalination method. The unit capital cost of the overall system is similar for every case since the cost contribution of the desalination plant is small compared to the Cu–Cl cycle.

The energy efficiencies of the Cu–Cl cycle, desalination plant, and the overall system, including the Cu–Cl cycle and the desalination plant, are shown in Fig. 7.27. The efficiency of the desalination plant for each case is evaluated based on the energy consumption given in Fig. 7.26. As observed earlier, the effect of the

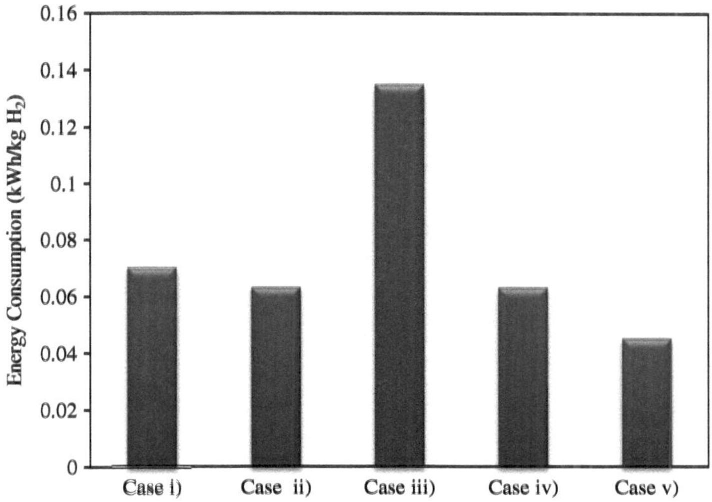

Fig. 7.26 Energy use of desalination process [data from Orhan et al. (2010)]

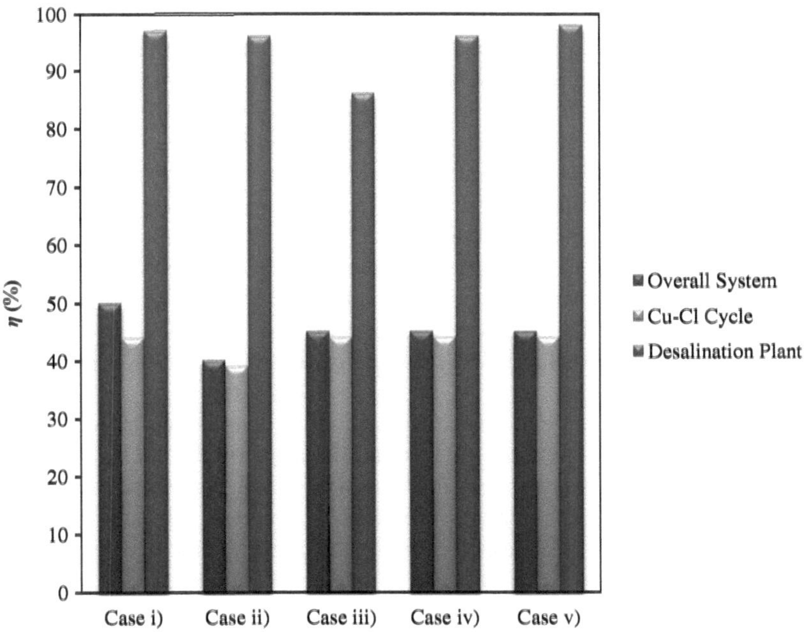

Fig. 7.27 Efficiency of the Cu–Cl cycle, desalination plant, and overall (Cu–Cl cycle and desalination plant) system

Cu–Cl cycle on the overall system is dominant as the desalination plant uses much less energy than the Cu–Cl cycle. Thus, the efficiency of the Cu–Cl cycle and the overall system are similar for each case. For the overall system, Case (i) exhibits a higher efficiency, since waste heat from the nuclear reactor (which is assumed to be "free") is utilized in this option. In Case (ii), recovered heat from the Cu–Cl cycle is used for desalination instead of internally, within the cycle. Thus, the Cu–Cl cycle and the overall system operate at lower efficiencies, since the effect of the cycle is very important on the overall system.

Based on the study, from results detailed in Orhan et al. (2010), the least expensive desalination method is HD at 7 \$/(kg H_2/day) and the technology with the highest initial cost (11.7 \$/(kg H_2/day)) is the multistage flash desalination. The capital costs of vapor compression and reverse osmosis desalination appear to be similar at 9 \$/(kg H_2/day), while the capital cost of multiple effect desalination system is about 10.8 \$/(kg H_2/day).

7.8 Conclusions

In this chapter, several integrated systems for nuclear hydrogen production were presented. In some systems, the only generated product at the output is hydrogen, while in some others, the products are hydrogen and electricity. Both systems are promising for large-scale conversion of nuclear energy in a sustainable manner. In systems that split water to generate hydrogen, there is a major factor of importance, which is the quality of water. Normally pretreated, clean water must be used for the process. Water pretreatment may be costly but it is necessary, especially in the regions where fresh water is not available. Systems that desalinate water or can handle brackish water are promising emerging technologies. Also, it appears economically justifiable to have integration of nuclear hydrogen pants with desalination systems which generate fresh water for two purposes: (1) hydrogen production via water splitting and (2) production of drinking water.

In the first part of the chapter, next generation nuclear reactors were discussed and their diagrams were shown. Next, the chapter studied a number of nuclear-based integrated systems including those with natural gas or coal conversion, high-temperature electrolysis, sulfur–iodine systems, hybrid sulfur systems, copper–chlorine cycle systems.

The cost of hydrogen generation for each option was indicated for each case. Regardless of the option chosen to generate hydrogen with nuclear energy, the production costs were very similar, at around US\$ 2/kg. This value is competitive with common methods of hydrogen production from coal and natural gas, especially at large production scales. If carbon taxes are implemented by authorities, then the economics becomes significantly in favor of the nuclear-driven process.

References

Abu-Khader MM (2009) Recent advances in nuclear power: A review. Prog Nuclear Energy 51: 225–235

Alemberti A, Carlsson J, Malambu E, Orden A, Struwe D, Agostini P, Monti S (2011) European lead fast reactor—ELSY. Nucl Eng Des 241:3470–3480

Delpech S, Merle-Lucotte E, Heuer D, Allibert M, Ghetta V, Le-Brun C (2009) Reactor physics and reprocessing scheme of innovative molten salt reactor system. J Fluorine Chem 130:11–17

Elder R, Allen R (2009) Nuclear hydrogen production: Coupling a very high/high temperature reactor to a hydrogen production plant. Prog Nuclear Energy 51:500–525

Granovskii M, Dincer I, Rosen MA, Pioro I (2008) Performance assessment of a combined system to link a supercritical water-cooled nuclear reactor and a thermochemical water splitting cycle for hydrogen production. Energ Convers Manag 49:1873–1881

Khartabil H (2009) SCWR: Overview. Generation IV International Forum, Paris, 9–10 September 2009

Kirchhoff R, Van Heek KH, Jüntgen H, Peters W (1984) Operation of a semi-technical pilot plant for nuclear aided steam gasification of coal. Nucl Eng Des 78:233–239

Naidin M, Mokry S, Baig F, Gospodinov Y, Zirn U, Pioro I, Naterer G (2009) Thermal design options for pressure channel SCWRs with cogeneration of hydrogen. J Eng Gas Turbines Power 131:012901

Ohashi H, Inaba Y, Nishihara T, Inagaki Y, Takeda T, Hayashi K, Katanishi S, Takada S, Ogawa M, Shiozawa S (2004) Performance test results of mock-up test facility of HTTR hydrogen production system. J Nuclear Sci Technol 41:385–392

Orhan MF, Dincer I, Naterer GF, Rosen MA (2010) Coupling of copper-chlorine hybrid thermochemical water splitting cycle with a desalination plant for hydrogen production from nuclear energy. *Int J Hydrogen Energy* 35:1560–1574

Rastoin J, Malherbe J, Pottier J, Lecoanet A (1979) Nuclear methane reforming for coal gasification. Int J Hydrogen Energy 4:535–540

Richards M, Shenoy A, Schultz K, Brown L, Harvego E, McKellar M, Coupey J-P, Reza SMM (2005) H_2-MHR conceptual designs based on SI Process and HTE. Nuclear production of hydrogen, third informational exchange meeting, Oarai, Japan, 5–7 October 2005

Sakaba N, Kasahara S, Onuki K, Kunitomi K (2007) Conceptual design of hydrogen production system with thermochemical water-splitting sulfur-iodine process utilizing heat from the high-temperature gas-cooled reactor HTTR. Int J Hydrogen Energy 32:4160–4169

Schrader L, Strauss W, Teggers H (1975) The application of nuclear process heat for hydrogasification of coal. Nucl Eng Des 34:51–57

Tsvetkova GV, Yang WS, Wade DC, Peddicord KL (2003) Reactor physics feasibility analysis of long-lived STAR-H_2 system. Trans Am Nuclear Soc 88:685–686

Van Rooijen WFG (2009) Gas-cooled fast reactor: a historical review and future outlook. Hindawi Publishing Corporation. Science and Technology of Nuclear Installations 965757

Varrin RD, Reifsneider K, Scott DS, Irving P, Rolfson G (2011) NGNP hydrogen technology down selection, Idaho National Laboratory

Verfondern K (2007) Nuclear energy for hydrogen production. Writings of Research Center Jülich, Energy Technology, Volume 58, Research Center Jülich GmbH: Jülich (Germany). http://juwel.fz-juelich.de:8080/dspace/bitstream/2128/2518/1/Energietechnik_58.pdf accessed in March 2012

Wang ZL, Naterer GF, Gabriel KS (2010) Thermal integration of SWCR nuclear and thermochemical hydrogen plants. Second Canada-China Joint Workshop on Supercritical Water Cooled Reactor, Toronto ON, 25–28 April 2010

Chapter 8
Future Trends and Emerging Opportunities with Nuclear Hydrogen Production

Abstract Nuclear hydrogen production includes technologies with both old roots and new emerging directions: nuclear power plants and hydrogen production processes. Hydrogen production from water or hydrocarbons has been established since the early twentieth century. Also, nuclear energy for commercial electricity generation at a large scale has been established over 50 years ago. Novel technologies of hydrogen production at a large scale have emerged recently such as high-temperature electrolysis and thermochemical cycles. A new generation of nuclear reactors is being developed through the Generation IV International Forum (GIF) with advanced designs and innovations for high-temperature process heat generation and advanced fuel cycles. Both fast and thermal neutron spectra are under development. In addition, nuclear fusion to generate high-temperature heat has achieved significant progress and is viewed as a major technology of the future. In this chapter, various trends and future emerging opportunities with nuclear hydrogen production are reviewed. One of the important issues is the development of advanced fuels for thermal and fast neutron spectra. TRISO fuel particles are one of the promising developments while pebble bed reactor designs are also significant. Intermediate heat exchangers will have a crucial role for heat transfer and radioactive particle isolation from the nuclear core to downstream processes. Advanced power generation cycles including combined and supercritical cycles represent an important development for effective power and hydrogen cogeneration.

8.1 Introduction

Many countries and regions such as the USA, Canada, Russian Federation, China, Japan, Korea, India, France, Switzerland, European Union, and others are actively engaged in the development of nuclear plants for power, process heat, hydrogen, and synthetic fuel production as part of their energy future. There are strong reasons to believe that nuclear reactor technology will expand and diversify further by

G.F. Naterer et al., *Hydrogen Production from Nuclear Energy*,
DOI 10.1007/978-1-4471-4938-5_8, © Springer-Verlag London 2013

including—beside electricity generation—process heat, hydrogen, and synthetic fuel production.

Sustainable development in the world is influenced by many constraints and factors. One major constraint is the continual diminishing of fossil fuel resources and conventional nuclear fuel reserves. Middle East countries which possess oil and natural gas in abundant quantity and excellent quality have intergenerational/sustainability issues because excessive resource exploitation today may lead to jeopardizing the access of future generations to valuable resources. Solar energy is also abundant in Gulf Region countries and it has not been fully harnessed. Similar situations can be found in North African countries which are relatively rich in hydrocarbons and have desert areas with abundant solar radiation and inexpensive land.

Another major global issue is related to rapid population growth at an accelerated pace which is also related to increased energy consumption. This is increasingly evident as more energy-consuming commodities (electronic and communication devices, various appliances, vehicles, etc.) are used today per capita than any time in the past. The worldwide energy demand spans from electricity to space heating, water heating, vehicle propulsion, and energy needs to produce materials from which all types of commodities are produced. Hydrogen has a key role because it is an energy carrier and valuable chemical feedstock at the same time.

Hydrogen is also needed for the fabrication of many types of materials and chemicals used by the modern world. For instance, as a chemical feedstock, hydrogen is a key element to produce synthetic fuels by combining hydrogen with carbon monoxide or carbon dioxide (which can be recovered from air or emission stacks of power plants). This yields a range of fuels such as synthetic natural gas, synthetic methanol, synthetic gasoline, synthetic diesel, etc. Another major use of hydrogen is for clean fertilizer production such as ammonia and urea. Furthermore, hydrogen is a crucial component in industrial processes, in the petrochemical industry, in the chlorine–sodium industry, for plastic material production, or in the food industry.

A major global concern is climate change and environmental pollution, which are interrelated. These have major importance in anthropogenic activity to not generate imbalances in the environment. From this perspective, production of hydrogen from water followed by its use and a clean energy carrier can become one of the major solutions. Through the use of hydrogen (or synthetic fuel produced with hydrogen and carbon dioxide), one can replace the polluting processes of fossil fuel combustion—largely used in industry—with clean processes which can generate high-temperature heat or electricity without environmental emissions.

Nuclear power together with solar energy will become two of the key pillars to generate hydrogen at a large scale. Other renewable energy sources such as wind, geothermal, OTEC, etc. can be used on a local or a regional scale. Solar energy has wider availability despite its intermittent nature. Solar energy illuminates most of the earth's surface. Technological challenges need to still be overcome to make large-scale, solar hydrogen generation economically competitive. This issue is related to the cost of land as well as the progress of light harnessing technology, as well as generating high-temperature heat to split water with thermochemical cycles. Photochemical and photovoltaic methods of solar hydrogen generation are also emerging and promising technologies.

As compared to solar energy-based methods and other renewable energy methods, production of hydrogen by nuclear energy is the most compact, in terms of production capacity per overall carbon footprint of the plant. Nuclear fuel is a concentrated source of energy. Nuclear reactor sites including the reactor building, power generation plant, and (in the future) hydrogen production plant occupy a relatively small amount of land with respect to the production capacity. For example, a nuclear plant generates about 1 MW power per acre (capacity factor >0.8), whereas concentrated solar power plants generate at most 25 kW per acre (capacity factor of 0.3) and wind farms only 5 kW per acre (capacity factor of 0.3).

Nuclear fuels (uranium, thorium) are not widely available but rather found in certain countries such as Australia, Kazakhstan, the USA, India, Russian Federation, Canada, Niger, Uzbekistan, Namibia, China, Brazil, Ukraine, and South Africa. Despite the regional availability, other countries are interested in the development of nuclear energy programs, although they have limited fuel reserves (e.g., France or Japan).

The future development of nuclear reactors will be related to improvement of safety issues, fuel cycles, and conversion efficiency (e.g., power or hydrogen production). Safety is crucial for political and public acceptance of nuclear technology. For better safety, the next generation of nuclear reactor designs has hydrogen production plants that are placed remotely from the nuclear reactor. More developments are planned in advanced fuel cycles and this chapter discusses these issues. For example, an emerging trend is to develop smaller capacity modular reactor units with long refueling periods of a few decades.

Although nuclear hydrogen production can be a major solution of large-scale hydrogen production at central locations, there are diverging views regarding the role of hydrogen in a future sustainable energy economy. On one side, the proponents of a hydrogen economy concept assume that hydrogen can be used as an energy carrier—or fuel—in vehicular applications in the future. However, it is well known that hydrogen is difficult to store and its energy density by volume is relatively low. So skeptics say that a hydrogen economy is not feasible, or at least hydrogen cannot be used to store energy, but rather a more compact way of energy storage is required. Production of synthetic fuels starting from hydrogen may become a more promising route for hydrogen generation at a large scale. In this concept, once produced, hydrogen is not stored but rather converted into a more compact synthetic fuel. It is important, however, that all of the production chain—hydrogen generation and fuel synthesis—is sustainable with reduced or no pollution to the environment. The first step in such a process is the sustainable production of hydrogen.

The previous chapters of this book discussed in detail the major paths of hydrogen production using thermal energy derived from nuclear reactors. Although there are many possible thermochemical cycles and other methods of hydrogen production, relatively few methods are economically feasible for nuclear hydrogen production. In summary, these include water electrolysis (alkaline or HTSE), thermochemical cycles, hybrid cycles, and coal or natural gas reformation using heat derived from nuclear energy. At this time, one purely thermochemical cycle appears closest to commercial viability: the sulfur–iodine cycle. Two major hybrid thermochemical cycles are under development: the hybrid sulfur cycle and the copper–chlorine cycle. Some other cycles have recently emerged, but they are in early conceptual phases of development and proof of concept. The limitations of the available technologies imply that system integration may be the key pathway towards improving overall system efficiency and reducing the hydrogen production cost. Multiproduct generation and hybridization can increase the economic attractiveness for large-scale hydrogen production.

This last chapter of this book aims to provide an overall perspective on future development of nuclear hydrogen generation technologies and research trends. In the chapter, future trends in the development of nuclear fuels and fuel cycles for fission reactors are presented. It is followed by a section referring to fusion reactor development for hydrogen generation.

8.2 New Emerging Systems and Advancements

Several design strategies and technological advances can be observed with nuclear fission reactors for high-temperature heat generation. The use of a helium coolant and graphite moderator represents a promising choice for improved safety and efficiency with high-temperature reactors. The nuclear fuel in new reactors must be packed in special materials, generally different than zirconia, which is customarily used in most commercial reactors currently in operation. It is important to maintain the coated fuel temperature below the safety limits. An annular core design is promoted since it offers a passive heat decay removal. In this system, the gaseous coolant flows in a high-temperature, high burn-up, porous ceramic core.

Tristructural isotropic (TRISO)-coated fuel particles were proven in test reactors to safely operate at high temperature. These fuel cartridges comprise a uranium oxide core coated with four layers of isotropic materials formed in porous carbon, permeable to fission products, which are embedded in pyrolytic carbon and a silicon carbide layer and an outer pyrolytic carbon shell. The outer layer of fuel cartridge acts as a pressure vessel since it is impermeable to any fission products up to temperatures higher than ~1,900 K.

TRISO particles are capable of retaining the fission fragments under any possible circumstances as they have a strong negative coefficient of reactivity. Moreover, the reactor core configurations are such that in case of an accident, there is no need of active cooling as decay heat is transferred by heat conduction from the core to the exterior. In pebble bed reactors, the TRISO particles are formed in 6 cm pebbles which are fed to the top and they advance slowly towards the bottom where they are extracted in a safe manner. Since the particles can be removed, any special reactor shutdown becomes unnecessary.

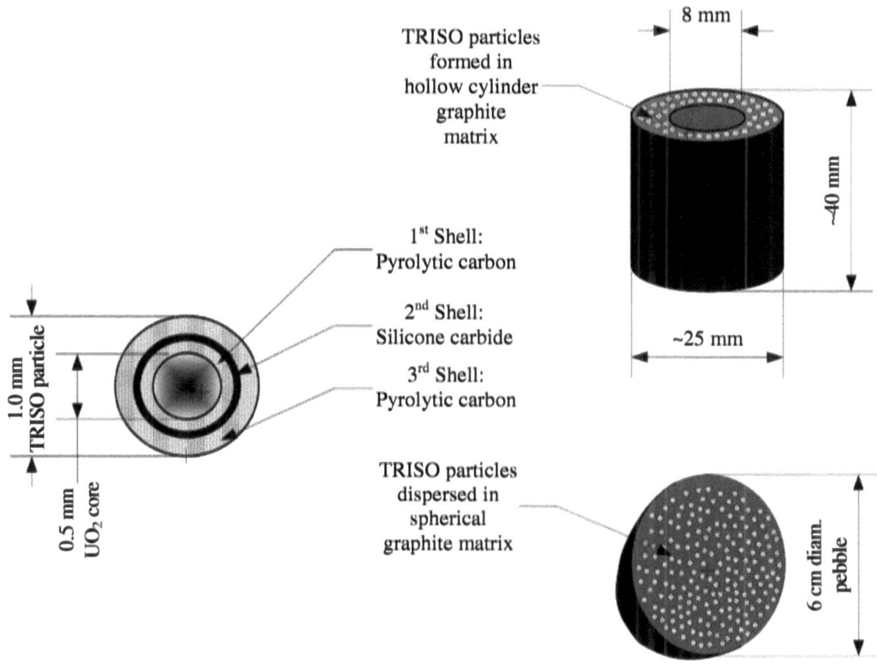

Fig. 8.1 Schematic of a TRISO fuel particle for fission reactors with enhanced safety

Figure 8.1 shows two options to assemble TRISO fuel particles in a fuel element. The dimensions of a TRISO particle are about ~1 mm diameter and there is a core of 0.5 mm uranium dioxide (fissile material). In a first option, the particles are formed in a hollow cylindrical rod with approximate dimensions of 40 mm length, 8 mm inner diameter, and 25 mm diameter. The cylindrical particles are placed in a prismatic graphite block of hexagonal form of ~50 cm width and ~70 cm height. The prismatic block constitutes a fuel element. Helium coolant is circulated such that it cools the fuel rods at their inner surface (Richards et al. 2006).

A second option is to assemble TRISO particles in a spherical matrix pebble as indicated in Fig. 8.1 (bottom right). In this case, the reactor core consists of a packed bed of pebbles and helium coolant is forced to flow through the void space between the spherical pebbles (Matzner and Wallace 2005).

A design challenge with linking the nuclear reactor to a downstream process such as hydrogen production is the possibility of tritium escape in the primary loop of coolant. Tritium is generated in the reactor core from side nuclear reactions involving lithium, helium, and boron atoms, whereas helium is the coolant. In some instances, tritium atoms—which are radioactive—can escape from the primary to secondary circuit through the intermediate heat exchanger (IHX) wall and reach the secondary helium circuit from where they contaminate radioactively the system by traversing the metallic wall of the IHX. As reported by Elder and Allen (2009), in a

Table 8.1 Main design parameters of IHX

Parameter			GA	JAEA	EURATOM	AREVA-NP
Capacity, $\dot{Q}$ (MW thermal)			600	170	613	608
LMTD (K)			24	154	48	50
Primary	Flow rate, $\dot{m}$ (kg/s)		321	324	210	240
	In	P (MPa)	7.03	5.0	5.5	5.5
		T (K)	1,223	1,223	1,223	1,123
	Out	P (MPa)	7.0	4.95	5.4	5.4
		T (K)	863	1,123	660	623
Secondary	Flow rate, $\dot{m}$ (kg/s)		321	80	219	614
	In	P (MPa)	7.10	5.15	5	5.5
		T (K)	838	773	623	573
	Out	P (MPa)	7.07	5.0	4.9	5.4
		T (K)	1,998	1,073	1,163	1,073

Source: Elder and Allen (2009)

VHTR, the release of tritium in the primary circuit is of order 60 μCi/s (Ci = Curie) whereas the tritium contamination of the hydrogen produced by a thermochemical plant connected to the IHX is of order 50 mCi per ton of hydrogen product. This is far below the maximum admissible limits of contamination, but it must be considered thoroughly for design purposes. One potential design measure for preventing tritium migration to the secondary circuit is by pressurizing the secondary circuit more than the primary circuit. The material selection of the IHX is also important; hastelloy XR is an option (Elder and Allen 2009). The IHX must be also a compact design with a high overall heat transfer coefficient, capable to transfer hundreds of megawatts of thermal energy from the primary to secondary side with a reduced temperature difference. Table 8.1 shows operating parameters of the IHX of four different concepts proposed in past literature.

Another measure to reduce contamination is by including helium purification systems in the primary and secondary coolant loops. Isolation vales must be included in the secondary loop lines to stop the coolant flow in case of pipe rupture and isolate the primary loop to prevent a release of coolant—which may be contaminated with fission products that are harmful to the environment. There are severe design constraints for the valve as it must operate at very high temperatures. Research activities are under way for finding suitable coating materials to increase the durability of the valves.

The Atomic Energy Commissariat (CEA) of France supports developments of AREVA's VHTR via several research projects including the design and testing of innovative concepts of crucial elements of helium circuits. As mentioned by Carre et al. (2010), a new concept of a compact IHX of high capacity (800 MW thermal energy) has been recently developed. The system is presented in a simplified diagram in Fig. 8.2. It shows the heat exchanger, primary isolation valve, secondary isolation valves, and the helium circulator of the primary loop; all elements are part of a compact assembly, denoted as IHX. Hot helium from the reactor enters the IHX assembly from the side via a tube that turns up inside the vessel and directs the gases

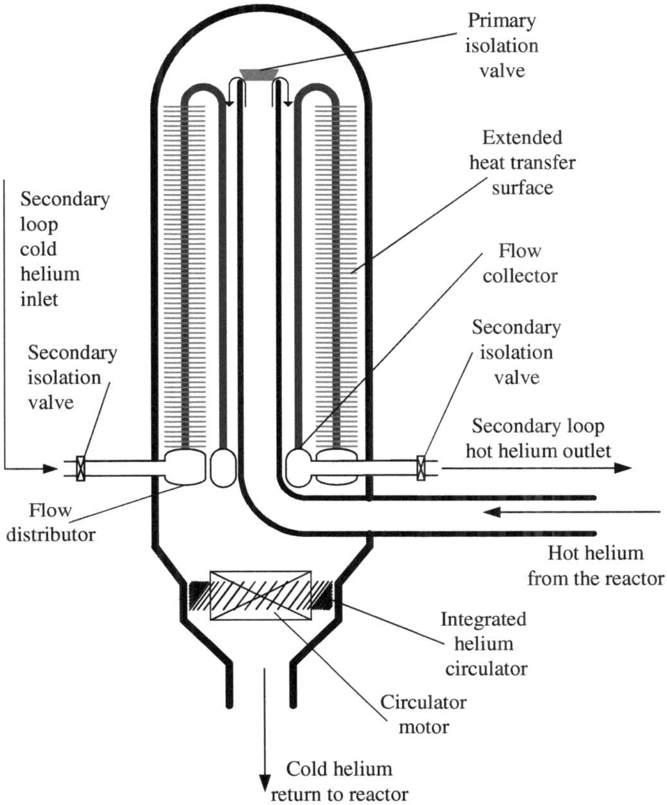

Fig. 8.2 IHX concept developed at AREVA-NP/CEA (modified from Carre et al. 2010)

towards the top of the reactor where it is placed with an isolation valve for the primary circuit. The hot gases from the top plenum are then forced downwards and across a finned tube surface for enhanced heat transfer with helium from the secondary loop. At the bottom of the vessel, the cooled gases are pressurized by the helium circulator which has a compact construction together with its motor. The secondary helium enters at lower temperatures from the side of the IHX vessel and it is fed to a distributor which diverts the flow in a number of vertical tubes with finned surfaces. Helium flows upwards as indicated and then it returns downwards. A the lower side of the heat exchanger module, there is a collector which collects the flow from all parallel tubes and directs it towards the exit port. A secondary isolation valve is placed at the exit port. The IHX vessel is made of industrial grade ferritic steel (9Cr1Mo) with 7.3 m inner diameter and a height of 20 m, weight of around 1,000 t, and transferring 0.8 kW thermal energy per kg with a corresponding power density of approximately 0.94 MW/m^3.

Another promising option for IHX is a printed circuit technology which consists of stacking steel plates via a diffusional bonding process. The plates have small

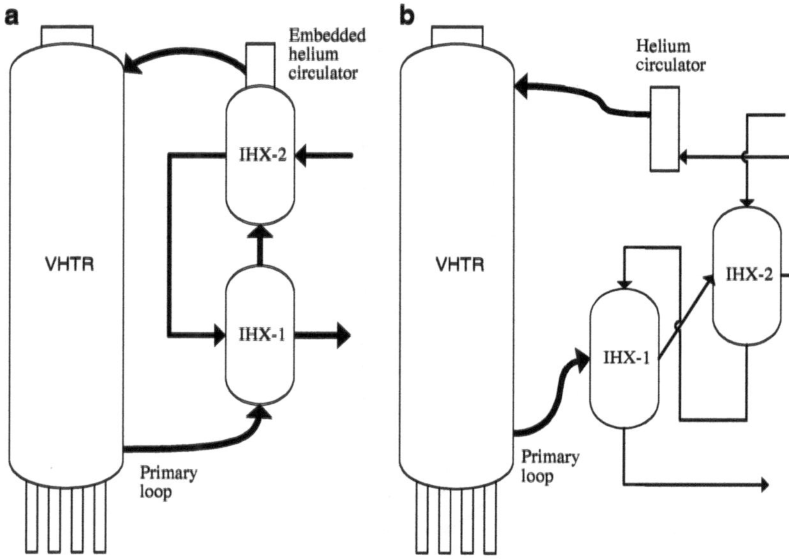

Fig. 8.3 Flow configuration for high-temperature and low-temperature IHX systems: (**a**) core-side configuration; (**b**) shell-side configuration

flow channels obtained by mechanical pressing or chemical etching. The advantage of this concept is approximately six times higher compactness than a conventional tubular IHX for the same effective heat transfer rate (Mizia 2008). Two types of printed circuit heat exchanger (PCHE) concepts exist, namely, the Toshiba concept and the Heatric Corporation concept. The IHX is formed with PCHE modules linked with flow distribution and collection lines and eventually embedded in an outer metallic shell. There are two IHX systems installed in series: one at the lower temperature side and with an extended lifetime of 60 years, and the other is a replaceable heat exchanger placed at the high-temperature end. This heat exchanger has a shorter lifetime due to the higher operating temperature and more severe conditions.

According to Mizia (2008), two types of configurations—core side and shell side—are considered for development. A core-side configuration is shown in Fig. 8.3a and the primary circuit has flow channels at the core side of the IHX. The hottest gas is circulated in the middle of the IHX, while the gas of the secondary loop comes from the outer site, which circulates through the shell. Inside the shell, there are PCHE modules. The low-temperature IHX (namely, IHX-2 in the figure) has an embedded helium circulator for the primary loop. The shell-side configuration is presented in Fig. 8.3b. In this case, the hot fluid of the primary helium loop is circulated through the heat exchanger's shell side from where it is forced to flow through the PCHE module and exits from the opposite part of the shell side. In the core, the secondary helium flows at lower temperature and it includes the shell distribution and collection systems for the secondary flow.

Table 8.2 Parameters of a 224 MW IHX for linking SCWR to the thermochemical cycle

Inner tube (supercritical water)		Annulus side (water → superheated steam)		Overall heat exchanger	
$\dot{m}$ (kg/s)	243	$\dot{m}$ (kg/s)	289	$\dot{Q}$ (kW/pipe)	77.7
P (MPa)	25	P (MPa)	5	$\dot{Q}$ (MW total)	224
T_{in} (K)	898	T_{in} (K)	557	No. of pipes	2,890
T_{out} (K)	678	T_{out} (K)	873	U (W/m^{2}K)	879
d_i (mm)	18.9	D_i (mm)	30.1	A (m^2/pipe)	1.93
d_o (mm)	26.7	D_i (mm)	33.4	L (m)	23
G (kg/m^{2}s)	300	G (kg/m^{2}s)	659	q'' (kW/m^2)	40.2
V_{max} (m/s)	4	V_{max} (m/s)	52	q''' (MW/m^3)	88

Source: Mokry et al. (2012)

Another type of IHX is for linkage of a supercritical water-cooled reactor (SCWR) to a thermochemical water-splitting plant. Mokry et al. (2012) have developed a conceptual design of a heat exchanger that uses supercritical water in the primary circuit and generates superheated steam at the secondary circuit. The design is based on experimental data which was correlated by the Buckingham Π theorem. The correlation for thermo-hydraulic calculations with 25 % uncertainty of the heat transfer coefficient has been developed. The IHX design is a double-pipe, counterflow type with a secondary circuit operating at 5 MPa.

The Mokry et al. (2012) correlation for the supercritical side of the heat exchanger with forced convection flow is described by

$$\text{Nu} = 0.0061\, Re^{0.904}\, Pr^{0.684} (\rho_w/\rho)^{0.564}, \tag{8.1}$$

where Re is Reynolds number and Pr is Prandtl number as calculated for the bulk flow.

For the Prandtl number, an average value is considered using the mean thermal conductivity along the heat exchanger pipe. The density of supercritical water is calculated for the bulk flow. The index w in (8.1) refers to flow conditions at the wall. The geometrical parameters of the IHX for the integrated SCWR/thermochemical plant are presented in Table 8.2. A very high compactness of the SCWR-IHX is obtained with respect to the VHTR-IHX, namely, ~88 MW/m^3 vs. ~1 MW/m^3 for a conventional tubular IHX or ~6 MW/m^3 for a PCHE-based design because of the significantly higher density of water.

The transfer of heat from the secondary loop to a hydrogen production process presents various challenges. In general—with high-temperature gas reactors—it implies the transfer of heat from flowing helium gas to a chemical reaction. In the case of SWCR, the heat transfer occurs from superheated steam at intermediate or low pressure to the endothermic chemical reactions of the hydrogen production process. Many thermochemical processes involve multiphase flows such as solids, liquids, gases, and often with corrosive media. Moreover, heat is transferred to a catalytic or non-catalytic reactor of various types including packed bed or fluidized bed, etc. A version of the IHX for SWCR that uses hot helium at the secondary side has been conceptually designed as a shell-and-tube configuration (whereas in a pipe, it flows supercritical water in the shell, it is helium at 2 MPa). This was reported in a recent paper by Abedin et al. (2012).

Significant advances were reported by the Japan Atomic Energy Agency (JAEA) in the development of the sulfuric acid decomposition unit for the S–I hydrogen production cycle, which reported an emerging design as shown in Fig. 8.4. The decomposition reactor recovers heat from the secondary helium loop and provides the required thermal energy input for the chemical reaction. A prototype was constructed at JAEA for testing at low-scale production as reported by Terada et al. (2005). Recent progress in the development of the JAEA reactor was reported by Ogawa et al. (2009). The main challenge has been to cope with a highly corrosive behavior of the reactant while facilitating a high heat transfer rate through the wall material. In a JAEA design, the reactor is made of two solid blocks of silicon carbide (SiC). The two blocks are piled vertically using a pure gold plate with holes as the sealant. Both the upper and lower blocks have two types of holes, namely, through-holes for the sulfuric acid decomposition side and shorter bores for the helium heat transfer fluid side. The rows of through-holes and bore-holes alternate such that heat is transferred from helium to the reacting stream.

In this reactor configuration, hot helium enters from above in a ring-type distributor which is detailed in a cross section at the upper left of Fig. 8.4. After its distribution, the hot helium flows downwards through the vertical tubes (see figure). Heat is transferred to the SiC block and eventually to the process stream. At the bottom of the lower SiC block, there is a collector for the colder helium stream which diverts the flow towards the annular space between the reactor block and outer mantel. Furthermore, helium flows upwards in the annular space.

The core reactor block is thermally insulated to impede heat transfer between the uprising helium and the process flow inside the SiC block. Cold helium is extracted from the upper side of the reactor's outer shell. The reactant—sulfuric acid—is fed at the bottom and distributed to each of the vertical tubes via a plenum-type distributor. After sufficient heating, the following reaction proceeds: $H_2SO_4(g) \rightarrow SO_3(g) + H_2O(g)$. In principle, the catalyst can be packed in the vertical tubes to enhance the reaction rate. The product stream is collected in an upper plenum (see the figure) and delivered at an exit port.

In a hybrid sulfur water-splitting cycle, there is a similar decomposition reaction of sulfuric acid, with a difference that the process is coupled with SO_3 conversion to SO_2 in the same reactor. The products are SO_2, O_2, and H_2O, according to $H_2SO_4 \rightarrow 0.5O_2 + SO_3 + H_2O$. An emerging version of hybrid sulfur process—denoted as once-through (Ot-HyS)—was proposed by Jung and Jeong (2010). As opposed to the HyS process, which splits water and recycles all other chemicals, in the Ot-HyS process, sulfur and water are consumed to generate two products: hydrogen and sulfuric acid. This process can be justified from ecological and economical perspectives wherein sulfur is a recovered product from desulfurization of fossil fuels. This is currently practiced in many refineries for processing of petroleum and natural gas. It is also a major global source of sulfur. Recovered sulfur is combusted within the Ot-HyS process with oxygen from air to form sulfur dioxide (SO_2).

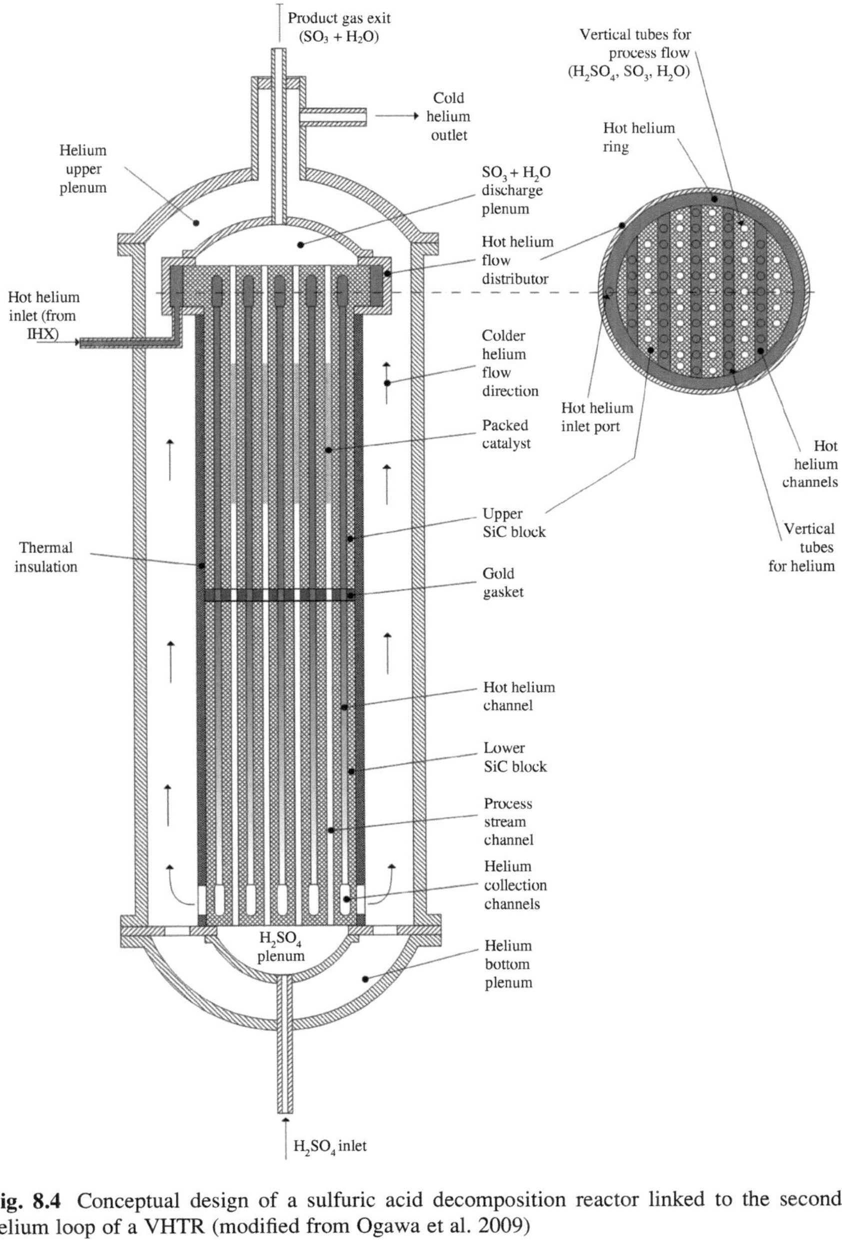

Fig. 8.4 Conceptual design of a sulfuric acid decomposition reactor linked to the secondary helium loop of a VHTR (modified from Ogawa et al. 2009)

The SO_2 is combined with water in a depolarized water electrolysis process—the same as in the HyS cycle. The electrolysis generates hydrogen and sulfuric acid as separate streams. There is some process heat required for sulfuric acid concentration which can be derived from intermediate-temperature nuclear reactors or sulfur combustion processes.

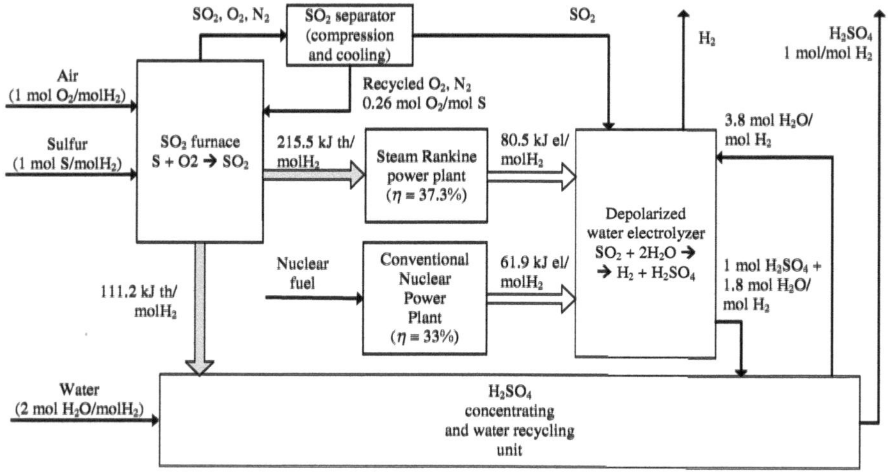

Fig. 8.5 Simplified diagram of Ot-HyS process (modified from Jung and Jeong 2010)

A simplified diagram of the Ot-HyS process is illustrated in Fig. 8.5. It involves a typical (present generation) nuclear power plant with an efficiency of 33 %, sulfur dioxide furnace, sulfur dioxide separation unit, steam Rankine power plant driven with heat generated by the sulfur dioxide furnace, sulfuric acid concentration and water recycling unit driven with heat recovered from the sulfur dioxide furnace, and a depolarized water electrolyzer unit driven by electricity generated jointly by the sulfur dioxide furnace and the nuclear power plant. The overall system generates 1 mol of hydrogen and 1 mol of sulfuric acid. It consumes 215.5 kJ thermal energy from the sulfur dioxide furnace and 185.7 kJ thermal energy from the nuclear fuel. Thus, the overall efficiency of the plant is estimated to be 47 % on an LHV basis for hydrogen. If the efficiency of the nuclear power plant is improved to 42 %, the LHV efficiency of the hydrogen production process becomes 50 %. This figure is compared to 34 % for the HyS system flow sheet simulation by Savannah River National Laboratory (SRNL) and 26–28 % nuclear based electrolytic hydrogen production efficiency (see Jung and Jeong, 2010).

8.3 Advanced Thermodynamic Cycles Coupled to Fission Reactors and Thermochemical Plants

Improvement of the energy and exergy efficiencies of integrated nuclear plants for hydrogen and power generation is crucially important in order to reach competitiveness with respect to fossil fuel-based processes such as coal plants, coal gasification, or natural gas reforming. Advanced combined cycles or other advanced cycles in conjunction with nuclear reactors represent a key method that assures a maximum efficiency of power generation. Various schemes were proposed in past

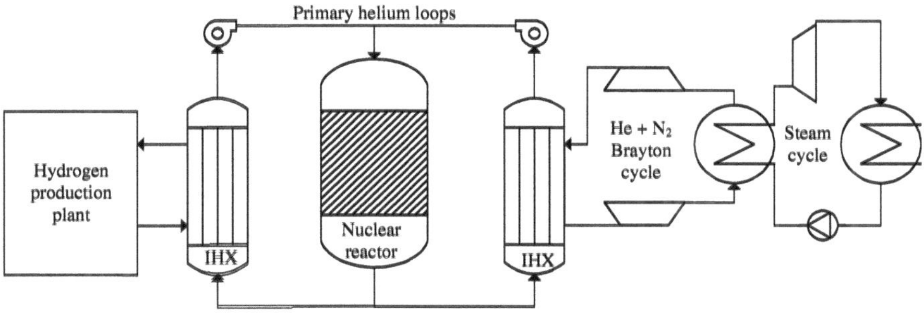

Fig. 8.6 Simplified diagram of nuclear hydrogen production system using a combined, indirect Brayton (He + N_2) and steam Rankine cycle

studies such as cascaded Brayton and Rankine cycles. The following working fluid combinations are the most common: helium–steam, helium and nitrogen mixture in a topping cycle and steam at the bottoming cycle, organic fluids in organic Rankine cycle (ORC), and CO_2 in a supercritical cycle. Two methods of coupling the cycle to the nuclear reactor will be considered: direct and indirect (via an IHX).

An example of a combined cycle indirectly coupled to the nuclear reactor and comprising a Brayton topping cycle which operates with a helium–nitrogen mixture and a steam Rankine cycle at the bottom was presented by Dumaz et al. (2007), see Fig. 8.6. The mixture of nitrogen–helium is attractive as a working fluid because it can be modified to have similar properties to air; therefore one can use existing turbomachinery technology for the Brayton cycle components. Moreover, the steam bottoming cycle benefits from widely available steam turbines and other specific equipment. According to a study by Dumaz et al. (2007), this cycle can reach a power generation efficiency of about 45 %.

A combined Brayton cycle with an ORC bottoming cycle of an indirect type was proposed by Pérez-Pichel et al. (2011). The working fluid for the Brayton cycle is helium gas, while the working fluid for ORC is R125. This design can benefit from the already existing turbomachinery technology for both the helium loop and ORC loop. The energy efficiency of power production, as predicted by Pérez-Pichel et al. (2011), is about 35 %. In Fig. 8.7, the IHX is the IHX which transfers heat from the primary helium loop to the Brayton cycle. In the figure, after the helium turbine, the expanded helium passes to a heat recovery heat exchanger and cools down while reheating the working fluid to the reentry temperature at the IHX. The HRVG is the heat recovery vapor generator for the ORC and designated "COOL." The precooler and intercooler reduce the temperature of the working fluid prior to the first and the second stages of compression, which take place in the low-pressure compressor (LPC) and high-pressure compressor (HPC), respectively. The ORC has a configuration with one turbine, one condenser ("COND"), and one pump.

A newly emerging power conversion technology with intermediate-temperature gas-cooled fast reactors was analyzed by Pope et al. (2009). It consists of a recompression cycle with supercritical carbon dioxide. In this cycle, carbon dioxide

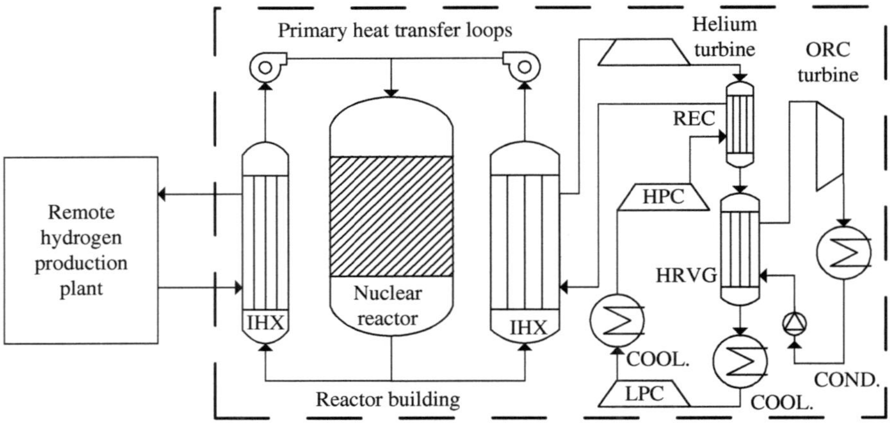

Fig. 8.7 Combined gas-turbine and organic Rankine cycle system for nuclear power and hydrogen production (modified from Pérez-Pichel et al. 2011)

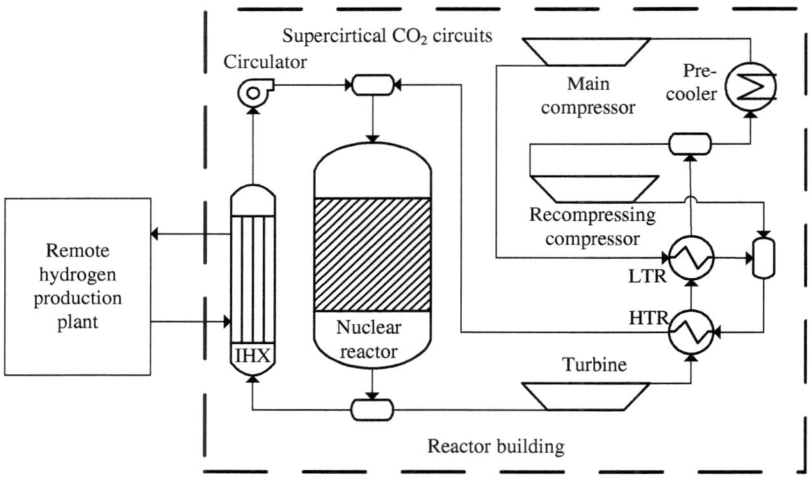

Fig. 8.8 Recompression cycle with supercritical carbon dioxide for gas-cooled nuclear reactor with hydrogen cogeneration (modified from Pope et al. 2009)

is maintained supercritical in all thermodynamic states. At the lowest temperature end of the cycle, the thermodynamic state of the working fluid is slightly above the critical point at 7.69 MPa. The compressibility is very low with a fluid density of about 600 kg/m^3, which increases to about 720 kg/m^3 during the recompression at 20 MPa. A schematic of the cycle is illustrated in Fig. 8.8. This cycle is a direct type: it is directly coupled to the nuclear reactor. In a system which couples the reactor to the power plant and nuclear plant, the supercritical carbon dioxide, which cools the nuclear core, is split into two directions: to the IHX which transfers heat to

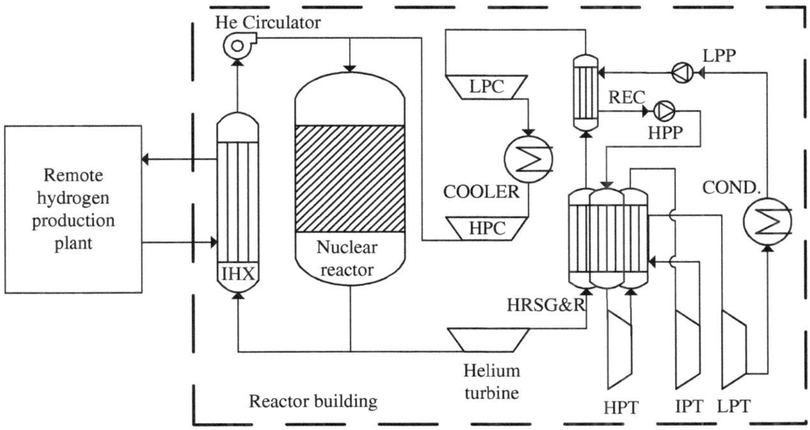

Fig. 8.9 Directly linked power plant with combined Brayton and double-reheat steam Rankine cycles for nuclear power and hydrogen production (modified from Elder and Allen 2009)

the hydrogen plant and the turbine which expands the working fluid and generates useful work. Heat is recovered from the expanded flow in two heat exchangers coupled in series—the high-temperature recuperator (HTR) and the low-temperature recuperator (LTR). Thereafter, a portion of 42 % of the working flow is recompressed while the main portion of 58 % is cooled to ~305 K and compressed at high efficiency. The main stream after the compressor is heated in LTR and it joins the stream from the recompressing compressor; thereafter, the overall flow is heated in HTR and returned at the prescribed parameters (pressure, temperature) for the reactor inlet. It was reported by Pope et al. (2009) that the net power generation efficiency of this system is 48 % assuming that the temperature at the turbine inlet is 923 K and the pressure is 19.45 MPa. The expansion ratio of the turbine is 2.4. If the power cycle is coupled indirectly to the reactor, about 2–3 % efficiency losses can be estimated, as reported by Hejzlar et al. (2006).

Directly linked combined cycles for power generation show the highest potential for efficient power generation in an integrated system for nuclear hydrogen and electricity production. Figure 8.9 illustrates a simplified diagram of a combined cycle that uses a helium-based Brayton cycle (topping) and a double-reheat steam Rankine cycle (bottoming cycle). The Brayton cycle has two stages of compression with intercooling. Heat is recovered from the expanded helium flow in the HRSG&R (heat recovery steam generator and reheater) to generate superheated steam for the high-pressure turbine (HPT) and steam reheating prior to the intermediate-pressure turbine (IPT) and low-pressure turbine (LPT). A recuperative heat exchanger is used to preheat water between the low-pressure pump (LPP) and high-pressure pump (HPP), avoiding steam extraction from IPT or LPT. As reported by Elder and Allen (2009), the efficiency of this cycle is in the range of 46–49 % when the reactor outlet temperature varies from 1,123 to 1,223 K. This efficiency is the highest amongst the following cycles: direct Brayton cycle (not

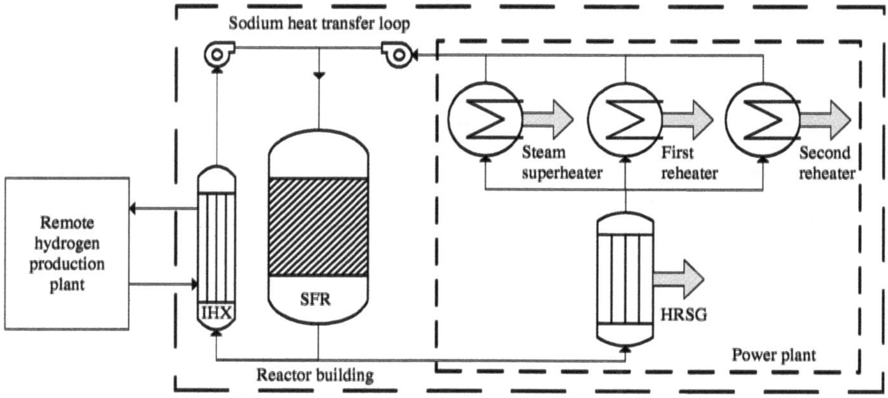

Fig. 8.10 Indirect linkage of a supercritical steam Rankine cycle to a nuclear plant for power and hydrogen generation

combined), combined cycles with indirect heat transfer and no intercooler, steam Rankine cycle with single reheat, steam Rankine cycle with no reheat, combined indirect Brayton–Rankine cycle, and Brayton indirect cycle.

Indirectly linked supercritical steam Rankine cycles have also been analyzed for power generation with nuclear reactors of intermediate temperature. Pérez-Pichel et al. (2011) analyzed a supercritical steam Rankine cycle with triple reheating and coupled to a sodium fast reactor (SFR) using an IHX. The hot-end temperature of the IHX is 800 K. The linkage manner of the indirect supercritical steam Rankine cycle with the nuclear reactor and hydrogen plant is illustrated in Fig. 8.10. The power generation efficiency reported by Pérez-Pichel et al. (2011)—assuming a secondary heat transfer loop with molten sodium—is about 45 %.

8.4 Nuclear Fusion-Based Hydrogen Generation

Practical realization of commercial reactors with nuclear fusion for power and process heat generation is a long-term goal of many research and development centers worldwide. As a process, nuclear fusion has been demonstrated as an uncontrolled reaction in a nuclear ("hydrogen") bomb. In this process, fusion of deuteron with triton occurs as follows (deuteron is the nucleus deuterium ${}^{2}_{1}H$ and triton is the tritium nucleus, ${}^{3}_{1}H$): ${}^{2}_{1}H + {}^{3}_{1}H \rightarrow {}^{4}_{2}He + n + 17.6\,MeV$.

This reaction requires ~10 keV for initiation. In the nuclear bomb, the associated temperature of 77×10^6 K has been obtained for fusion reaction initiation using an explosive fission reaction of uranium-235. The process duration in a thermonuclear bomb is about 1 μs. There has been considerable research throughout the world to develop a reactor which performs the fusion reaction in a controlled manner. Possible options are to use magnetic fields to keep the reaction in a confined space at high temperature, or inertial confinement through shock waves that are thermally induced by heating with laser beams.

It is interesting to note that about 1 Mt of coal has an equivalent energy with 1 t of nuclear fuel for fusion reactors. The fuel for a nuclear fusion reactor consists of deuterium and lithium in a mass proportion of 0.1:3. The main radioactive product of the fusion reaction is tritium, which has a half-life of 12 years, as compared to the fission products of common nuclear reactors which present radioactivity for hundreds or thousands of years.

In 2006, an international research project called ITER ("the way" in Latin) was established by Euratom, the USA, Russia, Japan, China, South Korea, and India with the aim to demonstrate a nuclear fusion reactor which is designed to generate 450 MW net power in pulses of at least 400 s each. The system generates about 500 MW of electricity but consumes 10 % to maintain the operation including the cryostats for superconducting magnets of the tokamak. It is estimated that fusion power technology can become commercially viable by 2050. Today, there are only a few theoretical concepts available for power plants supplied with thermal energy from controlled fusion reactions.

Using of nuclear fusion reactors to generate hydrogen by water splitting with high-temperature electrolysis, thermochemical cycles, or hybrid cycles has been investigated theoretically since the 1970s. In principle, a nuclear fusion reactor is surrounded by various cooling blankets with the role of maintaining a constant temperature of the plasma core. The blanket transfers heat released by the core to a useful process such as a power generation system or a thermochemical water-splitting cycle. The main elements of a hydrogen generation system with nuclear fusion were identified by Dang and Steinberg (1980), while more recent results were reported by Schultz (2003). It was claimed that the S–I cycle can be integrated with a nuclear fusion reactor. The main elements of an integrated system can be summarized as follows:

- Tokamak plasma reactor
- Cooling blanket shell
- Tritium extraction system
- A number of cascaded heat transfer loops (such as primary, secondary, and tertiary helium loops) which transfer the high-temperature heat to the downstream process
- Coolant purification systems
- Power production system
- Hydrogen production system

Figure 8.11 shows the general configuration of a fusion-based hydrogen generation facility which includes a thermochemical plant for water splitting. The only consumption within the plant is deuterium fuel which is extracted from seawater. The products are a relatively large quantity of deuterium-free water, hydrogen, oxygen, and electricity. Although very promising in the long term, there are still many challenges related to a fusion system. One important issue relates to the deposition of large quantities of deuterium-free water. Since the concentration of

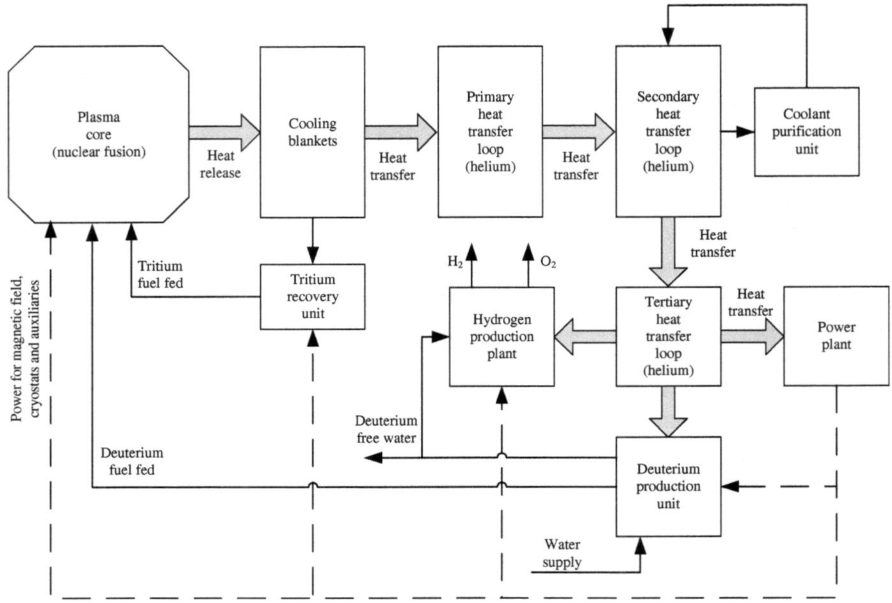

Fig. 8.11 Block diagram of a nuclear fusion plant for hydrogen production

deuterium in seawater is extremely low, large quantities of water must be circulated through the deuterium extraction system. The deposited water cannot be mixed with fresh water as they are not sources of deuterium for the process.

8.5 Conclusions

In this chapter, the future trends and potential opportunities with nuclear hydrogen production have been reviewed. Sustainable deployment of nuclear energy is closely related to the diversification of outputs, from energy to hydrogen, synthetic fuels, desalination, and/or process heat. Many countries have a development program of nuclear hydrogen production, as hydrogen is one of the most important constituents of many valued products and chemicals, while also a potentially major energy carrier in the future. This chapter discussed how a nuclear hydrogen economy can address—in the future—many global issues today, such as continuous population growth coupled with a sharp increase of energy and commodity demands, atmospheric pollution, and global warming. Nuclear hydrogen production will be a key method of hydrogen generation at a large scale in the future.

Several technological and scientific advances were reported in past literature and reviewed in this chapter. The development of TRISO-coated fuel particles represents one of the most remarkable advances in nuclear reactor technology with an intrinsic safety feature characterized by a negative reactivity coefficient and impermeability to fission fragments up to ~1,900 K. Pebble bed reactor configurations represent the safest nuclear reactor technology with regard to fuelling and simplicity of shutdown procedures.

Several concepts of IHX were discussed in this chapter. The IHX is used to decouple the primary coolant circuit from a downstream process of hydrogen production, and sometimes from a power generation plant. Helium is one of the most promising coolants due to its chemical stability in nuclear radiation media. A promising and compact IHX concept for helium-to-helium heat transfer includes a primary isolation valve and helium circulator that has been developed by AREVA NP and CEA. For SCWR, the IHX transfers heat from supercritical water at the primary side to a coolant at the secondary side for which there are two options—as published in past literature. One possibility is to have superheated steam generation at the secondary side. The other possibility is to heat helium as a heat transfer fluid in a secondary loop that links the nuclear facility to a remotely located thermochemical hydrogen production plant.

In the thermochemical hydrogen production plant, one of the crucial elements is the highest temperature reactor where an endothermic reaction is conducted. For example, in the S–I cycle, this reaction is the decomposition of sulfuric acid, while in the copper–chlorine cycle, it is the decomposition of copper oxychloride. JAEA developed a promising and compact concept of a chemical reactor for this reaction, which is made of solid blocks of silicon carbide sealed with pure gold gaskets, which is very resistant to highly corrosive fluids at elevated temperatures.

Another promising concept is nuclear hydrogen production with the once-through hybrid sulfur process. This process consumes sulfur, air, and water to generate hydrogen and sulfuric acid, two valuable products, at the expense of conventional nuclear electricity only. The system is very promising as it can use sulfur recovered from polluting processes.

There were several advanced power cycles discussed in this chapter that may improve the electricity generation efficiency with a nuclear reactor and thus hydrogen cogeneration effectiveness can be increased. Directly coupled cycles have slightly better efficiency than systems which are indirectly coupled to the nuclear reactor—via an IHX.

In the further distant future of nuclear hydrogen production—about 2050 or thereafter—it is envisaged that nuclear fusion systems may become commercial technologies for nuclear fission systems of electricity and hydrogen production. Nuclear hydrogen and electricity production can be achieved using nuclear fusion by consumption of relatively small amounts of deuterium fuel. The tritium is generated and used inside the process, thus not needed to be supplied externally. At the same time, the system generates enough power to maintain operation including its cryostats, maintaining the magnetic fields and tokamak plasma, tritium recovery, coolant purification, hydrogen and power production and their auxiliaries, heavy water extraction, and deuteron separation (by electrolysis). However, besides the technological challenges that must be overcome, the nuclear fusion process at a large scale will need to process very large amounts of water which can be deposited in separate locations such as deuterium-free water which cannot be used as a source of fuel in the future. This leads to potentially difficult and expensive operations.

Nuclear fission may be a safer way to advance nuclear technology in the future and achieve diversification of outputs. The generation IV nuclear roadmap of next-generation reactor development (for power, process heat, and hydrogen production) represents a pathway with fewer technological challenges than fusion reactors. The development of high-temperature components and special materials is a crucial element for the future development of nuclear technology at higher temperatures. Using these nuclear technologies, hydrogen generation and its derivatives (such as the production of synthetic fuels and materials) will become a key part of the nuclear energy future.

References

Abedin AH, Wang Z, Rosen MA (2012) Heat extraction form supercritical water-cooled nuclear reactors for hydrogen production plants. Int J Hydrogen Energy. 37:16527–16534

Carre F, Yvon P, Anzieu P, Chauvin N, Malo J-Y (2010) Update of the French R&D strategy on gas-cooled reactors. Nucl Eng Des 240:2401–2408

Dang VD, Steinberg M (1980) Application of the nuclear fusion reactor to thermochemical–electrochemical hybrid cycles and electrolysis for hydrogen production from water. Energy Conv Manage 20:85–101

Dumaz P, Allègre P, Bassi C, Cadiou T, Conti A, Garnier JC, Malo JY, Tosello A (2007) Gas cooled fast reactors—status of CEA preliminary design studies. Nucl Eng Des 237:1618–1627

Elder R, Allen R (2009) Nuclear heat for hydrogen production: coupling a very high/high temperature reactor to a hydrogen production plant. Prog Nucl Energy 51:500–525

Hejzlar P, Dostal V, Driscoll MJ, Dumaz P, Poullennec G, Alpy N (2006) Assessment of gas cooled fast reactor with indirect supercritical CO_2 cycle. Nucl Eng Technol 38:109–118

Jung YH, Jeong YH (2010) Development of the once through hybrid sulfur process for nuclear hydrogen production. Int J Hydrogen Energy 35:12255–12267

Matzner D, Wallace E (2005) The PBMR's progress. Nucl Eng Int 50:17–21

Mizia RE (2008) Next generation nuclear plant intermediate heat exchanger acquisition strategy. Idaho National Laboratory. Report INL/EXT-08-14054

Mokry S, Lukomski A, Pioro I, Gabriel K, Naterer G (2012) Thermalhydraulic analysis and heat transfer correlation for an intermediate heat exchanger linking a supercritical water-cooled reactor and a copper-chlorine cycle for hydrogen co-generation. Int J Hydrogen Energy. 37:16542–16556

Ogawa M, Hino R, Inagaki Y, Kunitomi K, Onuki K, Takegami H (2010) Present status of HTGR and hydrogen production development at JAFA. Nuclear Hydrogen Production. Fourth Information Exchange Meeting Oakbrook, IL. Nuclear Energy Agency 6805

Pérez-Pichel GD, Linares JI, Herraz LE, Moratilla BY (2011) Potential applications of Rankine and He-Brayton cycles to sodium fast reactors. Nucl Eng Des 241:2643–2652

Pope MA, Ik LJ, Hejzlar P, Driscoll MJ (2009) Thermal hydraulic challenges of gas cooled fast reactors with passive safety features. Nucl Eng Des 239:840–854

Richards M, Shenoy AS, Venneri F, Brown L, Fukuie M, Harvego E (2006) Modular helium reactor deployment for flexible energy demands. In: Third international topical meeting on high temperature reactor technology, Johannesburg

Schultz KR (2003) Production of hydrogen by fusion energy: a review and perspective. Fusion Sci Technol 44:393–399

Terada A, Imai Y, Noguchi H, Ota H, Kanagawa A, Ishikura S, Kubo S, Iwatsuki J, Onuki K, Hino R (2005) Experimental and analytical results on H_2SO_4 and SO_3 decomposer for I-S process pilot plant. Nuclear production of hydrogen, third information exchange meeting, Oarai, Japan, 5–7 October

MIX
Papier aus verantwortungsvollen Quellen
Paper from responsible sources
FSC® C105338

If you have any concerns about our products,
you can contact us on
ProductSafety@springernature.com

In case Publisher is established outside the EU,
the EU authorized representative is:
Springer Nature Customer Service Center GmbH
Europaplatz 3, 69115 Heidelberg, Germany

Printed by Libri Plureos GmbH
in Hamburg, Germany